Essential College Physics I

Ju H. Kim, Ph.D.

UNIVERSITY OF NORTH DAKOTA

Published by Linus Learning

Ronkonkoma, NY 11779

Copyright © 2014 Linus Learning

All Rights Reserved.

Editorial Director: Steven Darian
Cover Designer: Ruckman Perera
Layout Designer: Mark Walken

ISBN 10: 1-60797-457-6

ISBN 13: 978-1-60797-457-4

No part of this publication may be reproduced, stored in a retrieval system, or transmitted, in any form or by any means, electronic, mechanical, photocopying, recording, or otherwise, without the prior permission of the publisher.

Printed in the United States of America.

This book is printed on acid-free paper.

Print Number   5   4   3   2   1

DEDICATION

This book is dedicated to my parents and to my family.

Table of Contents

Acknowledgments ix

Preface xi

CHAPTER 1: UNITS AND PHYSICAL QUANTITIES 1

1.1 What is physics? 2
1.2 Physics at work 3
1.3 Scientific International Units: MKS system 5
1.4 Accuracy versus Precision 8
1.5 Dimensional Analysis 10
1.6 Basic Mathematics Needed 11
1.7 Pythagorean Theorem 14
Summary 15
More Worked Problems 16

CHAPTER 2: ONE DIMENSIONAL KINEMATICS 23

2.1 Kinematics in One Dimension 24
2.2 Free Fall Motion 34
2.3 Real figure of the Earth: the Geoid and Gravity 38
Summary 40
More Worked Problems 41

CHAPTER 3: TWO DIMENSIONAL KINEMATICS 49

3.1 Scalars versus Vectors 50
3.3 Projectile Motion 58
3.4 Relative Velocity 69
Summary 73
More Worked Problems 74

CHAPTER 4: LAWS OF MOTION 81

4.1	Motion and Forces	83
4.2	Newton's Three Laws of Motion	84
4.3	What causes motion?	85
4.4	Basic properties of force	88
4.5	Forces Affecting Dynamics	90
4.6	Newton's Second Law	96
4.7	Newton's Third Law	98
4.8	Universal Law of Gravitation	102
4.9	Normal Force, Tension, and Friction	106

Summary 117

More Worked Problems 118

CHAPTER 5: WORK AND ENERGY 125

5.1	Energy	126
5.2	Work	130
5.3	Energy Conservation	133
5.4	Kinetic Energy	134
5.5	Work-Energy Principle	135
5.6	Potential Energy	137
5.7	Classification of Forces	140
5.8	Effect of Dissipative Forces	146
5.9	Power	148

Summary 152

More Worked Problems 156

CHAPTER 6: MOMENTUM AND COLLISIONS 161

6.1	Linear Momentum	162
6.2	Law of Conservation of Momentum	165
6.3	Collision between Two Objects	169
6.4	Elastic Collision in Multi-dimensional Space	176
6.5	Center of Mass	178

| 6.6 | Motion of Center of Mass | 182 |

Summary ... 185

More Work Problems ... 186

CHAPTER 7: CIRCULAR MOTION ... 193

7.1	Rotational motion versus Circular motion	194
7.2	Rotational Kinematics	196
7.3	Rolling motion	200
7.4	Dynamics of uniform circular motion	206
7.5	Gravity and circular motion	216

Summary ... 223

More Worked Problems ... 224

CHAPTER 8: EQUILIBRIUM AND ROTATIONAL DYNAMICS ... 231

8.1	Rotational Dynamics	232
8.2	Rigid body in Equilibrium	237
8.3	Center of Gravity	243
8.4	Inertia for a rotating body	247
8.5	Energies of a moving rigid body	252
8.6	Angular momentum	255
8.7	Conservation of angular momentum	257

Summary ... 261

More Worked Problems ... 263

CHAPTER 9: SOLIDS AND FLUIDS ... 271

9.1	Effects of an applied force	273
9.2	Volume Deformation:	280
9.3	Fluids	283
9.4	Object in fluid	287
9.5	Pascal's Principle	290
9.6	Archimedes' Principle	293
9.7	Fluid Dynamics	297

9.8	Bernoulli's Principle	300
9.9	Real Fluids	305
Summary		308
More Worked Problems		309

CHAPTER 10: THERMAL PHYSICS ... 317

10.1	Zeroth Law of Thermodynamics	318
10.2	Temperature and Heat	319
10.3	Expansion of materials	319
10.4	Atomic theory of matter	322
10.5	Ideal Gas Law	330
10.6	Kinetic theory	336
10.7	Internal (or thermal) Energy	341
Summary		348
More Worked Problems		349

CHAPTER 11: ENERGY TRANSFER PROCESSES ... 355

11.1	Heat	356
11.2	Heat involved in Phase Change	363
11.3	Heat Transfer Mechanisms	369
Summary		381
More Worked Problems		382

CHAPTER 12: THERMODYNAMICS ... 389

12.1 Laws of Thermodynamics	390
12.2 Application of First Law of Thermodynamics	391
12.3 Application of Second Law of Thermodynamics	402
12.4 Entropy	414
Summary	418
More Worked Problems	419

APPENDIX ... 425
INDEX ... 441

Acknowledgments

I wish to thank all my College Physics students who provided valuable feedback that helped me improve my teaching and provided me with motivation for writing this book. I also wish to express my appreciation to my colleagues in the Department of Physics and Astrophysics at the University of North Dakota for their support, especially to Dr. Yen Lee Loh, who spent many hours reading the manuscript and provided many valuable suggestions and critical comments.

Also, I would like to acknowledge all contributors to Wikipedia, which provided a wealth of useful information on a number of topics presented in this book.

x

Preface

During my years of teaching algebra-based physics classes at the University of North Dakota, I noticed that many students, with and without high school physics backgrounds, started out with the false concern that they would not do well because physics was a difficult subject to comprehend. In fact, physics is not difficult and learning physics can be quite enjoyable, but bringing out these positive aspects requires quite a different approach from traditional college courses. I have received positive feedback from many students who started the college physics course with some level of fear, but completed it with an appreciation for physics. This has encouraged me to write a book that aims to reveal the beauty of how Mother Nature works in the real world through a one-year algebra-based physics course.

One of the challenges that I and other teachers have had to face is to dispel the common misconception among students that physics is merely a collection of equations that can be used blindly to solve problems. In fact, developing good problem-solving techniques and strategies starts with a firm grasp of physics concepts and how they fit together to provide a coherent description of the physical world. Therefore my first goal is to help students develop conceptual understanding and problem-solving skills. This book features Understanding-the-Concept Questions and Exercise Problems as a means towards this goal.

Through rigorous training, physicists know that the ability to reason in an organized manner is essential to solving problems. My second goal is helping students to improve their reasoning ability. The Problem-Solving Strategies for many worked-out problems in the book are meant to guide students in how to think about the problems. A strong reasoning ability, combined with a good conceptual understanding and problem-solving skills, will eventually lead to a much happier feeling and a deeper appreciation for physics (as illustrated below!) This is the legacy that I wish to leave to my students.

Worried about a problem? + **Think Physics and Problem solving strategy!** = **Be happy!**

Finally, I want students to become aware that physics principles play a hugely important role in their lives. Since it is always easier to learn something new if it is directly related to our lives, the book describes many direct applications of physics principles. I have chosen many real-world biomedical examples dealing with human physiology, but sometimes physics applications in futuristic fictional worlds are also discussed for fun! Physics should be fun to learn because it is a fascinating subject. I also hope that the in-depth discussions incorporated throughout the text and the many worked-out problems at the end of each chapter will serve as useful learning tools.

UNITS AND PHYSICAL QUANTITIES

CHAPTER 1

As we look at the physical world around us, we may notice that there are many phenomena occurring every day. These physical phenomena include, but are not limited to, falling leaves or snow flakes, moving automobiles, flowing of water, evaporation of water, airplane flying, and ice cubes floating in your favorite cold beverage. Because we are so familiar with these phenomena, they do not appear unusual to us. Hence, we may neither consider of these phenomena as interesting, nor ask why the physical world works as it does. However, if we are little curious and ask questions about why these phenomena always happen the same way, then we may find that there are underlying reasons for everything. This may be the first step in recognizing that the physical world around us, including ourselves, is governed by invisible laws. We may then ask that if there are such physical laws, how do we know what these laws are? These thought-provoking questions can be answered by physics.

1.1 WHAT IS PHYSICS?

A short answer to the question is that **physics is a study of the physical world around us.** We should note that everything around us is physical because the physical world around us is made of atoms, which serve as basic building blocks. Since biological organisms and complex chemical compounds are made of atoms, these are also considered physical objects. However, due to their complexity, describing biological organisms or complex chemical molecules in terms of basic physical laws may not be convenient.

Figure 1.1: Laws of physics are fundamental. When a miracle happens in the physical world, it too must happen within the laws of physics.

Regardless of complexity, all physical objects obey **laws of physics** because the physical laws are fundamental. As shown in Fig. 1.1, we may consider the work of God involves the laws of physics. Hence understanding of these laws through physics allows us to explain and predict how Mother Nature will behave in a given situation. Also, laws of physics help us comprehend various physical phenomena. For convenience of learning about the laws that govern physical world, we divide physics into several areas. Some these areas of physics include thermodynamics, fluid dynamics, waves, electricity and magnetism, and optics. In both part I and II of this book, we will discuss these topics.

Understanding the laws of physics is an important part of understanding how the physical world works, as the laws of physics are absolute. In other words, no physical phenomenon can violate any one of these fundamental laws. Sometimes, a physical phenomenon appears very strange to us and may seem to **violate** physical laws. However, if the phenomenon is observable, then there is no violation of the fundamental laws. Figure 1.2 shows that one group of monks continues to walk

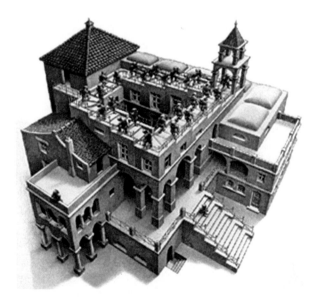

Figure 1.2: An illustration of a situation which violates the laws of physics. This situation appears real but cannot happen in the real world. (The image is from Gödel, Escher, and Bach.)

UNITS AND PHYSICAL QUANTITIES

up the stairs while another group continues to walk down. Of course, our common sense tells us that this phenomenon cannot happen in the real world. This common sense is based on our understanding of the physical world which is governed by the fundamental laws.

What is a consequence of violation any one of the fundamental laws? A small violation of physical laws may appear to us as being harmless. However, this may have a catastrophic consequence. For example, our universe may be entirely different than one we know. Also, all known physical phenomena may be completely unpredictable. In short, there will be unimaginable consequences as a result of violating the fundamental laws. Hence, it would be safe to say that everything that is physical around us will happen in complete accordance with the fundamental laws of physics.

1.2 PHYSICS AT WORK

Finding out the laws of physics fulfills our intellectual curiosity because it allows us to understand world around us, but it also provides enormous benefits to us. We may take advantage of our knowledge of the physical laws and manipulate the physical world through engineering. Since all engineering results from our understanding of the world around us, all of the modern day conveniences that we take for granted represent only a small fraction of what is possible. What is possible in the future is only limited by our imagination and the laws of physics. Hence, there is an inseparable relationship between physics and the

Figure 1.3: Automobile, airplane, smart phone, and tablet PC are all based on an understanding of physical laws.

modern conveniences shown in Fig. 1.3. The automobile involves mechanics and thermodynamics; the airplane involves fluid dynamics; the personal computer and smart phone involve electricity, magnetism and electromagnetic waves.

As we shall see below, physics helps us understand how these modern conveniences work. Also, it is the basis for future technologies. To understand how physics works, we need to be able to describe the physical world within the context of physical law. This raises an interesting question. We may ask "How do we know what are the physical laws?" More specifically, we need to answer what are these laws and how we know that these laws exist. As we may recognize by now that understanding the physical laws are important for describing how the physical world around us, but these laws, unlike our civil and criminal laws, are not written down by someone and enforced on us. We need to find out what these laws of physics are by carefully observing what physical phenomena are happening around us.

How to Describe the Physical world?

We can describe the physical world using the laws of physics. A more difficult task is to know what these laws are. One way to discover the laws of physics is through careful observations of physical phenomena. This requires lots of patience. For example, when we throw a baseball vertically as hard as we can, it will go up. However, if we wait a long enough time, then we will see that the baseball comes back down to the surface. If we are to connect this phenomenon to a law of physics, then we need to be confident that this is not an accidental event. If we repeat the experiment of throwing the baseball vertically up for many times and find the same final outcome, then we may conclude that a physical law governing an object is "what goes up must come down." This type of general understanding based on observations becomes a foundation for discovering the laws of physics. In Fig. 1.4, our knowledge on physical reality of falling is reflected in a cartoon illustrating that even in the virtual world nothing stays up in air forever, and everything eventually falls back down to the ground.

Once we know the qualitative nature of a law of physics governing a phenomenon, we need to be able to quantify it with a precise understanding which allows us to predict the future outcome of the similar event. So, to describe the physical phenomena accurately, we need to measure physical quantities associated with the phenomenon quantitatively. For the case of throwing a baseball upward, we may need to measure, for example, how high the baseball went up, how long it took to come back down, and how heavy it is. We need to quantify these observable quantities through measurements and to find relationships among measured quantities. For this purpose, the measurements must be accurate and reliable. For **accuracy** and **reproducibility**, we need to define the units (or dimensions) in which the measurements are made.

Figure 1.4: Laws of physics are applied equally at everywhere and to everyone. There are no exceptions. A comical illustration of this fact points out that even a cartoon character obeys the law of universal gravitation.

1.3 SCIENTIFIC INTERNATIONAL UNITS: MKS SYSTEM

In order to make measurements that are accurate and reproducible, it is important to have a system of measurement which is acceptable and agreeable to everyone. It would be meaningless to have one person measure a length, that is recognizable only to that person. Other people should also be able to measure the same length and agree on the value, within reason. For everyone's benefit, a measurement by one person should be easily understood by another person and be able to be confirmed. For this purpose, there are three basic units for measuring physical quantities. These basic quantities are length, mass, and time. There are many different units for measuring these quantities due to their purpose (i.e., commerce or science) and due to their historical origin. However, for scientific purposes, the basic unit of length is the **meter** (abbreviated m). This unit of length was the first international standard which was established by the French Academy of Sciences in the 1790's. The meter was intended to be one ten-millionth of the distance from the Earth's equator to the North Pole (at sea level), but its definition has been refined periodically refined due to our growing knowledge of metrology. After 1983, the meter is defined as the distance travelled by light in a vacuum during $1/(299,792,458)$ of a second. The standard unit of mass is the **kilogram** (kg). The mass of the international prototype of the kilogram is equal to the mass of one liter of water at 4.2 °C. Finally, the standard unit for time is the **second** (s). One second is defined as the time required for 9,192,631,770 periods of the radiation emitted by cesium atoms when they pass between to particular atomic states. These three standard units of measurement, as shown in Fig. 1.5, comprise the "MKS system". This means that no matter who measures, where one measures, and how one measures one meter, one kilogram, and one second, the measurements should be the same.

Figure 1.5: Scientific International standard for length, mass, and time: the meter, kilogram, and second (MKS).

It should be noted that when dealing with laws and equations of physics it is important to use a consistent set of units. Also, all of the physical quantities should be expressed within the context of this set of units (i.e., m, kg, and s) to avoid unnecessary confusions and complications. It should be noted that the most important system of units for both commercial and scientific purposes is the **Système International** (French for International System) or known as SI system. This system is most commonly used throughout the world. However, the **British engineering system** which takes the foot for length, the pound for force, and the second for time are used in United States of America. We adapt the SI system in this book. It should be noted that the MKS (with three basic units) is a subset of the SI system (with seven basic units).

Basic units for Physical Quantities

Physical Quantity	Symbol	S.I. Unit
Length (l)	m	meter
Mass (m)	kg	kilogram
Time (t)	s	second
Electric Current (I)	A	ampere
Temperature (T)	K	Kelvin
Amount of Substance (n)	mol	Mole
Luminous Intensity (Iv)	cd	candela

There are many physical quantities which may be described as a combination of the basic units. Throughout the book, we may represent these physical quantities by using either Greek or Roman letters. We should remember that because there are not enough letters to represent all physical quantities, the same symbol is often used to denote two very different physical quantities. For example, the letter T may be used to represent temperature in one equation, while it may be used to represent period in another. So, it is important to check to see how the physical quantity is represented by the symbol in equation.

Units for Physical Quantities Derived from the Basic Units

Physical Quantities	Relation to other Quantities	S.I. Unit
Area (A)	$A = l \times l$	m^2
Volume (V)	$V = l \times l \times l$	m^3
Molar mass (M)	$M = m/n$	kg/mol
Density	$\rho = m/V$	kg/m^3
Molarity (m)	$M = n/m_{solvent}$	mol/kg
Concentration (C)	$C = n/V$	mol/m^3
Acceleration (a)	$a = v/t$	m/s^2
Velocity (v)	$v = l/t$	m/s
Force (F)	$F =$ mass $\times$ acceleration	Newton (N) = kg/(m·s^2)
Pressure (P)	Pressure $= F/A$	Pascal (Pa) = kg·m/s^2
Energy (E)	$E =$ Force $\times$ distance	Joule (J) = kg·m^2/s^2
Power (P)	Power $= E/t$	Watt (W) = kg·m^2/s^3
Electric Potential (V)	$V = E/I$	Volt (V) = kg·m^2/(s3·A)
Electric Charge (Q)	$Q = I \times t$	Coulomb (C) = A·s
Electrical Resistance (R)	$R = V/I$	Ohm (Ω) = kg·m^2/(s^3·A^2)
Electrical Conductance (G)	$G = 1/R$	Siemens (S) = A/Ω

UNITS AND PHYSICAL QUANTITIES

Also, we should note that there are multiple systems of units. However, a physical quantity represented in one system may also be represented in another system. In the table below, the correspondence between SI units and English units are listed:

SI Units Compared to English Units

Quantity	S.I. Unit	Related Units	English Units
Length	meter (m)	cm or mm	foot (ft)
Mass	kilogram (kg)	gram (g) or mg	pound (lb)
Volume	cubic meter (m^3)	Liter (l) or ml	Quart (qt)
Temperature	Kelvin (K)	Celsius (°C)	Fahrenheit (°F)
Pressure	Pascal (Pa)	torr or atm	Pound per square inch (PSI)
Energy	Joule (J)	calorie (cal)	British thermal Unit (BTU)

The different systems of units are compatible. The compatibility between these two systems of units (i.e., units of meter and foot for length) or between two related units within the same metric system (i.e., units of meter and centimeter) means that there are conversion factors which allow measurements in one system to be converted to those in another. Conversion of a measurement from one unit to another can be done easily by multiplying by an appropriate conversion factor. For example, a speed limit of 72.0 miles per hour (mph) in US can be converted to 116 km/h which may be more useful to a person who is familiar with the metric system. Also, the value of 32.2 m/s, rather than 116 km/h, may be more useful for determining the distance traveled by an object if an equation is used. Some of the commonly used conversion factors for numerical calculation of physical quantities are listed below:

Conversion Factors

	English System	Metric System
Length units	1 foot = 12 inches =	0.3048 m = 30.48 cm
Length units	1 yard = 36 inches =	0.914 m = 91.44 cm
Length units	1 mile = 5,280 feet =	1.609 km = 1,609 m
Volume units	1 gallon = 4.0 quarts =	3.785 liters = 3.785x10^3 ml
Mass units	1 pound = 16 ounces =	0.4536 kg = 435.6 grams
Mass units	1 ton =	2,000 pounds (lbs)
Mass units	1 atomic mass unit (amu) =	1.6606 x 10^{-27} kg
Length units	1 Angstrom (Å) =	1.0 x 10^{-10} m
Volume units	1 pint =	16 fluid ounces
Time units	1 hour =	3600 seconds
Time units	1 year =	3.15576 x10^7 seconds
Temperature units	0 Celsius (°C) =	273.15 Kelvin (K)
Energy units	1 calorie (cal) =	4.184 Joules (J)

In converting a physical quantity from one unit to another, we may carry out the factor label method. This method may be executed in the following way:

1. Write the given number and unit.
2. Set up a conversion factor, which is a fraction used to convert one unit to another. This is done as follows:

i. Place the given units as the denominator of conversion factor.
   ii. Place the desired unit as numerator.
   iii. Place a "1" in front of the larger unit.
   iv. Determine the number of smaller units needed to make "1" of the larger unit.
3. Cancel units and solve the problem.

We should remember that a unit conversion may be considered as an algebraic calculation, and the conversion factors may be considered as different ways to write unity (i.e., 1). As an example we convert the speed limit of 75 mi/h into the corresponding value in the MKS system. The units for the speed in the MKS system is m/s. By following the algorithm above, we write

$$75\frac{mi}{h} \times \frac{1.609 Km}{1 mi} \times \frac{1000 m}{1 km} \times \frac{1 h}{3600 s} = 33.5 \frac{m}{s}.$$

Here, we used the conversion factors to change the units from the mile (mi) to the meter (m) for distance and from the hour (h) to the second (s) for time. Note this approach does not work for temperatures. For example, the unit conversion from the Celsius (°C) to Kelvin (K) scale is obtained by adding 273.15 to a value in degreeCelsius.

1.4 ACCURACY VERSUS PRECISION

Both careful and reliable observations are needed to understand the physical world around us. This means that we need to make accurate and precise measurements. It should be noted that although accurate measurements are an important part of discovering physical laws, this does not mean that they are precise. No measurement is absolutely precise. There is a technical difference between "precision" and "accuracy". **Precision** in a strict sense refers to the repeatability of the measurement using a given instrument. However, **accuracy** refers to how close a measurement is to the true value. To understand the difference between precision and accuracy, we consider an example of doing a target practice by using a toy gun. Suppose we make four shots. Figure 1.6 shows four different ways in whicn the shots may land on the target. If the four shots land right on the bulls-eye of the target and are close together, then the shots are both precise and accurate. However, if these shots land on the target close together but away from the bulls-eye, then the shooting is precise but not accurate.

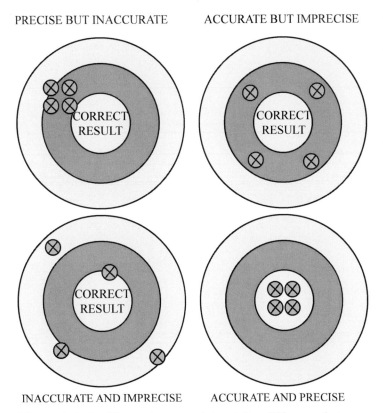

Figure 1.6: An illustration to point out the difference between accuracy and precision. It should be noted that an accurate measurement does not necessarily imply a precise measurement.

UNITS AND PHYSICAL QUANTITIES

When recording the result of observation, it is important to recognize that the precision of a measurement depends on the instrument used. Hence, it is important to state the precision of the instrument by providing an estimated uncertainty when presenting the result of a measurement. This gives an idea of the reliability of the data. For example, the height of an Apple iPhone 5 might be written as 123.8 ± 0.1 mm. The value of ±0.1 mm ("plus or minus 0.1 mm") represents the estimated uncertainty in the measurement. This uncertainty indicates that the actual value of the height is likely to be between 123.7 mm and 123.9 mm.

Significant Figures

There is another way to show the reliability of a number obtained from a calculation. As a physical quantity may be measured by using an instrument with a different level of precision and presented as a number, it is necessary to indicate the reliability of the measurement by using significant figures. The number of **significant figures** (sometimes called significant digits) is defined as the number of reliably known digits in a number. For example, a number written 23.21 has four significant figures, while a measured length of 0.062 cm has only two significant figures. For a rational number less than one, the zeros appearing immediately after the decimal point are not considered significant. For example, the number shown in Fig. 1.7 has three significant figures.

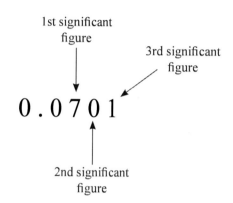

Figure 1.7: An illustration showing that significant figures are defined as the number of reliably known digits in a number.

As you may have suspected, there are rules for determining the number of significant digits in a number. In fact, there are three such rules. The **first rule** is that end zero(s) may or may not be significant if the number has no decimal point. For example, the number 9,000 has at least one significant figure, but the zeros may or may not be significant. To clarify the number of significant digits, we must express the number in scientific notation or add a decimal point to clarify the number of significant digit. If all zeros are significant, we may write the number either as 9,000. by adding a decimal point at the end or as 9.000×10^3 in scientific notation. The **second rule** is that all digits are significant except zeros at the beginning of a number and possibly zeros at the end of a number. For example, the number 0.0701 has three significant figures. Finally, the **third rule** is that the final result of a multiplication or division should have only as many digits as the number with the least number of significant figures used in the calculation. For example, a product of 11.2 cm in length and 6.7 cm in width will yield 75 cm² (i.e., 11.2 cm × 6.7 cm = 75 cm²), but not 75.04 cm² as we may obtain by using a calculator.

Powers of 10

Multiple	Symbol	Prefix Name	Its Value
10^{12}	T	tera-	one trillion times
10^9	G	giga-	one billion times
10^6	M	mega-	one million times
10^3	k	kilo-	one thousand times
10^2	h	hecto-	one hundred times
10^1	de	deka-	ten times
10^{-1}	d	deci-	one-tenth of

Multiple	Symbol	Prefix Name	Its Value
10^{-2}	c	centi-	one-hundredth of
10^{-3}	m	milli-	one-thousandth of
10^{-6}	μ	micro-	one-millionth of
10^{-9}	n	nano-	one-billionth of
10^{-12}	p	pico-	one-trillionth of
10^{-15}	f	femto-	one-quadrillionth of
10^{-18}	a	atto-	one-quintillionth of

It is interesting to note that there is a difference between the meaning of billion in American Englishand that in British English. A billion means 1,000,000,000 (i.e., 10^9) in American English, but it means 1,000,000,000,000 (i.e., 10^{12}) in British English.

1.5 DIMENSIONAL ANALYSIS

As we discussed above, the three units, meter, kilogram, and second, are the basic units in the SI system. These units are used to denote the fundamental quantities in physics. The fundamental quantities used in physical measurement are called **dimensions**. For example, length, mass, and time are dimensions. Dimensions for useful physical quantities may not be this simple. In either computing or describing a physical quantity, a mixture of dimensions may be involved since a number of different physical quantities are involved. For instance, the speed of an object involves two physical quantities, length and time. This means that when we compute the speed of a moving object, the final answer must have the dimension of [Length] over [Time]. Hence, this suggests that a dimensional analysis is a useful way to check the validity of an answer or relationship (i.e., formula.).

what is dimensional analysis?

We may consider the dimensional analysis as a procedure for checking the dimensional consistency of any equation. This analysis is used to verify the validity of an equation. Also, we can determine the dimension of an unknown constant in the correct equation. For example, the dimension of the constant G in the gravitational force equation $F = Gm1m^2/r^2$ is determined from the dimensional analysis, using the dimensions of F, m, and r. Since all physical quantities have dimensions, two sides of an equation must be equal not only in numerical value, but also in dimensions. For example, a product of two length measurements, 3.0 cm and 4.0 cm, will yield 12 cm² (i.e., 3.0 cm × 4.0 cm = 12 cm²). If the dimensions on one side of a formula do not equal to that on the other side, then the relation is incorrect. Sometime, we may become confused by using different units of the same dimensional quantity. For example, seconds and hours are two different units of time. So, in order to minimize confusion, when working on problems, we need to use the same unit for a given dimension throughout the problem.

Before doing a dimensional analysis, first, we need to do a unit analysis which allows us to check for mixed units. When two or more units are used to represent the same physical quantity, we need to use the conversion factors to express the quantities in the same unit. For example, when the speed of one object is given as 32 ft/s and the other object is given as 100 cm/s, we need to convert both

UNITS AND PHYSICAL QUANTITIES

values into units of meter per second (i.e., m/s). For this purpose, some useful conversion factors for length are 1 m = 100 cm, 1 ft = 12 in, and 1 in = 2.54 cm. Other useful conversion factors are listed in the conversion factor table in Sec. 1.3.

1.6 BASIC MATHEMATICS NEEDED

As many students who are taking physics for the first time may have guessed, some level of basic mathematics is needed because mathematics is the natural language for describing physics. Many peoples are frightened about using mathematics in physics. Just as our knowledge of the English language is needed to study English literature, we need to know some basic level of mathematics to study physics. Our mathematics knowledge will serve as a very important tool, not only for describing physical phenomena, but also for gaining a deeper understanding of physical principles. For both part I and II of this book, working knowledge of the following branches of mathematics is essential: Plane Geometry, College Algebra, and Trigonometry. Calculus is not required, but occasionally, some knowledge of Calculus may be helpful in understanding derivation of several important formulas. However, we should feel comfortable about carrying out algebraic manipulations. Armed with mathematical tools, we may focus on understand the basic physical principles and applying them to solve problems, rather than looking for the right equations. We need to develop problem-solving strategies because we may not always find the right equation to solve a problem. However, we may always apply appropriate principles and/or definitions to solve a problem. Our knowledge of mathematics will allow us to derive an answer starting from the mathematical equation(s) describing the basic physical principle(s) involved in the problem.

Algebra

Algebra is an important mathematical tool for solving problems. If your basic knowledge of Algebra is little bit rusty, then you will find it helpful to review College Algebra and pay careful attention to follow the basic algebraic manipulations discussed in the book. Throughout this book, we represent relationships between physical quantities as equations involving symbols (usually either Greek or Roman letters). Usually, the manipulation of these equations is done using algebra. Hence, we will use a great deal of algebra to do physics.

In solving any problem in this book, our first goal is to start with the simple equation(s) that describes the basic principle. As we know, an equation involves an equal sign. The equation tells us that the quantities on either side of the equals sign have the same value and dimensions. Examples of equation are $5 + 7 = 12$, $4x + 3 = 14$, and $ab^2 + c = 5$. The first equation involves only numbers and is called an arithmetic equation. The remaining two equations are algebraic since they involve symbols. The third equation contains the quantity ab^2 which means the product of a times b times b: $ab^2 = a \times b \times b$.

As a step in solving a problem, most likely, we need to solve for one or more symbols, which we treated as unknown. For example, in the equation $4x + 3 = 14$, x is the unknown. The solution to equation is $x = 11/4$. Determining what value (or values) the unknown(s) can have to satisfy the equation(s) is called solving the equation. To solve an equation, we need to use the following rule: *an equation will remain true if any operation performed on one side is also performed on the other side*. For example, operation include (a) addition or subtraction of either a number or symbol, (b) multiplication or division by either a number or symbol, (c) raising each side of the equation to the same power, or taking the same root (such as square root). In some problems, we may have two or more unknowns.

If we have two or more unknowns, If there is more than one unknown, one equation is not sufficient to find a solution. In general, if there are n unknowns, then n independent equations are needed to find a solution. For example, if there are two unknowns, we need two equations. If the unknowns are x and y, then a typical procedure is to solve one equation for x in terms of y, and then substitute this into the second equation to find the solution.

Sometimes, we may encounter equations that involve an unknown, say x, which appears not only to the first power, but also to the second power (i.e., squared) as well. This is called a quadratic equation and can be written in the form

$$a x^2 + b x + c = 0. \tag{1.1}$$

The quantities a, b, and c are typically numbers or constants that are given. The general solutions to quadratic equations are given by the quadratic formula:

$$x = \frac{-b \pm \sqrt{b^2 - 4ac}}{2a} \tag{1.2}$$

The ± sign indicates that there are two solutions for x: one solution uses the plus sign, and the other uses the minus sign.

Trigonometry

Another very useful branch of mathematics for this book is trigonometry. More specifically, we will use trigonometric functions. Trigonometric functions are defined for an angle α. Here, the angle α is defined by constructing a right triangle as shown in Fig. 1.8. A right triangle is defined as a triangle with one of the internal angles is 90° (i.e., right angle). The three sides of a right triangle about the angle α ($\neq 90°$) are named the adjacent side, the opposite side, and the hypotenuse. The opposite and adjacent sides are the lengths of the sides opposite and adjacent to the angle α, respectively and the hypotenuse is the diagonal length.

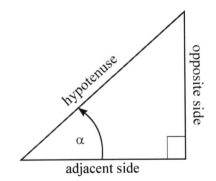

Figure 1.8: A triangle has three internal angles, and a right triangle has 90° as one of three internal angles. Three sides of the right triangle with respect to an angle α are defined.

Three trigonometric functions that we use through the book are sin α, cos α, and tan α. These three functions are defined by taking the ratios of the lengths of three sides of the right triangle, as follows:

$$\sin \alpha = \frac{opposite\ side}{hypotenuse} \tag{1.3}$$

$$\cos \alpha = \frac{adjacent\ side}{hypotenuse} \tag{1.4}$$

$$\tan \alpha = \frac{opposite\ side}{adjacent\ side} = \frac{\sin \alpha}{\cos \alpha} \tag{1.5}$$

UNITS AND PHYSICAL QUANTITIES

These functions can be either positive or negative. The signs (+ or -) that cosine, sine, and tangent take on for angle α depends on which one of the four quadrants (0° to 360°) the angle α falls in. The angle α is measured counterclockwise from the x axis; negative angles are measured from below the x axis, clockwise. Also, it is helpful to remember that sin 0° = 0, sin 90° = 1, cos 0° = 1, and cos 90° = 0. Some calculation may involve algebraic manipulations of trigonometric functions. Hence, the following trigonometric identities are very useful:

$$\sin^2 \theta + \cos^2 \theta = 1 \qquad (1.6)$$

$$\sin(A \pm B) = \sin(A) \cos(B) \pm \cos(A) \sin(B) \qquad (1.7)$$

$$\cos(A \pm B) = \cos(A) \cos(B) \mp \sin(A) \sin(B) \qquad (1.8)$$

Of course, there are other useful trigonometric identities. However, if we need to use other identities than these three shown above, then we will derive them and use them in calculation.

Exercise Problem 1.1

On a sunny day, a tall building casts a shadow that is 67.2 m long. The angle between the sun's ray and the ground is $\theta = 50.0°$. Determine the height of the building.

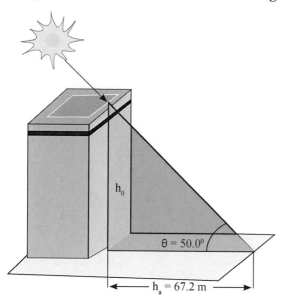

PROBLEM SOLVING STRATEGY

What do we know?

One side of a right triangle and one of the internal angles are given.

What concepts are needed?

We need to apply the definition of tangent function.

Why?

We need to determine the length of the opposite side from the information on the length of the adjacent side.

Solution:

Since we have the information about the length of the shadow (i.e., adjacent side) and the angle θ and need to determine the height of the building (i.e., opposite side), we use the definition of tangent

$$\tan q = \frac{\text{opposite side}}{\text{adjacent side}} = \frac{h_o}{h_a}$$

We may multiply both sides of the above equation by h_a to solve for h_o. We obtain

$$h_o = h_a \tan \theta = 67.2m \times \tan(50.0°) \cong 80.2 \text{ m}$$

This result indicates that, when the angle θ is smaller, the length of the shadow is longer.

Inverse Trigonometric Functions

In solving problems, you may need to use the inverse trigonometric functions to determine the angle α from the sides of the right triangle. The inverse trigonometric functions (which are occasionally called cyclometric functions) are the inverse functions of the trigonometric functions. These functions have suitably restricted domains. We often use the notations $\sin^{-1}$, $\cos^{-1}$, and $\tan^{-1}$ to denote arcsin, arcos, and arctan, respectively. However, we should not confuse this convention with the common convention for expression like $\sin^2(x)$ which refers to numeric power rather than function composition. Three inverse trigonometric functions that we will use often are defined as follows:

$$\alpha = \cos^{-1}\left(\frac{adjacent\ side}{hypotenuse}\right) \tag{1.9}$$

$$\alpha = \sin^{-1}\left(\frac{opposite\ side}{hypotenuse}\right) \tag{1.10}$$

$$\alpha = \tan^{-1}\left(\frac{opposite\ side}{adjacent\ side}\right) \tag{1.11}$$

You may see easily that Eq. (1.9) is obtained from the definition of the cosine function of (1.4) by taking the inverse on the both sides of Eq. (1.4) as shown below:

$$\cos\alpha = \frac{adjacent\ side}{hypotenuse} \Rightarrow \cos^{-1}\left(\cos\alpha = \frac{adjacent\ side}{hypotenuse}\right) \Rightarrow \alpha = \cos^{-1}\left(\frac{adjacent\ side}{hypotenuse}\right)$$

We may follow a similar approach for the derivation of the angle α from the remaining two trigonometric functions.

1.7 PYTHAGOREAN THEOREM

For a right triangle as shown in Fig. 1.9, there is the relationship among the lengths of three sides. Namely, the square of the length of the hypotenuse (the side opposite the right angle) is equal to the sum of the squares of the lengths of the other two sides. This relation is called the Pythagorean equation and is written as

$$(hypotenuse)^2 = (adjacent\ side)^2 + (opposite\ side)^2 \tag{1.12}$$

The theorem written as Eq. (1.12) has numerous proofs. It may possibly have the most number of proofs compared to any mathematical theorem. These proofs are very diverse, including both geometric proofs and algebraic proofs, with some dating back thousands of years.

The Pythagorean equation which is written as Eq. (1.12) is named after the Greek mathematician Pythagoras (570 BC - 495

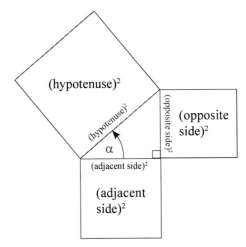

Figure 1.9: A schematic diagram illustrating the meaning of the Pythagorean equation. The theorem may be considered in terms of the area generated by the three sides of a right triangle.

UNITS AND PHYSICAL QUANTITIES

BC), who by tradition is credited with its discovery and proof. However, it is often argued that knowledge of the theorem predates him. There is evidence that Babylonian mathematicians understood the formula, but there is little surviving evidence that they used it in a mathematical framework, as we will need to do for our discussion. We will find Eq. (1.12) quite useful throughout the book. The first place where we need to use Eq. (1.12) is for determining the magnitude of a vector in Chapter 3.

SUMMARY

i. For scientific purposes, the most commonly used basic unit for measuring length is the **meter** (abbreviated **m**), the standard unit of mass is the **kilogram** (**kg**), and the standard unit of time is the **second** (**s**).

ii. To understand the physical world around us, we need to make accurate and precise measurements to discovering physical laws. However, there is a technical difference between "precision" and "accuracy". **Precision** refers to the repeatability of the measurement using a given instrument, and **accuracy** refers to how close a measurement is to the true value.

iii. One useful algebraic equation is the **quadratic equation** which may be written in the form

$$ax^2 + bx + c = 0,$$

where a, b, and c are constants. The general solutions to the quadratic equations are given by the quadratic formula:

$$x = \frac{-b \pm \sqrt{b^2 - 4ac}}{2a}$$

where the ± sign indicates that there are two solutions for x.

iv. Three useful **trigonometric functions** are defined by taking the ratio of the lengths of the sides of the right triangle. These functions are written as follows:

$$\sin \alpha = \frac{opposite\ side}{hypotenuse},$$

$$\cos \alpha = \frac{adjacent\ side}{hypotenuse},$$

$$\tan \alpha = \frac{opposite\ side}{adjacent\ side} = \frac{\sin \alpha}{\cos \alpha}.$$

The inverse trigonometric functions, $\sin^{-1}$, $\cos^{-1}$, and $\tan^{-1}$, may be used to determine the angle α from the sides of the right triangle.

v. The square of the length of the hypotenuse (the side opposite to the right angle) is equal to the sum of the squares of the lengths of the other two sides. This statement is called the Pythagorean theorem. This theorem may be written as

$$(hypothnuse)^2 = (adjacent\ side)^2 + (opposite\ side)^2$$

This equation is called the Pythagorean equation.

MORE WORKED PROBLEMS

Problem 1.1 :

Engineers often express the density of substance as a weight density, for example, in pounds per cubic foot. (a) What is the weight density of water? (b) What is the weight of one gallon of water?

Solution:

a. The density of water $\varrho = 1000$ kg/m³. Since 1.00 m is equal to 3.28 ft, one cubic meter is (i.e., 1 m³) is equal to 35.29 ft³. Also, 1.00 kg is equal to 2.2 lb. So the weight density of water in the MKS system may be written in the English units, by using the conversion factors, as

$$1000 \frac{kg}{m^3} \times \frac{1 m^3}{35.29 ft^3} \times \frac{2.2 lb}{1.0 kg} = 62.3 \frac{lb}{ft^3}$$

b. The weight of one gallon of water may be expressed in pounds by using conversion factors

$$1 gal \times \frac{0.134 ft^3}{1 gal} \times 62.3 \frac{lb}{ft^3} = 8.34 lb$$

Problem 1.2 :

A spring is hanging down from the ceiling, and an object of mass m is attached to the free end. The object is pulled down, thereby stretching the spring, and then released. The object oscillates up and down, and the time T required for one complete up-and-down oscillation is given by the equation $T = 2\pi(m/k)^{1/2}$, where the spring constant k is unknown. What must be the dimension of k for this equation to be dimensionally correct?

Solution:

The dimension of the spring constant k must be determined by solving the equation

$$T = 2\pi \left(\frac{m}{k}\right)^{1/2} \Rightarrow T^2 = 4\pi^2 \frac{m}{k} \Rightarrow k = 4\pi^2 \frac{m}{T^2}$$

Here, the term $4\pi^2$ is a numerical factor which does not have a dimension. So it can be ignored in this analysis. Since the dimension for mass is [M] and that for time is [T], the dimension of the spring constant k has mixed dimensions which may be written as

$$\text{Dimension of } k = \frac{[M]}{[T]^2}$$

UNITS AND PHYSICAL QUANTITIES

Problem 1.3:

A student wants to determine the distance of a small island from the lakeshore. He first draws a 50.0-m line parallel to the shore. Then he goes to the ends of the line and measures the angles of the lines of sight from the island relative to the line he has drawn. The angles are 30° and 40°. How far is the island from the shore?

PROBLEM SOLVING STRATEGY

What do we know?

We are given the internal angles of two right triangles.

What concepts are needed?

We need to apply the definition of tangent function.

Why?

We need to determine the length of the opposite side of the right triangle.

Solution:

Here it is noted that the angles $\theta = 30°$ and $\theta' = 40°$. For the both right triangle with the angle θ and θ', the distance d is the opposite side. This suggests that we may use the definition of tangent to write

$$d = x \tan \theta = (50 \text{ m} - x) \tan \theta'$$

We rewrite this equation as

$$x \tan \theta = (50 \text{ m}) \tan \theta' - x \tan \theta'$$

We now solve for x by collecting the terms with the variable x on the left side of the equation and leaving the remaining term on the right hand side of equation. We obtain

$$x(\tan \theta + \tan \theta') = (50 \text{ m}) \tan \theta'$$

We solve for x and obtain

$$x = \frac{(50 \text{ m}) \tan \theta'}{\tan \theta + \tan \theta'} = \frac{50 \text{ m} \times \tan 40°}{\tan 30° + \tan 40°} = 29.6 \text{ m}$$

We now use the value of x that we computed above and use the definition of tangent once more to obtain

$$d = x \tan \theta = 29.6 \text{ m} \times \tan 30° = 17 \text{ m}$$

Problem 1.4:

The drawing below shows the face-centered cubic structure of sodium chloride (common table salt) crystal. Each unit cell of the crystal has two atoms: one sodium atom and one chloride atom. Sodium and chloride ions position at the corners of a cube that is part of the crystal structure. The edge of the cube is 0.281 nm (1 nm = 1 nanometer = 10-9 m) in length. Find the distance (in nanometers) between the sodium ion located at point P and the chloride ion located at point Q.

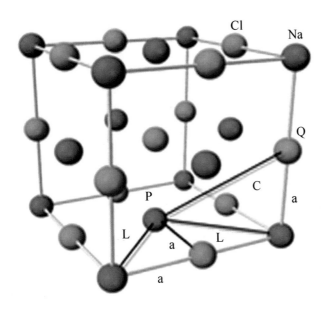

PROBLEM SOLVING STRATEGY

What do we know?

We are given the length of the edge of the cube, which is part of a sodium chloride crystal.

What concepts are needed?

We need to apply the Pythagorean equation.

Why?

We need to determine the length of the hypotenuse of a right triangle from the lengths of two known sides.

Solution:

Suppose we consider the above view of the cube. The length L of the diagonal of the bottom face of the cube can be found using the Pythagorean equation. Applying this equation, we obtain the length L as

$$L^2 = a^2 + a^2 = 2a^2 = 2(0.281 \text{ nm})^2 = 0.158 \text{ nm}^2 \Rightarrow L = 0.397 \text{nm}$$

Once again, we find the required distance C by using the Pythagorean equation. We obtain

$$C^2 = L^2 + a^2 = (0.397 \text{ nm})^2 + (0.281 \text{ nm})^2 = 0.237 \text{ nm}^2 \Rightarrow C = 0.487 \text{ nm}$$

PROBLEMS

1. many seconds are there in (a) two hour and thirty-five minutes and (b) three days?

2. A sailor tells you that if his ship is traveling at 23 knots (nautical miles per hour), it is moving faster than the 25 miles per hour your car travels. How can this be? [Note: 1 knot is 1.15078 miles.]

3. A commuting student wants to buy 15 gal of gas, but the gas station has installed new pumps that are measured in liters. How many liters of gas (round off to a whole number) should he ask for?

4. Common product label indicates that a bottle of soft drink contains 2 fl. Oz. From the units on the labels, find (a) the number of milliliters in 2 fl. Oz and (b) the number of ounces in 100 g.

UNITS AND PHYSICAL QUANTITIES

5. Engineers often express the density of substance as a weight density, for example, in pounds per cubic foot. **(a)** What is the weight density of water? **(b)** What is the weight of one gallon of water?

6. What is the difference between the volume of one kilogram of lead and that of gold? Note that the density of lead and gold is has 11.000kg/m³ and 19.000kg/m³, respectively.

7. The weight of in a cubical container is measured to be 56.5 pounds. (a) How many liter of water is in this container? (b) What is the dimension of the container?

8. Express each of the following numbers to only three significant figures: (a) 11.072 m; (b) 0.0520 mm; (c) 0.002559 kg; (d) 93001000 mi.

9. The diameter of a sphere is 5.36 in. Find (a) the radius of the sphere in centimeters, (b) the surface area of the sphere in square centimeters, and (c) the volume of the sphere in cubic centimeters.

10. (a) How many seconds are there in a year? (b) If one micrometeorite (a sphere with a diameter on the order of 10^{-6} m) struck each square meter of the Moon each second, estimate the number of years it would take to cover the Moon with micrometeorites to a depth of one meter. (*Hint*: Consider a cubic box, 1 m on a side, one the Moon, and find how long it would take to fill the box.)

11. The diameter of a sphere is measured to be 5.36 in. Find (a) the radius of the sphere in centimeters, (b) the surface area of the sphere in square centimeters, and (c) the volume of the sphere in cubic centimeters.

12. If an automobile is traveling at the speed of 75 miles per hour, what is its speed in meters per second?

13. An automobile tire is rated to last for 50,000 miles. Estimate the number of revolutions the tire will make in its lifetime.

14. The amount of water in reservoirs is often measured in acre-ft. One acre-ft is a volume that covers an area of one acre to a depth of one foot. An acre is 43,5600 ft². Find the volume in SI units of a reservoir containing 50.0 acre-ft of water.

15. Soft drinks are commonly sold in aluminum containers. (a) To an order of magnitude, how many such containers are thrown away or recycled each year by U.S. consumers? (b) How many tons of aluminum does this represent? In your solution, state the quantities your measure or estimate and the values you take for them.

16. According to the information provided by World Health Organization in 2013, the average life expectancy of a male living in the United States is 79.8 years. Estimate how many times the heart will beat during the lifetime of an average male.

17. Bacteria and other prokaryotes are found deep underground, in water, and in the air. One micron (10^{-6} m) is a typical length scale associated with these microbes. (a) Estimate the total number of bacteria and other prokaryotes in the biosphere of the Earth. (b) Estimate the total mass of all such microbes. (c) Discuss the relative importance of humans and microbes to the ecology of planet Earth. Can *Homo sapiens* survive without them?

18. The correct equation for the volume of a sphere is $V = 4pr^3/3$, where r is the radius of the sphere. What should be the equation when expressed in terms of the diameter d? (F05)

19. The base of a pyramid covers an area of 13.0 acres (1 acre = 43,560 ft²) and has a height of 481 ft. If the volume of a pyramid is given by the expression $V = bh/3$, where b is the area of the base and h is the height, find the volume of this pyramid in cubic meters.

20. Einstein's famous mass-energy equivalence is expressed by the equation $E = mc^2$, where E is energy, m is mass, and c is the speed of light. (a) What are the SI base units of energy? (b) Another

equation for energy is $E = mgh$, where m is mass, g is the acceleration due to gravity, and h is height. Does this equation give the same units as in (a)?

21. A shape that covers an area A and has a uniform height h has a volume $V = Ah$. (a) Show that $V = Ah$ is dimensionally correct. (b) Show that the volumes of a cylinder and of a cubical box can be written in the form $V = Ah$, identifying A in each case. (Note that A, sometimes called the "footprint" of the object, can have any shape and that the height can, in general, be replaced by the average thickness of the object.)

22. Newton's law of universal gravitation is represented by

$$F = G \frac{Mm}{r^2}$$

where F is the gravitational force, M and m are masses, and r is a length. Force has the SI units of kg·m/s². What are the SI units of the proportionality constant G?

23. The radius of a circle is measured to be (10.5 ± 0.2) m. Calculate (a) the area and (b) the circumference of the circle, and give the uncertainty in each value.

24. The radius of a solid sphere is measured to be (21.5 ± 0.1) m. Calculate (a) the volume and (b) the surface area of the sphere, and give the uncertainty in each value.

25. The side of a cube is measured to be (5.67 ± 0.03) m. Calculate (a) the volume and (b) the surface area of the cube, and give the uncertainty in each value.

26. You are driving into Washington DC, and in the distance you see the famous Washington Monument. This monument rises to a height of 169 m. You estimate your line of sight with the top of the monument to be 1.5° above the horizontal. Approximately how far (in kilometers) are you from the base of the monument?

27. On takeoff, an airplane climbs with a speed of 190 m/s at an angle of 35° above the horizontal. The speed and direction of the airplane constitute a vector quantity known as the velocity. The sun is shining directly overhead. How fast is the shadow of the plane moving along the ground? (That is, what is the magnitude of the horizontal component of the plane's velocity?)

28. A corner construction lot has the shape of a right triangle. If the two sides perpendicular to each other are 37 m long and 42.3 m long, respectively, what is the length of the hypotenuse?

29. Two chains of length 2.0 m are used to support a lamp. The distance between the two chains is 1.0 m along the ceiling. What is the vertical distance from the lamp to the ceiling?

30. You and your friend live in different buildings that are next to each other. You live in the building on the left, and a friend lives in the other building. The two of you are having a discussion about the height of the buildings, and your friend claims that his building is half as tall as yours. To resolve the issue you climb to the roof of your building which makes an angle of 21° above the horizontal (when viewed from your friend's building), while your line of sight to the base of the other building makes an angle of 52° below the horizontal. Determine the ratio of the height of the taller building to the height of the shorter building. State whether your friend is right or wrong.

31. A high fountain of water is located at the center of a circular pool as shown in figure below. Not wishing to get his feet wet, a student walks around the pool and measures its circumference to be 15.0 m. Next, the student stands at the edge of the pool and uses a protractor to gauge the angle of elevation at the bottom of the fountain to be 48.0°. How high is the fountain?

32. Suppose the mountain height is y, the woman's original distance from the mountain is x, and the angle of elevation she measures from the horizontal to the top of the mountain is θ. If she moves a distance d closer to the mountain and measures an angle of elevation ϕ, what is a general

equation for the height of the mountain y in terms of d, ϕ, and θ? Neglect the height of her eyes above the ground.

33. Cube 1 has surface area A_1 and volume V_1, and cube 2 has surface area A_2 and volume V_2. If the side of cube 2 is triple the side of cube 1, what is the ratio of (a) the areas, A_2/A_1 and (b) the volumes V_2/V_1? (F12)

ONE DIMENSIONAL KINEMATICS

CHAPTER 2

Have you seen a snail move? If you have, then have you wonder about how it moves? Most snails move by gliding along on their muscular foot, which is lubricated with mucus. To help them move, snails secrete a stream of slime (mucus) from a gland located at the front of their foot. Snails produce mucus in order to aid locomotion by reducing friction. This slime enables them to glide smoothly over many different types of surface and helps to form a suction that helps them cling to vegetation and even hang upside down. Also the mucus helps reduce the snail's risk of mechanical injury from sharp objects. A snail is able to push against a surface using its muscular foot and propel its body forward by creating an undulating wave-like motion along the length of the foot. This motion is powered by successive waves of muscular contraction which move along the bottom surface of the foot. This muscular action is clearly visible when a snail is crawling on the glass of a window or aquarium. This is a rather complex description of a snail's motion. A much shorter description would be that snails move slowly. Snails move at a proverbially low speed (1 mm/s is a typical speed). Their progress is slowed by the weight of their shell. In proportion to its body size, the shell is quite a load for a snail.

If we are asked to describe the motion of a snail, how would we describe it precisely? In other words, if we are asked to state the position of the snail as a function of time, what information do we need? We may not think that describing the motion of a snail as something special. However, it is special because once we know how to do this we may apply it to all moving objects since they may be described in the same way, regardless whether that object is a slow moving snail or a fast moving racing car. In describing motion of an object, we need to consider the two aspects of its motion. One aspect is *how* it moves and other is *why* it moves. These two aspects must be described together in order to have a complete account of motion. These two aspects of motion combined are called *mechanics*. To have a better understanding of how to describe the motion of an object, we need to start with the simplest case: motion in one dimension. It should be noted that much more complex motion of an object may still be expressed in a similar way.

2.1 KINEMATICS IN ONE DIMENSION

Mechanics is the study of motion of objects and related concepts of force and energy. As the concepts of force and energy are more abstract, we discuss them in Chapters 4 and 5, respectively, once we have a better grasp of the concept of motion. The study of motion of objects is divided into two parts: *kinematics* and *dynamics*. It is noted that kinematics deal with describing how objects move, while dynamics deal with force and describe why objects move as they do. We will first study kinematics of an object in Chapters 2 and 3, and then study its dynamics later in Chapter 4.

Also, it should be noted that, in general, movement of any object may be described in terms of two types of motion. These two types are i) translational and ii) rotational motion. A complex motion involves a combination of translational and rotational motion. For example, when we throw a baseball horizontally, it will have translational motion. Also, it will have rotational motion if it is thrown with spin. This means that we may always decompose the motion of an object into translational and rotational motion. It should be noted that rotational kinematics is in many ways very similar to translational kinematics. We will discuss rotational kinematics once we understand how to formulate translational kinematics.

Translational motion

As the simplest case, we study translational motion of an object in one dimension. In describing kinematics in one dimension (i.e., along the x-axis), we need to know the position of an object at any given moment. We may use a mathematical function to express the relationship between these two physical quantities. However, for describing the position x as a function of time t from the initial position x_1 at t_1 to the final position x_2 at t_2, there are two apparently similar, but, different quantities that we need to define. These two quantities are distance and displacement.

ONE DIMENSIONAL KINEMATICS

What is the difference between the distance and displacement?

Displacement is the change in the position of an object (i.e., $x_2 - x_1$). If there is no change in direction while going from the initial to the final position, then the magnitude of displacement and distance are the same. However, when there is a change in the direction, the two quantities are different. For example, if we start from Grand Forks and drive to Fargo, then the distance we travel is 72 miles which is the distance between the two cities. Also, the displacement is 72 miles due south from Grand Forks. However, if we travel from Grand Forks to Fargo and return back to Grand Forks, which involves changing the direction at Fargo, then the two quantities are different. For this case, the distance we traveled is 144 miles while the displacement is 0 miles. One important difference between these two is the direction. For displacement, both the magnitude and direction are needed, while for distance only the magnitude is needed. For this reason, displacement is a vector while distance is a scalar. In describing a vector, we need to specify both the magnitude and direction. So, do we know where we will be if we travel 72 miles from Grand Forks? The answer is "NO!" because we are not given information about the direction.

Another quantity we need to describe an object's motion is its speed or velocity. Both speed and velocity describe how fast the object is moving. However, speed is a scalar, while velocity is a vector (see Chapter 3). It should be noted that, when there is no change in the direction of motion, the average speed and the average velocity are the same. **Average speed** is defined as the distance traveled divided by the time to travel that distance and is written as

$$\bar{v} = \text{average speed} = \frac{\text{distance}}{\text{time elapsed}}, \tag{2.1}$$

where the bar over the symbol v denotes average. On the other hand, average velocity is defined as the displacement divided by the time elapsed and is written as

$$\bar{\vec{v}} = \text{average velocity} = \frac{\text{displacement}}{\text{time elapsed}} = \frac{\vec{x}_2 - \vec{x}_1}{t_2 - t_1} = \frac{\Delta \vec{x}}{\Delta t}. \tag{2.2}$$

The units for both speed and velocity are **m/s**. It should be noted we denote a vector by using either a bold faced symbol or a symbol with an arrow on top. The average velocity may depend on the time interval chosen to evaluate $\bar{v}$ because the object may change either its speed or direction, or both. Figure 2.1 illustrates this point using a plot of position versus time for a car. We can read off a lot of information from the plot. For example, at $t = 0$ min, the car is 10 km to the right of the origin. The value of x decreases until $t = 30$ min, indicating that the car is moving to the left. The car stops for 10 minutes at a position 20 km to the left of the origin. The car starts moving back to the right at $t = 40$ min. Finally, the car reaches the origin at $t = 80$ min. The position versus time curve plot in Fig. 2.1 also illustrates how the speed of the car is changing. From the initial position $x_i = 10$ km at $t_i = 0$ min to the intermediate position $x_1 = -20$ km at $t_1 = 30$ min, the average speed during this time interval is

$$\bar{v} = \frac{\Delta x}{\Delta t} = \frac{x_1 - x_i}{t_1 - t_i} = \frac{(-20 \text{ km}) - 10 \text{ km}}{30 \text{ min} - 0 \text{ min}} \times \frac{60 \text{ min}}{1 \text{ h}} = -60 \frac{\text{km}}{\text{h}},$$

where the minus sign indicates that the car is traveling backward. From the intermediate time $t_1 = 30$ min to $t_2 = 40$ min, the car is not moving at all since the position does not change. From the intermediate position $x_2 = -20$ km at $t_2 = 40$ min to the final position $x_f = 0$ km at $t_f = 80$ min, the average speed is

$$\vec{v} = \frac{\Delta x}{\Delta t} = \frac{x_f - x_2}{t_f - t_2} = \frac{0 \text{ km} - (-20 \text{ km})}{80 \text{ min} - 40 \text{ min}} \times \frac{60 \text{ min}}{1 \text{ h}} = 30 \frac{\text{km}}{\text{h}}.$$

Here, the plus sign indicates that the car is moving forward (i.e., along the +x direction).

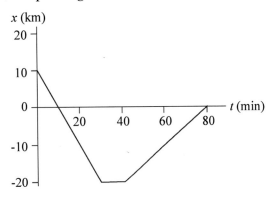

Figure 2.1: A plot of the position versus time for a car may be used to determine the average velocity during different time intervals and for the entire trip. These may not be the same.

Since the car changes direction at $t = 40$ min, we may apply the definition of average velocity to the entire trip and obtain

$$\vec{v} = \frac{\Delta \vec{x}}{\Delta t} = \frac{\vec{x}_f - \vec{x}_i}{t_f - t_i} = \frac{0 \text{km} - 10 \text{km}}{80 \text{min} - 0 \text{min}} \times \frac{60 \text{min}}{1 \text{h}} = -7.5 \frac{\text{km}}{\text{h}}.$$

Here, the result of -7.5 km/h indicates that the magnitude of the average velocity is 7.5 km/h, and that the direction is backward. The result contains information about how fast the car is moving and its direction, indicating that the average velocity is a vector. In general, a vector should be distinguished from a scalar, which has no information about direction.

There is another point we should recognize about our calculation of the average velocity from the plot in Fig. 2.1. The average velocity only describes an average over 80 minutes, but it does not tell us information about the velocity at any instant of time. As we have computed above, this is not 7.5 km/h along the -x direction but depends on the time interval. For accurate information about how fast the car is moving at a given time, we need to know the instantaneous velocity.

What is the difference between average velocity and instantaneous velocity?

Instantaneous velocity is defined in a similar way as the average velocity except that the time elapsed in Eq. (2.2) is taken to be much shorter than before. More precisely, it is defined over an infinitesimally short time period. The definition of instantaneous velocity is

$$\vec{v} = \lim_{\Delta t \to 0} \frac{\Delta \vec{x}}{\Delta t} \qquad (2.3)$$

where the units for instantaneous velocity is **m/s**. Let us consider the following example to understand the difference between average and instantaneous velocity. Suppose we are driving from Grand Forks to Fargo, which is about 72 miles south. If it takes us one hour to get there then we may say that our average speed is 72 miles per hour (mph). However, this is not what we would see on the speedometer of the automobile. For example, when we first enter the I-29 freeway we may find that the speed indicated on the speedometer is 50 mph. When there is no car nearby, we may decide to speed up a little by stepping

ONE DIMENSIONAL KINEMATICS

on the accelerator petal a little harder. Now, we find that the speed has increased to 80 mph. However, when we see a state trooper ahead of us, we would most definitely reduce the speed to below 75 mph so that we would not get a ticket for speeding. Throughout the trip, we may notice that the speed indicated on the speedometer is rarely 72 mph. So, what is the difference between the two values of speed? One value of speed is obtained by using the travel distance and time, while other is the one that indicated on the speedometer. As we may speculate, 72 mph is the average speed because the time interval we used to determine this value is one hour, which is too long to be instantaneous. However, the speed on the speedometer is closer to the instantaneous speed. Technically, this speed too is an average speed even though the time interval of few seconds is much shorter than one hour. In the context of measuring speed of an automobile, the time interval of few seconds is good enough to be considered as an infinitesimal time interval. Note that the time interval of few seconds may not be considered as infinitesimal in a different physical phenomenon such as the flapping of a hummingbird's wings. A hummingbird hovers in mid-air by rapidly flapping its wings 12 – 80 times per second (depending on the species). A few seconds is a long time for a phenomenon which happens on time scales shorter than one second.

It is noted that a **speedometer** is a gauge that measures and displays the instantaneous speed of a land vehicle. Although this device is now universally fitted to motor vehicles, the mechanical speedometers became available in the 1900s. Since about 1910, they have become available as standard equipment in all motor vehicles. Speedometers for other vehicles have specific names, such as pitometer log for a ship and airspeed gauge for an airplane. Also, many modern speedometers sense speed of the vehicle by using eddy-current, optical, or magnetic sensor.

Exercise Problem 2.1:

You are driving home from school steadily at 65 mph for 130 miles. It then begins to rain and you slow to 55 mph. You arrive home after driving 3 hours and 20 minutes. (a) How far is your hometown from school? (b) What was your average speed?

Problem Solving Strategy

What do we know?

We have information on the driving distance, time, and speed.

What concepts are needed?

We need to apply the definition of average speed.

Why?

Since the direction of motion does not change during the entire trip, the average velocity is the same as the average speed.

Solution:

First, we write the total time taken to drive from school to home as

$$T_{total} = T_{65mph} + T_{55mph} = 3\frac{1}{3}.$$

(1a)

By using the definition of average speed, we obtain the driving time for traveling 130 mi

$$\bar{v} = \frac{\Delta x}{\Delta t} = \frac{\Delta x}{T_{65mph}} \Rightarrow T_{65mph} = \frac{\Delta x}{\bar{v}} = \frac{130\,mi}{65\,mi/h} = 2\,h, \tag{1b}$$

Using Eq. (1a), we determine T_{55mph} and obtain $T_{55mph} = T_{tot} - T_{65mph} = (4/3)$h. Also, we compute the average speed by applying the definition once more and obtain

$$\bar{v} = \frac{\Delta x}{\Delta t} = \frac{\Delta x}{T_{55mph}} \Rightarrow \Delta x = \bar{v} T_{55mph} = 55\frac{mi}{h} \times \frac{4}{3}h = 73\,mi,$$

The total distance traveled is

$$\Delta x_{tot} = 130\,mi + 73\,mi = 203\,mi$$

Hence, the average speed for the entire trip is

$$\bar{v}_{total} = \frac{\Delta x_{total}}{T_{total}} = \frac{203\,mi}{10\,h/3} = 61\frac{mi}{h}$$

As indicated by the speedometer, the car's speed may be changing constantly throughout the trip. Hence, the car is accelerating since an object whose velocity is changing is said to be accelerating. **Acceleration** specifies how rapidly the velocity of an object is changing. Note that the acceleration a is *a* vector. We may define the average acceleration as

$$\vec{\bar{a}} = \text{average acceleration} = \frac{\vec{v}_2 - \vec{v}_1}{t_2 - t_1} = \frac{\Delta \vec{v}}{\Delta t}, \tag{2.4}$$

where the units for acceleration is **m/s²**. Note that acceleration is a vector. This means that both magnitude and direction are needed to specify it. Similar to the relationship between average velocity and instantaneous velocity, we may define the instantaneous acceleration as

$$\vec{a} = \lim_{\Delta t \to 0} \frac{\Delta \vec{v}}{\Delta t}. \tag{2.5}$$

We should remember that when the average acceleration does not change in time, it is the same as the instantaneous acceleration. For describing the kinematics of an object, we will first consider the case of constant acceleration. When the acceleration is constant, a plot of position versus time is described by a parabolic function as shown in Fig. 2.2. The curve shows that the displacement during the 0.2 s time interval almost right after the start of motion (i.e., small *t*) is small, indicating that the velocity is small. However, the displacement over a later 0.2 s time interval is greater, indicating that the velocity is increasing with time.

ONE DIMENSIONAL KINEMATICS

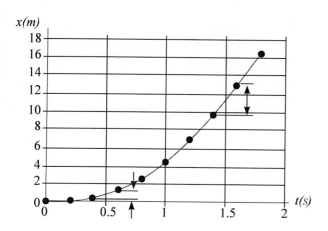

Figure 2.2: A plot of the displacement versus time indicates that the displacement in a later time interval is greater than that in an earlier time interval because the velocity is increasing, reflecting a constant positive acceleration.

There is one clear way to know whether the acceleration of an object remains constant or is changing. We need to examine a velocity-versus-time (i.e., v versus t) plot. If the acceleration is constant then, as we shall discuss below, the plot will show a straight line. However, if the acceleration changes with time, then the plot may show a curved line as shown in Fig. 2.3. From the velocity-versus-time plot, we can also obtain other interesting information about motion, such as the displacement of an object as a function of time.

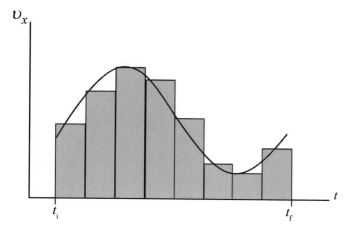

Figure 2.3: A curved line in the velocity versus time plot shows that the acceleration changes in time. Also, the distance traveled may be computed from the plot as the area under the curve.

The total displacement Δx is the sum of the areas of all the rectangles, which is approximately equal to the area under the curve. If the object starts from an initial position x_i with velocity v_i at $t = t_i$ and arrives at a final position x_f with velocity v_f at $t = t_f$, then the total displacement of the object (i.e., $x_f - x_i$) is the area under the velocity versus time curve from t_i to t_f. This area may be approximated by summing the areas of the shaded rectangles shown in Fig. 2.3. Of course, the approximation becomes improved when the width of each rectangle becomes narrower, increasing the number of rectangles. It is noted that the approximation becomes exact when the width approaches zero while the number of rectangles increases to infinity. In this limiting case, we are basically doing calculus which is beyond the scope of this book. So, we do not consider the case of a non-linear curve in the velocity-versus-time plot. From now on we will restrict our discussion of kinematics to cases where the velocity-versus-time plot consists of straight lines only.

Understand-the-Concept Question 2.1:

If the velocity of an object is zero, does it mean that the acceleration is zero?

If the acceleration is zero, does it mean that the velocity is zero?

Explanation:

Velocity and acceleration are related to each other, but they are two different quantities. This means that the acceleration does not have to be zero when the velocity is zero. An example of this case is a drag racing car standing at the start of a race. Also, velocity may not be zero, even when the acceleration is zero. A good example of this point is a car traveling at a constant speed of 65 mph under cruise control on a highway. In this case, the acceleration is zero since the speed is not changing, but the velocity is non-zero.

Drag racing is a competition in which specially prepared automobiles or motorcycles compete. The race follows a straight course from a standing start to a finish over a measured distance of, most commonly, a quarter mile (roughly 400 m). Usually, two cars compete at a time to be first to cross a set finish line. At a drag race, the automobiles quickly increase their speed from rest, reflecting their large acceleration.

Uniformly Accelerated Motion

Let us now derive a set of equations relating displacement, velocity, and time in the case of a uniformly accelerating object. As we noted above, when the acceleration is uniform, the average acceleration is the same as the instantaneous acceleration. This indicates that we may use the definition of both velocity and acceleration to derive the kinematic relationships. These relationships include the position and velocity as a function of time. First, we derive the position versus time relationship. In deriving this relation, we take $t_1 = 0$ and $t_2 = t$, for simplicity. Also, we assume that the initial position x_1 and velocity v_1 of an object are given by $x_1 = x_o$ and $v_1 = v_o$. Moreover, we assume that the direction of motion does not change and start with the definition of average velocity,

$$\bar{v} = \frac{\Delta x}{\Delta t} = \frac{x_2 - x_1}{t_2 - t_1} = \frac{x - x_o}{t} \Rightarrow x = x_o + \bar{v}t, \qquad (2.6)$$

and the definition of average acceleration,

$$a = \frac{\Delta v}{\Delta t} = \frac{v - v_o}{t}. \qquad (2.7)$$

ONE DIMENSIONAL KINEMATICS

From Eq. (2.7), we may obtain the velocity v as

$$v = v_0 + at. \tag{2.8}$$

The velocity versus time relationship of Eq. (2.8) is one of the four equations we will use for solving kinematics problems. The linear relation between the velocity and time may be seen easily from the graph shown in Fig. 2.4. The intercept at $t = 0$ represents the initial velocity v_0 and the slope of this linear curve is the acceleration.

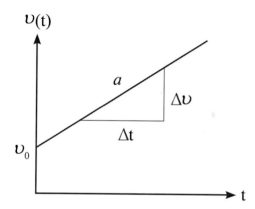

Figure 2.4: A straight line in the velocity versus time plot indicates that the acceleration $a = \Delta v / \Delta t$, which is the slope of the line, is constant. The intercept at $t = 0$ corresponds to the initial velocity v_0.

Additional information may be extracted from the linear relationship between velocity and time. Since the velocity increases at a uniform rate, the average velocity may be evaluated as the midpoint value between the initial and final velocities. Hence, the average velocity, according to Fig. 2.4, may be written as

$$\bar{v} = \frac{v_0 + v}{2}. \tag{2.9}$$

Equation (2.9) is the second of the four kinematics equations. We now obtain the third kinematics equation by using equations (2.6), (2.8), and (2.9). We rewrite Eq. (2.6) by using the average speed equation of Eq. (2.9) and by using the speed-versus-time equation of Eq. (2.8). With a little bit of algebraic calculation, we may derive the relationship between position and time as

$$x = x_0 + \bar{v}t \;\Rightarrow\; x = x_0 + v_0 t + \frac{1}{2} a t^2. \tag{2.10}$$

Finally, the fourth kinematics equation may be obtained by removing the time variable from Eq. (2.6). To do this, we use Eq. (2.8) and obtain the time t as

$$v = v_0 + at \;\Rightarrow\; t = \frac{v - v_0}{a}. \tag{2.11}$$

Now, we substitute equations (2.9) and (2.11) into Eq. (2.6) and obtain

$$x = x_0 + \frac{v_0 + v}{2} t = x_0 + \frac{v^2 - v_0^2}{2a}. \tag{2.12}$$

Again, with a simple algebraic calculation, we may rewrite Eq. (2.12) in a useful form

$$x = x_o + \frac{v^2 - v_o^2}{2a} \Rightarrow v^2 = v_o^2 + 2a(x - x_o). \tag{2.13}$$

Now we have a set of four kinematics equations (see Table 2.1 below) to describe the motion of an object. These four equations are equations (2.8), (2.9), (2.10), and (2.13). Note that these four kinematics equations do not represent any physical principles. They are just the definition of average velocity and acceleration which are rewritten in a more convenient way to describe motion of an object. This means a good starting point for solving a kinematics problem is one of these four equations. However, it is helpful to identify which equation (out of four) is a useful starting point. A useful starting point for approaching a kinematics problem may depend on information provided in the problem and the physical quantity needs to be determined.

With the four kinematics equations we derived above, we are ready to tackle any one dimensional problem. However, the question is "how do we do it?" To get an idea about how to apply the four kinematics equations, we consider a Saturn V rocket lifting off from a launch pad as shown in Fig. 2.5. There are seven red dots representing seven vertical positions measured at seven different times. The vertical position y of the rocket is plotted as a function of time t. From the plot, we may notice that the each measurement was taken at a roughly 0.3 second interval with the first measurement at $t = 0.6$ s. The vertical position is increasing faster than linearly in time t. According to Eq. (2.10), this means the rocket is accelerating (i.e., $a \neq 0$) upward. The position-versus-time graph of Fig. 2.5 for the rocket is a parabola which implies that it is being constantly accelerated. However, the value of acceleration may not be extracted directly from the curve. This task may be accomplished conveniently from a velocity-versus-time plot, as shown in the lower panel. A linear fit to the data indicates that the speed of the rocket is increasing linearly in time. According to Eq. (2.8) and Fig. 2.4, the slope of this linear curve is the acceleration. From the data in Fig. 2.5, the estimated acceleration is $a = \Delta v_y/\Delta t = (27 \text{ m/s})/1.5 \text{ s} = 18 \text{ m/s}^2$. Also, the initial velocity of the rocket may be obtained from the plot by extrapolating the curve to $t = 0$. Since the linear line goes through the origin, the y-intercept (which measures the initial velocity) is zero.

In our discussion of the Saturn V rocket launch, we used equations (2.8) and (2.10) to analyze both the position and vertical velocity data. If these position and velocity data are not available to us, would we be able to determine the acceleration? The answer is "Yes!" Of course, for this case, we need to use Eq. (2.13) which does not involve the time t variable. By solving Exercise Problem 2.2 below, we shall see how this velocity-versus-position relation may be applied in solving a kinematics problem.

Table 2.1: Four kinematics equations derived from the definition of average velocity and acceleration. These equations are useful for solving one-dimensional kinematics problems with a uniform acceleration.

$u = u_0 + at$	speed-versus-time
$\bar{u} = \dfrac{u_0 + u}{2}$	average speed under constant acceleration
$x = x_o + u_0 t + \dfrac{1}{2}at^2$	position-versus-time
$u^2 = u_0^2 + 2a(x - x_o)$	speed-versus-position

ONE DIMENSIONAL KINEMATICS

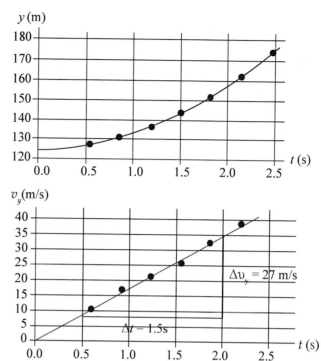

Figure 2.5: The vertical position y (upper right panel) and speed v_y (lower right panel) of a Saturn V rocket is plotted as a function of time t to illustrate the kinematics of the rocket which has a constant acceleration.

Exercise Problem 2.2:

You are designing an airport for small planes. One kind of airplane that might use this airfield must reach a speed before takeoff at least 27.8 m/s (100 km/h), and can accelerate at 2.00 m/s². (a) If the runway is 150 m long, can this plane reach proper speed to take off? (b) If not, what minimum length must the runway have?

Problem Solving Strategy

What do we know?

We know the takeoff velocity at the end of the runway. Also, the airplane has a constant acceleration during takeoff.

What concepts are needed?

We need to apply the velocity-versus-position relation of Eq. (2.13).

Why?

In applying Eq. (2.13), the unknown quantity is the length of runway. So, we need to compute the distance and compare with the 150 m runway.

> **Solution:**
>
> Note that we are given the final speed and acceleration of the airplane as $v = 27.8$ m/s and $a = 2.00$ m/s², respectively. For simplicity, we set the initial position $x_o = 0$ and initial velocity $v_o = 0$. We need to find the final position x by using the information provided in the problem and compare it with 150 m. We start with the equation
>
> $$v^2 = v_0^2 + 2a(x - x_0) \Rightarrow v^2 = 2ax.$$
>
> We solve for the distance x and obtain
>
> $$x = \frac{v^2}{2a} = \frac{\left(27.8 \,\text{m}/\text{s}\right)^2}{2 \times 2.0 \,\text{m}/\text{s}^2} = 193 \,\text{m}.$$
>
> Our calculation indicates that a 193 m long runway is needed to reach the proper speed. Therefore, the 150 m runway is too short.

As you may have guessed the takeoff speed of airplanes is not always the same. The takeoff speed for commercial airliners ranges from 250 km/h (150 mph) for a 737 weighing 100,000 pounds to 360 km/h (225 mph) for a Concorde, which is no longer in service, weighing 400,000 pounds. The takeoff speed required varies with air density, aircraft gross weight, lift coefficient, and aircraft configuration (flap or slat position, as applicable). Air density is affected by factors such as field elevation and air temperature. In a single-engine or light twin-engine aircraft, the pilot calculates the length of runway required to take off and clear any obstacles to ensure sufficient runway to use for takeoff. In most cases, a safety margin can be added to provide the option to stop on the runway in case of a failure to takeoff.

2.2 FREE FALL MOTION

Free fall motion is the most common example of uniform motion in one dimension. A free falling object is an object that is falling under the sole influence of gravity. The **acceleration due to gravity** is g ≈ **9.81 m/s²**. This value is called Earth's gravity and refers to the acceleration of objects due to Earth on or near its surface. We should note that this approximate value ignores the effects of air resistance and means that the speed of an object falling freely near the Earth's surface will increase by about 9.81 meters (about 32.2 ft) per second every second. The concept of free fall was first introduced by Galileo Galilei. For free fall of an object near the Earth's surface, Galileo (1564 – 1642) postulated that all objects would fall with the same constant acceleration in the absence of air or other

Figure 2.6: Galileo showed that all objects near the surface of Earth fall with the same acceleration, regardless its weight and size, if the effects of air resistance are negligible.

ONE DIMENSIONAL KINEMATICS

resistance. As schematically illustrated in Fig. 2.6, Galileo's classic experiment led to the finding that all objects free fall at the same rate, regardless of their mass. According to legend, Galileo dropped balls of different mass from the Leaning Tower of Pisa to support his idea that their time of descent was independent of their mass. This was contrary to what Aristotle had taught: that heavy objects fall faster than lighter ones, in direct proportion to weight.

As noted above, the main assumption of Galileo for performing the experiment from the Leaning Tower of Pisa was that free-falling objects do not encounter air resistance. However, for most objects moving near the surface of Earth, it is difficult to avoid the effects of air resistance. How does air resistance affects falling motion? Let us discuss briefly the effects of air resistance.

Effects of Air Resistance

Air resistance is the slowing effect which air creates on an object as it moves through the atmosphere. (This is also called wind resistance.) The effects of air resistance may be easily observable. For example, objects or people in free fall will find their descent slowed by air. This effect depends on a number of different factors. For the motion of **aircraft** or any vehicle, speed is one such factor. Also, the air resistance depends on the shape of the object. A larger object tends to experience greater air resistance. For instance, a larger object must push more air out of the way in order to move through it, so a larger area means more air resistance. This is the reason that fast cars and airplanes need to be streamlined.

Air resistance has a variety of other effects. Some of them are easily observable as discussed above. Humankind has always been able to observe the effects of air resistance, but the physical factors involved were not understood until the seventeenth century. Trying to understand the principle of gravity, Galileo used experiments to test Aristotle's thesis that heavier objects fall faster than lighter ones. He was able to prove this was not true by showing that the gravitational force affects every object in the same way. However, he realized the lighter objects were slowed by resistance from the air, and the heavier objects had enough weight to counter this factor. We now know that air resistance is caused by the collision of a solid object with gas molecules in the atmosphere. When there is agreater number of air molecules, the resistance is greater. In practice, this means an object with a wider surface encounters greater resistance as illustrated in Fig. 2.7. Also, a faster object has greater

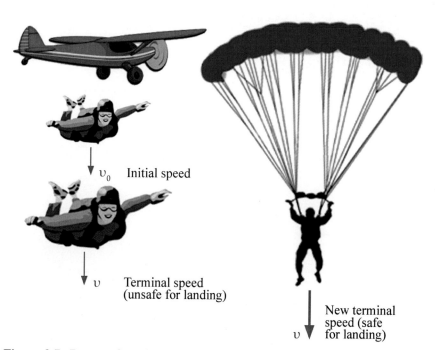

Figure 2.7: Due to air resistance, all free-falling objects will reach the terminal velocity. The value of this terminal velocity depends on the amount of air resistance experienced by the object.

air resistance because it contacts more air molecules in a given span of time. When the resistance of an object in free fall equals the pull of gravity on the object, it no longer accelerates. This is called **terminal velocity**, and it varies depending on factors such as weight, surface area, and speed. These effects can be observed by watching skydivers in action as shown in Fig. 2.7. Before activating her **parachute**, the skydiver falls at terminal velocity, seemingly held aloft by the air. If she draws her limbs in and points her body downward, her speed will increase as the resistance is decreased. By positioning her body parallel to the ground and spreading her arms and legs, she can slow her descent. Once she opens her parachute, the air resistance will increase, slowing her further. The terminal velocity of the opened parachute is low enough that she will strike the ground at a survivable speed.

Exercise Problem 2.3:

Suppose a ball is thrown downward with an initial velocity of 3.00 m/s from a tower 70.0 m high. (a) What would be its position after 1.00 s and 2.00 s? (b) What would be its speed after 1.00 s and 2.00 s?

Solution:

We use Eq. (2.10) to compute the position of the ball from the top of the tower ($y_o = 0$):

$$y = y_0 + v_0 t + \frac{1}{2}at^2 \Rightarrow y = v_0 t + \frac{1}{2}at^2.$$

At $t = 1.00$ s, the ball's vertical position from the top is

$$y = 3.00\frac{m}{s} \times 1.00s + \frac{1}{2} \times 9.80\frac{m}{s^2} \times (1.00s)^2 = 7.90 m.$$

At $t = 2.00$ s, the ball's vertical position from the top is

$$y = 3.00\frac{m}{s} \times 2.00s + \frac{1}{2} \times 9.80\frac{m}{s^2} \times (2.00s)^2 = 25.6 m.$$

For computing the speed, we use the speed versus time equation,

$$v = v_0 + at.$$

We obtain that, at $t = 1.00$ s, the downward speed is

$$v = 3.00\frac{m}{s} + 9.80\frac{m}{s^2} \times 2.00s = 12.8\frac{m}{s}$$

and, at $t = 2.00$ s, the downward speed is

$$v = 3.00\frac{m}{s} + 9.80\frac{m}{s^2} \times 2.00s = 22.6\frac{m}{s}$$

ONE DIMENSIONAL KINEMATICS

In solving Exercise Problem 2.3, we chose the top of the tower as the reference point and the downward direction as positive. We should note that our choice of the reference point is arbitrary. Also, our choice of the coordinate adapted for the problem is arbitrary. However, the sign of the gravity depends on the choice of the +y direction because the gravitational acceleration always points toward the center of Earth (assuming that Earth is a perfect sphere with a uniform weight distribution). Our choice of the downward direction as the +y direction indicates that the gravitational acceleration is $a = g \approx 9.80$ m/s². However, if the +y direction is upward, then the acceleration is $a = -g \approx -9.80$ m/s². Here, the sign indicates the direction of the gravitational acceleration relative to the chosen coordinate.

Exercise Problem 2.4:

A stone is thrown vertically upward with a speed of 12.0 m/s from the edge of a cliff which is 75.0 m high. (a) How much later does it reach the bottom of the cliff? (b) What is its speed just before hitting? (c) What total distance did it travel?

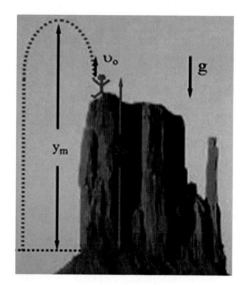

PROBLEM SOLVING STRATEGY

What do we know?

A stone is thrown vertically upward with an initial speed. What goes up must come down!

Concepts are needed?

We need to apply the kinematics equations for the upward and downward motion, separately.

Why?

Although the acceleration due to gravity always points towards the center of the Earth, it slows the stone moving upward and it increases the speed of the stone moving downward.

Solution:

When the stone is thrown vertically upward, the vertical speed becomes zero when it reaches the maximum height. We use this condition to rewrite Eq. (2.8) and obtain

$$v = v_o + at \Rightarrow 0 = v_o - gt_u \tag{4a}$$

Using Eq. (4a), we solve the time t_u taken by the stone to reach the maximum height and obtain

$$t_u = \frac{v_o}{g} = \frac{12.0 \, \text{m/s}}{9.8 \, \text{m/s}^2} = 1.22 \, \text{s} \tag{4b}$$

Now, we compute the distance the stone goes up from the cliff by using equations (2.10) and (4b). We obtain

$$y_m - y_0 = v_0 t - \frac{1}{2}gt^2 = 12.0 \frac{\text{m}}{\text{s}} \times 1.22 \text{s} - \frac{1}{2} \times 9.8 \frac{\text{m}}{\text{s}^2} \times (1.22 \text{s})^2 = 7.35 \text{m} \tag{4c}$$

The total distance traveled by the stone is

$$d = y_m + (y_m - y_o) = 2(y_m - y_o) + y_o = 2 \times 7.35 + 75.0\,\text{m} = 89.7\,\text{m},$$

Where y_o is the initial vertical position of the stone. Once again, we use Eq. (2.10) to compute the time the stone takes to fall a distance of y_m, the maximum height. We use the condition that $y = 0$ at the bottom of the cliff and obtain

$$y = y_o + v_o t + \frac{1}{2}at^2 \Rightarrow 0 = y_m - \frac{1}{2}gt_d^2. \quad \text{(4d)}$$

Using Eq. (4d), we solve for t_d and obtain

$$t_d = \sqrt{\frac{2y_m}{g}} = \sqrt{\frac{2(75.0\,\text{m} + 7.35\,\text{m})}{9.8\,\text{m}/\text{s}^2}} = 4.1\,\text{s} \quad \text{(4e)}$$

The total time that T the stone is in the air is

$$T = t_u + t_d = 1.2\,\text{s} + 4.1\,\text{s} = 5.3\,\text{s}$$

The speed just before hitting the ground may be computed by using Eq. (2.8). We obtain

$$v = v_o + at \Rightarrow v = -gt_d = -9.8\,\frac{\text{m}}{\text{s}^2} \times 4.1\,\text{s} = -40.2\,\frac{\text{m}}{\text{s}}$$

In solving this type of problem, it may be useful for us to remember that assuming the +y-axis is upwards the condition for determining this maximum height is when the vertical speed is zero. If the vertical speed is positive, then the object continues to move upward and the maximum height is not yet reached. However, if the vertical speed is negative, then the object is coming back down towards Earth.

As we know, gravity is important for motion of an object on or near the surface of Earth. For example, due to gravity, any object thrown vertically upward with a speed less than the escape velocity will reach a maximum height and will return back to the surface. Here the escape speed is defined as the speed needed to "break free" from the gravitational attraction on the Earth without further propulsion, which depends of the value of gravitational acceleration. However, the value of gravity is not the same everywhere on or near the surface of Earth.

2.3 REAL FIGURE OF THE EARTH: THE GEOID AND GRAVITY

As we discussed in Sec. 2.2, the gravitational acceleration of Earth is an important quantity for the kinematics equations, but its precise strength depends on location as well as many other factors. It should be noted that the nominal "average" value at the Earth's surface is known as standard gravity and this value, by definition, is 9.80665 m/s² (32.1737 ft/s²). The precise strength of Earth's gravity g at the equator of Earth is 9.78033 m/s², indicating that g at other places may be higher. So, what are the reasons for the variation in g?

One reason for the variation in g is that Earth is not perfectly symmetrical. If Earth is a perfect sphere of spherically uniform density (i.e., density varies only on the distance from center) then Earth would produce a gravitational field of uniform magnitude at all points on its surface, yielding the same g everywhere and always pointing directly towards the center of sphere. However, as illustrated in Fig. 2.8, the Earth is not a perfect sphere. It deviates slightly from this ideal. Consequently, there are slight deviations in both the magnitude and direction of gravity across its surface, and the weight, which is the net force exerted on the object by Earth, varies due to this deviation. Also, the presence of other factors, including inertial response to the Earth's rotation, affects g. Hence, the gravity measured by using a scale or plumb bob is called either "effective gravity" or "apparent gravity". Other parameters affecting the apparent or actual strength of Earth's gravity include i) latitude, ii) altitude, and iii) the local topography and geology, iv) other factors. Now, we examine the effects of these parameters on gravity.

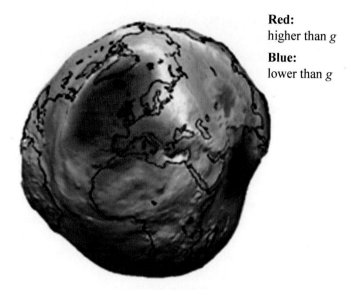

Red: higher than g

Blue: lower than g

Figure 2.8: A schematic picture illustrating that Earth is not a perfect sphere. This is one of the reasons the value of gravity is not uniform on the surface of Earth. In the diagram, a deviation from a perfect sphere is exaggerated.

i. **Latitude:** There are two major reasons for the dependence of g on latitude. The first reason is that the inertia produced by Earth's rotation at latitudes nearer the Equator is stronger than at polar latitudes. This counteracts the Earth's gravity to a small degree and reduces the downward acceleration of falling objects, up to a maximum of 0.3% at the Equator. The second reason is that the Earth's equatorial bulge, which is also caused by inertia, causes objects at the Equator to be farther from the planet's center than objects at the poles. These two effects cause a change in both the magnitude and direction of the true gravitational force relative to an ideal spherical Earth. More specifically, an object at the Equator experiences a weaker gravitational pull than an object at the poles. In combination, the equatorial bulge and the effects of the Earth's inertia mean that sea-level gravitational acceleration increases from about 9.780 m/s² at the Equator to about 9.832 m/s² at the poles, indicating that an object will weigh about 0.5% more at the poles than at the Equator. These two factors also influence the direction of the effective gravity. The effective gravity at anywhere on Earth, away from either the Equator or poles, points not exactly toward the center of the Earth. Roughly half of this deviation is due to inertia while other half is due to the extra mass around the Equator.

ii. **Altitude:** The value of gravity varies relative to the height of an object. Gravity decreases with altitude, since greater altitude means greater distance from the Earth's center. This dependence on altitude may be seen from Newton's universal law of Gravitation which we will discuss in Chapter 4. Assuming that all other things are equal, an increase in altitude from sea level to the top of Mount Everest (8,848 meters) causes a weight to decrease about 0.28%. An additional factor affecting apparent weight is the decrease in air density at altitude, which lessens an object's buoyancy (see Chapter 9). It is a worth noting a common misconception that astronauts in orbit are weightless because they have flown high enough to "escape" the Earth's gravity. This is not true. In fact, at an altitude of 400 kilometers (250 miles), equivalent to a typical orbit of the Space Shuttle, gravity is still nearly 90% as strong as at the Earth's surface, and apparent weightlessness actually occurs because orbiting objects are in free-fall.

iii. **Local topography and geology:** On the surface, local variations in topography, such as the presence of mountains, and geology, such as the density of rocks in the vicinity, cause fluctuations in the Earth's gravitational field. This is known as gravitational anomalies. Some of these anomalies can be very extensive, resulting in bulges in sea level and throwing pendulum clocks out of synchronization. The fluctuations are measured with highly sensitive gravimeters because these are small. When the effects of topography and other known factors are subtracted, the resulting data used to draw conclusions. These techniques are used by either prospectors or surveyors to find oil and mineral deposits. Denser rocks which often contain mineral ores cause higher than normal local gravitational fields on the Earth's surface, while less dense sedimentary rocks cause lower local gravitational fields.

iv. **Other factors:** As we will discuss in Chapter 9, objects in fluid, such as air, experience a supporting buoyant force which reduces the apparent strength of gravity. As measured by an object's weight, the magnitude of this buoyancy depends on air density and, hence, air pressure yields a smaller apparent weight. The concept of *apparent weight* will be discussed in Chapter 4. Also, the Moon and Sun affect the gravity of Earth. The gravitational effects of the Moon and the Sun which also cause the tides have a very small effect on the apparent strength of Earth's gravity. However, this depends on their relative positions. Typical variations are 2 μm/s² over the course of a day.

SUMMARY

i. Movement of any object may be described in terms of **two types of motion.** These two types are
   1. translational motion, and   2. rotational motion.

ii. **Displacement** is the change in the position of the object (i.e., $x_2 - x_1$). If there is no change in the direction of motion from the initial to final position, then the displacement and distance are the same. However, if there is a direction change between the initial and final positions, these two quantities are different, sine displacement is a vector while distance is a scalar.

iii. **Average speed** is defined as the distance traveled by an object divided by the time it takes to travel that distance. According to this definition, we may write

$$\bar{v} = \text{average speed} = \frac{\text{distance}}{\text{time elapsed}}.$$

Similarly, average velocity is defined as the displacement divided by the time elapsed

$$\bar{\vec{u}} = \text{averge velocity} = \frac{\text{displacement}}{\text{time elapsed}} = \frac{\vec{x}_2 - \vec{x}_1}{t_2 - t_1} = \frac{''\vec{x}}{''t}.$$

The units of velocity are **m/s**. Velocity is a vector while speed is a scalar.

iv. An object whose velocity is changing is accelerating. **Acceleration** specifies how rapidly the velocity of an object is changing and is defined as

$$\bar{\vec{a}} = \text{average acceleration} = \frac{\vec{v}_2 - \vec{v}_1}{t_2 - t_1} = \frac{\Delta \vec{v}}{\Delta t}$$

where the units of acceleration are **m/s²**. Acceleration is a vector, indicating that both magnitude and direction are needed to specify the quantity. When the average acceleration is constant than the instantaneous acceleration

ONE DIMENSIONAL KINEMATICS

$$\vec{a} = \lim_{\Delta t \to 0} \frac{\Delta \vec{v}}{\Delta t}$$

is identical to the average acceleration.

v. Four kinematics equations that are useful for solving one-dimensional kinematics problems are:

$$v = v_0 + at \text{ (speed-versus-time)}$$

$$\bar{v} = \frac{v_0 + v}{2} \text{ (average speed under constant acceleration)}$$

$$x = x_o + v_o t + \frac{1}{2}at^2 \text{ (position-versus-time)}$$

$$v^2 = v_0^2 + 2a(x - x_o) \text{ (speed-versus-position)}$$

These equations can all be derived from the definition of average velocity and acceleration.

vi. A free falling object is an object that is falling under the influence of gravity. For free fall, Galileo postulated that all objects would fall with the same constant acceleration in the absence of air or other resistance. The acceleration due to gravity is $g \approx 9.81 \text{ m/s}^2$ which is also known as the **Earth's gravity**.

MORE WORKED PROBLEMS

Problem 2.1:

A train normally travels at a uniform speed of 72 km/h on a long stretch of straight level track. On a particular day, the train must take a 2.0-min stop at a station along this track. If the train decelerates at a uniform rate of 1.0 m/s² and, after the stop, accelerates at a rate of 0.50 m/s², how much time is lost because of stopping at the station?

Solution:

According to the problem, the train will first decelerate, and then it will stop for 2.0 min. The train will accelerate to 72 km/h. First, we convert the speed from 'km/h' to 'm/s' and obtain

$$72 \frac{\text{km}}{\text{h}} \times \frac{1000 \text{m}}{1 \text{km}} \times \frac{1 \text{h}}{3600 \text{s}} = 20 \frac{\text{m}}{\text{s}}$$

Using the kinematics equation, $v = v_0 + at$, we find the time lost during both deceleration and acceleration. During deceleration, the time to stop and the distance travelled during this time is

$$\Delta t_1 = \frac{\Delta v}{a_1} = \frac{0 \frac{\text{m}}{\text{s}} - 20 \frac{\text{m}}{\text{s}}}{-1.0 \frac{\text{m}}{\text{s}^2}} = 20 \text{s} \text{ and } \Delta x_1 = \bar{v} \Delta t = \frac{20 \frac{\text{m}}{\text{s}} + 0 \frac{\text{m}}{\text{s}}}{2} \times 20 \text{s} = 200 \text{m}$$

It would have taken the train $t = d/v = 10$ s to travel 200 m. So, the train lost only 10 s (= 20 s − 10 s) during deceleration. However, during acceleration, the time to obtain the cruising speed and the distance travelled is

$$\Delta t_2 = \frac{\Delta v}{a_2} = \frac{20\frac{m}{s} - 0\frac{m}{s}}{0.5\frac{m}{s^2}} = 40\,s \quad \text{and} \quad \Delta x_2 = \bar{v}\,\Delta t = \frac{0\frac{m}{s} + 20\frac{m}{s}}{2} \times 40\,s = 400\,m$$

It would have taken the train $t = d/v = 20$ s to travel 400 m. So, it lost only 20 s (= 40 s − 20 s) during acceleration. Now, we combine the result to find that the total time lost by the train is 150 s (= 10 s + 2.0 min + 20 s).

Problem 2.2:

A student at a window on the second floor for a dorm sees his math professor walking on the sidewalk besides the building. He drops a water balloon from 18.0 m above the ground when the professor is 1.00 m from the point directly beneath the window. If the professor is 170 cm tall and walks at a rate of 0.450 m/s, does the balloon hit her? If not, how close does it come?

Solution:

We need to find the time it takes for the water balloon to reach the level of the professor's head. We note that the vertical distance between the initial position of the balloon and the professor's head given in the problem is $y - y_0 = -(18.0\,m - 1.70\,m) = -16.3\,m$. The initial velocity is zero (i.e., $v_0 = 0$).

By using the kinematics equation $y - y_0 = v_0 t - \frac{1}{2}gt^2$, we find the time t as

$$t = \sqrt{\frac{-2(y - y_0)}{g}} = \sqrt{\frac{-2(-16.3\,m)}{9.80\,m/s^2}}$$

During the time t, the professor advances a distance equal to

$$d = vt = 0.450\,m/s \times 1.824\,s = 0.821\,m < 1.00\,m$$

Note that: That $y - y_0$ is a vertical displacement of the water balloon and d is a horaizontal displacement. This means that the water balloon does not hit her. We now calculate the time it takes for the balloon to hit the ground

$$t = \sqrt{\frac{-2(y - y_0)}{g}} = \sqrt{\frac{-2(-18.0\,m)}{9.80\,m/s^2}} = 1.917\,s$$

During this time, the professor advances a distance of $d = vt = 0.450\,m/s \times 1.917\,s = 0.862\,m$ which is less than 1.00 m. So the water balloon hits 1.00 m − 0.862 m = 0.14 m in front of the professor.

ONE DIMENSIONAL KINEMATICS

Problem 2.3:

A locomotive is accelerating at 1.6 m/s². It passes through a 20.0-m wide crossing (along the direction of the locomotive's motion) in a time 2.4 s. After the locomotive leaves the crossing, how much time is required until its speed reaches 32 m/s?

Solution:

First, we need to determine the initial speed u_o of the locomotive by using the kinematics equation, $x = x_o + v_o t + \frac{1}{2} a t^2$, and obtain

$$d = x - x_o = v_o t + \frac{1}{2} a t^2 \Rightarrow v_o = \frac{d - \frac{1}{2} a t^2}{t} = \frac{20\,\text{m} - \frac{1}{2} \times 1.6 \frac{\text{m}}{\text{s}^2} \times (2.4\,\text{s})^2}{2.4\,\text{s}} \cong 6.4 \frac{\text{m}}{\text{s}}$$

Now, to determine the time required to reach the speed $v = 32$ m/s, we use another kinematics equation

$$v = v_{after\,crossing} + at \Rightarrow t = \frac{v - v_{after\,crossing}}{a} = \frac{32 \frac{\text{m}}{\text{s}} - 10.2 \frac{\text{m}}{\text{s}}}{1.6 \frac{\text{m}}{\text{s}^2}} \cong 14\,\text{s}$$

We should note that the velocity of the locomotive after it leaves the crossing is

$$v_{after\,crossing} = v_o + at = 6.4 \frac{\text{m}}{\text{s}} + 1.6 \frac{\text{m}}{\text{s}^2} \times 2.4\,\text{s} = 10.2 \frac{\text{m}}{\text{s}}$$

Problem 2.4:

A motorist is driving at 20 m/s when she sees that a traffic light 200 m ahead has just turned red. She knows that this light stays red for 15 s, and she wants to reach the light just as it turns green again. It takes her 1.0 s to step on the brakes and begin slowing at a constant deceleration. What is her speed as she reaches the light at the instant it turns green?

Solution:

First, we compute the distance traveled during the reaction time. The reaction time is 1.0 s and the motion during this time is

$$x_1 = x_o + v_o (t_1 - t_o) = 0\,\text{m} + 20 \frac{\text{m}}{\text{s}} \times 1.0\,\text{s} = 20\,\text{m}$$

Knowing the final position x_2 is 200 m ahead (i.e., $x_2 = 200$ m), we compute the deceleration during slowing down from the position-versus-time equations of

$$x_2 = x_1 + v_1 (t_2 - t_1) + \frac{1}{2} a_1 (t_2 - t_1)^2 = 200\,\text{m}$$

and obtain the deceleration as

$$a_1 = \frac{x_2 - x_1 - v_1(t_2 - t_1)}{\frac{1}{2}(t_2 - t_1)^2} = \frac{200\,\text{m} - 20\,\text{m} - 20\frac{\text{m}}{\text{s}} \times (15\,\text{s} - 1.0\,\text{s})}{\frac{1}{2} \times (15\,\text{s} - 1.0\,\text{s})^2} \cong -1.02\,\frac{\text{m}}{\text{s}^2}$$

Now, by using the velocity-versus-time relation, the final speed v_2 can be obtained as

$$v_2 = v_1 + a_1(t_2 - t_1) = 20\,\frac{\text{m}}{\text{s}} + \left(-1.02\,\frac{\text{m}}{\text{s}^2}\right) \times (15\,\text{s} - 1.0\,\text{s}) = 5.7\,\frac{\text{m}}{\text{s}}$$

PROBLEMS

1. Due to continental drift, the North American and European continents are drifting apart at an average speed of about 3 cm per year. At this speed, how long (in years) will it take for them to drift apart by another mile?

2. Alan leaves Los Angeles at 8:00 A.M. to drive to San Francisco, 400 mi away. He travels at a steady 50 mph. Beth leaves Los Angeles at 9:00 A.M. and drives a steady 60 mph. (a) Who gets to San Francisco first? (b) How long does the first to arrive have to wait for the second?

3. In a 10.0 km race, one runner runs at a steady 11.0 km/h and another runs at 14.0 km/h. How far from the finish line is the slower runner when the faster runner finishes the race?

4. A bicyclist makes a trip that consists of three parts, each in the same direction (due east) along a straight road. During the first part, she rides for 25 minutes at an average speed of 7.5 m/s. During the second part, she rides for 40 minutes at an average speed of 5.5 m/s. Finally, during the third part, she rides for 8.0 minutes at an average speed of 15 m/s. (a) How far has the bicyclist traveled during the entire trip? (b) What is her average velocity for the trip?

5. In reaching her destination, a backpacker walks with an average velocity of 1.40 m/s, due west. This average velocity results because she hikes for 6.85 km with an average velocity of 2.66 m/s, due west, turns around, and hikes with an average velocity of 0.447 m/s, due east. How far east did she walk?

6. An insect crawls along the edge of a rectangular swimming pool of length 30 m and width 20 m. If it crawls from one corner to another which is across in the diagonal direction in 60 min, (a) what is its average speed, and (b) what is the magnitude of its average velocity?

7. A person takes a trip, driving with a constant speed of 90.0 km/h, except for a 30.0-min rest stop. If the person's average speed is 77.8 km/h, (a) how much time is spent on the trip and (b) how far does the person travel?

8. An earthquake releases two types of traveling waves called "transverse" and "longitudinal." (See in Chapter 13.) The average speeds of transverse and longitudinal seismic waves in rock are 9.2 km/s and 5.1 km/s, respectively. A seismograph records the arrival of the transverse waves 70 s before that of the longitudinal waves. Assuming that the waves travel in straight lines, how far away is the center of the earthquake from the seismograph?

9. The driver of a pickup truck going 100 km/h applies the brakes, giving the truck a uniform deceleration of 6.50 m/s² in 10.0 seconds. (a) How far does it travel during this time? (b) What is the speed of the truck in kilometers per hour at the end of this distance?

10. In a 10.0 km race, one runner runs at a steady 12.5 km/h and another runs at 14.5 km/h. How long does the faster runner have to wait at the finish line to see the slower runner cross?

11. Figure P2.11 shows a plot of velocity versus time for an object in linear motion. (a) What are the instantaneous velocities at $t = 30.0$ s and $t = 40.0$ s? (b) Compute the final displacement of the object. (c) Compute the total distance the object travels.

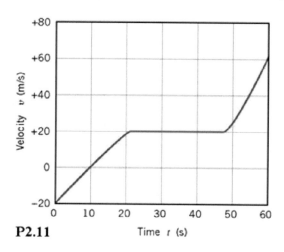

P2.11

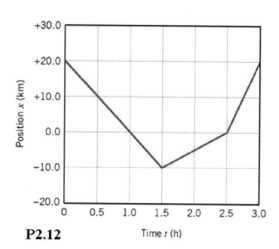

P2.12

12. Figure P2.12 shows the position graph of a particle. (a) Draw the particle's velocity graph for the interval $0 \text{ h} \leq t \leq 3.0 \text{ h}$. (b) Does this particle have a turning point or points? If so, at what time or times does this point or these points occur?

13. Consider the velocity graph of an object in Figure P2.13. Draw the object's acceleration graph for the interval $0 \text{ s} \leq t \leq 10 \text{ s}$. Give both axes an appropriate numerical scale.

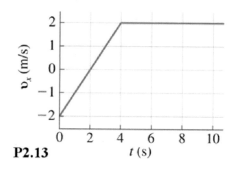

P2.13

14. A car starts from $x_i = 10$ m at $t_i = 0$ s and moves with the velocity graph shown in Figure P2.13. (a) What is the object's position at $t = 2$ s, 3 s, and 4 s? (b) Does this car ever change direction? If so, at what time does it change?

15. Runner A is initially 4.0 mi east of a flagpole and is running with a constant velocity of 6.0 mi/h due west. Runner B is initially 3.0 mi west of the flagpole and is running with a constant velocity of 5.0 mi/h due east. How far are the runners from the flagpole when they meet?

16. In getting ready to slam-dunk the ball, a basketball player starts from rest and sprints to a speed of 6.2 m/s in 1.4 s. Assuming that the player accelerates uniformly, determine the distance he runs.

17. A speed skater moving across frictionless ice at 8.5 m/s hits a 6.0-m-wide patch of rough ice. She slows steadily, and then she continues on at 6.5 m/s. What is her acceleration on the rough ice?

18. An object moving with uniform acceleration has a velocity of 15.0 cm/s in the positive x-direction when its x-coordinate is 3.00 cm. If its x coordinate 2.00 s later is -5.00 cm, what is its acceleration?

19. Two cars travel in the same direction along a straight highway, one at a constant speed of 55 mi/h and the other at 70 mi/h. (a) Assuming they start at the same point, how much sooner does the faster car arrive at a destination 10 mi away? (b) How far must the faster car travel before it has a 15-min lead on the slower car?

20. Two runners start one hundred meters apart and run toward each other. Each runs ten meters during the first second. During each second thereafter, each runner runs ninety percent of the distance he ran in the previous second. Thus, the velocity of each person changes from second to second. However, during any one second, the velocity remains constant. Make a position-time graph for one of the runners. From this graph, determine (a) how much time passes before the runners collide and (b) the speed with which each is running at the moment of collision.

21. Figure P2.19 is a somewhat simplified velocity graph for Olympic sprinter Carl Lewis starting a 100 m dash. Estimate his acceleration during each of the intervals A, B, and C.

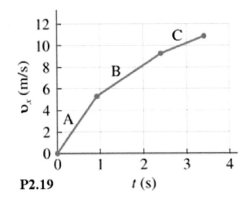

P2.19

22. A jet aircraft being launched from an aircraft carrier is accelerated from rest along a 94-m track for 2.5 s. This takeoff from a short runway is assisted by an aircraft catapult. (a) What is the acceleration of the aircraft, assuming it is constant? (b) What is the launch speed of the jet?

23. A jet plane is cruising at 250 m/s when suddenly the pilot turns the engines up to full throttle. After traveling an additional 5.0 km, the jet is moving with a speed of 320 m/s. (a) What is the jets acceleration, assuming it to be a constant acceleration? (b) Is your answer reasonable? Explain.

24. A train 400 m long is moving on a straight track with a speed of 82.4 km/h. The engineer applies the brakes at a crossing, and later the last car passes the crossing with a speed of 15.2 km/h. Assuming constant acceleration, determine how long the train blocked the crossing. Disregard the width of the crossing.

25. A speedboat increases its speed uniformly from u_i = 20.0 m/s to u_f = 30.0 m/s in a distance of 2.00×10^2 m. (a) Draw a coordinate system for this situation and label the relevant quantities, including vectors. (b) For the given information, what is single equation is most appropriate for finding the acceleration? (c) Solve the equation selected in part (b) symbolically for the boat's acceleration in terms of u_i, u_f, and Dx. (d) Substitute given values, obtaining that acceleration. (e) Find the time it takes the boat to travel the given distance.

26. You're driving down the highway late one night at u_i = 20 m/s when a deer steps onto the road d = 35 m in front of you. Your reaction time before stepping on the brakes is 0.50 s, and the maximum deceleration of your car is 10 m/s². (a) How much distance is between you and the deer when you come to a stop? (b) What is the maximum speed you could have and still not hit the deer?

ONE DIMENSIONAL KINEMATICS

P2.26

27. The minimum stopping distance for a car traveling at a speed of 30 m/s is 60 m, including the distance traveled during the driver's reaction time of 0.50 s. (a) What is the minimum stopping distance for the same car traveling at a speed of 40 m/s? (b) Draw a position-versus-time graph for the motion of the car in part a. Assume the car is at $x_i = 0$ m when the driver first sees the emergency situation ahead that calls for a rapid halt.

28. A car starts from rest and travels for 5.0 s with a uniform acceleration of +1.5 m/s². The driver then applies the breaks, causing a uniform acceleration of -2.0 m/s². If the brakes are applied for 3.0 s, (a) how fast is the car going at the end of the braking period, and (b) how far has the car gone?

29. A speedboat starts from an initial speed of 0.200 m/s and accelerates at +2.01 m/s² for 7.00 s. At the end of this time, the boat continues for an additional 6.00 s with an acceleration of +0.518 m/s². Following this, the boat accelerates at -1.49 m/s² for 8.00 s. (a) What is the velocity of the boat at $t = 21.0$ s? (b) Find the total displacement of the boat.

30. A steam catapult launches a jet aircraft from the aircraft carrier *George H. W. Bush*, giving it a speed of 175 mi/h in 2.50 s. (a) Find the average acceleration of the plane. (b) Assuming the acceleration is constant, find the distance the plane moves.

31. A football player, starting from rest at the line of scrimmage, accelerates along a straight line for a time of 1.5 s. Then, during a negligible amount of time, he changes the magnitude of his acceleration to a value of 1.1 m/s². With this acceleration, he continues in the same direction for another 1.2 s, until he reaches a speed of 3.4 m/s. What is the value of his acceleration (assumed to be constant) during the initial 1.5-s period?

32. A certain cable car in San Francisco can stop in 10 s when traveling at maximum speed. On one occasion the driver sees a dog a distance d in front of the car and slams on the brakes instantly. The car reaches the dog 8.0 s later, and the dog jumps off the track just in time. If the car travels 4.0 m beyond the position of the dog before coming to a stop, how far was the car from the dog? (*Hint*: You will need three equations.)

33. An ice sled powered by a rocket engine starts from rest on a large frozen lake and accelerates at +40 ft/s². After some time t_1, the rocket engine is shut down and the sled moves with constant velocity v for a time t_2. If the total distance traveled by the sled is 17,500 ft and the total time is 90 s, find (a) the times t_1 and t_2 and (b) the velocity v. At the 17,500-ft mark, the sled begins to accelerate at -20 ft/s². (c) What is the final position of the sled when it comes to rest? (d) How long does it take to come to rest?

34. A penny is dropped from rest from the top of the Empire State Building in New York City. Considering that the height of the building is 381 m and ignoring air resistance, what is the speed with which the penny strikes the ground?

35. A log is floating on swiftly moving water. A stone is dropped from rest from an 80-m-high bridge and lands on the log as it passes under the bridge. If the log moves with a constant speed of 4.5 m/s, what is the horizontal distance between the log and bridge when the stone is released?

36. In an action movie, the villain is rescued from the ocean by grabbing onto the ladder hanging from a helicopter. He is also intent on gripping the ladder that he lets go of his briefcase of counterfeit money when he is 150 m above the water. If the briefcase hits the water 7.0 s later, what was the speed at which the helicopter was ascending?

37. A cement block accidentally falls from rest from the ledge of a 60.0-m-high building. When the block is 14.0 m above the ground, a man, 2.00 m tall, looks up and notices that the block is directly above him. How much time, at most, does the man have to get out of the way?

38. Two stones are released from the edge of a cliff, one a short time after the other. (a) As they fall, the first is always going faster than the second. Does the difference between their speeds get larger, get smaller, or stay the same? Explain. (b) Does their separation increase, decrease, or stay the same? Explain. (c) Will the time interval between the instants at which they hit the ground be smaller than, equal to, or larger than the time interval between the instants of their release? Explain.

39. While standing on a bridge 15.0 m above the ground, you drop a stone from rest. When the stone has fallen 3.20 m, you throw a second stone straight down. What initial velocity must you give the second stone if they are both to reach the ground at the same instant? Take the downward direction to be the negative direction.

40. A mountain climber stands at the top of a 50.0-m cliff that overhangs a calm pool of water. She throws two stones vertically downward 1.00 s apart and observes that they cause a single splash. The first stone had an initial velocity of -2.00 m/s. (a) How long after release of the first stone did the two stones hit the water? (b) What initial velocity must the second stone have had, given that they hit the water simultaneously? (c) What was the velocity of each stone at the instant it hit the water?

41. A tennis player tosses a tennis ball straight up and then catches it after 2.00 s at the same height as the point of release. (a) What is the acceleration of the ball while it is in flight? (b) What is the velocity of the ball when it reaches its maximum height? Find (c) the initial velocity of the ball and (d) the maximum height it reaches.

42. An arrow is shot vertically upward. Three seconds later, it is at a height of 35 m. (a) What was the initial speed of the arrow? (b) How long is the arrow in flight from launch to returning to the initial height? (Assume the arrow falls vertically downward.)

43. In a physics demonstration, a physics professor throws a tennis ball at rest vertically upward and walks towards a table which is 2.00 meters away at a constant speed of 1.50 m/s to pick up a physics book. Neglecting air resistance, what is the minimum height he must throw the ball to catch it?

44. A model rocket is launched straight upward with an initial speed of 50.0 m/s. It accelerates with a constant upward acceleration of 2.00 m/s² until its engines stop at an altitude of 150 m. (a) What can you say about the motion of the rocket after its engines stop? (b) What is the maximum height reached by the rocket? (c) How long after liftoff does the rocket reach its maximum height? (d) How long is the rocket in the air?

TWO DIMENSIONAL KINEMATICS

CHAPTER 3

This is a picture of two opposing armies fighting for a territory in a medieval battle scene. As we can see from the picture, the main long-range weapon is the catapult which was quite common in a medieval battlefield. This weapon is a large-scale version of the simple sling. The simple sling, in experienced hands, was considered as the most effective personal projectile weapon until the fifteen century. This weaponry surpassed the accuracy and deadliness of the bow and even early firearms. Both catapults and simple slings are types of projectile weapons. A projectile weapon may be understood as any weapon that lobs an object at the enemy. Technically, this definition covers most siege engines and cannon. However, projectile weapons that we commonly think about are bows, slings, throwing weapons, and other similar weaponry. As we know, such projectile weapons were of paramount importance in medieval warfare because the farther away one army could deliver a hit and take out their enemy on the battlefield, the more likely that army would have the winning advantage in the battle and, possibly, in the war. The ranged units in an army were responsible for using projectile weapons to send showers of missiles into enemy ranks to thin the line and break up the opposing shield wall. Also, the barrage of projectiles also had psychological effects, causing confusion and demoralizing the enemy. This approach has been one of the most important military strategies ever since caveman picked up rocks to throw at approaching enemies.

The technology of projectile weaponry increased at an incredible pace during the medieval era. Over the course of this martial evolution, new technologies and military tactics began to relegate the role of the slinger to that of an auxiliary soldier and ultimately removed it from the battlefield of medieval Europe. From simple bows and slings in the early medieval times to advanced gunpowder weapons of the late medieval times, the range of projectiles increased enormously. Along with its range, the importance of projectile weapons grew as cannon and firearms were adapted on a widespread basis on the battlefields of the sixteenth century and onward. From this perspective, the modern warfare had not changed much since the medieval period. Many modern projectile weapons have a long firing range which is well beyond visual range. The range of some weapons has become so long that the effects of Earth's rotation and variations in Earth's gravity may need to be taken into account for accurate targeting.

As the main focus of this Chapter, we discuss how to describe the projectile motion of an object in two dimensions by extending the set of four kinematic equations we derived in Chapter 2. Also, we discuss relative motion in one and two dimensions to describe the motion of an object in a coordinate system that is moving with respect to a fixed reference frame.

3.1 SCALARS VERSUS VECTORS

The motion of objects in two dimensions can be described by using the same terminology as in Chapter 2. A lay person, without much physics knowledge, might describe the motion of objects using phrases such as, *going fast, stopped, slowing down, speeding up,* and *turning*. These words may provide a sufficient vocabulary for a lay person, but we use these words and many more so that we may describe motion of an object more precisely in physics. In Chapter 2, we expanded a lay person's vocabulary list with words such as *distance, displacement, speed, velocity,* and *acceleration*. We have seen that although these words appear to describe the similar quantities, they are in fact associated with strict definitions of different mathematical quantities. The mathematical quantities that are used to describe the motion of objects can be divided into two categories. These quantities are either **scalars** or **vectors**. In physics, measurable physical quantities as well as many abstract quantities are categorized as either scalar or vector quantities.

What are scalars?

Scalars are quantities that are fully described by a magnitude (or numerical value) alone. This is a quantity that can be described with a single number. Examples of scalar quantities are speed of an automobile, wind speed, and distance.

TWO DIMENSIONAL KINEMATICS

What are vectors?

Vectors are quantities that are fully described by both a magnitude and a direction. Examples of vector quantities are velocity of an automobile, wind velocity, and displacement. These quantities require both magnitude and direction for complete definition.

In other words, scalar quantities can be manipulated using the usual laws of algebra. However, these laws do not apply to vector quantities! Hence, we need to understand the rules for adding and subtracting vectors.

Addition and Subtraction of Vectors

Let us discuss how to add or subtract vectors using a graphical approach as shown in Fig. 3.1. To add, we need to relocate one vector with respect to the other. Since a vector is specified by its magnitude and direction, the vector **D** must be relocated with respect to the vector **E** while preserving its original orientation and size. (Note that a vector is denoted by using either a bold-faced letter or a letter with an arrow on top.) Vectors are said to be invariant under translation.

There are two graphical approaches for adding vectors: i) tip-to-tail rule and ii) parallelogram rule. These two approaches are illustrated in Fig. 3.2. First, let us discuss the tip-to-tail rule. If we think of vectors **D** and **E** as two successive displacement vectors which describe a person walking, then it is clear that the resultant vector **F** = **D** + **E**, which represents the total displacement, can be found by placing **E** such that its tail is at the same point as the head of **D**. The resultant vector **F** then has its tail at the tail of **D** and its head at the head of **E** as shown in Fig. 3.2. Now, we discuss the parallelogram rule. We move **E** so that its tail coincides with the tail of **D**. The resultant vector is seen to lie along the diagonal of the parallelogram formed by **D** and **E**, with the tails of all three vectors coinciding. The same resultant vector **F** is obtained when we shift **D** such that its tail coincides with the head of **E**. Now, it is obvious from these ways of adding vectors, by using the parallelogram rule, that **F** = **D** + **E** = **E** + **D**. The equivalence between these two ways shows the commutative law holds for vector addition.

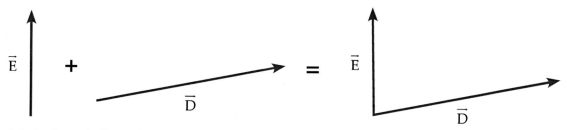

Figure 3.1: A schematic illustration of adding two vectors, **E** and **D**, by joining the ends of the vectors. Vector **D** may be transported anywhere as long as both the magnitude and direction remain the same.

Figure 3.2: Two ways to add vectors are schematically illustrated. For the tip-to-tail rule (left panel), the tail of **E** is moved to meet the tip of **D**. For the parallelogram rule (right panel), the diagonal of the parallelogram formed by **D** and **E** is found to compute the resultant vector **F**.

Since we may write **E − D = E + (−D)**, vector subtractions and additions are similar. The orientation of **−D** is opposite to the orientation of **D**, but the magnitude is the same.

Adding two vectors by decomposing them into components...

As we have seen from the application of the tip-to-tail and parallelogram rule, both the direction and magnitude of two vectors must be considered. These two rules provide nice qualitative ways to obtain the resultant vector, but we may need more precise accounts of both the direction and magnitude of the resultant vector. To obtain this, we first need to **decompose the vectors into components**. Then, we add the vector components that are in the same direction. To decompose a vector in two dimensions, first, we need to visualize the x- and y-component of an arbitrary vector **A**. The diagrams in Fig. 3.3 show that the **x-component** *of vector* **A** is A_x and the **y-component** of **A** is A_y.

In Fig. 3.3, the decomposition of vector **A** allows us to write **A** as the sum of the two component vectors, A_x and A_y,

$$\vec{A} = \vec{A}_x + \vec{A}_y = A_x \hat{x} + A_y \hat{y}, \tag{3.1}$$

where $A_x = A\cos\theta$ and $A_y = A\sin\theta$ denote the magnitude of the component vectors $\mathbf{A}_x$ and $\mathbf{A}_y$ respectively, representing projections along the x- and y-axis. It is noted that $\hat{x}$ and $\hat{y}$ denote the unit vector along the x and y direction, respectively. Here, a **unit vector** is a vector (often a spatial vector) whose length is 1 (the unit length). We may think of a unit vector as a pointer for direction. Also, a unit vector is often denoted by a lowercase letter with a "hat", like this: $\hat{i}$ (pronounced "i-hat"). Since the component vectors $\mathbf{A}_x$ and $\mathbf{A}_y$ are perpendicular to each other, the magnitude of the vector **A** is (from the Pythagorean equation)

$$A = \sqrt{A_x^2 + A_y^2} \tag{3.2}$$

and the direction of **A** with respect to the horizontal axis is described by the angle

$$\theta = \tan^{-1}\left(\frac{A_y}{A_x}\right). \tag{3.3}$$

With two vectors decomposed and written in terms of their component vectors, we will now add the vectors by using the **component method**. The two vectors we will add here are displacement vectors, but the method is applied in the same for any other type of vectors, such as velocity, acceleration, or force vectors. So, how do we add two vectors **A** and **B**? To add these two vectors, first, we break up

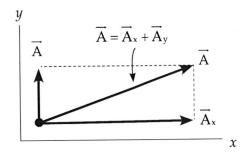 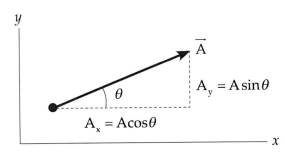

Figure 3.3: A vector may be represented as the sum of x- and y-components. The x-component is parallel to the x-axis and the y-component is parallel to the y-axis. The magnitude of the components corresponds to its projection.

TWO DIMENSIONAL KINEMATICS

each vector into its components, A_x and A_y for A (i.e., $A = A_x + A_y$), and B_x and B_y for B (i.e., $B = B_x + B_y$), as shown on the Fig. 3.4. From the rules for the equality of vectors, the vector B in Fig. 3.4 (left) may be relocated. This vector B, before relocation, is equal to the vector B, after relocation, in Fig. 3.4 (right) because it has equal length and equal direction. Here, the tip-to-tail rule is applied to illustrate the vector addition procedure graphically. Now, for the analytic approach, we can add the **x-components** (i.e., A_x and B_x) to generate the **x-component** of the resultant vector R

$$R_x = A_x + B_x \tag{3.4}$$

since the components of vectors A and B in the same direction may be considered scalars. Similarly, we can add the **y-components** of A and B (i.e., A_y and B_y) to obtain the y-component of the resultant vector R (i.e., R_y)

$$R_y = A_y + B_y \tag{3.5}$$

The new component vectors, R_x and R_y, completely define the resultant vector R by specifying both the magnitude and the direction. Looking carefully at the diagrams in Fig. 3.4, we can see that adding two vectors produces the resultant vector which is **not** in the direction of either of the original vectors A and B. Also, the magnitude is **not** equal to the sum of the magnitudes of the original vectors.

As we have seen above, vector algebra is very different from scalar algebra because we must account for both magnitude and direction. However, vector algebra has many useful applications in the real world. For example, suppose we imagine a situation where we are in a boat or a plane, and we need to plot a course. We will need to plan our navigation on a map, but there are no streets or signs along the way. To plot a course, we would need to know where we are starting and where we want to be. However, the main problem is how to get there. To solve this problem, we need to use a couple of vectors. As we move along our course, we may swerve a bit off course because of wind or water currents. In this case, we just go back to the map, find our current location, and plot a new vector that will take us to our destination. We need to draw the vector between the two points (i.e., the current location and the destination) and continue on our way. Captains of ships or airplanes use vectors (they know the speed and direction) to plot their courses.

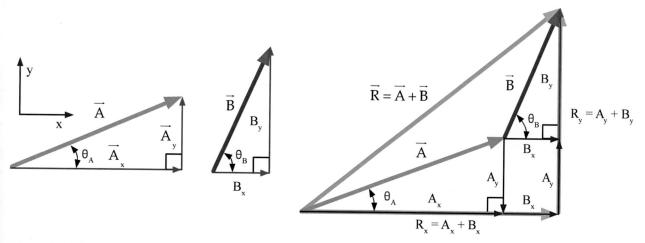

Figure 3.4: The separate vectors A and B decomposed into their components. Both the x- and y-components of two vectors are added, separately, to obtain the resultant vector R.

Understand-the-Concept Question 3.1:

Consider the following three vectors: vector **A** = (1,3), vector **B** = (3,0), and vector **C** = **A** + **B**. What is the magnitude of **C**?

a. 3   b. 4   c. 5   d. 7

Explanation:

The magnitude of vector **C** may be computed by using the Pythagorean equation. To do this, we must find the components of the resultant vector by adding the components of vectors **A** and **B**. The x- and y-components of **C** are given by $C_x = 1 + 3 = 4$ and $C_y = 3 + 0 = 3$, respectively. Applying the Pythagorean equation, we find that $C = (C_x^2 + C_y^2)^{1/2} = (16 + 9)^{1/2} = 5$.

Answer: c

Exercise Problem 3.1:

Three vectors are shown in the figure. Their magnitudes are given in arbitrary units. Determine the sum of three vectors. Give the resultant vector in terms of (a) components, (b) magnitude and angle with x-axis. (Note: The magnitude and the direction of vectors are not drawn to scale.)

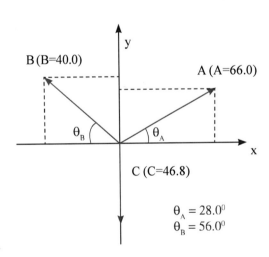
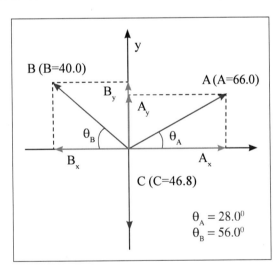

Solution:

First, we need to decompose the three vectors, **A**, **B**, and **C**, into their components:

$$\vec{A} = \vec{A}_x + \vec{A}_y \text{ where } A_x = A\cos\theta_A \text{ and } A_y = A\sin\theta_A$$

$$\vec{B} = \vec{B}_x + \vec{B}_y \text{ where } B_x = B\cos\theta_B \text{ and } B_y = B\sin\theta_B$$

$$\vec{C} = \vec{C}_y \text{ where } C_y = C$$

TWO DIMENSIONAL KINEMATICS

To add these three vectors and to obtain the resultant vector **R**, we need to add each component separately (i.e., add the x-components of three vectors together and also add the y-components of three vectors together). By following this approach, we obtain

$$\vec{R} = \vec{A} + \vec{B} + \vec{C} = \vec{R}_x + \vec{R}_y = (\vec{A}_x + \vec{B}_x) + (\vec{A}_y + \vec{B}_y + \vec{C}_y)$$

Therefore, the x- and y-component of **R** are

$$R_x = A_x - B_x = A\cos\theta_A - B\cos\theta_B = 35.9,$$

and

$$R_y = A_y + B_y - C_y = A\sin\theta_A + B\sin\theta_B - C = 17.3,$$

respectively. The magnitude of the resultant vector **R** is

$$R = \left(R_x^2 + R_y^2\right)^{1/2} = 39.9.$$

The direction of **R** may be computed by applying the definition of the tangent function. We obtain the direction of the resultant vector **R** as

$$\tan\theta = R_y/R_x \Rightarrow \theta = \tan^{-1}\left(R_y/R_x\right) = 25.7°.$$

For describing the direction of any vector, there are a variety of conventions. The two conventions that we will use are the following: 1) the direction of a vector may be expressed as an angle of rotation of the vector about its "tail" from east, west, north, or south; 2) the direction of a vector may be expressed as a counterclockwise angle of rotation of the vector about its "tail" from due east (i.e., the positive x-axis).

3.2 Kinematics in Two Dimensions

One of the reasons we discuss vectors and vector algebra is that there are few real examples of one-dimensional motion as we have seen in Chapter 2, but there are many more examples of two-dimensional motion. For example, an apple falling from a tree, a car moving on a straight road, and an elevator rising in a shaft are examples of one-dimensional motion which do not require vectors for their description. However, there are many more examples of motion in two dimensions which require vector quantities to describe motion. For example, a batter hitting a home run, a car driven to work, a ship sailing to Europe, a dog wagging its tail, a caribou migrating, and a hunter shooting an arrow are example of two-dimensional motion. Fortunately, the rules of motion in one dimension may be used to describe motion in most two-dimensional situations if they are applied properly and are decomposed into two one-dimensional motions.

In most situations, these two dimensions lie in a horizontal plane. In these cases, we may label them as north/south and east/west. However, sometimes we may use a vertical plane and label the two dimensions as up/down and forward/backward. As we know, in life, situations are not always as simple as just dropping rocks from the roof of a building or pushing a block

Figure 3.5: The motion of a biker jumping off a ramp may be decomposed into their motion along the horizontal direction and along the vertical direction. These two motions are independent of each other.

across a table at constant speed. If these were the only things happening in the real world, our lives would be very boring. Luckily, the real world is much more interesting because life takes on a more three-dimensional aspect. Actually, we are living in a four-dimensional world because time may be considered to be another dimension. However, within the context of kinematics, the word "dimension" is used to refer to spatial dimensions and not temporal dimensions. Setting aside these points, we consider motion in two dimensions. When we say that an object is moving in a "plane" it usually means that it is moving on a flat two-dimensional coordinate axis like the x- and y-axes (i.e., as if it was moving on a smooth flat surface).

A few things to remember...

We may ask "In what kind of situation is an object considered to move in two dimensions?" Well, in one-dimensional motion, the object moves in exclusively one direction, either a horizontal "x" or vertical "y" direction. So, in two-dimensional motion, the object moves in *both* the x and y directions. Under most ordinary circumstances, motion of an object in two different directions (i.e., x and y) is independent. This means that if we are describing motion of a fighter jet as shown in Fig. 3.6 then its position and velocity in the horizontal direction does not depend on its position or velocity in the vertical direction. Therefore, it is helpful to decompose the problem into two one-dimensional problems. Even after the decomposition of a problem, some features, such as the time of flight or the starting position, may still be considered common to both directions. In decomposing the problem, one important thing to remember is to choose the coordinate axes carefully because, sometimes, the calculation may be significantly simplified by picking one coordinate system over another.

For describing two-dimensional kinematics, we use the following physical quantities: i) displacement, ii) velocity, and iii) acceleration. These vector quantities can be decomposed each into components. The decomposition may be done to express each vector as a sum of the component vectors as shown in Fig. 3.7. Displacement, velocity, and acceleration vectors are expressed as

TWO DIMENSIONAL KINEMATICS

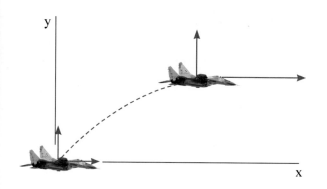

Figure 3.6: The motion of an object in the x-direction is independent from that in the y-direction. In the left panel, a fighter jet's movement may be controlled by using its vector thruster. In the right panel, a vector thruster of a military fighter plane may be used to maneuver the plane.

$$\vec{r} = x\,\hat{x} + y\,\hat{y} \quad \text{(Displacement)}, \tag{3.6a}$$

$$\vec{v} = v_x\hat{x} + v_y\hat{y} \quad \text{(Velocity)}, \tag{3.6b}$$

$$\vec{a} = a_x\hat{x} + a_y\hat{y} \quad \text{(Acceleration)}, \tag{3.6c}$$

respectively. We should remember that vectors in two dimensions have two components and those in three dimensions have three components.

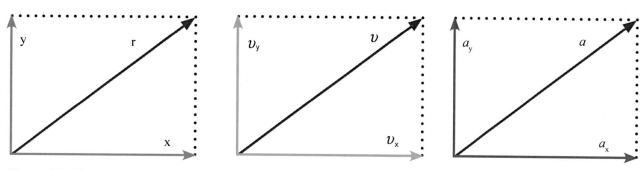

Figure 3.7: Three quantities, displacement, velocity, and acceleration, which are useful for describing kinematics of an object in two dimensions are decomposed and written as a sum of two component vectors.

As shown in Fig. 3.7, the decomposition of displacement, velocity, and acceleration vectors into x- and y-components in two dimensions indicates that the kinematics may also be decomposed. One important advantage of doing this decomposition is that we can treat the motions of an object along the x-axis and the y-axis as independent of each other.

Kinematic Equations in Two Dimensions

Noting that the kinematics of an object in two dimensions can be described by decomposing it into the x- and y-component, we need four kinematic equations for each direction. The mathematical relations between the quantities that are used to describe an object's motion are the following:

	x-component	y-component
Velocity-versus-time	$v_x = v_{ox} + a_x t$	$v_y = v_{oy} + a_y t$
Position-versus-time (using average speed)	$x = \frac{1}{2}(v_{ox} + v_x)t$	$y = \frac{1}{2}(v_{oy} + v_y)t$
Position-versus-time	$x = v_{ox}t + \frac{1}{2}a_x t^2$	$y = v_{oy}t + \frac{1}{2}a_y t^2$
Velocity-versus-position	$v_x^2 = v_{ox}^2 + 2a_x x$	$v_y^2 = v_{oy}^2 + 2a_y y$

We should remember that these two sets of four kinematic equations assume that an object is moving with constant acceleration and that it travels different distances in the x- and y-direction in equal time intervals.

3.3 PROJECTILE MOTION

A special case of two-dimensional kinematics near the surface of Earth is projectile motion. **Projectile motion** is defined as a form of motion where a particle, called a projectile, is thrown obliquely near the Earth's surface, and it moves along a curved path under the action of gravity. The path followed by a projectile motion is called its trajectory. Projectile motion only occurs when there is one force applied at the beginning of the trajectory after which there is no interference other than that from gravity. Air resistance is often important for objects moving through the air, but this effect can be ignored in many cases. In the cases in which air resistance is not negligible, it provides notable interference to the trajectory of a projectile. However, we will not take this effect into account formally in our description of two-dimensional kinematics. Examples of projectile motion are a golf ball, a thrown or batted baseball, kicked footballs, and an athlete doing a long jump. However, any objects that have any source of propulsion are not considered as projectiles because acceleration other than the Earth's gravity needs to be considered.

Understand-the-Concept Question 3.2:

Are these projectile motions?

TWO DIMENSIONAL KINEMATICS

Explanation:

When an object leaves a reference point with some initial speed at an angle with respect to the horizontal direction and moving in the air only under the influence of gravitational acceleration, we consider it as a projectile. This indicates that water leaving a drinking fountain and an artillery shell leaving a howitzer are examples of projectiles. Here, we are assuming the effects of air resistance are negligible.

In comparing the trajectory of the two projectiles in the Understanding-the-Concept Question 3.2, we may notice that the shape of the curved path can be described as a parabola. Is there any reason for this shape? The answer is "Yes!" We shall see below that all projectiles follow parabolic trajectories.

Under the action of gravity:

To understand projectile motion, let us first consider a ball projected horizontally from the top of a table as shown in Fig. 3.8. In the absence of gravity, the ball would continue its horizontal motion at a constant velocity. This is consistent with the law of inertia (see Chapter 4). However, if the ball was merely dropped from rest in the presence of gravity, it would accelerate downward, gaining speed at a rate of 9.8 m/s every second. This is consistent with the concept of free-falling objects accelerating at a rate known as the **acceleration of gravity** since an object in the air experiences a vertically downward acceleration g. Therefore, the horizontally projected red ball is expected to reach the ground in the same time as the blue ball which is dropped vertically.

Let us continue our thought experiment. We project the red ball horizontally in the presence of gravity. The red ball has a constant horizontal velocity and would maintain the same horizontal motion as before. Furthermore, the force of gravity will act upon the red ball to cause the same vertical motion as before and yields a downward acceleration. The ball will fall the same amount of distance as it did when it was merely dropped from rest (as for the blue ball in Fig. 3.8). The presence of gravity does not affect the horizontal motion of the projectile. To alter the horizontal motion, there must be a horizontal force to cause a horizontal acceleration. It should be noted that we introduced the concept of force here to describe the action of push or pull on an object. An extensive discussion of this concept is presented in Chapter 4. In terms of force, we know that there is only one vertical force acting upon projectiles since there is only the force of gravity. This vertical

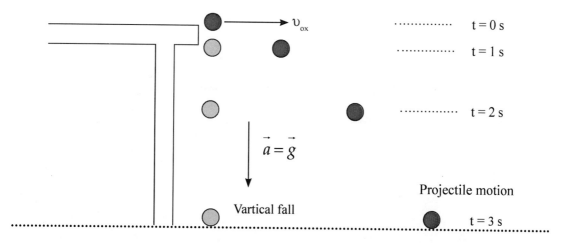

Figure 3.8: The effect of gravity on two balls is illustrated. The vertical motion of both the red ball which is projected horizontally with the initial speed v_{ox} and the blue ball which is dropped from rest are the same.

force acts perpendicular to the horizontal motion. Hence, this force will not affect the horizontal motion of the ball since perpendicular components of motion are independent of each other. Thus, the projectile such as the red ball travels with a constant horizontal velocity v_{ox} and a downward vertical acceleration g as shown in Fig. 3.8.

Kinematic Equations for Projectile Motion

Since the vertical motion is independent of the horizontal motion, we can simply break up the motion of an object into two parts, each of which follows the familiar one-dimensional equations as shown below.

	x-component	y-component
Velocity-versus-time	$v_x = v_{ox}$	$v_y = v_{oy} - gt$
Position-versus-time	$x = x_o + v_{ox}t$	$y = y_o + v_{oy}t - \frac{1}{2}gt^2$
Velocity-versus-position	$v_x^2 = v_{ox}^2$	$v_y^2 = v_{oy}^2 - 2g(y - y_o)$
Average speed	$v_x = v_{ox}$	$\bar{v}_y = \frac{v_{oy} + v_y}{2}$

We should note that these two sets of kinematic equations are obtained by assuming that the positive y-axis is upward, $a_x = 0$, and $a_y = -g$. Another interesting point to remember is that the kinematic equations for the x-component are independent of those for the y-component, but both the x- and y-direction motions share one common factor: **time!**

With the two sets of four kinematic equations, we are ready to describe projectile motion of objects. However, before we proceed further, some helpful hints on how to use these equations to solve kinematic problems are in order. To solve two-dimensional problems, we should remember that it is always helpful to understand what is going on in the problem by visualizing the situation. It may be helpful to organize our thoughts in the following way:

1. Read the problem carefully
2. Draw a schematic diagram representing the problem
3. Write down what we know in both directions separately
4. Figure out what we need to determine
5. Check the equations of one-dimensional kinematics to see if we can turn what we know into what we want.

One general strategy we should remember is that we may need to use information about motion in one dimension to gain information about that in other dimension. For example, we may need to use motion in the y-direction to find the time of travel, so that we can add time to the list of known quantities in the x-direction to find displacement.

TWO DIMENSIONAL KINEMATICS

Understand-the-Concept Question 3.3:

Suppose a child sits upright in a wagon which is moving to the right at constant speed. The child extends her hand and throws an apple straight upward (from her own point of view), while the wagon continues to travel forward at constant speed. If air resistance is neglected, will the apple land

  a. behind the wagon   b. in the wagon   c. in front of the wagon.

Explanation:

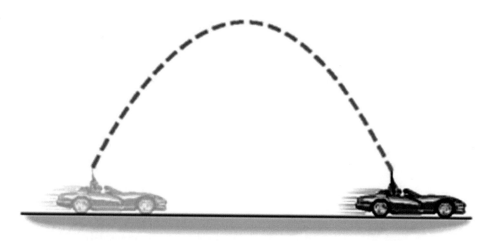

As we may notice that if the car is completely closed (all windows shut and all vents turned off) and traveling at a constant speed, then a ball would not move backward when it is thrown vertically. It would appear to rise and fall in a perfectly vertical path, relative to the car and the passengers inside. Similar to this case, when either a ball or an apple, respectively, is thrown vertically from a moving convertible car (as shown schematically in the figure above) or a wagon, we will find the same result if there is no air resistance.

Answer: b

  We may easily recognize that the situation described in the Understanding-the-Concept Question 3.3 may not happen easily in the real world because of air resistance and non-zero acceleration. For instance, if the ball is subject to moving air (such as if it was thrown upward from a convertible with an open top) or if the car accelerated after the ball was thrown, then the ball would indeed move backward, relative to the car and its passengers. On the other hand, to an outside observer, the ball would move in the same direction as the car. However, the ball would be moving at a different horizontal speed than the car. This is in contrast to the case of no air resistance and no acceleration in which both the ball and car move with the same horizontal speed.

Trajectory of a Projectile

Now, let us consider the shape of a trajectory. The **trajectory** of a projectile is defined as the path that a thrown or launched projectile will take under the action of gravity, neglecting all

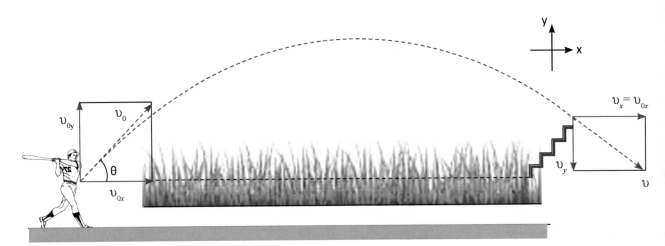

Figure 3.9: An illustration of a batter hitting a home run. The trajectory of the baseball obtained by treating the vertical and horizontal motion separately is parabola when the effects of air resistance on the ball are negligible.

other forces, such as friction from air resistance, without propulsion. As noted earlier, many projectiles undergo a vertical motion as well as a horizontal motion. This means that as they move vertically (i.e., upward or downward) they are also moving horizontally. In other words, there are two components to the projectile's motion: horizontal and vertical motion. Moreover, these two components of motion can be discussed separately since the perpendicular components of motion are independent of each other. We now follow the horizontal and vertical components of a projectile's motion and show that the path followed by any projectile is a parabola if we can ignore air resistance and can assume that g is constant.

Before we do this, let us first think about how a projectile appears to us as we follow it along its path. With negligible air resistance, if a projectile, such as a baseball, travels in a parabolic path and if we are positioned so as to catch it as it descends, we will see its angle of elevation increase continuously throughout its flight. The tangent of the angle of elevation is proportional to the time since the ball was sent into the air because it was struck with a bat. Even when the ball is descending, near the end of its flight, its angle of elevation seen by us continues to increase. Therefore, we will see the ball as if it was ascending vertically at a constant speed. Finding the place from which the ball appears to rise steadily will help us to position ourselves correctly to catch it. If we are too close to the batter who hit the ball, then the ball will appear to rise at an accelerating rate. However, if we are too far from the batter, then the ball will appear to rise slowly, and descends.

With this in mind, we derive the mathematical expression which describes the trajectory of the baseball in Fig. 3.9. For simplicity, we set $x_o = y_o = 0$ and decompose the initial velocity v_o of the projectile into two components as $v_o = v_{ox} + v_{oy} = v_{ox}\hat{x} + v_{oy}\hat{y}$. The magnitude of the x- and y-components of the initial velocity of the projectile are $v_{ox} = v_o \cos\theta$ and $v_{oy} = v_o \sin\theta$, respectively. Here, the angle θ is the launch angle (the angle that the initial velocity make with the x-axis). Along the x-axis, the position of the ball is given by

$$x = x_o + v_{ox}t \Rightarrow x = v_{ox}t \Rightarrow t = \frac{x}{v_{ox}} \tag{3.7}$$

Similarly, along the y-axis, the vertical position of the ball is given by

$$y = y_o + v_{oy}t - \frac{1}{2}gt^2 \Rightarrow y = v_{oy}t - \frac{1}{2}gt^2 \tag{3.8}$$

TWO DIMENSIONAL KINEMATICS

We now substitute Eq. (3.7) into Eq. (3.8) to obtain

$$y = v_{oy}\left(\frac{x}{v_{ox}}\right) - \frac{1}{2}g\left(\frac{x}{v_{ox}}\right)^2 = \frac{v_{oy}}{v_{ox}}x - \frac{g}{2v_{ox}^2}x^2 = ax - bx^2 \qquad (3.9)$$

where the coefficients $a = v_{oy}/v_{ox}$ and $b = g/(2v_{ox}^2)$ are constants. We may easily recognize that Eq. (3.9) is an equation for parabola. In simple parabolic projectile motion, the horizontal speed is assumed to be constant and only the vertical speed changes. At the top of the arc (i.e., the apex of the parabola) the vertical speed is zero as the projectile changes from upward to downward motion. Conversely, if we assume that the initial and final position of the projectile are at the same level, then the maximum speed will be at the point of launch or landing since there is a constant acceleration downward during the entire flight due to gravity. Mathematically, the instantaneous speed of the projectile is the slope of the tangent of the parabola. If the parabola is of the form $y = ax - bx^2$, then the slope is $a - 2bx$, which has the minimum magnitude when $x = a/2b$ and has the maximum magnitude when $x = +\infty$ or $-\infty$.

Air resistance (sometimes called drag) is one influence which may change the trajectory of projectiles dramatically since objects moving through air are slowed down. This air resistance affects a spacecraft when it re-enters the Earth's atmosphere. Also, it affects the path of a projectile such as a bullet or a ball since the trajectory of a projectile is changed when air resistance is taken into account.

Contrary to what we know now about projectile motion, the medieval scientists believed that a projectile went upwards at an angle along a straight path and then it went through a short curved section before falling vertically back to the ground again. Both the range of a projectile and the maximum height that it reaches are affected by air resistance. The mathematics of describing the motion with air resistance without the use of calculus may be quite complicated us, especially if we consider the change in the shape and/or surface of a projectile and the variation of the density of the air with height. However, as we can see from the red curve in Fig. 3.10, the schematic diagram illustrates generally how air resistance affects both the trajectory and the velocity of a projectile (i.e., slope of the trajectory). The blue line represents the trajectory of the projectile with no air resistance. The maximum height, the range and the velocity of the projectile represented by the red line are all reduced.

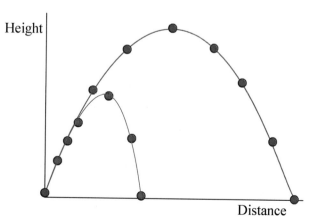

Figure 3.10: A long jumper with a hang time is a projectile. The term "hang time" is the length of time an athlete can jump off the ground. In sports such as long-jump hang time and air resistance are critical. The longer the hang time the greater the distance that the athlete can jump (blue curve), but air resistance (red curve) reduces it.

CHAPTER 3

The effects of air resistance depend on the speed of projectiles. The resistance is often taken as being proportional to either the velocity of the object or the square of the velocity of the object. When an object is moving at a low speed the strength of air resistance may be approximated as being proportional to the speed. In fluid mechanics, low speed motion is characterized by low Reynolds number. Here, the **Reynolds number** is a dimensionless number defined as a measure of the ratio of inertial forces to viscous forces and consequently quantifies the relative importance of these two types of forces for given flow conditions. The approximation is done so that the equations describing the particle's motion are easily solved. However, the assumption that air resistance may be taken to be in direct proportion to the velocity of the particle is not correct for a typical projectile in air with a velocity above a few tens of meters per second. When the velocity is above few tens of meters per second, the projectile is considered to move at high speed. At higher speed the air resistance is proportional to the square of the particle's velocity. High speed motion is characterized by high Reynolds number.

As we noted above, the Reynolds number is used to characterize different flow regimes, such as laminar or turbulent flow. Laminar flow occurs at low Reynolds numbers, where viscous forces are dominant, and is characterized by smooth, constant fluid motion. However, turbulent flow occurs at high Reynolds numbers and is dominated by inertial forces, which tend to produce chaotic eddies, vortices and other flow instabilities. Also, for a fluid in relative motion to a surface, this number can be defined for a number of different situations which include the fluid properties, such as density and viscosity, a velocity, and a characteristic length. We will discuss about fluid dynamics in Chapter 9.

Exercise Problem 3.2:

A football is kicked at an angle $\theta_o = 37.0°$ with a velocity of 20.0 m/s. Calculate (a) the maximum height, (b) the time of travel before the football hits the ground, (c) how far away it hits the ground, (d) the velocity vector at maximum height, and (e) the acceleration vector at maximum height. Assume the ball leaves the foot at ground level.

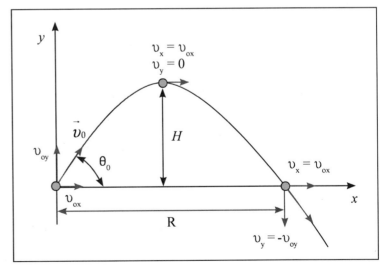

Solution:

The initial velocity vector may be written as the sum of the x- and y-component of velocity $\vec{v}_o = \vec{v}_{ox} + \vec{v}_{oy}$, where the magnitude of each component is

TWO DIMENSIONAL KINEMATICS

$$v_{ox} = v_o \cos\theta_0 = 20.0 \, \text{m}/\text{s} \times \cos(37.0°) = 16.0 \, \text{m}/\text{s}.$$

and

$$v_{oy} = v_o \sin\theta_0 = 20.0 \, \text{m}/\text{s} \times \sin(37.0°) = 12.0 \, \text{m}/\text{s}.$$

At the maximum height H, the vertical speed is $v_y = 0$. We use this condition to simplify the kinematics and obtain

$$v_y = v_{oy} - gt \Rightarrow 0 = v_{oy} - gt.$$

We solve for the time t to reach the maximum height and obtain

$$t = \frac{v_{oy}}{g} = \frac{12.0 \, \text{m}/\text{s}}{9.8 \, \text{m}/\text{s}^2} = 1.22 \, \text{s}.$$

We now use the time $t = 1.22$ s to compute the maximum height reached by the football by using the kinematic equation for vertical position versus time and obtain

$$y = y_o + v_{oy}t - \frac{1}{2}gt^2 \Rightarrow y = 12.0 \frac{\text{m}}{\text{s}} \times 1.22 \, \text{s} - \frac{1}{2} \times 9.8 \frac{\text{m}}{\text{s}^2} \times (1.22 \, \text{s})^2 = 7.35 \, \text{m}.$$

We may obtain the same result by using another kinematic equation

$$v_y^2 = v_{oy}^2 - 2g(y - y_o) \Rightarrow y = \frac{v_{oy}^2}{2g} = \frac{(12.0 \, \text{m}/\text{s})^2}{2 \times 9.8 \, \text{m}/\text{s}^2} = 7.35 \, \text{m}.$$

The time of travel may be determined by noting that the final vertical position is $y = 0$. We compute the time of travel from the position-versus-time kinematic equation and obtain

$$y = y_o + v_{oy}t - \frac{1}{2}gt^2 \Rightarrow 0 = v_{oy}t - \frac{1}{2}gt^2 \Rightarrow t = \frac{2v_{oy}}{g} = \frac{2 \times 12.0 \, \text{m}/\text{s}}{9.8 \, \text{m}/\text{s}^2} = 2.45 \, \text{s}.$$

During the time $t = 2.45$ s, the horizontal speed of the football does not change. Hence, the horizontal distance traveled by the football is

$$x = x_o + v_{ox}t \Rightarrow x = v_{ox}t = 16.0 \frac{\text{m}}{\text{s}} \times 2.45 \, \text{s} = 39.2 \, \text{m}.$$

At the maximum height, the x-component of the velocity is the same but the y-component is zero. So, we may write

$$v_x = v_{ox} \text{ and } v_y = 0.$$

The acceleration vector at maximum height, as well as during the entire flight of the football, is 9.8 m/s² downward.

In football, one of the most important skills for playing it well is being able to throw or kick a football farther and accurately in order to score. One way of accomplishing this is to throw or kick a football with a high velocity/speed at an optimum angle of attack. For example, a field goal kick, which will travel end-over-end, presents an interesting challenge to a kicker. Accounting the effects of drag and lift during the flight of the ball is not an easy task because one side of the ball may not be constantly facing the air. This consideration may change the optimum angle. It turns out that for many objects travelling in the air or any fluid with viscosity effects will travel farther at angle of attack greater than 45° or less than 45°. This angle depends entirely on the relative strength of lift to drag, indicating that the flight path of the football will not be a symmetric parabola.

Range of a projectile

As we have noted above, a **trajectory** is an imagined trace of positions followed by an object moving through space. Also, a particular trajectory may be described mathematically either by the geometry (i.e., the set of all positions taken by the object) or by the position as a function of time. The range of an object may be determined from its trajectory. However, computing the range may not be straightforward in real situations. Suppose we consider a simple case in which a projectile is launched under the influence of a uniform gravitational force only. For example, a rock thrown on the nearly airless surface of the Moon is a good approximation. In this case, the trajectory takes the shape of

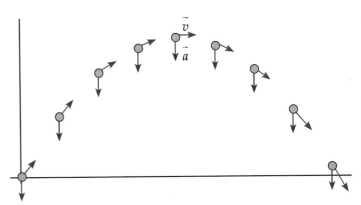

Figure 3.11: The ideal case of motion of a projectile in a uniform gravitational field, in the absence of other forces (such as air drag), exhibits a symmetric parabolic curve. The velocity changes constantly throughout trajectory but the acceleration remains the same, pointing toward the ground.

a parabola as in Fig. 3.11, provided the rock is not thrown too far. However, the precise trajectory of a projectile on Earth requires taking into account non-uniform gravitational forces and other forces such as drag and wind. A projectile is influenced by both gravity and aerodynamics when it is thrown in air. These influences may change the trajectory and thereby changing the range, the distance travelled during the time the object is in air.

Let us ignore all of these complications and consider an ideal projectile motion. In the ideal projectile motion, there is no air resistance and no change in gravitational acceleration. This assumption simplifies the mathematics greatly and is a good approximation to actual projectile motion in cases in which the distances travelled are small. Ideal projectile motion also serves as a good introduction to the calculation of range, before adding the complications of air resistance. For simplicity, we examine the case where y_o is zero. The horizontal position of the projectile with respect to the initial position $x_o = 0$ is

$$x(t) = v\cos\theta \cdot t. \tag{3.10}$$

In the vertical direction, the projectile's position is described by

$$y(t) = v\sin\theta \cdot t - \frac{1}{2}gt^2. \tag{3.11}$$

Here, we are interested in the time when the projectile returns to the same height of its origination (i.e., $y = 0$). So, we impose the condition of $y = 0$ in Eq. (3.11) and obtain

TWO DIMENSIONAL KINEMATICS

$$0 = t\left(v\sin\theta - \frac{1}{2}gt\right). \tag{3.12}$$

By factoring, we find that the two solutions are

$$t = 0 \quad \text{and} \quad t = \frac{2v\sin\theta}{g}. \tag{3.14}$$

The first solution corresponds to the time when the projectile is launched. The second solution is useful for determining the range of the projectile since it corresponds to the time when the projectile lands on the ground after the flight. Plugging the second solution for t into the horizontal equation of Eq. (3.10), we obtain the range of the projectile as

$$R = \frac{2v^2 \cos\theta \sin\theta}{g}. \tag{3.15}$$

We may apply the trigonometric identity $sin(\theta+\phi) = sin\theta \cdot cos\phi + sin\phi \cdot cos\theta$ to simplify Eq. (3.15). With the relation $sin(2\theta) = 2cos(\theta) \cdot sin(\theta)$, we may simplify Eq. (3.15) to

$$R = \frac{v^2 \sin 2\theta}{g}. \tag{3.16}$$

where R denotes the horizontal distance travelled. We should note that when θ is 45°, the solution becomes $R = v^2/g$. This means that an object launched at a 45° angle had the greatest range. More sophisticated arguments for why the 45° launch angle yields the greatest range exist, but we will not pursue them here because they involve calculus.

Exercise Problem 3.3:

Romeo is chucking pebbles gently up to Juliet's window, and he wants the pebble to hit the window with only a horizontal component of velocity. He is standing at the edge of a rose garden 8.0 m below her window and 9.0 m from the base of the wall. How fast are the pebbles going when they hit her window?

PROBLEM SOLVING STRATEGY

What do we know?

We know both the horizontal and vertical distance that the pebbles must travel. The pebbles have only the horizontal speed when they reach the window.

What concepts are needed?

We need to use the kinematic equations for the y-component to compute the time. We need to use this time to compute the horizontal speed.

Why?

When the pebbles reach the Juliet's window, they are at the maximum height of the trajectory.

Solution:

When the pebbles hit the window with only horizontal component of velocity, the vertical component of velocity is zero. Using this condition, we compute the y-component of the initial speed of pebbles leaving Romeo's hand:

$$v_y^2 = v_{oy}^2 - 2g(y-y_o) \quad \Rightarrow \quad 0 = v_{oy}^2 - 2gd_y.$$

We solve for v_{oy} and obtain

$$v_{oy} = \sqrt{2gd_y} = \sqrt{2 \times 9.8 \, \text{m}/\text{s}^2 \times 8.0 \, \text{m}} = 12.5 \, \frac{\text{m}}{\text{s}}.$$

Now, we compute the time pebbles take to reach Juliet's window and obtain

$$v_y = v_{oy} - gt \quad \Rightarrow \quad t = \frac{v_{oy}}{g} = \frac{12.5 \, \text{m}/\text{s}}{9.8 \, \text{m}/\text{s}^2} = 1.28 \, \text{s}.$$

The horizontal component of velocity is given by

$$x = x_o + v_{ox} t \quad \Rightarrow \quad v_{ox} = \frac{x - x_o}{t} = \frac{d_x}{t} = \frac{9.0 \, \text{m}}{1.28 \, \text{s}} = 7.0 \, \frac{\text{m}}{\text{s}}.$$

It should be noted that since there is no horizontal acceleration the horizontal speed of the pebbles remains the same. When the pebbles hit the window, the two components of the velocity are

$$v_x = v_{ox} \quad \text{and} \quad v_y = 0 \quad \Rightarrow \quad v = v_x = 7.0 \, \frac{\text{m}}{\text{s}}.$$

As we have noted above, determining the range of a projectile in the real world may not be simple. It is complicated by the fact that there are many factors such as volume, shape, surface roughness, and mass, influencing its trajectory. In general, a larger projectile (i.e., with greater volume) tends to face greater air resistance, which reduces the range of the projectile. The amount of air resistance faced by the projectile can be modified by its shape. For example, a tall and wide, but short, projectile will face greater air resistance than a low and narrow, but long, projectile of the same volume. The surface of the projectile must be considered. A smooth projectile will face less air resistance than a rough-surfaced one. Irregularities on the surface of the projectile may change its trajectory if they create more drag on one side of the projectile than on the other. Also, mass is important since a more massive projectile will have more motional energy and will be less affected by air resistance. The distribution of mass within the projectile can be important since an unevenly weighted projectile may spin undesirably, causing irregularities in its trajectory due to the Magnus effect. The Magnus effect generates sidewise force on a spinning cylindrical or spherical solid immersed in a fluid (liquid or gas) when there is relative motion between the spinning body and the fluid. This effect is responsible for the "curve" of a served tennis ball or a driven golf ball and affects the trajectory of a spinning artillery shell. Sometimes, if a projectile is given rotation along its axes of travel, then irregularities in the projectile's shape and weight distribution tend to be canceled out.

TWO DIMENSIONAL KINEMATICS

3.4 RELATIVE VELOCITY

In our discussion of motion up to this point, we considered kinematics of an object with respect to a fixed point of reference. Sometimes, it may be convenient for us to describe the motion of an object with respect to a reference frame that is moving with respect to another reference frame which may be either fixed or moving. This means that description of motion in these two reference frames is different due to the relative motion of these reference frames. Hence, when considering how observations are made in different reference frames, the relative velocity is used to relate different perspectives. We examine below the relative motion of reference frames without any acceleration.

Relative motion in one dimension

Most people find relative velocity to be a somewhat difficult concept. Let us consider relative motion in one dimension since it is reasonably straightforward. Suppose we are ground-based observers watching a passenger walking on a train which is moving with a speed of 9.0 m/s with respect to the ground as shown in Fig. 3.12. If he is walking along at a speed of 2.0 m/s with respect to the train in the same direction as the train, then he is moving at the speed of 11.0 m/s to us, ground-based observers. In this case, both the passenger and train move in the same direction, and hence both speeds are added to determine the passenger's speed with respect to the ground. However, if the passenger is walking in the opposite direction, then we would see him still moving in the same direction as the train but at a slower speed of 7 m/s.

Let us consider another example to grasp the concept of relative motion. Here, we will use the vector notation to account for the direction of relative motion formally. Suppose we are walking along a road at 8 km/h west. A train track which runs parallel to the road and a train is passing by, traveling at 40 km/h west. There is also a car driving by on the road, going 30 km/h east. How fast is the train traveling relative to us? How fast is the car traveling relative to us? And how fast is the train traveling relative to the car? One way to look at the problem is that the train will be 40 km west of where we are now in an hour, but we will be 8 km west. So the train will be 32 km further west than us in an hour. Relative to us, then, the train has a velocity of 32 km/h (West). Similarly, relative to the train, we have a velocity of 32 km/h (East). Using a subscript U for us, T for the train, and G for the ground, we can say that our velocity relative to the ground is $v_{UG} = 8$ km/h (West) and the velocity of the train relative to the ground is $v_{TG} = 40$ km/h (West). Note that if we flip the order of the subscripts, to get the velocity of the ground relative to us, for example, then we would get a vector which is equal in magnitude but opposite in direction. We can write this equal and opposite vector by flipping the sign (i.e., reversing the direction). The velocity of the ground relative to us is $v_{GU} = -v_{UG} = -8$ km/h (West) = 8 km/h (East). The velocity of the train relative

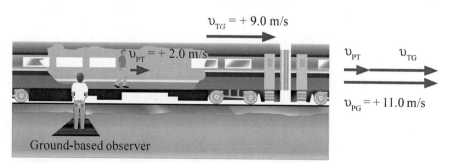

Figure 3.12: A schematic illustration of relative motion in one dimension. The motion of a person moving in the same direction as the moving train is seen by an observer on the ground.

to us, v_{TU}, can be found by adding vectors appropriately. Hence, we should pay attention to the order of the subscripts in this equation: $v_{TG} = v_{TU} + v_{UG}$. Here, two subscripts are used to represent a relative velocity. The first refers to the object and the second to the reference frame in which it has this velocity.

Relative motion in two dimensions

When the velocities are either in the same direction or in the opposite direction as in one dimension, we can find relative velocities without using vector subtraction. Relative motion in two dimensions is little more complex since velocities are not always in the same or opposite directions. Hence, we need to use vector subtraction and decompose the vectors into rectangular components (i.e., vector decomposition) to add or subtract vectors. The relative velocity equations in two dimensions look similar to those in one dimension. Suppose we consider the effect of the river current upon the motorboat as shown in Fig. 3.13. If a motorboat heads straight across a river (i.e., if the boat points its bow straight towards the other side), it would not reach the shore directly across from its starting point due to the river current. The river current influences the motion of the boat and carries it downstream. The motorboat may be moving with a velocity of 4.0 m/s directly across the river, but its resultant velocity will be greater than 4.0 m/s and at an angle in the downstream direction. While the speedometer of the boat may read 4.0 m/s, its speed with respect to an observer on the shore will be greater than 4.0 m/s.

The resultant velocity of the motorboat can be determined as the vector sum of the boat velocity v_{BW} and the river velocity v_{WS}. Here, we use the subscript B, W, and S to denote boat, water, and shore, respectively. Since the boat heads straight across the river and the current is always directed straight downstream, the two vectors are at right angles to each other. Thus, the Pythagorean equation can be used to determine the resultant velocity. Suppose that the river was moving with a velocity of 2.0 m/s, East and the motorboat was moving with a velocity of 4.0 m/s, North. What would be the resultant velocity of the motorboat (i.e., the velocity relative to an observer on the shore)? The magnitude of the resultant can be found as follows:

$$v_{BS} = \sqrt{v_{BW}^2 + v_{WS}^2} = \sqrt{\left(4.0\frac{m}{s}\right)^2 + \left(2.0\frac{m}{s}\right)^2} = 4.5\frac{m}{s}.$$

The direction is the counterclockwise angle of rotation that the resultant vector makes with due East. This angle can be determined using a trigonometric function as shown below:

$$\theta = \tan^{-1}\left(\frac{v_{BW}}{v_{WS}}\right) = \tan^{-1}\left(\frac{4.0\,m/s}{2.0\,m/s}\right) = 63°.$$

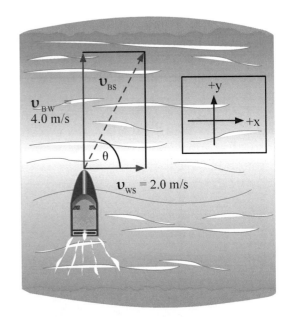

Figure 3.13: A schematic illustration of relative motion in two dimensions. To move directly across a stream of water along the y-direction, the boat must account for the speed of the waters flowing along the −x-direction.

The motorboat problem that we discussed here is a typical example of relative motion in two dimensions. These problems are typically accompanied by three separate questions. First,

TWO DIMENSIONAL KINEMATICS

what is the resultant velocity (both magnitude and direction) of the boat? Second, if the width of the river is x meters wide, then how much time does it takes the boat to travel shore to shore? Finally, what distance downstream does the boat reach the opposite shore? The first of these three questions was answered above; the resultant velocity of the boat can be determined using the Pythagorean equation (magnitude) and a trigonometric function (direction). The second and third of these questions can be answered using the average speed equation (and a lot of logic).

Understand-the-Concept Question 3.4:

Suppose a man in a rowboat is trying to cross a river that flows due west with a strong current. The man starts on the south bank and is trying to reach the north bank directly north from his starting point. He should

   a.   head due north,
   b.   head due west,
   c.   head in a northwesterly direction,
   d.   head in a northeasterly direction?

Explanation:

This question is similar to the motorboat problem that we discussed above. To describe the relative motion, we use the subscripts B, W, and S to denote boat, water, and shore, respectively. The velocity of the boat with respect to the shore is given as the vector sum of the velocity of the boat with respect to the river v_{BW} and the velocity of water with respect to the shore v_{WS}: $v_{BS} = v_{BW} + v_{WS}$. This means that the boat (i.e., v_{BW}) must head in a northeasterly direction to reach the north bank directly north from the sting point. A schematic diagram illustrating the situation of the man in a rowboat is shown below.

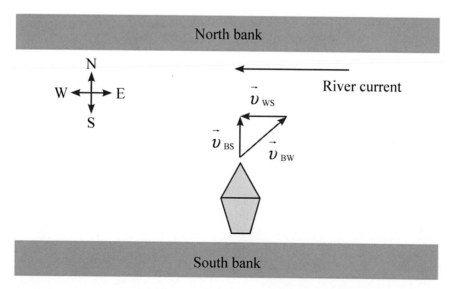

As we can see from this problem that relative motion is important for navigation of a ship and airplane, which is affected by water and air currents, respectively. Accounting for the direction of these currents is important, but it may not always be perpendicular with respect to the direction of motion for the vessels as in the problem.

Answer: d

Exercise Problem 3.4:

Two boats are heading away from shore. Boat 1 is heading due north at a speed of 3.00 m/s relative to the shore. Relative to Boat 1, Boat 2 is moving 30.0° northeast at a speed of 1.60 m/s. A passenger on Boat 2 walks due east across the deck at a speed of 1.20 m/s relative to boat 2. What is the speed of the passenger relative to the shore?

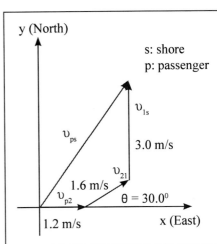

Solution:

The velocity of the passenger relative to the shore v_{ps} is related to the velocity of the passenger relative to boat 2, v_{p2}, and the velocity of boat 2 relative to the shore v_{2s} by

$$\vec{v}_{ps} = \vec{v}_{p2} + \vec{v}_{2s}. \tag{5a}$$

However, v_{2s} is related to v_{21} and v_{1s} by

$$\vec{v}_{2s} = \vec{v}_{21} + \vec{v}_{1s}, \tag{5b}$$

where v_{21} is the velocity of boat 2 relative to boat 1 and v_{1s} is the velocity of boat 1 relative to the shore. We may substitute Eq. (5b) into Eq. (5a) and obtain

$$\vec{v}_{ps} = \vec{v}_{p2} + \vec{v}_{21} + \vec{v}_{1s} = (v_{ps} + v_{21}\cos\theta)\hat{x} + (v_{21}\sin\theta + v_{1s})\hat{y}$$

$$= (1.2 + 1.6 \times \cos 30°)\frac{m}{s}\hat{x} + (1.6 \times \sin 30° + 3.0)\frac{m}{s}\hat{y} = 2.59\frac{m}{s}\hat{x} + 3.8\frac{m}{s}\hat{y}.$$

Finally, we compute the magnitude of the velocity v_{ps} by using the Pythagorean equation and obtain

$$v_{ps} = \sqrt{\left(2.59\frac{m}{s}\right)^2 + \left(3.80\frac{m}{s}\right)^2} \cong 4.60\frac{m}{s}.$$

We may further appreciate the importance of relative motion when we recognize that there is no such thing as absolute motion in the Universe. An object is only at rest or in motion relative to some reference frame. In other words, relative motion is motion with respect to an arbitrarily selected object,

TWO DIMENSIONAL KINEMATICS

which may or may not have actual or true motion. A mountain on Earth may be at rest relative to the Earth, but it is in motion relative to the Sun. Although all motion is relative, motion is usually defined in terms of its direction and rate of movement relative to the Earth. If the object we want to describe is a ship, then this motion is defined in terms of the true course and speed. The motion of an object also may be defined in terms of its direction and rate of movement relative to another object which may also be in motion. The motion of one ship relative to the motion of another ship is defined in terms of the direction and speed of relative movement. Each form of motion may be depicted by a velocity vector, which represents direction and rate of movement. So we should not be surprised to know that, for decades, the maneuvering board chart has been a mainstay for solving relative motion problems. The bearing and distance marks from the center and the convenient scales at the sides allow one to plot both course and speed vectors easily and quickly. Vector addition allowed navigators to provide course and speed for rendezvous (or the necessary changes to avoid collision). For a submariner, relative motion is almost indispensable if he/she wants to fire a torpedo and expects to hit a target.

SUMMARY

i. **Scalars** are quantities that are fully described by their magnitude alone, (i.e., by a single number). **Vectors** are quantities that are fully described by both magnitude and direction.

ii. **Projectile motion** is a form of motion where a particle (called a projectile) is thrown obliquely without propulsion near the Earth's surface. Projectiles move along a curved path under the action of gravity. The path followed by a projectile is called its trajectory and is described by a parabola.

iii. The **trajectory** of a projectile, neglecting all other forces including friction from air resistance, can be described by breaking up the motion of an object into two components, each of which follows the familiar one-dimensional equations of motion as shown below:

	x-component	y-component
Velocity-versus-time	$v_x = v_{ox}$	$v_y = v_{oy} - gt$
Position-versus-time	$x = x_o + v_{ox} t$	$y = y_o + v_{oy} t - \frac{1}{2} g t^2$
Velocity-versus-position	$v_x^2 = v_{ox}^2$	$v_y^2 = v_{oy}^2 - 2g(y - y_o)$
Average speed	$\overline{v_x} = v_{ox}$	$\overline{v_y} = \dfrac{v_{oy} + v_y}{2}$

iv. The laws of physics are the same when they are applied in any reference frame which is moving at a constant velocity with respect to the Earth. However, the motion may appear differently when viewed from a different frame of reference. The difference in the description of the motion can be explained by including relative motion of the reference frame.

MORE WORKED PROBLEMS

Problem 3.1:

A 2.05-m-tall basketball player takes a shot when he is 6.02 m from the basket (at the three –point line). If the launch angle is 25° and the ball was launched at the level of the player's head, what must be release speed o the ball or the player to make the shot? The basket is 3.05 m above the floor.

Solution:

For simplicity, we set $x_o = y_o = 0$. To obtain the time of flight, we use the position-versus-time kinematic equation

$$x = x_o + v_{xo}t = v_o \cos\theta\, t \Rightarrow t = \frac{x}{v_o \cos\theta}.$$

We use the result and the position-versus-time kinematic equation for the y-axis to obtain

$$y = y_o + v_{yo}t - \frac{1}{2}gt^2 = v_o \sin\theta \frac{x}{v_o \cos\theta} - \frac{1}{2}g\left(\frac{x}{v_o \cos\theta}\right)^2 = x\tan\theta - \frac{gx^2}{2v_o^2 \cos^2\theta}.$$

We rewrite the position-versus-time equation for the y-axis to solve for v_o and obtain

$$y - x\tan\theta = -\frac{gx^2}{2v_o^2 \cos^2\theta} \Rightarrow v_o^2 = \frac{gx^2}{2\cos^2\theta(x\tan\theta - y)}.$$

Now, we compute the speed v_o by substituting in the value of the known quantities and obtain

$$v_o = \frac{x}{\cos\theta}\sqrt{\frac{g}{2(x\tan\theta - y)}} = \frac{6.02\,\text{m}}{\cos 25°}\sqrt{\frac{9.80\,\text{m/s}^2}{2\times(6.02\,\text{m}\times\tan 25° - 1.0\,\text{m})}} = 10.9\,\frac{\text{m}}{\text{s}}.$$

Problem 3.2:

An airplane is flying with a velocity of 240 m/s at an angle of 30.0° with the horizontal, as the drawing shows. When the altitude of the plane is 2.4 km, a flare is released from the plane. The flare hits the target on the ground. What is the angle θ?

TWO DIMENSIONAL KINEMATICS

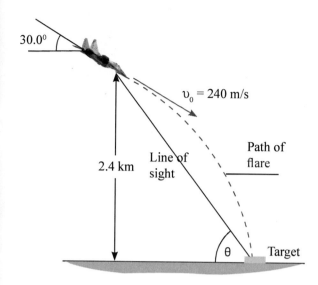

PROBLEM SOLVING STRATEGY

What do we know?

Both the speed and angle of attack for the airplane are given. Also, the altitude of the plane is known.

What concepts are needed?

We need to decompose the motion of the flare into the horizontal and vertical components. Then, we apply the two sets of kinematic equations for the projectile.

Why?

The initial velocity of the flare is the same as that of the plane. The horizontal motion of the flare is independent of its vertical motion.

Solution:

According to the schematic diagram, the angle θ can be found from

$$\tan\theta = \frac{2400\,\text{m}}{x} \Rightarrow \theta = \tan^{-1}\frac{2400\,\text{m}}{x},$$

where x is the horizontal displacement of the flare. Since $a_x = 0$ for the horizontal motion of the flare (i.e., projectile), we may write the position-versus-time equation as

$$x = v_{ox}t = v_o \cos\phi\, t,$$

with $\phi = 30°$. The flight time is determined by consider the vertical motion. Now, assuming that upward is the positive direction, we rewrite the position-versus-time equation as

$$y = -v_{oy}t + \frac{1}{2}a_y t^2 = -v_o \sin\phi\, t + \frac{1}{2}a_y t^2.$$

Noting the condition that the flare travelled the vertical distance of 2.4 km (i.e., $y = 2.4$ km) when it hits the ground, we rearrange the position-versus-time equation to obtain

$$\frac{1}{2}a_y t^2 - v_o \sin\phi\, t - y = 0.$$

Noting that $y = 2400$ m and $a_y = 9.80$ m/s², we solve for the time t and obtain

$$t = \frac{v_o \sin\phi \pm \sqrt{(v_o \sin\phi)^2 - 4\left(\frac{a_y}{2}\right)(-y)}}{2\left(\frac{a_y}{2}\right)}$$

$$= \frac{240\,\text{m/s} \times \sin 30° \pm \sqrt{\left(240\,\text{m/s} \times \sin 30°\right)^2 + 2 \times 9.8\,\text{m/s}^2 \times 2400\,\text{m}}}{-9.8\,\text{m/s}^2}.$$

We obtain $t = 13\,\text{s}$ as a solution to the quadratic formula. We should note that between the two solutions of the quadratic formula only this solution is physically sensible one. We may now use this result to find that the flare travels the horizontal distance of

$$x = v_o \cos\phi\, t = 240\,\frac{\text{m}}{\text{s}} \times \cos 30° \times 13\,\text{s} = 2700\,\text{m}.$$

So, knowing the value of x, we may easily determine the angle θ as

$$\theta = \tan^{-1}\left(\frac{2400\,\text{m}}{2700\,\text{m}}\right) = 42°.$$

Problem 3.3:

The punter on a football team tries to kick a football so that it stays in the air for a long "hang time". If the ball is kicked with an initial velocity of 25.0 m/s at an angle of 60.0° above the ground, what is the "hang time"?

Solution:

First, we compute the x- and y-component of the initial velocity v_o and obtain

$$v_{ox} = v_o \cos\theta = 25.0\,\frac{\text{m}}{\text{s}} \times \cos 60° = 12.5\,\frac{\text{m}}{\text{s}}.$$

$$v_{oy} = v_o \sin\theta = 25.0\,\frac{\text{m}}{\text{s}} \times \sin 60° = 21.7\,\frac{\text{m}}{\text{s}}.$$

respectively. The "hang time" can be determined by solving the vertical position-versus-time equation of

$$y = y_o + v_{yo} t - \frac{1}{2} g t^2.$$

We set both $y_o = 0$ and $y = 0$ because both the initial and final position of the football is on the ground. We use these conditions to write

$$0 = v_{oy} t - \frac{1}{2} g t^2 = t\left(v_{oy} - \frac{1}{2} g t\right).$$

We can see easily that there are two solutions. One solution, $t = 0$, corresponds to the initial vertical position $y = 0$. Other solution corresponds to the hang time T, which is obtained by solving

$$v_{oy} - \frac{1}{2} g T = 0 \Rightarrow T = \frac{2 v_{oy}}{g} = \frac{2 \times 21.7\,\text{m/s}}{9.80\,\text{m/s}^2} = 4.42\,\text{s}.$$

TWO DIMENSIONAL KINEMATICS

Problem 3.4:

You are in a hot-air balloon that, relative to the ground, has a velocity of 6.0 m/s in a direction due east. You see a hawk moving directly away from the balloon in a direction due north. The speed of the hawk relative to you is 2.0 m/s. What are the magnitude and direction of the hawk's velocity relative to the ground? Express the directional angle relative to due east?

Solution:

It is noted that the velocity of the hawk with respect to the ground is

$$\vec{v}_{HG} = \vec{v}_{BG} + \vec{v}_{HB}$$

where v_{BG} = 6.0 m/s is the velocity of balloon with respect to the ground, and v_{HB} = 2.0 m/s is the velocity of the hawk with respect to the balloon. The magnitude of v_{HG} is given by

$$v_{HG} = \sqrt{v_{BG}^2 + v_{HG}^2} = \sqrt{\left(6.0\frac{m}{s}\right)^2 + \left(2.0\frac{m}{s}\right)^2} = 6.3\frac{m}{s}.$$

The direction of v_{HG} is computed as

$$\tan\theta = \frac{v_{HB}}{v_{BG}} \Rightarrow \theta = \tan^{-1}\frac{v_{HB}}{v_{BG}} = \tan^{-1}\left(\frac{2.0\,m/s}{6.0\,m/s}\right) = 18°.$$

PROBLEMS

1. Three displacements are **A** = 200 m due south, **B** = 250 m due west, and **C** = 150 m at 30.0° east of the following possible ways of adding these vectors $R_1 = A + B + C$; $R_2 = B + C + A$; $R_3 = C + B + A$. (b) Explain what you can conclude from comparing the diagrams.

2. The route followed by a hiker consists of three displacement vectors **A**, **B**, and **C**. Vector **A** is along a measured trail and is 1550 m in a direction 25.0° north of east. Vector **B** is not along a measured trail, but the hiker uses a compass and knows that the direction is 41.0° east of south. Similarly, the direction of vector **C** is 35.0° north of west. The hiker ends up back where she started, so the resulting displacement is zero, or **A** + **B** + **C** = 0. Find the magnitude of vector **B** and (b) vector **C**.

3. The Earth moves around the Sun in a nearly circular orbit of radius 1.50×10¹¹ m. During the three summer months (an elapsed time of 7.89×10⁶ s), the Earth move one-fourth of the distance around the Sun. (a) What is the average speed of the Earth? (b) What is the magnitude of the average velocity of the Earth during this period?

4. An airplane flies 200 km due west from city A to city B and then 300 km in the direction of 30.0° north of west from city B to city C. (a) In straight-line distance, how far is city C, from city A? (b) Relative to city A, in what direction is city C? (c) Why is the answer only approximately correct?

5. An ocean liner leaves New York City and travels 18.0° north of east for 155 km. How far east and how far north has it gone? In other words, what are the magnitudes of the components of the ship's displacement vector in the directions (a) due east and (b) due north?

6. You friend has slipped and fallen. To help her up, you pull with a force **F**, as the drawing shows. The vertical component of this force is 130 N, and the horizontal component is 150 N. Find (a) the magnitude of **F** and (b) the angle θ.

7. Three forces are applied to an object, as indicated in the drawing. Force $\mathbf{F}_1$ has a magnitude of 21.0 N and is directed 30.0° to the left of the +y axis. Force $\mathbf{F}_2$ has a magnitude of 15.0 N and points along the +x axis. What must be the magnitude and direction (specified by the angle θ in the drawing) of the third force $\mathbf{F}_3$ such that the vector sum of the three forces is 0 N?

8. An automobile travels, at a constant speed of 60 km/h, 800 m along a straight highway that is inclined 5.0° to the horizontal. An observer notes only the vertical motion of the car. What is the car's (a) vertical velocity magnitude and (b) vertical travel distance?

9. A golf ball rolls off a horizontal cliff with an initial speed of 11.4 m/s. The ball falls a vertical distance of 15.5 m into a lake below. (a) How much time does the ball spend in the air? (b) What is the speed v of the ball just before it strikes the water?

10. A motorcycle daredevil is attempting to jump across as many buses as possible (see the drawing in the book). The takeoff ramp makes an angle of 18.0° above the horizontal, and the landing ramp is identical to the takeoff ramp. The buses are parked side by side, and each bus is 2.74 m wide. The cyclist leaves the ramp with a speed of 33.5 m/s. What is the maximum number of buses over which the cyclist can jump?

11. A horizontal rifle is fired at a bull's-eye. The muzzle speed of the bullet is 670 m/s. The barrel is pointed directly at the center of bull's-eye, but the bullet strikes the target 0.025 m below the center. What is the horizontal distance between the end of the rifle and the bull's-eye?

12. A small plane takes off at a constant velocity of 150 km/h at an angle of 37°. At 3.00 s, (a) how high is the plane above the ground, and (b) what horizontal distance has the plane traveled from the liftoff point?

13. An airplane with a speed of 97.5 m/s is climbing upward at an angle of 50.0° with respect to the horizontal. When the plane's altitude is 732 m, the pilot releases a package. (a) Calculate the distance along the ground, measured from a point directly beneath the point of release, to where the package hits the earth. (b) Relative to the ground, what is the angle of the velocity vector of the package just before impact?

14. A pitcher throws a fastball horizontally at a speed of 140 km/h toward home plate, 18.4 m away. (a) If the batter's combined reaction and swing times total 0.350 s, how long can the batter watch the ball after it has left the pitcher's hand before swinging? (b) In traveling to the plate, how far does the ball drop from its original horizontal line?

15. Two cannons are mounted as shown in the drawing and rigged to fire simultaneously. They are used in a circus act in which two clowns serve as human cannonballs. The clowns are fired toward each other and collide at a height of 1.00 m above the muzzles of the cannons. Clown A is launched at a 75.0° angle, with a speed of 9.00 m/s. The horizontal separation between the clowns as they leave the cannons is 6.00 m. Find the launch speed u_{0B} and the launch angle θ_B (> 45.0°) for clown B.

16. A golf ball with an initial speed of 50.0 m/s lands exactly 240 m downrange on a level course. (a) Neglecting air friction, what two projection angles would achieve this result? (b) What is the maximum height reached by the ball, using the two angles determined in part (a)?

17. A rock is thrown upward from the level ground in such a way that the maximum height of its flight is equal to its horizontal range R. (a) At what angle θ is the rock thrown? (b) In terms of the original range R, what is the range R_{max} the rock can attain if it is launched at the same speed

> # TWO DIMENSIONAL KINEMATICS

but at the optimal angle for maximum range? (c) Would your answer to part (a) be different if the rock is thrown with the same speed on a different planet? Explain.

18. A dart gun is fired while being held horizontally at a height of 1.00 m above ground level and while it is at rest relative to the ground. The dart from the gun travels a horizontal distance of 5.00 m. A college student holds the same gun in a horizontal position while sliding down a 45.0° incline at a constant speed of 2.00 m/s. How far will the dart travel if the student fires the gun when it is 1.00 m above the ground?

19. Starting from rest, several toy cars roll down ramps of differing lengths and angles. Rank them according to their speed at the bottom of the ramp, from slowest to fastest. Car A goes down a 10 m ramp inclined at 15°, car B goes down a 10 m ramp inclined at 20°, car C goes down at 8.0 m ramp inclined at 20°, and car D goes down a 12 m ramp inclined at 12°.

20. One of King Arthur's knights fires cannon from the top of the castle wall. The cannonball is fired at a speed of 50 m/s and an angle of 30°. A cannonball that was accidentally dropped hits the moat below in 1.5 s. How far from the castle wall does the fired cannonball hit the ground?

21. A tennis player hits a ball 2.0 m above the ground. The ball leaves his racquet with a speed of 20 m/s at an angle 5.0° above the horizontal. The horizontal distance to the net is 7.0 m, and the net is 1.0 m high. Does the ball clear the net? If s, by how much? If not, by how does it miss?

22. At some airports there are speed ramps to help passengers get from one place to another. A speed ramp is a moving conveyor belt that you can either stand or walk on. Suppose a speed ramp has a length of 105 m and is moving at speed of 2.0 m/s relative to the ground. In addition, suppose you can cover this distance in 75 s when walking on the ground. If you walk at the same rate with respect to the speed ramp that you walk with respect to the ground, how long does it take for you to travel the 105 m using the speed ramp?

23. When the moving sidewalk at the airport is broken, as it often seems to be, it takes you 50 s to walk from your gate to the baggage claim. When it is working and you stand on the moving sidewalk the entire way, without walking, it takes 75 s to travel the same distance. How long will it take you to travel from the gate to baggage claim if you walk while riding on the moving sidewalk?

24. A swimmer maintains a speed of 0.15 m/s relative to the water when he swims directly toward the opposite shore of a river. The river has a current that flows at 0.75 m/s. (a) How far downstream is he carried in 1.5 min? (b) What is his velocity relative to an observer on shore?

25. A river has a steady speed of 0.500 m/s. A student swims upstream a distance of 1.00 km and swims back to the starting point. (a) If the student can swim at a speed of 1.20 m/s in still water, how long does the trip take? (b) How much time is required in still water for the same length swim? (c) Intuitively, why does the swim take longer when there is a current?

26. A person looking out the window of a stationary train notices that raindrops are falling down at a speed of 5.0 m/s relative to the ground. When the train moves at a constant velocity, the raindrops make an angle of 25° when they move past the window, as the drawing shows. How fast is train moving?

27. You are in a hot-air balloon that, relative to the ground, has a velocity of 6.0 m/s in a direction due east. You see a hawk moving directly away from the balloon in a direction due north. The speed of the hawk relative to you is 2.0 m/s. What are the magnitude and direction of the hawk's velocity relative to the ground? Express the directional angle relative to due east?

28. A jet liner can fly 6.00 hours on a full load of fuel. Without any wind it flies at a speed of 2.40×10^2 m/s. The plane is to make a round-trip by heading due west for a certain distance, turning around, and then heading due east for the return trip. During the entire flight, however, the

plane encounters a 57.8-m/s wind from the jet stream, which blows from west to east. What is the maximum distance that the plane can travel due west and so that it will just be able to return home?

29. A Coast Guard cutter detects an unidentified ship at a distance of 20.0 km in the direction 15.0° east of north. The ship is traveling at 26.0 km/h on a course at 40.0 east of north. The Coast Guard wishes to send a speedboat to intercept and investigate the vessel. (a) If the speedboat travels at 50.0 km/h, in what direction should it head? Express the direction as a compass bearing with respect to due north. (b) Find the time required for the cutter to intercept the ship.

30. An assembly line has a staple gun that rolls to the left at 1.0 m/s while parts to be stapled roll past it to the right at 3.0 m/s. The staple gun fires 10 staples per second. How far apart are the staples in the finished part?

LAWS OF MOTION

4 CHAPTER

In the name of Love, two hamsters are barely hanging on. What is holding them in place so they would not fall? Why would we expect them to fall if one of the two hamsters cannot hold on for a whatever reason? The answer to these questions is "force". Although we may be familiar with this abstract concept because we use the word quite commonly, let us consider few examples we are familiar with in order to have a better grasp of what force is. One example is a rock climber who is hanging on a cliff, similar to two hamsters hanging on as shown in the picture above. In both cases, the rock climber and two hamsters are fighting against the force of gravity.

A rock climber has to use precarious holds and balancing movements to make it up a mountain face. These movements are possible because the human body functions with precision. Also, equipment is used to make some of the more difficult climbs possible in the presence of force of gravity which is always pulling the rock climbers down. One part of rock climbing has to do with the way the climber's hands work. There are two different grips that climbers must use when doing rock climbing: the power grip and the precision grip. The power grip uses the entire hand to grip an object so that it is pressed between the fingers and the palm of the hand using the muscles of the forearm. This allows the grip to be tight and powerful as opposed to the precision grip.

The precision grip utilizes fine motor skills. It grips the object between the fingers. Using the power of the muscles in the hand rather than the larger muscles of the forearm. When climbing, this techniques comes into play when ensuring safety with a chock wedged into the mountainside. The precision grip is used to grip the chock and to shove it into a crack in the mountain. A karabiner is attached to the chock and the rope is threaded through it. If the climber falls, rock climbing science says that the rope would bring the climber to a stop by briefly exerting a force much larger than the force exerted on the body by gravity.

Another perspective on the concept of force is found in George Lucas' fictional universe of "Star Wars". In the Star Wars movie, the **Force** is a binding, metaphysical, and ubiquitous power. In the original Star Wars film, the Force is described as an energy field created by all living things. The Force surrounds and penetrates living beings and binds the galaxy together. Also, the Force has a "Dark side", which feeds off emotions such as anger, jealousy, fear, lust, and hate, but the Jedi are only supposed to use the Force for peaceful purposes. The Force can enhance natural, physical, and mental abilities, including strength (such as during a "Force jump" or to slow a fall from an otherwise dangerous height) and accuracy. A number of other Force powers include telekinesis, telepathy, levitation, deep hypnosis, enhanced empathy, reflexes, precognition, and enhanced speed. Also, the Force gives enhanced skills in light-saber combat as well as a number of other powers, such as the ability to heal or drain the life force of others, increase resistance to attack, warp space and to dissipate energy attacks. Those people who possess the discipline and subtlety of mind can sense the disturbances in the Force. Since the Force is "an energy field created by all living things", a disturbance can be felt when there is death or suffering on a massive scale. Force-sensitivity is a condition in the Star Wars universe where a life form possesses a natural connection to the Force.

Is there any connection between the mythical Force in the Star Wars universe and the concept of force in physics in our own universe? Perhaps! If the Force has its counterpart in physics, then, most likely, it may be associated with the concept of force and energy. Now, let us direct our attention to the concept of force in this chapter and defer our discussion on energy to Chapter 5.

In general, **force** is defined as any influence that causes an object to undergo a certain change in its movement, orientation, or geometrical construction. In physics, there are fundamental interactions, sometimes called **fundamental forces,** which govern the way universe evolves. These interactions are used to describe fundamental physics as patterns of relations in physical systems which evolve over time. However, these fundamental forces are not reducible to more basic entities. The known fundamental interactions are gravitation, electromagnetism, strong nuclear, and weak nuclear which are all non-contact forces. In principle, these forces allow us to understand all of physical phenomena around us as well as any changes occurring in the universe, from the subatomic to cosmic scales.

Forces that we will discuss in this book are derived from these fundamental forces and are the underlying reason for motion of objects. A force can be described in terms of intuitive concepts such as a push or pull. In other words, a force can cause an object with mass to change its velocity (which includes from a state of rest to begin moving) or to accelerate. Also, it can cause a flexible object to deform. A force is a vector, having both magnitude and direction. The SI unit for measuring force is Newton (N) and typically is represented by the symbol **F**. The effects of force on an object may depend on a situation. Some forces lead to translational motion, while others lead to rotational motion. For translational motion, if the mass of the object is constant, then the acceleration is directly proportional to the net force acting on the object. The acceleration is in the direction of the net force, and is inversely proportional the mass of the object.

4.1 MOTION AND FORCES

In Chapters 2 and 3, we looked at kinematics which describes how an object moves. Now, in Chapters 4 and 5, we look at why an object moves. As we have seen from the quantities determining motion, the concept of force is plays a central role for understanding why an object moves. Also, the effects of force may be described in terms of laws of motion.

What makes an object at rest begin to move and why do objects move as they do?

The answer is "force"! Sir Isaac Newton (1642 – 1727) stated that forces lead to changes in a property of matter that he called momentum. The momentum of an object is equal to its mass multiplied by its velocity (see Chapter 6). A greater force leads to a faster rate of change of momentum. For an object with a fixed mass, that rate of change can be simplified to a change in velocity, which is known as acceleration. The concept of acceleration was not recognized by Aristotle (384 BC – 322 BC) because he lacked the mathematics needed to deal with it. It took many centuries before, Newton invented calculus to deal with acceleration and led directly to one of the most famous equations in all science, **F** = m**a**. This equation relates the force **F** acting on a mass m to the acceleration a it produces.

What is force?

As we discussed above, a force is any kind of push or pull. The concept of force was originally defined by Newton in his three laws of motion. Force is a quantitative description of the interaction between two physical bodies, such as an object and its environment. There are contact and non-contact forces. Contact force is defined as the force exerted when two physical objects come in direct contact with each other. On the other hand, non-contact forces, such as gravitation and electromagnetic force, can exert themselves even across the empty vacuum of space. For these forces, no contact is necessary.

It is generally understood that there are four fundamental forces governing physics in the universe. These four fundamental forces are gravitational force, Coulomb force, weak force, and strong force. The **strong interaction** is very strong, but very short-ranged. It acts only over ranges of order 10^{-15} m and is responsible for holding atomic nuclei together. Basically, it is an attractive force, but it can be an effective repulsive force in some circumstances. The **electromagnetic force** causes electric and magnetic effects such as the repulsion between like electrical charges or the interaction of bar magnets. It is long-ranged but is much weaker than the strong force. It can be either attractive or repulsive and acts only between pieces of matter carrying electrical charge. The **weak force** is responsible for radioactive decay and neutrino interactions. It has a very short range and it is very weak, as indicated by its name. The **gravitational force** is weak, but very long ranged. Furthermore, it is always attractive and acts between any two pieces of matter in the Universe since mass is its source. However, these four fundamental forces are found to be related to each other. One of the most challenging problems in physics is to find a way to unify of all of these four fundamental forces.

Why does force not always give rise to motion?

According to Newton, there will be no motion if and only if net force acting on an object at rest is zero. Since force is a vector, several forces may act on an object in different directions. The sum of forces acting on a particle is called the total force or net force. The net force is defined as a single force that replaces the effect of the original forces on the particle's motion. As described by Newton, the net force gives the same acceleration for the particle as all of those actual forces combined together.

4.2 NEWTON'S THREE LAWS OF MOTION

There are basic governing principles for motion of objects. In his book entitled "Philosophiae Naturalis Principia Mathematica" (Mathematical Principles of Natural Philosophy), which was first published in 1687, Newton formulated the governing principles as laws of motion. Newton's laws of motion are three physical laws that form the basis of classical mechanics. They describe the relationship between the forces acting on a body and its motion due to those forces. Let us now state the laws of motion.

The three laws of motion can be summarized as follows. The **first law** states that if an object experiences no net force, then its velocity is constant: the object is either at rest (if its velocity is zero), or it moves in a straight line with constant speed (if its velocity is nonzero). Here, no net force may mean that there are no forces acting on the object or that there are several forces acting on the object but they all cancel out. Also since velocity is a vector quantity which expresses both the object's speed and the direction of its motion, the statement that the velocity is constant implies that both its speed and direction are constant. This is called "law of inertia." This law predicts the behavior of objects for which all existing forces are balanced. The **second law** states that the acceleration of an object is directly proportional to the net force acting on it and is inversely proportional to its mass. The direction of the acceleration is in the direction of the net force acting on the object.

$$\vec{a} = \frac{1}{m}\sum_i \vec{F}_i \Rightarrow \sum_i \vec{F}_i = m\vec{a}. \tag{4.1}$$

Newton's second law pertains to the behavior of objects when all forces are not balanced. Equation (4.1) states that the acceleration of an object depends on two variables: i) the net force acting upon the object and ii) the mass of the object. As the force acting upon an object is increased, the acceleration of the object is increased. However, as the mass of an object is increased, the acceleration of the object is decreased. The **third law** states that whenever one object exerts force on a second object, the second exerts an equal and opposite force on the first. The third states that all forces exist in pairs. If object A exerts force $\mathbf{F}_A$ on object B, then B simultaneously exerts force $\mathbf{F}_B$ on A. These two forces are equal in magnitude but opposite in direction: $\mathbf{F}_A = -\mathbf{F}_B$. The third law means that all forces are interactions between different bodies, and thus there is no such thing as a unidirectional force or force that acts on only one body. Also, this law states that to every action there is an equal and opposite reaction. It does not matter which is called the action and which is called reaction; both forces are part of a single interaction, and neither force exists without the other. From a conceptual point of view, Newton's third law is realized when a person pulls a rope attached to a wall as shown in Fig. 4.1, the wall pulls the rope against the person with the equal strength. Similarly, the tires of a car push against the road when the road pushes back on the tires, indicating that the tires and road simultaneously push against each other. In swimming, a person interacts with the water in a swimming pool, pushing the water backward, while the water simultaneously pushes the person forward. Again, both the person and water push against each other. The reaction forces in these examples account for the motion. Another useful example is

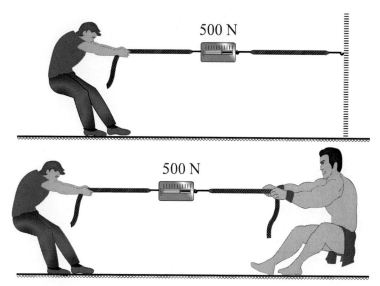

Figure 4.1: A schematic illustration of Newton's third law. According to this law, for every action there is an equal (in size) and opposite (in direction) reaction. Here, a fictitious weight lifter representing the wall is providing the reaction force.

forces that depend on friction. When friction is absent, a person or car on ice is unable to exert the action force to produce the needed reaction force.

These three laws of motion are used to explain and to investigate motion of many physical objects and systems. For example, in the third volume of the Philosophiae Naturalis Principia Mathematica, Newton showed that these laws of motion can be combined with his universal law of gravitation to derive the Kepler's three laws of planetary motion, which were inferred from careful observations. Kepler's three laws are the following: 1) the orbit of every planet is an ellipse with the Sun at one of the two foci; 2) a line joining a planet and the Sun sweeps out equal areas during equal intervals of time; and 3) the square of the orbital period of a planet is proportional to the cube of the semi-major axis of its orbit. (See Chapter 7 for more on this law of planetary motion.)

4.3 WHAT CAUSES MOTION?

To grasp the concept of force better, it will be helpful for us to see how our understanding of mechanics was developed and improved through the ages. According to Aristotle, the natural state of an earthly object is to be at rest. This viewpoint was developed based on observations. Aristotle provided a philosophical discussion on the concept of a force as an integral part of Aristotelian cosmology. In Aristotle's view, the natural world held four elements that existed in "natural states". Aristotle believed that the natural state of objects with mass on Earth, such as the elements water and earth, is to be motionless on the ground and that they tended towards that state if left alone. He distinguished between the innate tendency of objects to find their "natural place" (e.g., for heavy bodies to fall), which led to "natural motion", and unnatural or forced motion, which required continued application of a force. This theory is based on the everyday experience of how objects move, such as the constant application of a force needed to keep a cart moving. However, this view had conceptual trouble accounting for the behavior of projectiles, such as the flight of arrows. The place where forces were applied to projectiles was only at the start of the flight. While the projectile sailed through the air, there is no discernible force acting on it. Aristotle was aware of this problem and proposed that the air displaced through the projectile's path provided the needed force to continue the projectile moving. This explanation demands that air is needed for projectiles. Based on the theory, no projectile would be able to move in a vacuum after the initial push. Additional problems with the explanation include the fact that air resists the motion of the projectiles. Aristotelian physics began facing criticism in the medieval period in the sixth century. Of course, we now know this simple viewpoint is incomplete!

The shortcomings of Aristotelian physics would not be fully corrected until the work of Galileo Galilei (1564 – 1642) in the seventeenth century. Galileo was influenced by the late medieval idea that objects in forced motion carried an innate force of impetus. As discussed in Chapter 2, Galileo constructed an experiment in which stones and cannonballs were both dropped at the same time to disprove the Aristotelian theory of motion. He showed that the bodies were accelerated by gravity and the acceleration was independent of their mass. He argued that objects retain their velocity unless acted on by a resistive force such as friction. Also Galileo suggested that an object is at rest due to nature's influences that are resistive to motion of the object. The natural state of an object is uniform motion with constant velocity when it is free of external influences. We should note that being "at rest" is a special case of uniform motion with zero velocity.

As an example of the resistive influences, we consider a sled sliding on different surfaces as shown in Figure. 4.2. Influences that lead to a deviation from uniform motion are known as "**forces**"; hence, force is defined as an influence tending to change the direction of motion of a body, produce motion, or stress in a stationary body. As its effect, force can change the direction of motion of body that is already in motion or can lead to the motion of a stationary body. For example, when a car moving on a road is suddenly hit from behind by a truck, the car's velocity increases due to the force applied by the truck. (Similarly, the truck's velocity decreases due to the reaction force from the car.) A collision between a moving and stationary object causes the velocity (and position) of the stationary object to change. Another example, shown in Fig. 4.3, illustrates that an external influence can change the car's uniform motion. Depends on the direction, force affects the velocity of an object differently. A force acting in the same direction as the velocity changes only the speed of the object.

Figure 4.2: The effect of a resistive force known as friction on the motion of a child on a sled is illustrated schematically. Here, all kinds of resistive forces including friction cause moving objects to slow down or stop.

LAWS OF MOTION

However, as we will see in Chapter 5 the force acting at right angles to the velocity does not change the speed but only changes the direction of the velocity. An example of such force is the gravitational force exerted by the Earth on a satellite moving in a circular orbit.

As shown in Fig. 4.3, the force of impact changes the object's motion. Does this force depend on the mass of the object? To answer this question, let us think about the force of impact on the child in a frontal crash. We assume the child has a small mass compared to the car. Because of the mass of the child is small compared to that of the car, the child does not affect the speed of the car significantly. So, we can calculate the approximate force of impact. Before the impact, the relative speed of the child with respect to the car is zero. However, as the car stops abruptly at the impact, the child will accelerate rapidly with respect to the car due to rapidly decreasing speed of the car. The distance at which the child accelerates to achieve this speed is roughly equal to the width of the child. If we assume an average child is 25 cm wide (2.8 inches), then using the velocity-versus-position equations

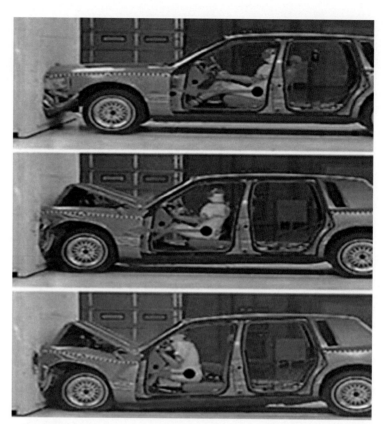

Figure 4.3: Whenever a car is involved in a crash, intense forces are exerted on the car to stop it. A given amount of force is presented during any crash. Note that the green arrows are velocity, not force! The magnitude of force varies based on the speed and mass of the car as well as the speed and mass of whatever it hits.

$$v^2 = v_o^2 + 2ad \implies a = \frac{v^2 - v_o^2}{2d} \tag{4.2}$$

we may estimate $a = 246.4$ m/s² as the acceleration. If the mass of the child is 35 kg (77 pounds), then according to Eq. (4.1) the total force of $F = ma = 8624$ N is directed upon the child at the impact. This is equivalent to the weight of 880 kg object or 1939 pounds (almost an English ton). Most certainly, this is enough force to injure anyone seriously. A slightly higher initial speed of the car would yield an even larger impact force.

To reduce the impact of crash, a crumple zone was invented. Crumple zones are designed to absorb the energy of motion from the impact during a traffic collision by controlled deformation. The concept of energy is discussed in Chapter 5. This energy is much greater than is commonly realized. A 2000 kg car travelling at 60 km/h (16.7 m/s), before crashing into a thick concrete wall, is subject to the same impact force as a front-down drop from a height of 14.2 m crashing on to a solid concrete surface. Increasing the speed by 50% to 90 km/h (25 m/s) compares to a fall from 32m which is an increase of 125%. The reason is that the energy stored in a moving object increases as the square of the impact velocity. Typically, crumple zones are located in the front part of the vehicle to absorb the impact of a head-on collision, but they may be found on other parts of the vehicle as well. According to a British Motor Insurance Repair Research Center, the study of where on the vehicle impact damage occurs indicates that 65% were front impacts, 25% were rear impacts, 5% were left side impacts, and 5% were right side impacts. To safeguard from a deadly collision, some racing cars use aluminum or composite/carbon fiber honeycomb to form an impact attenuator to dissipate crash energy by using a much smaller volume and lower weight crumple zones than those for the road cars. Also, impact attenuators have also been introduced on highway maintenance vehicles in some countries.

4.4 BASIC PROPERTIES OF FORCE

Newton's first law states that an object in motion subject to no force will continue to move in a straight line. Newton's second law states that, force can change the motion of an object. In general, force is defined as a type of push or pull that causes an object to change its velocity or direction. We may ask "What are the common properties of force?" There are five properties that are shared by all forces as shown in Fig. 4.4. The basic properties of forces are the following: 1) a force is a push or a pull on an object, 2) a force acts on an object, 3) a force requires an agent, 4) a force is a vector, and 5) a force can be either a contact force or a long-range force.

The last property is useful for classification of forces. Forces may be separated into two categories: i) forces that act by direct contact and ii) forces that act at a distance, with no physical contact between the objects. Direct contact forces seem to make sense to us because we encounter this type of force all

A force is a push or a pull on an object.

A force acts on an object. It must have an object to act on.

A force requires an agent to deliver it.

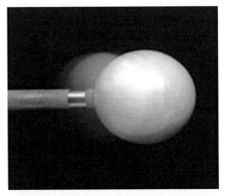

A force is a vector. It has both direction and magnitude.

LAWS OF MOTION

A force can be either a contact force or a non-contact force (i.e. a long-range force).

Figure 4.4: All forces share some common property. All forces lead to influence that causes an object to undergo a certain change, either concerning its movement, direction, or geometrical construction.

the time. For example, when we push on a door to open it, we apply a direct force. Hence, it seems natural that if a moving object smashes into another object, then it will cause the second object to move in the same direction, as explained by Newton's laws of motion. On the other hand, it is not as easy to grasp non-contact forces (or acting over a distance) such as gravitational and electric. It is interesting to note that Newton also struggled with the concept of action-at-a-distance. So, we may ask the following questions: "How can a force act on another object at a distance?" and "What is 'pushing' on the second object to make it move?" These questions seem natural for the case of two magnets of like poles near each other. From a broader examination of non-contact forces, it may appear that most forces that act at a distance seem to "pull" more than they push. What is this mystical force of gravity that pulls an object to the earth from a distance? There is no really good explanation until we can examine the physics behind these forces at a much deeper level than we are prepared to do. So, we will simply have to accept the idea that forces can act at a distance, even though this answer may not really satisfy to our curiosity.

when several forces are present...

More often than not, a body will be subjected to the simultaneous action of several forces. The question we may now ask is "How can we calculate the simultaneous effect of two or more forces?" The answer is supplied by the principle of superposition for forces. The **superposition principle** states that, for all linear systems, the net response at a given place and time caused by two or more stimuli is the sum of the responses which would have been caused by each stimulus individually. This principle applies to any linear system and has many applications in physics because many physical systems can be modeled as linear systems. According to this principle, when several forces, F_1, F_2, F_3, ..., and F_N, are exerted on an object, they combine to form the **net force**

$$\vec{F}_{net} = \vec{F}_1 + \vec{F}_2 + \vec{F}_3 + \cdots + \vec{F}_N = \sum_{i=1}^{N} \vec{F}_i. \tag{4.3}$$

This summation is called **superposition of forces.** Here, the single force that has the same effect as the combination of the individual forces is called the **net force** or the **resultant force**. According to Eq. (4.3), the net force is the vector sum of the individual forces. In terms of the net force, Newton's second law becomes $\vec{F}_{net} = m\vec{a}$. Thus, the principle of superposition of forces is equivalent to the assertion that each force produces acceleration independently of the presence or absence of other forces. We must emphasize that this principle is a law of nature which has the same status as Newton's laws. The most precise empirical test of this principle emerges from the study of planetary motion. There, we find that the net force on a planet is indeed the vector sum of all the gravitational pulls exerted by the Sun and by other planets. Somewhat less precise tests of this principle can be performed in a laboratory by pulling on a body with known forces in known directions.

4.5 FORCES AFFECTING DYNAMICS

For the purpose of describing motion of objects that we see around us, we do not need to apply the fundamental forces (gravitational, electromagnetic, weak, and strong forces) directly because it would be too complicated and unnecessary. Instead, we work with the effective forces that are derived from the gravitational and electromagnetic interactions. These forces are 1) weight, 2) spring force, 3) tension, 4) normal force, 5) friction, 6) drag, 7) thrust, and 8 electric) and magnetic forces. This will allow us to solve most dynamics problems.

Weight is defined as the gravitational pull of the Earth on or near its surface. This is a non-contact force. The **weight** of an object is usually taken to be the force on the object due to gravity as in Fig. 4.5. Its magnitude W is the product of the mass m of the object and the magnitude of the local gravitational acceleration g. Thus, we write $W = mg$. The term weight and mass are often confused with each other in everyday discourse, but they are distinct quantities. For example, an object with a mass of 1 kg has a weight of about 9.8 N (Newton = kg·m/s^2) on the surface of the Earth, which is about one-sixth as much on the Moon. In this sense, a body can be weightless only if it is far away from any gravitating mass. So, if an object with a mass of 1kg floats freely in outer space, its weight is zero.

Spring force is a contact force exerted on an object by a spring. Springs are typically metal coils, but other flexible objects such as rubber bands, leaf springs, and diving boards may also be represented by springs. An ideal spring is taken to be massless, frictionless, unbreakable, and infinitely stretchable. Such springs exert forces that push when contracted or pull when extended as shown in Fig. 4.6, in proportion to the displacement of the spring from its equilibrium position. This is an elastic force which acts to restore the spring to its natural length. The spring force arises from the deformation of the solid body which depends only on the body's instantaneous deformation and not on its previous history.

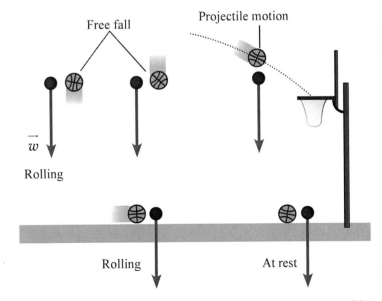

Figure 4.5: Weight is the force on an object due to gravity. Any object with mass near the surface of Earth will have weight W. The magnitude of this force is the object's mass times the gravitational acceleration (i.e., $W = mg$).

LAWS OF MOTION

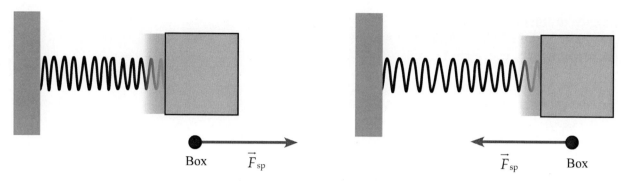

Figure 4.6: A force exerted on an object by either a stretched spring or compressed spring is described by Hooke's law. Hooke's law is a first order linear approximation to the real response of springs and other elastic bodies to applied forces. The direction of this force is opposite to the direction of spring deformation from equilibrium.

Tension force is a contact force exerted on an object by a string, rope, or wire as shown in Fig. 4.7. The direction of the tension force *is always along the string or rope*. Tension forces can be modeled by using ideal strings which are massless, frictionless, unbreakable, and inextensible (not stretchable.) Ideal strings transmit tension forces instantaneously in action/reaction pairs so that if two objects are connected, any force directed along the string by the first object is accompanied by a force directed along the string in the opposite direction by the second object. Ideal strings can be combined

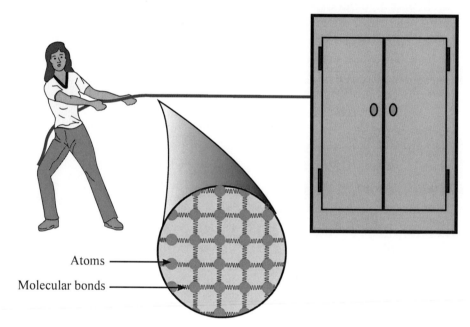

Figure 4.7: Tension is the pulling force exerted by a string, cable, chain, or similar solid object on another object. It results from the net electrostatic attraction between the particles in a solid when it is deformed so that the particles are further apart from each other than when at equilibrium.

with ideal pulleys to change the physical direction of tension. Also, the tension force on a load can be multiplied by connecting the same string multiple times to the same object through the uses of movable pulleys. Such machines allow for an increase in force, but there is a corresponding increase in the length of string that must be displaced in order to move the load.

Normal force is a contact force exerted by a surface against an object that is pressing against the surface. Here, the surface is acting as the agent. The normal force is due to repulsive forces of interaction between atoms at close contact. Microscopically, when an object and surface are in contact then their electron clouds overlap, resulting in a force acting in the direction normal to the surface interface between two objects. (Note that "normal" and "perpendicular" mean the same.) If a table rests on a floor, it does not sink into the floor because the floor provides a normal force that counteracts the weight of the table. Also, the normal force responds when an external force pushes on a solid object.

Examples of the normal force in action are the impact force on an object crashing into an immobile surface and a person's hand pressing on a wall as in Fig 4.8.

The normal force $\mathbf{F}_N$ (occasionally N) acting on a body is generally associated with the force that the surface of one body exerts on the surface of another body in the absence of any frictional forces between the two surfaces. The normal force is always perpendicular to the surfaces of contact. This is the origin of its name - normal to the surface. The normal force is an action/reaction force. A surface will not exert a normal force on an object in contact with it unless some other external force pushes the object into the surface. It is noted that a surface exerts a force perpendicular to itself as the molecular spring press outward. This is a real force arising from real compression of molecular bonds. Hence, this force may be viewed as the surface of a floor or wall preventing the object from penetrating the surface. For example, considering a person standing still on the ground, the normal force is one of the components of the ground reaction force. In another common situation, when an object hits a surface with some speed, the surface can withstand it because the normal force provides for a rapid deceleration, which will depend on the flexibility of the surface.

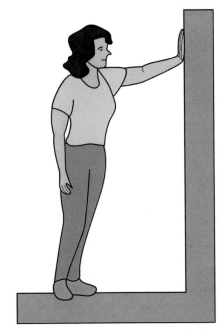

Figure 4.8: When she leans on a wall by pressing her hand against the wall, the compressed molecular springs in the wall presses outward against her hand. This response of the wall is an example of the normal force.

Friction is a contact force similar to the normal force but it is always parallel to the surface. It is directly related to the normal force which acts to keep two solid objects separated at the point of contact. On a microscopic level, friction arises as atoms from the object and atoms on the surface run over each other due to the relative motion. Friction originates from the electromagnetic forces between atoms.

There are two broad classifications of frictional forces: static friction and kinetic friction as shown in Fig. 4.9. Friction opposes the relative motion or tendency of such motion of two

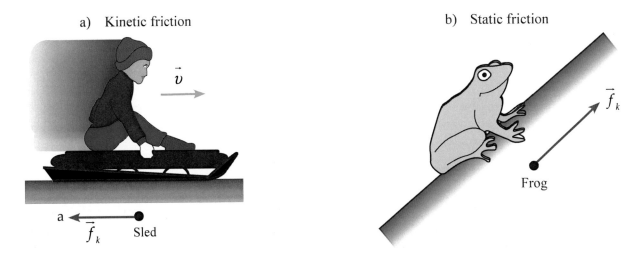

Figure 4.9: a) Kinetic friction opposes sliding and static friction opposes slipping. b) For static friction, noting that the frog would slip downhill if there was no friction; the static friction points uphill to oppose slipping.

surfaces in contact. Friction is an extremely important force. For example, it allows us to walk on the ground without slipping, it helps propel automobiles and other ground transport, and it is involved in holding nails, screws, and nuts. On the other hand, friction also causes wear and tear on the materials in contact. In situations where the surfaces in contact are moving relative to each other, the friction between the two objects converts energy of motion into other forms energy such as heat (i.e., atomic vibrations), sound, and radiation. Friction between solid objects and fluids (i.e., gases *or* liquids) is called fluid friction.

Drag is a resistive force experienced by objects moving through fluids. Drag arises as a result of collision between fluid molecules and the moving solid object. Hence, it points opposite to the direction of motion as shown in Fig. 4.10 for falling leaves. When the fluid is a gas like air, it is called aerodynamic drag (or air resistance). However, when the fluid is a liquid like water, it is called hydrodynamic drag.

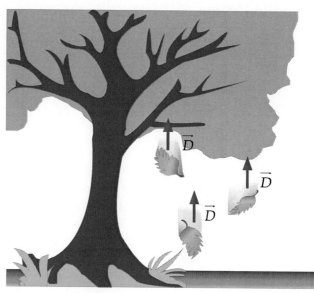

Figure 4.10: Air resistance is a significant force on falling leaves. It points in the opposite direction to motion of an object. Air resistance, which is also called drag, is the forces that are opposite to the relative motion of an object through the air. Drag forces act opposite to the oncoming flow velocity. Drag depends directly on velocity.

In fluid dynamics, **drag** refers to forces which act on a solid object in the direction of the fluid flow velocity. Unlike friction, which is nearly independent of velocity, drag forces depend on velocity! Drag forces tend to decrease fluid velocity relative to the solid object in the path of fluid. Examples of drag include the component of the net aerodynamic or hydrodynamic force acting in the opposite direction to the movement of the object, such as a car, aircraft and boat, relative to the Earth. In the case of viscous drag of fluid in a pipe, drag force exerted by the immobile pipe causes fluid flow to be slower near the (inner) surface of the pipe.

Thrust is a force that propels an object forward by expelling the exhaust gas in the direction opposite to the direction of the force as shown in Fig. 4.11. When a system expels or accelerates mass in one direction, the accelerated mass will cause a force of equal magnitude but opposite direction on that system. Hence, thrust is a reaction force which is described quantitatively by Newton's second and third laws. The "strength" of a rocket engine is called its **thrust**. Thrust is measured in "pounds of thrust" in the U.S. and in Newton under the metric system (4.45 N of thrust equals 1 pound-force of thrust). A pound force of thrust is the amount of thrust it would take to overcome an one pound object against the

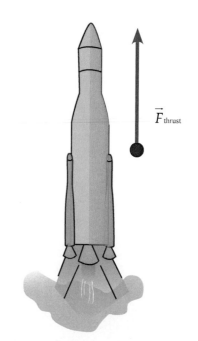

Figure 4.11: Thrust force is exerted on a rocket by exhaust gases. Thrust is a reaction force described quantitatively by Newton's second and third laws. When a system expels or accelerates mass in one direction, the accelerated mass will cause a force of equal magnitude but opposite direction on that system.

force of gravity on Earth. According to this definition, the acceleration of gravity is 32 feet per second per second (21 mph per second) on Earth. If we were floating in space with a bag of baseballs and we threw one baseball per second away from us at 21 mph, our baseballs would be generating the equivalent of 1 pound of thrust. Similarly, if we threw the baseballs at 42 mph, then we would be generating 2 pounds of thrust. Of course, the direction of thrust is opposite to the direction of the baseballs. If we throw them at 2,100 mph (perhaps by shooting them out of some sort of baseball gun), then we are generating 100 pounds of thrust, and so on. This is the same principle used to provide thrust to a rocket as illustrated in Fig. 4.11.

Electric and magnetic forces are long range (i.e., non-contact) forces which act on charged particles. The electrostatic force was first described in 1784 by Charles Coulomb as a force that existed intrinsically between two charges. As we will discuss in Chapter 15, the electrostatic force obeys an inverse square law and is directed in the radial direction. This force can be both attractive and repulsive depending on the sign of electric charges, but it is independent of the mass of the charged objects and follows the superposition principle. We can also think of this force as the force exerted on a charged particle in an electric field. Similarly, the magnetic force is a force exerted on a moving charge in a magnetic field. The electromagnetic force, which is a unified description of electric and magnetic forces, is one of the four fundamental interactions in nature. **Electromagnetism** is concerned with the forces that occur between electrically charged particles. In electromagnetic theory, these forces are explained using electromagnetic fields. With the exception of gravity, electromagnetism is the interaction responsible for almost all the phenomena encountered in daily life. As we know, ordinary matter takes its form as a result of intermolecular forces between individual molecules. The intermolecular forces are derived from the electromagnetic force between electrons. Electrons are bound by electromagnetic forces in orbitals around atomic nuclei to form atoms, which are the building blocks of molecules. This governs the processes involved in chemistry. Hence, all processes in chemistry arise from interactions between the electrons of neighboring atoms, which are in turn determined by the interaction between electromagnetic force and the momentum of the electrons.

Also, there are fictitious forces which depend on a frame of reference, meaning that these forces appear due to the adoption of non-inertial reference frames. Examples of these fictitious forces include the centrifugal force and the Coriolis force. These forces are fictitious because they appear in the accelerating frames of reference. The physics in an accelerating frame of reference are modified by the appearance of fictitious forces. In the theory of general relativity (see Chapter 26), gravity may be considered a fictitious force in situations in which the space-time deviates from a flat geometry. As an extension of the concept of curvature of the space-time, Kaluza-Klein theory and string theory attributes the origin of electromagnetism and the other fundamental forces to the curvature of differently scaled dimensions, which would ultimately imply that all forces may be fictitious.

What do forces do?

Let us discuss the effect of forces. We know that a physical quantity changes due to the action of a force. For example, force tends to change a body's state of rest or of uniform motion in a straight line. In particular, when the forces acting on a body are unbalanced, they produce a net force, which causes the velocity of the body to change. Forces may be balanced or unbalanced. For example, an iron ball suspended from a hook by a wire has two forces acting on it: its weight, acting vertically downward, and the tension acting on the wire vertically upward. Since these forces are equal in magnitude but opposite in direction, their resultant is zero. As in this case, when the resultant of a group of forces acting on the same object is zero, the forces are balanced. When the resultant force is not equal to zero, the forces are unbalanced.

LAWS OF MOTION

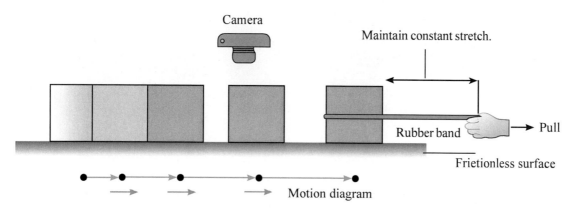

Figure 4.12: A schematic illustration for measuring the effect of applied force on an object. A rubber band is used to apply a force to the object with known mass, and a camera is used to measure its increasing speed with time.

When there is an unbalanced force, a non-zero net force is experienced by the object. In Fig. 4.12, an object is being pulled with a rubber band. The rubber band exerts a constant force. According to Newton's second law, when an object is pulled with a constant force, it will move with a constant acceleration. The magnitude of force exerted on the object is directly proportional to the number of rubber band used. If the object is pulled with the three identical rubber bands, then three times the force of a single rubber band is applied. The three lines shown in Fig. 4.13 indicate that the magnitude of acceleration is directly proportional to the size of the force. If mass m is constant, then force F is directly proportional to acceleration a. When the force is tripled, then acceleration will triple.

What about the dependence of acceleration on the mass of the object? According to Newton's second law, acceleration is inversely proportional to the mass on which the force acts. If the acceleration a is constant, then force **F** is inversely proportional to the mass m. When the force is doubled, then the mass is halved. Similarly, if the mass is increased by a factor of 2 then the force becomes 1/2 of its original setting as shown in Fig. 4.14. Here, the acceleration is inversely proportional to the number of blocks. Both force and acceleration are vector quantities, indicating that they have a direction associated with their magnitude. However, the mass can change only the magnitude of the acceleration but not the direction

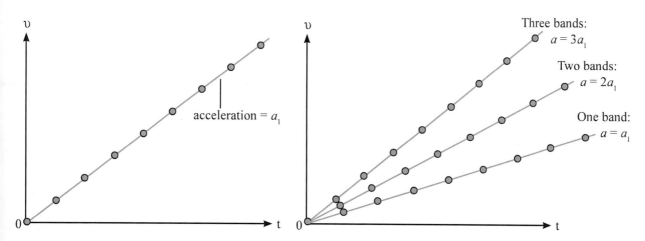

Figure. 4.13: The velocity-versus-time plot for a single and multiple rubber bands are shown in the left and right panel, respectively. The slopes of the graph, reflecting the acceleration, are directly proportional to the applied force.

Mass is the property of an object which determines how it accelerates in response to an applied force. More specifically, inertial mass is a quantitative measure of an object's resistance to acceleration. In addition to this, **gravitational mass** is defined as a measure of magnitude of the gravitational force which is i) exerted by an object (active gravitational mass) or ii) experienced by an object (passive gravitational force) when interacting with a second object. The inertial mass of an object determines its acceleration in the presence of an applied force. Macroscopically, mass is associated with matter even though matter is poorly defined in science. In normal situations, the weight of an object is proportional to its mass. So, usually, we would not consider using the same unit for both concepts as a problem. However, the distinction between mass and weight becomes important for measurements with a precision better than a few percent because of slight differences in the strength of the Earth's gravitational field at different places. Also, this distinction is more apparent for places far from the surface of the Earth, such as in space or on other planets where the gravity is notably different.

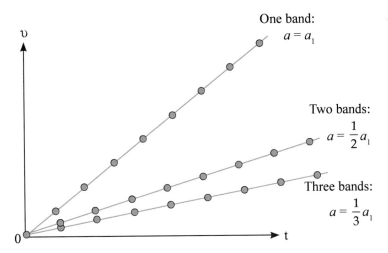

Figure 4.14: Acceleration is inversely proportional to the number of blocks. The relationship between force and mass may be understood as the following: the greater the mass, the greater amount of force is needed to make that mass change speed or direction.

4.6 NEWTON'S SECOND LAW

We now focus on the second law of motion which may be used to describe motion of an object. Newton's second law states that the acceleration of an object is directly proportional to the net force acting on it and is inversely proportional to its mass. In general, a precise definition of force is the change in momentum per unit time. **Momentum**, as we discuss later in Chapter 6, is a characteristic of a moving body, determined by the product of the body's mass and velocity. To determine the change in momentum per unit time, Newton developed differential calculus. In this section, we consider a simple case of constant mass. For an object with a constant mass, Newton's original equation for force looked something like $\mathbf{F} = m\, \Delta v/\Delta t$. Because the acceleration is defined as the change in velocity in an instant of time (i.e., $a = \Delta v/\Delta t$), the equation for force is often rewritten as $\mathbf{F} = m\mathbf{a}$. The equation describing Newton's second law allows us to specify a unit of measurement for force. Because the standard unit of mass is the kilogram (kg) and the standard unit of acceleration is meters per second squared (m/s^2), the unit for force must be a product of the two which is kg·m/s^2. This is a little awkward, so a decision was made to use a Newton (N) as the official unit of force. One Newton is equivalent to 1 kg·m/s^2. There are 4.448 N in 1 pound-force. So what do we do with Newton's second law? As it turns out, $\mathbf{F} = m\mathbf{a}$ allows us to quantify motion of every variety by determining the acceleration a. Once the constant acceleration of an object is known, we may use the kinematics equations to describe how the object moves. This is one way in which dynamics is related to kinematics.

LAWS OF MOTION

Exercise Problem 4.1:

Very small forces have tremendous effects on the motion of very small objects. Consider a single electron, with a mass of 9.1×10^{-31} kg, subject to a single force equal to the weight of a penny, 2.5×10^{-2} N. What is the acceleration of the electron?

Solution:

Applying Newton's second law **F** = m**a**, we obtain the acceleration as

$$a = \frac{F}{m} = \frac{2.5 \times 10^{-2} \, \text{N}}{9.1 \times 10^{-31} \, \text{kg}} = 2.7 \times 10^{28} \, \frac{\text{m}}{\text{s}^2}$$

This is a huge acceleration, as expected. This huge acceleration is due to the small mass of an electron. So, this means that the force of 2.5×10^{-2} N is a huge force for an electron, even though it does not appear much of force to us.

Free-body Diagram

In analyzing a dynamics problem, it is important to account for all of the forces acting on an object and determine both the direction and magnitude of the net force. This may be done easily by drawing a free-body diagram.

What is a free-body diagram?

Since the net external force acting on the object must be obtained in order to apply Newton's second law, we use a free-body diagram to represent our knowledge about force and motion. A free-body diagram is a sketch of an object of interest with all the surrounding objects stripped away, showing only the forces acting on the body. Drawing a free-body diagram is an important step in solving mechanics problems since it helps us visualize all the forces acting on the object. This diagram represents an object as a particle and shows all forces acting on it. We may dare say that drawing free-body diagrams is an essential tool. In fact, it is the most important step. We should always

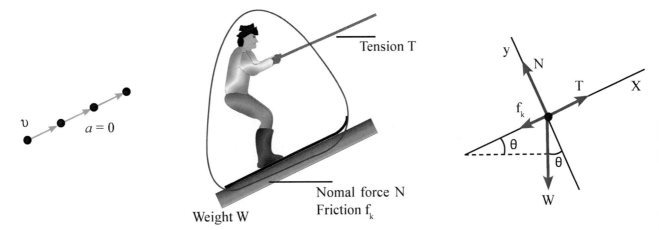

Figure 4.15: Free-body diagrams are a tool for solving problems with multiple forces acting on a single body. The diagram shown here can be used for the summation of forces. The free-body diagram reduces the complexity of situation for easy analysis and is used as a starting point for a mathematical model of the forces acting on an object.

begin by drawing a free-body diagram. As shown in Fig. 4.15, a free-body diagram can be used as a convenient way to keep track of forces acting on a system. Here, a skier is being pulled up on an inclined plane by a rope. There are the following four forces acting on the skier in different directions: weight, normal force, friction, and tension. Ideally, a free-body diagram is drawn with the angles and relative magnitudes of the force vectors preserved so that graphical vector addition can be done to determine the net force. In Fig. 4.14, the angle θ between **W** and the negative y-axis is the same as the angle of the incline.

A **free-body diagram** is not meant to be a scaled drawing. Rather it is a working sketch which is open to modification as we work through the problem. Typically, we need to have seen through the problem before we arrive at a satisfactory diagram. There is an element of art and an inherent flexibility in the whole process. There is no hard and fast algorithm. Not only how it is drawn but also how it is interpreted depends crucially on how a body is modeled in a free-body diagram. Since a free-body diagram is a pictorial device which represents a rough working sketch of a force diagram, engineers and physicists use this diagram to analyze the forces acting on a body. The body itself may consist of multiple components (as in an automobile) or just a part of a component (as in a short section of a beam). In analysis of structures, free-body diagrams for a component of a structure or whole structure are used in determining shear forces and bending moments. In a complex problem, a whole series of such diagrams may be necessary to analyze forces. The free body in a free-body diagram is not free of constraints. The constraints have been replaced by arrows representing the forces. Also, drawing a free-body diagram can help us determine the unknown forces and equations of motion of the body and thus can help us to analyze a problem in statics or dynamics.

4.7 NEWTON'S THIRD LAW

Among the three laws of motion stated in the "Philosophiae Naturalis Principia Mathematica" in 1686, the third law may appear more abstract than the other two laws, but it is essential for describing motion. Newton's third of motion plays an important role in the derivation of a fundamental physics principle: conservation of momentum. We will discuss this principle in Chapter 6. Newton's third law states that for every action (i.e., force) in nature there is an equal and opposite reaction. In other words, if object A exerts a force on object B, then object B also exerts an equal and opposite force on object

LAWS OF MOTION

A (i.e., **to every action there is an equal and opposite reaction**). Notice that the action and reaction forces are exerted on different objects. Let us see how Newton's third law may be used in analysis of a dynamics problem. For an aircraft, the principal of action and reaction is very important. It helps to explain the generation of lift from an airfoil. In this problem, the air is deflected downward by the action of the airfoil, and in reaction the wing is pushed upward. Similarly, for a spinning ball, the air is deflected to one side, and the ball reacts by moving in the opposite direction. A jet engine also produces thrust through action and reaction. The engine produces hot exhaust gases which flow out the back of the engine. In reaction, a thrusting force is produced in the opposite direction.

Understanding-the-Concept Question 4.1:

What makes a car go forward?

PROBLEM SOLVING STRATEGY

What do we know?

A force is needed to move the car forward.

What concepts are needed?

We need to apply Newton's third law.

Why?

The force exerted on the car is reaction to the force that the car exerted on the ground, which is action.

Explanation:

Newton's third law applies when we drive an automobile. The action force is the pushing against the road and the reaction is the road pushing against the tires. There are many other examples of action/reaction forces. For example, when a car hits a person, the person hits the car. When a car hits telephone pole, the pole hits the car. When a car hits a wall, the wall hits the car. When a person's foot pushes the gas pedal, the gas pedal pushes the person's foot. These arguments indicate that a car moves forward due to the friction force exerted by the ground on the tires, and this force is the reaction to the force exerted on the ground by the tires. It is difficult to drive on an icy road because the tire cannot grip the road because there is less fiction. So, when the car tries to push on the road it cannot because the road is icy. Hence, if the car cannot push on the road, the road cannot push on the car. As a result the car slides on the ice.

Understanding-the-Concept Question 4.2:

Michelangelo's assistant has been assigned the task of moving a block of marble using a sled. He says to his boss, "When I exert a forward force on the sled, the sled exerts an equal and opposite force backward. So how can I ever start it moving? No matter how hard I pull, the backward reaction force always equals my forward force, so the net force must be zero. I will never be able to move this load." Is this a case of a little knowledge being dangerous?

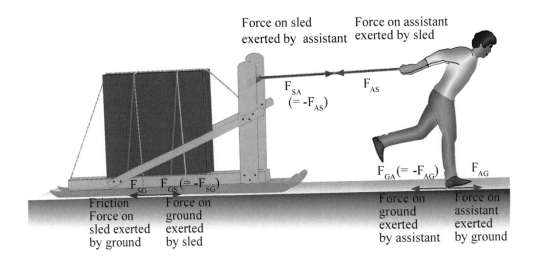

Explanation:

To answer this question, we examine how both Newton's second and third law may be applied. Suppose we imagine two people are on roller skates or ice skates and facing each other. One person pushes on other person with a certain force, then the two people will be accelerated backwards according to $\mathbf{F} = m\mathbf{a}$, in the opposite direction to each other. This is due to Newton's third law: one person's push causes an equal opposite push on other person. So, both people will accelerate backwards. The force one person applies on other person is $\mathbf{F} = m\mathbf{a}$. So, the acceleration a of other person's motion is dependent on the force $\mathbf{F}$ one person pushes and his/her mass m. However, the same force is applied on both people. Now, suppose we are pushing on a wall with a certain force. Since the wall is stationary, our force is reflected back against our hands. The force we apply is the same force that is applied to our hands. If we are wearing roller skates, and we push on the wall with a certain force, we will move backward as if that same force was applied to you. Suppose we pull on a large box that is on the floor. If the friction between the box and the floor is greater than the force we are applying, then the equal force will be pulling on our hands. However, if the resistive force of friction is less than our force, the box will slide along the floor. The force on the box will be the force we applied minus the force of friction. Now, we can add Newton's second law to the third law and conclude that the box will slide only when the net force acting on it is non-zero. This means the force on the sled exerted by the assistant has to be large enough to overcome the friction force on the sled exerted by the ground.

It is important to remember that we need to consider only the forces that are acting on the object under consideration when we apply Newton's second law. So, in order to determine if the assistant moves or not, we must consider only the forces acting on the assistant and then apply

$$\sum_i \vec{F}_i = m\vec{a}$$

This is important because Newton's second law is concerned with the net force. We could rewrite the law to say that when a non-zero **net force** acts on an object, the object accelerates in the direction of the net force. To exemplify this point further, let us consider a pack of six dogs, three on the left side and other three on the right side, pulling a sled. The sled is moving straight. Now, one of the dogs on the left breaks free and runs away. Suddenly, the force pulling to the right is larger than the force pulling to the left, so the sled accelerates to the right. The net force does not have the same effect on the movement of the object as the original system forces as we will discuss

LAWS OF MOTION

in Chapter 8. This suggests that the point of application of the net force determine the effect of the resultant force. The resultant force can lead to rotational motion, rather than linear motion if the line of action does not pass through center of mass (see Chapter 8). It is always possible to determine a point of application of a net force so that it maintains the movement of the object under the original system of forces.

Exercise Problem 4.2:

A weight lifter stands up from a squatting position while holding a heavy barbell across his shoulders. Identify all of the third-law pairs of forces, and then draw free-body diagrams for the weight lifter and the barbell. Use dotted lines to connect the members of all action/reaction pairs.

Barbell (BB)

Weight lifter (WL)

Problem Solving Strategy

What do we know?

The weight lifter is holding a heavy barbell, but he is being supported by the ground.

What concepts are needed?

We need to apply Newton's third law and identify the action/reaction forces. We need to draw free-body diagrams.

Why?

There are three parts to the system: i. barbell, ii. weight lifter, and iii. ground. The sum of the action/reaction forces is zero.

Solution:

First, we separate the system into three parts: i) barbell, ii) weight lifter, and iii) ground. Then, we identify the action/reaction forces acting on each part of the system as shown below. We should note that since we have the closed system, the sum of all of the action/reaction forces is zero.

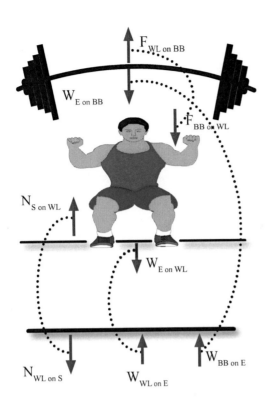 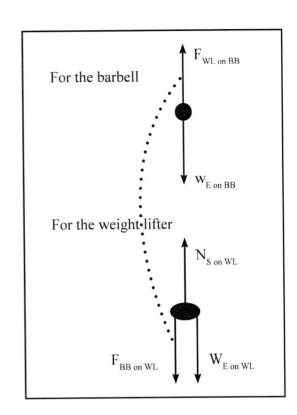

Two separate free-body diagrams are needed to analyze the problem. One diagram is for the barbell and other one is for the weight lifter. Accounting for the forces on each free-body diagram, we may apply Newton's second law.

4.8 UNIVERSAL LAW OF GRAVITATION

The **weight** of an object is defined as the force on the object due to gravity. An object of mass m on the surface of Earth will experience a gravitational force $\mathbf{F} = mg$ where g is the acceleration due to gravity.

What is the difference between mass and weight?

Although weight and mass are often used synonymously by a lay person, these two quantities are not the same. Weight is the magnitude of the gravitational force on an object, and mass is an intrinsic property of an object. In scientific contexts, mass refers loosely to the amount of "matter" in an object even though "matter" may be difficult to define. However, weight refers to the force experienced by an object due to gravity. In other words, an object with a mass of 1.0 kilograms will weigh 9.8 Newton on Earth. Here, newton is the unit of force, while kilogram is the unit of mass. Weight is its mass multiplied by the gravitational field strength. So, weight changes depending on the place where it was measured. Weight of an object, compared to its value on Earth, will be less on Mars (where gravity is weaker), more on Saturn, and negligible in space when far from any significant source of gravity, but it will always have the same mass. While the **weight** of a mass is a function of the strength of gravity, the **mass** of an object is constant for any given observer, so long as no energy or matter is added to the object. This raises another question: "If the weight of an object is determined by the strength of gravity, then what determines the gravity?" This question can be answered by universal law of gravitation.

LAWS OF MOTION

Newton's law of universal gravitation states that every point mass in the universe attracts every other point mass with a force that is directly proportional to the product of their masses and inversely proportional to the square of the distance between them. Separately, it was shown that large spherically symmetrical masses attract each other as if all their mass were concentrated at their centers (i.e., point masses). This general physical law was derived from empirical observations. According to the universal law of gravitation, every particle in the universe exerts an attractive force on every other particle. This means that all particles pull toward each other as illustrated in Fig. 4.16. The force that each particle exerts on other is directed along the line joining the particles. Here, we use the word particle very loosely and consider a planet as a particle. The magnitude of this force of attraction is given by

Figure 4.16: The fall of the Moon around the Earth is the same kind of motion as the fall of an apple to the Earth. Both are described by the same three laws of motion, and both feel a gravitational force described by the same, universal force law.

$$F = G \frac{m_1 m_2}{r^2} \qquad (4.4)$$

where G is the universal gravitational constant $G = 6.673 \times 10^{-11}$ N·m²/kg². The value of the constant G was first accurately determined from the result of the experiment conducted by Cavendish in 1798. For his experiment, Cavendish constructed a torsion balance which is made of a six-foot (1.8 m) wooden rod suspended from a wire, with a 2-inch (51 mm) diameter 1.61-pound (0.73 kg) lead sphere attached to each end. Two 12-inch (300 mm) 348-pound (158 kg) lead balls were located near the smaller balls, about 9 inches (230 mm) away, and held in place with a separate suspension system. The experiment measured the faint gravitational attraction between the small balls and the larger ones. This experiment was the first test of Newton›s theory of gravitation between masses in the laboratory. It took place 111 years after the publication of Newton's Principia and 71 years after Newton's death. So, Newton's calculations could not use the numerical value of G. Instead, he could only calculate the magnitude of a gravitational force relative to another force.

Equation (4.4) indicates that the force is proportional to the product of the two masses and inversely proportional to the square of the distance between them. Here, the distance between the two masses is the center-to-center distance as shown in Fig. 4.17. The attractive nature of gravitation is important for understanding the evolution of the universe. From a cosmological perspective, gravitation causes dispersed matter to coalesce and the coalesced matter to remain intact. This picture accounts for the existence of planets, stars, galaxies and most of the macroscopic objects in the universe. Also, gravitation is responsible for keeping the Earth and the other planets in their orbits around the Sun, for keeping the Moon in its orbit around the Earth, for the formation of tides, for natural convection, for heating the interiors of forming stars and planets to very high temperatures, and for various other phenomena observed on Earth and throughout the universe.

On the surface of Earth, we may easily observe the effects of gravitation. For example, fluid flow occurs under the influence of a density gradient and gravity. Also, an object with mass has weight on Earth and stays on the surface. The weight of an object with mass m is

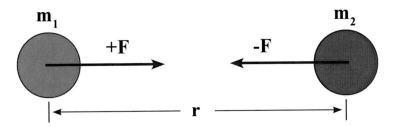

Figure 4.17: Newton's law of universal gravitation states that every point mass in the universe attracts every other point mass with a force that is directly proportional to the product of their masses and inversely proportional to the square of the distance between them.

$$W = G\frac{M_E m}{r_E^2} = mg \qquad (4.5)$$

where $M_E = 5.98 \times 10^{24}$ kg and $r_E = 6.38 \times 10^6$ m is the mass and radius of the Earth, respectively. A simple calculation of g using these values yields $g = GM_E/r_E^2 = 9.80$ m/s^2. We now can compute our weight (i.e., gravitational force acting on us due to the force of pull by the Earth). Knowing our weight an interesting question that we can ask is "Do we always feel this force?" In the operational definition, the weight of an object is the force measured by the operation of weighing it, which is the force it exerts on its support. This can make a considerable difference, depending on the details. For example, an object in free fall exerts almost no force on its support. This situation is commonly referred to as weightlessness. However, being in free fall does not affect the weight according to the gravitational definition. Therefore, the operational definition is sometimes refined by requiring that the object be at rest.

Sensation of Weight

The weight of an object is the force of gravity on that object, but gravity is not the force that we can feel or sense directly. Our sensation of weight is due to contact forces pressing against us. In other words, we feel apparent weight! **Apparent weight** is defined as the magnitude of the contact force that supports the object. Apparent weight is the property of an object that corresponds to how heavy it "feels". The apparent weight of an object will differ from its weight whenever the force of gravity acting on the object is not balanced by an equal and opposite normal force. Remember that the weight of an object is "defined" as the (magnitude of the) gravitational force acting on it. This means that even a "weightless" astronaut in low Earth orbit has almost the same weight as he would have while standing on the ground. An object that rests on the ground is subject to a normal force exerted by the ground. The normal force acts only on the boundary of the object that is in contact with the ground. This force is transferred into the body; the force of gravity on every part of the body is balanced by stress forces acting on that part. A "weightless" astronaut feels weightless due to the absence of these stress forces. By defining the apparent weight in terms of normal forces, we can capture this effect of the stress forces. In Fig. 4.18, a man feels the normal force pressing against his feet when he is on the ground. However, he feels heavier than normal while accelerating upward.

In real world situations, the act of weighing may produce a result that differs from the ideal value of weight provided by the definition. A common example of this is the effect of buoyancy.

LAWS OF MOTION

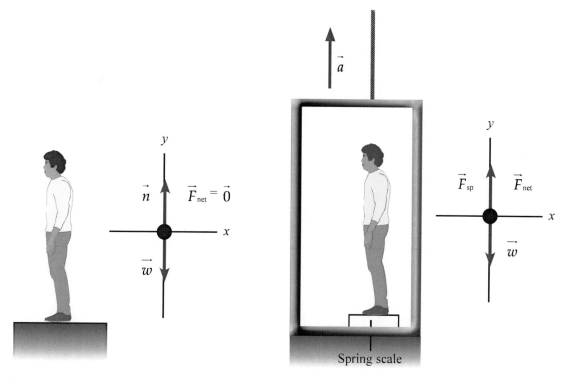

Figure 4.18: When a person stands on a scale, the reading on the scale is actually the normal force that the scale exerts back towards him to support his weight. When the elevator accelerates upward, the scale has to push upward with extra force on the person to accelerate his mass upward.

When an object is immersed in a fluid, the displacement of the fluid will cause an upward force on the object, making it appear lighter when weighed on a scale. The apparent weight may be similarly affected by levitation and mechanical suspension. When the gravitational force definition of weight is used, the operational weight measured by an accelerating scale in Figure 4.18 is often also referred to as the apparent weight.

Understanding-the-Concept Question 4.3:

What is his apparent weight?

Problem Solving Strategy

What do we know?

He is falling.

What concepts are needed?

We need to apply the definition of apparent weight.

Why?

The weight and apparent weight of a falling object is not the same.

Explanation:

The person in the picture is happy because he is feeling weightless. Weightlessness is a phenomenon experienced by people during free-fall. Although the term zero gravity is often used as a synonym, weightlessness in orbit is not the result of the force of gravity being eliminated or even significantly reduced. In fact, the force of the Earth's gravity at an altitude of 100 km is only 3% less than that on the Earth's surface. Weightlessness typically occurs when an object or person is falling freely, in orbit, in deep space (far from a planet, star, or other massive body), or in an airplane following a particular parabolic flight path. Weightlessness occurs whenever all forces applied to a person or object are uniformly distributed across the object's mass (as in a uniform gravitational field), or when the object is not acted upon by any force. This is in contrast with typical human experiences in which a non-uniform force is acting, such as standing on the ground and sitting in a chair on the ground. In a place where gravity is countered by the reaction force of the ground, a person will have a sensation of weight which is the same as the gravitational force. In general, the sensation of weight differs from the gravitational force. This difference is due to the acceleration of a supporting vehicle. For example, when flying in a plane, a reaction force is transmitted from the lift that provided by the wings. Hence, the sensation of weight is modified by the airplane. Also, during atmospheric reentry or during the use of a parachute, atmospheric drag decelerates a vehicle. During an orbital maneuver in a spacecraft or during the launch phase of a rocket, the rocket engines provide thrust.

4.9 NORMAL FORCE, TENSION, AND FRICTION

In section 4.5, we listed the forces we will use in solving dynamics problems. Among these forces, in addition to weight, the following forces are used quite often in the analysis: i) normal force, ii) tension, and iii) friction. So, it would be helpful for us to learn how these forces are applied in analysis of the dynamics problems. In this section, we discuss these three forces further and work through some exercise problems involving these forces.

The **normal force** F_N is the component of the contact force exerted on an object by the surface. The direction of this force is perpendicular to the surface of contact. For example, the surface of a floor or wall which prevents the object from penetrating the surface exerts a normal force. The normal force is one of the components of the ground reaction force. If we consider an object resting on a base, which is on the ground as shown in Fig. 4.19, then the base exerts normal force on the object and the ground exerts normal force on the base as the reaction force. Another common situation is an object hitting a surface. In this case, if the object hits the surface with some speed and the surface can withstand it, then the normal force provides for a rapid deceleration which will depend on the flexibility of the surface.

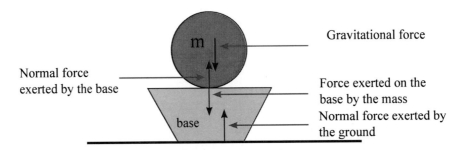

Figure 4.19: In many common situations, the normal force is just the weight of the object which is sitting on some surface, but if an object is on an incline or has components of applied force perpendicular to the surface, then it is not equal to the weight.

LAWS OF MOTION

The magnitude of normal force may change depending on situations. If a passenger were to stand on a "weighing scale", such as a conventional bathroom scale, while riding an elevator, the scale will be reading the normal force it delivers to the passenger's feet. When the elevator is either stationary or moving at constant velocity, the normal force on the person's feet balances the person's weight. However, the reading on the scale will be different than the person's ground weight if the elevator cab is accelerating up or down. This is because the scale measures the normal force, which varies as the elevator cab accelerates, and not gravitational force, which does not vary as the cab accelerates. In an elevator that is accelerating upward, the normal force is greater than the person's ground weight and so the person's perceived weight increases (making the person feel heavier). In an elevator that is accelerating downward, the normal force is less than the person's ground weight and so a passenger's perceived weight decreases. This example indicates that it is impossible for us to measure true gravitational force without knowledge of the motion of our immediate environment.

Exercise Problem 4.2:

A friend has given you a special gift, a box of mass 10.0 kg with a mystery surprise inside. It's a reward for writing your physics final examination. The box is resting on the smooth (frictionless) horizontal surface of a table. (a) Determine the weight of the box and normal force acting on it. (b) Now your friend pushes down on the box with a force of 40.0 N. Again determine the normal force acting on the box. (c) If your friend pulls upward on the box with a force of 40.0 N. What now is the normal force on a box?

Solution:

According to Newton's second law $\vec{F}_{net} = m\vec{a}$, the net force $\mathbf{F}_{net}$ is a vector. So, we may decompose the net force $\mathbf{F}_{net}$ into the x- and y-components and obtain

$$\sum_i F_{ix} = ma_x \quad \text{and} \quad \sum_i F_{iy} = ma_y.$$

The weight of the box is

$$F_g = mg = 10.0 \text{kg} \times 9.8 \, \text{m}/\text{s}^2 = 98.0 \text{N}.$$

a. When the box is resting on the table, the normal force may be obtained from

$$\sum_i F_{iy} = ma_y = 0 \Rightarrow F_N = F_g.$$

The computed value of the normal force is

$$\therefore F_N = F_g = 98.0 \text{N}.$$

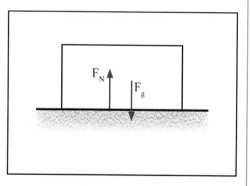

b. When you press on the top of the box, we need to account for the force pushing down and write

$$\sum_i F_{iy} = ma_y = 0 \Rightarrow F_N - F_g - F_d = 0.$$

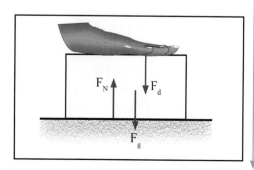

Solving for the normal force, we obtain

$$\therefore F_N = F_g + F_d = 98.0\,\text{N} + 40.0\,\text{N} = 138.0\,\text{N}.$$

c. Finally, when a string is attached to the box and being pulled upward, the net force on the box is

$$\sum_i F_{iy} = ma_y = 0 \Rightarrow F_N - F_g + F_u = 0.$$

Solving for the normal force, we obtain

$$\therefore F_N = F_g - F_u = 98.0\,\text{N} - 40.0\,\text{N} = 58.0\,\text{N}.$$

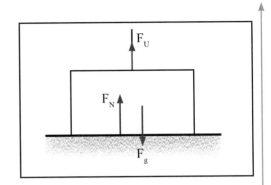

In this problem, it is useful to remember that i) normal face can only be zero or repulsive, not attractive and ii) it is always normal to surface.

Tension is the pulling force exerted by a string, cable, chain, or similar solid object on another object. Tension is the opposite of compression. Slackening reduces tension. In single-particle dynamics, tension is defined as the force exerted on an object by a rope or string. The magnitude of tension is measured in Newton (or sometimes pounds) and the direction is always measured parallel to the string on which it applies. There are two basic possibilities for systems of objects held by strings: i) acceleration is zero and the system is in equilibrium and ii) acceleration is non-zero and a net force is present. Usually a string is assumed to have negligible mass as a good simplifying approximation.

How does a string exert force?

Tension results from the net electrostatic attraction between the particles in a solid. When an applied force deforms the solid so that the particles in the solid are farther apart from each other and farther away from the equilibrium position, this force is balanced by repulsion due to electron shells and the resulting pulling force is exerted by a solid trying to restore its original, more compressed shape. As shown in Fig. 4.20, when the rope is stretched slightly, each piece of the rope pulls its on a neighboring piece. The hand (H) pulls on the rope which is attached to a wall (W) with a force $\mathbf{F}_{\text{H-on-rope}}$. The rope pulls back with a force $\mathbf{T}_{\text{rope-on-W}}$. Here, two forces $\mathbf{F}_{\text{H-on-rope}}$ and $\mathbf{T}_{\text{rope-on-W}}$ form an action/reaction pair so the tension in the rope is $\mathbf{T}_{\text{rope-on-W}} = \mathbf{F}_{\text{H-on-rope}}$. At an imaginary cut anywhere in the rope, each half of the rope pulls on the other half with a force of equal magnitude. The magnitude of this force is the tension in the rope. Finally, at the end of the rope, the rope pulls on the wall and the wall pulls back on the wall. These are an action/reaction pair. As the result, the tension in a stationary rope pulls equally hard at both ends of the rope whenever object is attached.

As we discussed above, tension is a force that stretches something. This indicates that tensile force is a force applied by an object being stretched, and tensional force is applied on the objects that are stretching it. Tensile force is directed along the stretched object. It pulls equally on the objects on either end of the rope that is being stretched. Tension is always directed parallel to the string on which it applies. This is the reason why strings, cords, wire, and rope appear difficult to handle. We need to have a distinct sense of direction when a rope is being pulled because the direction of tension is the direction of the force on the rope as shown in Fig. 4.21.

LAWS OF MOTION

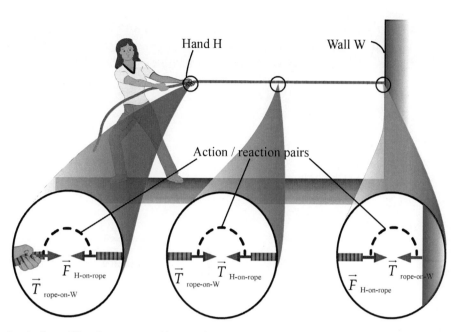

Figure 4.20: Tension is the pulling force exerted by a string, cable, chain, or similar solid object on another object. It results from the net electrostatic attraction between the particles in a solid when the particles are farther apart from each other than when they are at equilibrium.

The dynamics problems involving a rope may also involve a pulley. For massless cords passing over frictionless pulleys, the whole rope is characterized by a single tension T. If a rope is in tension, then at any cross section we might choose, the left part pulls on the right by a force T and the right side pulls on the left by a force, T. So all we have to remember is that i) there is a single tension, T characterizing an "ideal" cord and ii) a rope can only pull along its length. It never pushes and it never exerts a force perpendicular to its length. Also, we may use the following two rules for determining tension: 1) set the magnitude of the forces produced by a cord and 2) determine the direction of the force produced on an object in contact with the cord.

A combination of fixed and movable pulleys, shown in Fig. 4.21, forms a block and tackle and can yield the mechanical advantage, requiring less force to lift the same object. A block and tackle can have several pulleys mounted on the fixed and moving axles, further increasing the mechanical advantage. A rope and pulley system, that is a block and tackle, is characterized by the use of a single continuous rope to transmit a tension force around one or more pulleys to lift or move a load. Here, the rope may be a light line or a strong cable. If the rope and pulley system does not dissipate or store energy, then its mechanical advantage is the number of parts of the rope that act

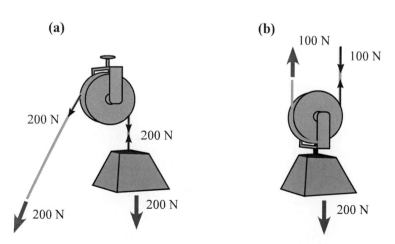

Figure 4.21: a) A fixed pulley has an axle mounted in bearings attached to a supporting structure. A fixed pulley changes the direction of the force on a rope that moves along its circumference. b) A movable pulley has an axle in a movable block. A single movable pulley is supported by two parts of the same rope.

on the load. The mechanical advantage of a pulley system can be analyzed using free-body diagrams which balance the tension force in the rope with the force of gravity on the load. In an ideal system, the massless and frictionless pulleys do not dissipate energy and allow for a change of direction of a rope that does not stretch or wear.

Exercise Problem 4.3:

A 6500 kg helicopter accelerates upward at 0.60 m/s² while lifting a 1200 kg car. (a) What is the lift force exerted by the air on the rotors? (b) What is the tension in the cable (ignore its mass) that connects the car to the helicopter?

PROBLEM SOLVING STRATEGY

What do we know?

Both a helicopter and a car are accelerating upward due to a lift force exerted by the air on the rotors.

What concepts are needed?

We need to draw a free-body diagram representing the system. Also, we apply Newton's second law.

Why?

Unbalanced lift force leads to acceleration of the helicopter and car, which may be described by Newton's laws.

Solution:

Since both the helicopter and car accelerate upward, the net force acting the system is non-zero. To analyze the problem, we need to draw a free-body diagram (shown above). Using this diagram, we write Newton's second law as

$$\sum_i F_{yi} = ma_y \Rightarrow F_{lift} - m_h g - F_T + F_T - m_c g = ma_y \tag{3a}$$

We simplify Eq. (3a) and obtain

$$(m_h + m_c)a_y = F_{lift} - (m_h + m_c)g \tag{3b}$$

Here the total mass m is the sum of the mass of helicopter and car. We solve for the lift force F_{lift} and obtain

$$F_{lift} = (m_h + m_c)(g + a_y)$$
$$= (6500\,\text{kg} + 1200\,\text{kg}) \times \left(9.8\,\text{m}/\text{s}^2 + 0.60\,\text{m}/\text{s}^2\right) = 8.01 \times 10^4\,\text{N} \tag{3c}$$

LAWS OF MOTION

To compute the tension on the rope, we apply Newton's second law for the car by accounting for the two forces acting on it. We obtain

$$\sum_i F_{yi} = m_c a_y \Rightarrow F_T - m_c g = m_c a_y \qquad (3d)$$

Solving for the tension from Eq. (3d), we obtain

$$F_T = m_c a_y + m_c g = m_c (a_y + g) = 1200\,\text{kg} \times \left(9.80\,\text{m}/\text{s}^2 + 0.60\,\text{m}/\text{s}^2\right) = 1.25 \times 10^4\,\text{N}$$

Friction arises from the electromagnetic forces between the charged particles constituting the two contacting surfaces. The complexity of these interactions makes the calculation of friction from first principles impossible. Hence, it is necessary to use empirical methods for analysis and to develop the theory of friction. Friction exists between two solid surfaces because even the smoothest looking surfaces are quite rough on the microscopic scale as shown in Fig. 4.22.

Friction is defined as the force resisting the relative motion of solid surfaces. Among several types of friction, such as dry friction, fluid friction, lubricated friction, skin friction, and internal friction, we consider dry friction. Dry friction resists relative lateral motion of two solid surfaces in contact. The two regimes of dry friction are 'static friction' between non-moving surfaces, and kinetic friction (sometimes called sliding friction or dynamic friction) between moving surfaces. We characterize the size of these types by using the coefficient of friction which is often symbolized by the Greek letter μ. The coefficient of friction is a dimensionless scalar value which describes the ratio of the force of friction between two bodies and the force pressing them together. The coefficient of friction depends on the materials used. For example, ice on steel has a low coefficient of friction, while rubber on pavement has a high coefficient of friction. The coefficients of friction range from near zero to greater than one. For example, a tire on concrete may have a coefficient of friction of 1.7 under good conditions.

Kinetic friction is sliding friction. This friction occurs when two objects are moving relative to each other and rub together (like a sled on the ground). For surfaces in relative motion $\mu = \mu_k$,

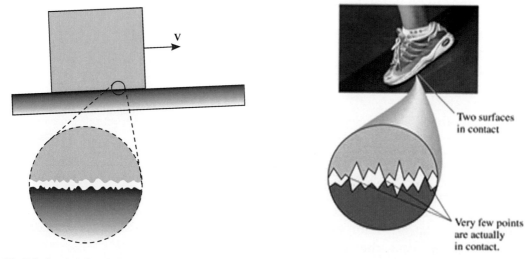

Figure 4.22: Friction is a force that is created whenever two surfaces move or try to move across each other. Friction always opposes the motion or attempted motion of one surface across another surface. Friction is dependent on the texture of both surfaces. Also, friction depends on the amount of contact force pushing the two surfaces together.

where μ_k is the coefficient of kinetic friction. The elementary properties of sliding (kinetic) friction were discovered by experiment in the fifteenth to eighteenth centuries and were expressed as three empirical laws: 1) the force of friction is directly proportional to the applied load (Amontons' first law); 2) the force of friction is independent of the apparent area of contact (Amontons' second law); and 3) kinetic friction is independent of the sliding velocity (Coulomb's law of friction). The Coulomb's law of friction indicates that the frictional force on each surface is exerted in the direction opposite to its motion relative to the other surface. The friction force is proportional to the normal force between two surfaces and is defined as

$$F_f = \mu_k F_N. \tag{4.6}$$

The direction of $\mathbf{F}_f$ is parallel to the two surfaces. On the other hand, **static friction** is friction between two or more solid objects that are not moving relative to each other. For surfaces at rest relative to each other $\mu = \mu_s$, where μ_s is the coefficient of static friction. Static friction is stationary friction which is defined as

$$F_{f,max} = \mu_s F_N. \tag{4.7}$$

The coefficient of static friction is usually higher than the coefficient of kinetic friction. For example, static friction can prevent an object from sliding down a sloped surface. The static friction force must be overcome by an applied force before an object can move, indicating that

$$F_f \leq \mu_s F_N. \tag{4.8}$$

Therefore, we may write that

$$\mu_k \leq \mu_s. \tag{4.9}$$

The maximum value of static friction, when motion is impending, is sometimes referred to as **limiting friction**, but this term is not used universally. For a tire, it is also known as traction. The interaction of different substances is modeled with different coefficients of friction. This means that certain substances have higher resistance to movement than other for the same normal force between them. These coefficients are listed in many places and depend on the state of the material like how smooth it is. Each value is experimentally determined.

When there is no sliding occurring, the friction force can have any value from zero up to F_{max}. As shown in Fig. 4.23, any applied force smaller than F_{max} attempting to slide one surface over the other is opposed by a frictional force of equal magnitude and opposite direction. Any force larger than F_{max} overcomes the force of static friction and causes sliding to occur. At the instant when sliding occurs, static friction is no longer applicable. In this case, the friction between the two surfaces is then called kinetic friction. For example, when we push a box on the ground, it does not move at first because

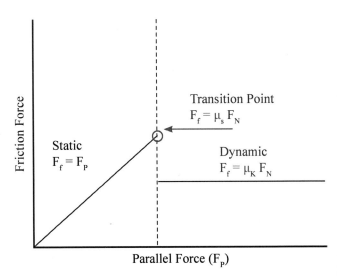

Figure 4.23: The relation between friction force and applied parallel force is illustrated. This explains why a stronger force is needed to start the motion of an object at rest than keep it in motion.

LAWS OF MOTION

Table 4.1: Coefficient of friction is inserted. (Note: Values are approximate and intended only as guide.)

Surfaces	Coefficient of static friction (m_s)	Coefficient of kinetic friction (m_k)
Wood on wood	0.4	0.2
Ice on ice	0.1	0.03
Graphite on graphite	0.1	
Metal on metal (lubricated)	0.15	0.07
Steel on steel (un-lubricated)	0.7	0.6
Rubber on dry concrete	1.0	0.8
Rubber on wet concrete	0.7	0.5
Rubber on other solid surfaces	1-4	1
Car tire on asphalt	0.72	
Car tire on glass	0.35	
Paper on cast iron	0.20	
Teflon on Teflon in air	0.04	0.04
Teflon on steel in air	0.04	0.04
Lubricated ball bearings	≤ 0.01	≤ 0.01
Synovial joints (in human limbs)	0.01	0.01

the force that we apply is not strong enough to overcome the force of friction. Mathematically, this is because the net force is zero or the applied force is equal to the friction force. When the box starts to move the force of friction becomes kinetic friction and can either be constant with any velocity or can change with velocity. Both static and kinetic frictions are proportional to the normal force exerted between the two solid bodies. The normal force between the ground and an object is the force of gravity of that object when the ground is level. In all cases, friction works against the motion or inclination towards motion.

Can the coefficient of friction become negative? Usual answer to this question is "No" because if the coefficient of friction was negative then the work done by the friction force would generate energy and a perpetual motion machine, which violates the second law of thermodynamics becomes possible. As of 2012, a single study done by Cannara and coworkers has demonstrated the potential for a negative coefficient of friction, meaning that a decrease in force leads to an increase in friction. This was reported in the scientific research journal *Nature* in October 2012. The reported research work involved the friction encountered by an atomic force microscope stylus when dragged across a graphene sheet in the presence of graphene-adsorbed oxygen. This contradicts the everyday experience that an increase of normal force increases friction.

Exercise Problem 4.4:

The skier has just begun descending the 30° slope. Assuming the coefficient of kinetic friction is 0.10, (a) first draw the free body diagram, then calculate (b) her acceleration, and (c) the speed she will reach after 4.0 s.

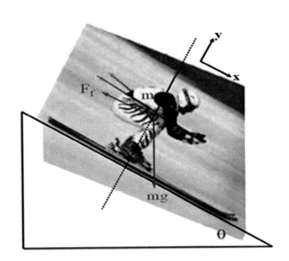

PROBLEM SOLVING STRATEGY

What do we know?

A skier is descending an incline plane which has a non-zero value of kinetic friction.

What concepts are needed?

We need to draw a free-body diagram and apply Newton's law of motion. Also, we need to orient the coordinate axes to simplify the problem.

Why?

This is a dynamics problem requiring application of Newton's second law.

Solution:

From the above free-body diagram, we need to decompose the weight vector into two components. The weight vector can be written as

$$m\vec{g} = mg\sin\theta\,\hat{x} - mg\cos\theta\,\hat{y}. \tag{4a}$$

We apply Newton's second law $F = ma$. Using Eq. (4a), we write the y- and x-component of $F = ma$ as

$$\sum_i F_{yi} = F_N - mg\cos\theta = 0 \;\Rightarrow\; F_N = mg\cos\theta \tag{4b}$$

and

$$\sum_i F_{xi} = ma_x \;\Rightarrow\; ma_x = mg\sin\theta - F_f \;\Rightarrow\; ma_x = mg\sin\theta - \mu_k F_N, \tag{4c}$$

respectively. In Eq. (4c), we used the relation $F_f = \mu_k F_N$. We now use Eq. (4c) to compute the acceleration along the incline plane and obtain

$$\begin{aligned}a_x &= \frac{1}{m}(mg\sin\theta - \mu_k mg\cos\theta) = g(\sin\theta - \mu_k\cos\theta) \\ &= 9.8\,\text{m}/\text{s}^2(\sin 30° - 0.10\times\cos 30°) = 4.0\,\text{m}/\text{s}^2.\end{aligned} \tag{4d}$$

Once we have computed the value of acceleration, we may use this to calculate the speed by using one of the kinematics equations. The speed that she will reach after t = 4.0s is determined by using $v_x = v_{ox} + a_x t = a_x t$. Noting that initial speed is $v_{ox} = 0$, we obtain (4.0m/s²)(4.0s) = 16.0m/s.

LAWS OF MOTION

Exercise Problem 4.5:

A block is given an initial speed of 3.0 m/s up the 22.0° plane shown in the figure. (a) How far up the plane will it go? (b) How much time elapses before it returns to its starting point? Ignore friction.

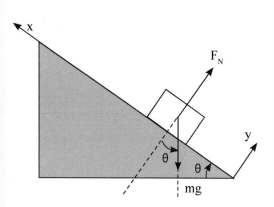

PROBLEM SOLVING STRATEGY

What do we know?

A block is moving up a frictionless inclined plane with an initial speed.

What concepts are needed?

We need apply Newton's second law to determine the acceleration and use the kinematic equations to solve the problem.

Why?

To use the kinematic equations, we need to have information on the acceleration of the object, which is obtained from application of Newton's second law.

Solution:

From the free-body diagram shown, we apply Newton's second law along the x- and y-direction separately. We obtain

$$\sum_i F_{xi} = ma_x \Rightarrow -mg\sin\theta = ma_x \Rightarrow a_x = -g\sin\theta \tag{5a}$$

and

$$\sum_i F_{yi} = ma_y \Rightarrow F_N - mg\cos\theta = 0. \tag{5b}$$

Note that we have chosen a coordinate system such that there is no y-component of the acceleration (i.e., $a_y = 0$). Using Eq. (5a), we use the kinematic equation (with $x_o = 0$) relating the speed and position of the block. We use

$$v_x^2 = v_{ox}^2 + 2a_x(x - x_o) \Rightarrow 0 = v_{ox}^2 - 2g\sin\theta \cdot x. \tag{5c}$$

We now use Eq. (5c) to solve x by noting that when the block goes up the maximum distance its speed is zero and obtain

$$x = \frac{v_{ox}^2}{2g\sin\theta} = \frac{(3.0\,\text{m/s})^2}{2 \times 9.8\,\text{m/s}^2 \times \sin 22.0°} = 1.23\,\text{m} \quad \text{up the plane.} \tag{5d}$$

Also, we may apply the same condition to compute the time needed to reach the maximum height. For this purpose, we use the kinematic equation which yields information on the speed as a function of time

$$v_x = v_{ox} + a_x t \Rightarrow 0 = v_{ox} - g\sin\theta \cdot t. \tag{5e}$$

We solve Eq. (5e) and obtain

$$t = \frac{v_{ox}}{g\sin\theta} = \frac{3.0\,\text{m/s}}{9.8\,\text{m/s}^2 \times \sin 22.0°} = 0.82\,\text{s}.$$

Therefore, the total elapsed time before it returns to its starting point is $2t = 1.6$ s.

Exercise Problem 4.6:

A small mass m is set on the surface of a sphere. If the coefficient of static friction is $\mu_s = 0.60$. At what angle ϕ would the mass start sliding?

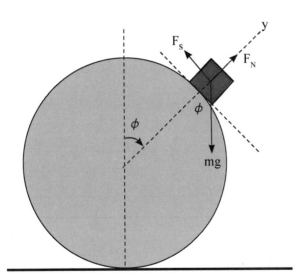

PROBLEM SOLVING STRATEGY

What do we know?

A small mass is place on a sphere due to static friction force.

What concepts are needed?

We need apply Newton's first law using a free-body diagram for the small mass. We need to impose the condition that at the moment the mass begins to slide its acceleration is still zero.

Why?

We may assume the mass will move so slowly at the moment it begins to slide. So the acceleration of the mass is negligible.

Solution:

First, we draw a free-body diagram and choose a coordinate system for the small mass. We, then, apply Newton's first law on the mass resting on the surface of the sphere. For y-component, Newton's law may be written as

$$\sum_i F_{y,i} = ma_y \Rightarrow F_N - mg\cos\phi = 0 \Rightarrow F_N = mg\cos\phi. \tag{6a}$$

Similarly, for x-component, Newton's law may be written as

$$\sum_i F_{x,i} = ma_x \Rightarrow mg\sin\phi - F_s = ma_x. \tag{6b}$$

LAWS OF MOTION

To determine the angle ϕ, we used the condition that the acceleration $a_x = 0$ when the mass starts sliding. Noting that the friction force may be written as $F_s = \mu_s F_N$, we use Eq. (6a) to rewrite Eq. (6b) as

$$mg \sin \phi = \mu_s F_N \implies mg \sin \phi = \mu_s mg \cos \phi. \tag{6c}$$

Using Eq. (6c), we solve for ϕ and obtain

$$\tan \phi = \mu_s \implies \phi_{\max} = \tan^{-1} \mu_s = \tan^{-1} 0.60 = 31°.$$

In physics, dynamics and kinematics are considered as the two pillars of classical mechanics. Hence, dynamics is included in mechanical, aerospace, and other engineering curriculums because of its importance in machine design, the design of land, sea, air, and space vehicles and other applications. Up until the beginning of the twentieth century, Newton's theory of motion was thought to constitute a *complete* description of all types of motion occurring in the Universe. We now know that this is not the case. The modern view is that Newton's theory is only an approximation which is valid under certain circumstances. The theory breaks down when the velocities of the objects under investigation approach the speed of light in vacuum, and must be modified in accordance with Einstein's *special theory of relativity*. The theory also fails in regions of space which are sufficiently curved that the propositions of Euclidean geometry (i.e., flat space-time geometry) do not hold to a good approximation, and must be augmented by Einstein's general *theory of relativity*. Finally, the theory breaks down on atomic and subatomic length-scales, and must be replaced by *quantum mechanics*. We shall neglect the effects both relativity and quantum mechanics in parts I and II of this book. This means we must restrict our investigations to the motions of *large* (compared to an atom) and *slow* (compared to the speed of light) objects moving in *Euclidean* space (i.e., flat space). Fortunately, virtually all of the motions which we commonly observe in the world around us fall into this category.

SUMMARY

i. **Newton's laws of motion** are three physical laws that form the basis for classical mechanics. **The First law** states that if an object experiences no net force, then its velocity is constant: the object is either at rest (if its velocity is zero), or it moves in a straight line with constant speed (if its velocity is nonzero). **The Second law** states that the acceleration of an object is directly proportional to the net force acting on it and is inversely proportional to its mass. The direction of the acceleration is in the direction of the net force acting on the object.

    $$\vec{a} = \frac{1}{m} \sum_i \vec{F}_i \implies \sum_i \vec{F}_i = m\vec{a}.$$

    The Third law states that whenever one object exerts a force on a second object, the second exerts an equal and opposite force on the first.

ii. **Newton's law of universal gravitation** states that every point mass in the universe attracts every other point mass with a force that is directly proportional to the product of their masses and inversely proportional to the square of the distance between them. (Separately it was shown that large spherically symmetrical masses attract and are attracted as if all their mass were concentrated at their centers.) The force that each exerts on the other is directed along the line joining the particle and has a magnitude given by

$$F = G\frac{m_1 m_2}{r^2}.$$

where G is the universal gravitational constant $G = 6.673 \times 10^{-11}$ N·m²/kg².

iii. **Kinetic friction** is sliding friction. This friction force is approximately proportional to the normal force between two surfaces.

$$F_f = \mu_k F_N.$$

where μ_k is the coefficient of kinetic friction. The direction of $\mathbf{F}_f$ is parallel to the two surfaces. **Static friction** is stationary friction

$$F_{f,\max} = \mu_s F_N.$$

where μ_s is the coefficient of static friction. The static friction force must be overcome by an applied force before an object can move. It is noted that

$$F_f \leq \mu_s F_N.$$

Therefore, the two coefficients of friction are related as

$$\mu_k \leq \mu_s.$$

The maximum value of static friction, when motion is impending, is sometimes referred to as **limiting friction**, but this term is not used universally.

MORE WORKED PROBLEMS

Problems 4.1:

Bonnie and Clyde are sliding a 300 kg bank safe across the floor to their getaway car. The safe slides with a constant speed if Clyde pushes from behind with 385 N of force while Bonnie pulls forward on a rope with 350 N of force. What is the safe's coefficient of kinetic friction on the bank floor?

Solution:

We apply Newton's first law separately in the vertical and horizontal directions. For the horizontal direction, we may write

$$F_{net,x} = \sum_i F_{x,i} = ma_x = 0 \Rightarrow F_B + F_C - f_k = 0.$$

since the safe slides with a constant speed. We solve for the friction force and obtain

$$f_k = F_B + F_C = 350\,\text{N} + 385\,\text{N} = 735\,\text{N}.$$

For the vertical direction, we may write

$$F_{net,y} = \sum_i F_{y,i} = ma_y = 0 \Rightarrow F_N - mg = 0.$$

LAWS OF MOTION

and obtain the normal force as

$$F_N = mg = 300\,\text{kg} \times 9.80\frac{\text{m}}{\text{s}^2} = 2940\,\text{N}.$$

Now, we compute the coefficient of kinetic friction from

$$f_k = \mu_k F_N \Rightarrow \mu_k = \frac{f_k}{F_N} = \frac{735\,\text{N}}{2940\,\text{N}} = 0.25.$$

Problems 4.2:

An Airbus A320 jetliner has a takeoff mass of 75,000 kg. It reaches its takeoff speed of 82 m/s (180 mph) in 35 s. What is the thrust of the engines? You can neglect air resistance and but not need to include rolling friction.

Solution:

Using the definition of acceleration, we obtain

$$a = \frac{\Delta v}{\Delta t} = \frac{82\,\text{m/s} - 0\,\text{m/s}}{35\,\text{s}} = 2.34\frac{\text{m}}{\text{s}^2}.$$

Now, we apply Newton's second law and write

$$\sum_i F_{x,i} = ma_x \Rightarrow F_{thrust} - f_r = ma \Rightarrow F_{thrust} = f_r + ma.$$

For rubber rolling on concrete, the coefficient of rolling friction is $\mu_r = 0.02$, and since the runway is horizontal

$$\sum_i F_{y,i} = ma_y \Rightarrow F_N - mg = 0 \Rightarrow F_N = mg.$$

Therefore, we compute the thrust of the engines as

$$F_{thrust} = \mu_r mg + ma = m(\mu_r + a) = 75000\,\text{kg} \times \left(0.02 \times 9.80\frac{\text{m}}{\text{s}^2} + 2.34\frac{\text{m}}{\text{s}^2}\right) = 1.9 \times 10^5\,\text{N}.$$

Problems 4.3:

The fastest pitched baseball was clocked at 46 m/s. If the pitcher exerted his force (assumed to be horizontal and constant) over a distance of 1.0 m, and a baseball has mass of 145 g. (a) Draw a free-body diagram of the ball during the pitch. (b) What force did the pitcher exert on the ball during this record-setting pitch? (c) Estimate the force in (b) as a fraction of the pitcher's weight.

Solution:

a. A free-body diagram for this problem may be drawn as

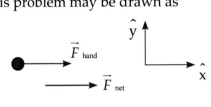

b. We solve the equation for the acceleration a_x using the kinematics equation

$$v_{x,f}^2 = v_{x,i}^2 + 2a_x(x_f - x_i) \Rightarrow a_x = \frac{v_{x,f}^2 - v_{x,i}^2}{2(\Delta x)} = \frac{\left(46\,\text{m/s}\right)^2 - \left(0\,\text{m/s}\right)^2}{2 \times 1.0\,\text{m}} = 1058\,\frac{\text{m}}{\text{s}^2}.$$

The force is

$$F_x = ma_x = 0.145\,\text{kg} \times 1058\,\frac{\text{m}}{\text{s}^2} = 150\,\text{N}.$$

c. If we assume that a typical pitcher weighs 170 pound-force (lbf) then this may be converted to Newton as

$$170\,\text{lbf} \times \frac{4.45\,\text{N}}{1\,\text{lbf}} = 760\,\text{N}$$

We now divide the force from part b) by this weight to see the fraction 150 N/760 N = 1/5. So the force the pitcher exerts on the ball is about 1/5 his weight.

Problems 4.4:

In the ideal set up shown below, $m_1 = 3.0$ kg and $m_2 = 2.5$ kg. (a) What is the acceleration of the masses? (b) What is the tension in the string? (Note: Assume the surface is frictionless.)

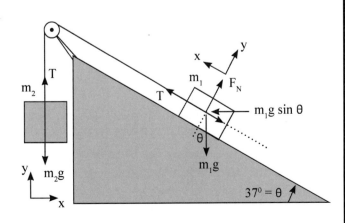

PROBLEM SOLVING STRATEGY

What do we know?

Two masses are connected by a string. Both the mass on the inclined plane and the hanging mass are accelerating.

What concepts are needed?

We need to apply Newton's second law by using a free-body diagram for both masses, separately.

Why?

Tension on the string connects the two masses. The direction of acceleration of these masses is different.

LAWS OF MOTION

Solution:

We apply Newton's second law for mass m_1 and obtain

$$\sum_i F_{x,i} = ma_x \Rightarrow T - m_1 g \sin\theta = m_1 a_x. \qquad (1)$$

We apply Newron's second law for mass m_2 and obtan

$$\sum_i F_{y,i} = ma_y \Rightarrow T - m_2 g = -m_2 a_y. \qquad (2)$$

We note that $a_x = a_y = a$ in magnitude. We subtract Eq. (2) from Eq. (1) and obtain

$$(T - m_1 g \sin\theta) - (T - m_2 g) = m_1 a - (-m_2 a) \Rightarrow (m_2 - m_1 \sin\theta)g = (m_1 + m_2)a.$$

We solve for acceleration a to obtain

$$a = \frac{m_2 - m_1 \sin\theta}{m_1 + m_2} g = \frac{2.5\,\text{kg} - 3.0\,\text{kg} \times \sin 37°}{2.5\,\text{kg} + 3.0\,\text{kg}} \times 9.80 \frac{m}{s^2} = 1.2 \frac{m}{s^2}.$$

For the direction, m_1 is moving up while m_2 is moving down. Now, we compute the tension T by using Eq. (2) and obtain

$$T - m_2 g = -m_2 a \Rightarrow T = m_2(g-a) = 2.5\,\text{kg} \times \left[9.80\frac{m}{s^2} - 1.2\frac{m}{s^2}\right] = 21\,\text{N}.$$

PROBLEMS

1. A person with a black belt in karate has a fist that has a mass of 0.80 kg. Starting from rest, this fist attains a velocity of 8.5 m/s in 0.13 s. What is the magnitude of the average net force applied to the fist to achieve this level of performance?

2. The motion of a very massive object is hardly affected by what would seem to be substantial force. Consider a supertanker, with a mass of 3.0×10^8 kg. If it is pushed by a rocket motor with 5.0×10^5 N- thrust and is subject to no other forces, what will be the magnitude of its acceleration?

3. A 0.20-kg ball is released from a height of 10 m above the beach; the impression the ball makes in the sand is 5.0 cm deep. What is the average force acting on the ball by the sand?

4. A 5.0-g bullet leaves the muzzle of a rifle with a speed of 320 m/s. What force (assumed constant) is exerted on the bullet while it is traveling down the 0.82-m-long barrel of the rifle?

5. A man pulling an empty wagon causes it to accelerate at 1.4 m/s². What will the acceleration be if he pulls with the same force when the wagon contains a child whose mass is three times that of the wagon?

6. Sam, who has mass of 75 kg, takes off across level snow on his jet-powered skis. The skis have a thrust of 200 N and a coefficient of kinetic friction on snow of 0.10. Unfortunately, the skis run out of fuel after only 10 s. (a) What is Sam's top speed? (b) How far has Sam traveled when he finally coasts to a stop?

7. A loaded Boeing 747 jumbo jet has a mass of 2.0×10^5 kg. What net force is required to give the plane an acceleration of 3.5 m/s² down the runway for takeoffs?

8. In the amusement park ride known as Magic Mountain Superman, powerful magnets accelerate a car and its riders from rest to 45 m/s (about 100 mi/h) in a time of 7.0 s. The combined mass of the car and riders is 5.5×10^3 kg. Find the average net force exerted on the car and riders by the magnets.

9. In an Olympic figure-skating event, a 60-kg male skater pushes a 45-kg female skater, causing her to accelerate at a rate of 2.0 m/s². At what rate will the male skater accelerate? What is the direction of his acceleration?

10. At a time when mining asteroids has become feasible, astronauts have connected a line between their 3500-kg space tug and a 6200-kg asteroid. Using their tug's engine, they pull on the asteroid with a force of 490 N. Initially the tug and the asteroid are at rest, 450 m apart. How much time does it take for the tug and the asteroid to meet?

11. A mountain climber is hanging from a rope in the middle of a crevasse. The rope is vertical. Identify the forces on the mountain climber.

12. A beach ball is thrown straight up, and sometime later it lands on the sand. Is the magnitude of the net force on the ball greatest when it is going up or when it is on the way down? Or is it the same in both cases? Explain. Air resistance should not be neglected for a large, light object.

13. Two forces F_A and F_B are applied to an object whose mass is 8.0 kg. The larger force is F_A. When both forces point due east, the object's acceleration has a magnitude of 0.50 m/s². However, when F_A points due east and F_B points due west, the acceleration is 0.40 m/s², due east. Find (a) the magnitude of F_A and (b) the magnitude of F_B.

14. (a) A 65-kg water skier is pulled by a boat with a horizontal force of 400 N due east with a water drag on the skis of 300 N. A sudden gust of wind supplies another horizontal force of 50 N on the skier at an angle of 60° north of east. At that instant, what is the skier's acceleration? (b) What would be the skier's acceleration if the wind force were in the opposite direction to that in part (a)?

15. A duck has a mass of 2.5 kg. As the duck paddles, a 0.10 N force acts on it in a direction due east. In addition, the current of the water exerts a force of 0.20 N in a direction of 52° south of east. When these forces begin to act, the velocity of the duck is 0.11 m/s in a direction due east. Find the magnitude and direction (relative to due east) of the displacement that the duck undergoes in 3.0 s while the forces are acting.

16. At an airport, luggage is unloaded from a plane into the three cars of a luggage carrier, as the drawing shows. The acceleration of the carrier is 0.12 m/s², and friction is negligible. The coupling bars have negligible mass. By how much would the tension in *each* of the coupling bar A, B, and C change if 39 kg of luggage were removed from car 2 and placed in (a) car 1 and (b) car 3? If the tension changes, specify whether it increases or decreases.

17. Two particles are located on the x axis. Particle 1 has a mass m and is at the origin. Particle 2 has a mass $2m$ and is at $x = +L$. A third particle is placed between particles 1 and 2. Where on the x axis should the third particle be located so that the magnitude of the gravitational force on both particle 1 and particle 2 doubles? Express your answer in terms of L.

18. On Earth, two parts of a space probe weigh 11,000 N and 3,400 N. These parts are separated by a center-to-center distance of 12 m and may be treated as uniform spherical objects. Find the magnitude of the gravitational force that each part exerts on the other out in space, far from any other objects.

19. Saturn has an equatorial radius of 6.00×10^7 m and a mass of 5.67×10^{26} kg. (a) Compute the acceleration of gravity at the equator of Saturn. (b) What is the ratio of a person's weight on Saturn to that on Earth?

LAWS OF MOTION

20. The mass of a robot is 5,450 kg. This robot weighs 3,620 N more on planet A than it does on planet B. Both planets have the same radius of 1.33×10^7 m. What is the difference $M_A - M_B$ in the masses of these planets?

21. The distance between two telephone poles is 50.0 m. When a 1.00-kg bird lands on the telephone wire midway between the poles, the wire sags 0.200 m. Draw a free-body diagram of the bird. How much tension does the bird produce in the wire? Ignore the weight of the wire.

22. A 23 kg child goes down a straight slide inclined 38° above horizontal. The child is acted on by his weight, the normal force from the slide, and kinetic friction. (a) Draw a free-body diagram of the child. (b) How large is the normal force of the slide on the child.

23. A stuntman is being pulled along a rough road at a constant velocity, by a cable attached to a moving truck. The cable is parallel to the ground. The mass of the stuntman is 109 kg, and the coefficient of kinetic friction between the road and him is 0.870. Find the tension in the cable.

24. A 20 kg loudspeaker is suspended 2.0 m below the ceiling by two cables that are each 30° from vertical. What is the tension in the cables?

25. A stalled 1500-kg automobile is pushed toward a gas station by a man and a woman on a level road. The applied horizontal forces are 200 N for the woman and 300 N for the man. (a) If there is an effective force of friction of 300 N on the car as it moves, what is its acceleration? (b) Once the car is moving appreciably, what would be an appropriate combined applied force, and why?

26. A 23 kg child goes down a straight slide inclined 38° above horizontal. The child is acted on by his weight, the normal force from the slide, kinetic friction, and a horizontal rope exerting a 30 N force. How large is the normal force of the slide on the child?

27. A 4000 kg truck is parked on a 15° slope. How big is the friction force on the truck?

28. An ice skater is gliding horizontally across the ice with an initial velocity of +6.3 m/s. The coefficient of kinetic friction between the ice and the skate blades is 0.081, and air resistance is negligible. How much time elapses before her velocity is reduced to +2.8 m/s?

29. A damp washcloth is hung over the edge of a table to dry. Thus, part (mass = m_{on}) of the washcloth rests on the table and part (mass = m_{off}) does not. The coefficient of static friction between the table and the washcloth is 0.40. Determine the maximum fraction $[m_{off}/(m_{on}+m_{off})]$ that can hang over the edge without causing the whole washcloth to slide off the table.

30. Objects with masses $m_1 = 10.0$ kg and $m_2 = 5.00$ kg are connected by a light string that passes over a frictionless pulley as in Figure P4.36. If, when the system starts from rest, m_2 falls 1.00 m in 1.20 s, determine the coefficient of kinetic friction between m_1 and the table.

31. A sled weighing 60.0 N is pulled horizontally across snow so that the coefficient of kinetic friction between sled and snow is 0.100. A penguin weighing 70.0 N rides on the sled, as in Figure P4.78. If the coefficient of static friction between penguin and sled is 0.700, find the maximum horizontal force that can be exerted on the sled before the penguin begins to slide off.

32. A softball player is throwing the ball. Her arm has come forward to where it is beside her head, but she hasn't yet released the ball. Identify all the third-law pairs of forces, and then draw free-body diagrams for the ballplayer and the ball. Use dotted lines to connect the members of all action/reaction pairs.

WORK AND ENERGY

CHAPTER 5

What was the cost of launching a space shuttle? There are two ways to answer the question. One way is to think about the financial cost and other way is to think about the energy cost. First, let us think about the financial cost. Each space shuttle launch cost American taxpayers between several hundred million dollars and more than a billion dollars. The figure may depend on who does the estimating. The National Aeronautics and Space Administration (NASA) reported the average shuttle launch cost is roughly $450 million. However, science policy researchers estimated the each shuttle launch cost as much as three times more. Pielke and Byerly, at the University of Colorado in Boulder, reported that the U.S. government and NASA spent a total of nearly $200 billion on the space shuttle program.

The space shuttle program began in 1981, with the launch of Columbia, and concluded with the launch of Atlantis in July 2011. Over those 20 years, there have been 134 missions. According to Pielke and Byerly, the average shuttle launch cost is about $1.5 billion. They argued that each shuttle launch cost about $1.2 billion from the time Columbia launched. However, if we include the lifetime costs of the shuttle program dating back to the early 1970s, the average shuttle launch cost rose to about $1.5 billion per flight. According to NASA, Endeavour cost about $1.7 billion. This is a measure of just how expensive it is to get into space.

What is the real cost of getting into space? In physics, this is measured in terms of energy. As we shall see in this chapter, the total energy cost is computed as the sum of kinetic and potential energy. An orbiting shuttle has potential energy by virtue of its height above the ground. It also has kinetic energy by virtue of its very considerable orbital speed. Once we grasp the concept of energy and look up some numbers, it is a simple calculation to estimate how much energy is needed. The potential energy (P.E.) is P.E. = mgh, which for typical shuttle takeoff weights and operating altitudes gives roughly 2.13×10^{11} J. The acceleration g due to gravity does change with altitude, but at shuttle altitudes the variation is small enough that we can overlook it for our rough calculation. The kinetic energy (K.E.) is the usual K.E. = $mv^2/2$. Plugging some more fairly typical figures, we can come up with a kinetic energy of 3.27×10^{12} J. The total energy obtained by adding the two energies is about 3.48×10^{12} J. The kinetic energy dominates the total energy. We now make an interesting comparison with the cost of electricity because electrical energy is also energy and energy is the common currency in physics. In Texas, electricity costs about ten cents per kilowatt hour (i.e., $0.10/kWh). How much would a space shuttle launch cost at that rate of energy cost? Well, there is 3.60×10^6 J in a kilowatt hour (kWh), so we see that 3.27×10^{12} J of energy costs about $90,750. Obviously this is a grossly unfair comparison. Even simple technology like an incandescent light bulb converts only a very small fraction of its electricity into visible light. Launching a shuttle is not just a matter of just turning electricity into altitude and speed. However, this estimate does suggest that there is some room for improvement. We may think the ideal method would be a space elevator system, where we would be getting pretty close to turning electricity into an orbit. Unfortunately, this approach is a long way away if it ever becomes practical. However, one thing is clear. The concept of energy allows us to examine practical problems in a completely different perspective.

5.1 ENERGY

In previous chapters, we have examined motion of objects in terms of both their kinematics and dynamics. As the underlying cause of motion and as for understanding the quantities which determine motion, the concept of force (i.e., laws of motion) played a central role. However, the force concept was somewhat inconvenient because force is a vector. We have to be mindful of both the magnitude and direction of all of the forces acting on an object, to assess its dynamics, and thereby kinematics, correctly. As a way to search for another conceptual framework to go beyond "kinematics" and "dynamics", we ask the following question:

Is there an alternative way to analyze the motion of an object?

The simple answer to this question is "Yes!" It is possible to approach and to solve both kinematics and dynamics problems by introducing new physical quantities and new concepts. Our hope is that by reformulating the problems in the context of these new quantities and new concepts, we may be able to solve kinematics and dynamics problems much more easily. If this is the case, then we may ask another question.

WORK AND ENERGY

How?

The problems we examined in Chapters 2, 3, and 4 may be reformulated and solved easily using physical quantities such as energy and momentum. There is a lot of overlap between the kinematics and energy problems, so it may be helpful to check whether both energy and momentum, as newly introduced physical quantities, are useful for reformulating the problem. Although the problems may be reformulated in terms of "energy" and "momentum", we need to find out why this new approach would work better. So, we may ask another question.

Why?

The answer to this question is simple but very powerful: **The total energy and momentum of an isolated system is conserved!** In other words, they remain constant. The laws of conservation of energy and momentum are among the most fundamental and useful laws in physics. These conservation principles can be used to find the solution of many mechanics problems, and they come up frequently in many fields of science. What these laws say is that if there are no external or internal forces acting in or on a system, then the energy of that system will remain constant. Newton's first law is hidden in these conservation laws. In addition, if there are no net forces on a system, then that system will have the same momentum at all times. We will discuss the concept of energy and energy conservation in this chapter but postpone our discussion on the concept of momentum and momentum conservation to Chapter 6.

The concepts of work and energy are closely tied to the concept of force because an applied force can do work on an object and cause a change in energy. Hence, loosely speaking, **energy** is defined as the ability to do work. Every process in the natural world involves energy. There is a saying "Money makes the world go round!" In reality, it is energy which makes the world go round. Energy may be considered as "natural money" because it resembles currency in many ways.

Money can be defined as: any kind of object or secure verifiable record that is generally accepted as payment for goods and services and repayment of debts in a given socio-economic context or country, But we know that there is no free lunch, and we have to pay for everything, knowingly or unknowingly. The role of energy in nature is very similar to that of money in an economy. As Nobel-prize winning physicist Richard Feynman put it: energy is the currency of the universe. If we want to speed it up, slow it down, change its position, make it hotter or colder, bend it, break it, whatever, we will have to pay for it (or be paid to do it). This is the role of energy in changes a system. Similar to money, energy may be i) converted from one form to another and ii) transferred from one object to another. To do this, we need to identify the types of energy a body can possess and to identify how energy is transformed.

Some important forms of energy

Similar to money, energy has a number of different forms. All forms of energy measure the ability of an object or system to do work on another object or system. In other words, there are different ways an object or a system can possess energy. Sometimes, it is difficult to define the energy in any given system precisely. However, in the physical sciences, several **forms of energy** have been defined. These include thermal energy, chemical energy, electric energy, the energy of electromagnetic radiation, nuclear energy, magnetic energy, elastic energy, sound energy, and mechanical energy. We will use some of these energies in describing both kinematics and dynamics.

Figure 5.1a: A wrecking ball is a heavy steel ball, usually hung from a crane. By using kinetic energy stored in the ball, the wrecking ball is used for demolishing large buildings. Wrecking balls range from about 450 kg to 5400 kg. The ball is made from forged steel. It is formed under very high pressure while the steel is red hot (soft but not molten) to compress and to strengthen it.

Figure 5.1b: A roller coaster car builds up gravitational potential energy as it climbs hill, moving away from Earth's surface.

Figure 5.1c: Elastic energy can be stored mechanically in a compressed gas or liquid, a coiled spring, or a stretched elastic band. On an atomic scale, the basis for the energy is a reversible strain placed on the bonds between atoms, meaning there's no permanent change to the material.

Figure 5.1d: Geothermal energy is thermal energy generated and stored in the Earth. Thermal energy is the energy that determines the temperature of matter. The geothermal energy of the Earth's crust originates from the original formation of the planet (20%) and from radioactive decay of minerals (80%).

The list of the known possible forms of energy in Fig. 5.1 is not necessarily complete. Whenever physical scientists discover that a certain phenomenon appears to violate the law of energy conservation, new forms may be added, as is the case with dark energy, a hypothetical form of energy that permeates all of space and tends to increase the rate of expansion of the universe.

Energy may be transformed between different forms at various efficiencies. Devices that transform between these forms are called transducers. Each form of energy can be transformed into any of the other forms, but energy is neither destroyed nor created. However, it may be lost to the surroundings. Losses of energy can always be accounted for by small transformations to other types of energy, like sound and heat. Power plants convert potential energy or kinetic energy into electrical energy. Electrical energy can in turn be converted back into other forms of energy

WORK AND ENERGY

Figure 5.1e: Chemical energy is stored in the bonds between atoms. This stored energy is released and absorbed when bonds are broken and new bonds are formed – chemical reactions. Chemical reactions change the way atoms are arranged.

Figure 5.1f: Nuclear energy is the stored potential of the nucleus, or center, of an individual atom. Most atoms are stable on Earth; they retain their identities as particular elements, like hydrogen, helium, iron, and carbon, as identified in the Periodic Table of Elements. Nuclear reactions change the fundamental identity of elements.

such as heat in an oven or light from a lamp. Energy is transformed from one form to another via a transfer mechanism.

Work and heat are transfer mechanisms

Work and heat are two important energy transfer mechanisms. As we can see in Fig. 5.2, an athlete is converting elastic energy stored in his muscle to kinetic energy of the shot. Similarly, a boy is transforming elastic energy stored in a sling into kinetic energy of a slingshot. In both cases, both the athlete and boy are doing work. Also, we can convert chemical energy stored in a match into radiation energy by using heat as an energy transfer mechanism.

Work is the energy transferred by a force to a moving object. Work is a scalar quantity, but it can be positive or negative. Work is a mode of transferring energy from one object to another object by means of an external force. Work performed on or by an object will change the object's mechanical energy. Mechanical energy is a mode of storing energy in an object. Machines provide a mechanical advantage to perform work. Another definition of work is the amount of energy transferred into or out of a system, not counting energy transferred by heat conduction. The conduction of heat is to be distinguished

Figure 5.2a: An athlete is throwing a shot in a shot put competition. The shot put is a track and field event involving "throwing" a heavy spherical object, which is called the shot, as far as possible. The shot put competition for men has been a part of the modern Olympics since their revival in 1896. Women's competition began in 1948.

Figure 5.2b: A match is a tool for starting a fire under controlled conditions. A typical modern match is made of a small wooden stick or stiff paper. One end is coated with a material that can be ignited by frictional heat generated by striking the match against a suitable surface.

Figure 5.2c: A slingshot is an example of how mechanical energy of an object can do work on another object. When a sling is loaded and pulled, the rubber bands are stretched, giving them mechanical energy. The mechanical energy of the stretched sling gives the rubber bands the ability to apply a force to the shot later on.

from heating by friction. This definition is based on the first law of thermodynamics which we will discuss in Chapter 12. In addition to work, heat is another important energy transfer mechanism. Heat is not a property of a system or body, but it is always associated with some kind of process and is synonymous with heat flow and heat transfer. So, **heat** is defined as energy transferred from one body to another by thermal interactions. This form of energy transfer can occur in a variety of ways, among them conduction, radiation, and convection. Heat energy is transferred as a result of a temperature difference. More specifically, energy is transferred as heat passes from a warm body with higher temperature to a cold body with lower temperature.

5.2 WORK

The concepts of work and energy are closely tied to the concept of force because an applied force can do work on an object and cause a change in energy. Since **energy** is defined as the ability to do work, the energy of an object may be changed by either the work done on or by the object. The "**work**" done on an object by a constant force (in both magnitude and direction) as shown in Fig. 5.3 is defined as the product of the magnitude of the displacement and the component of the force parallel to the displacement

$$W = \vec{F} \cdot \vec{d} = F_{paralleld} = (F\cos\theta)d \tag{5.1}$$

The unit used to describe work is the Joule (J) which is equal to the product of the Newton and the meter (i.e., N·m). Equation (5.1) indicates that work is a scalar quantity obtained as a product between the force vector **F** and displacement vector **d**. This type of multiplication operation between two vectors is called a "dot product". As indicated in Eq. (5.1), the result of a dot product depends on the cosine of the angle between the two vectors. This means that if the applied force is perpendicular to the displacement of an object (i.e., $\theta = 90^0$), then there is no work done on the object by the force.

WORK AND ENERGY

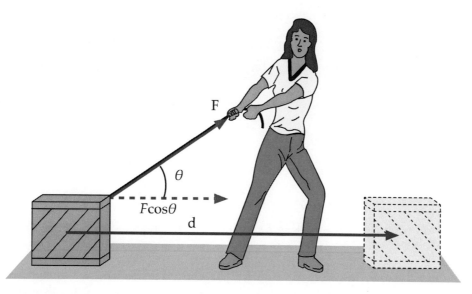

Figure 5.3: When a rope is used to pull a crate box along a horizontal floor, the force F exerted on the box moves it over a distance d. This displacement is due to the work done on the box by the person who is pulling it.

Exercise Problem 5.1:

Determine (a) the work a hiker must do on a 15.0 kg backpack to carry it up to a hill of height $h = 10.0$ m. Determine also (b) the work done by gravity on the backpack, and (c) the net work done on the backpack. For simplicity, assume the motion is smooth and at constant velocity (i.e., negligible acceleration.)

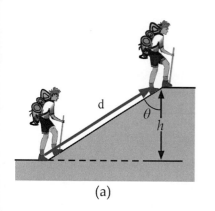
(a)

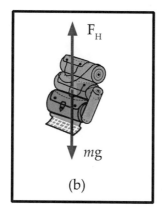
(b)

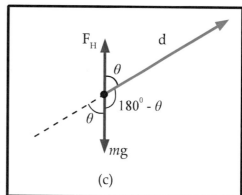
(c)

Solution:

According to the free-body diagram shown above, there are two forces acting on the backpack. Since the backpack is not accelerating, the net force acting on it is zero. We apply Newton's first law and write

$$\sum_i F_{i,y} = ma_y \Rightarrow F_H - mg = 0$$

The force exerted by the hiker F_H is the same as the weight of the backpack

$$F_H = mg = 15.0 \text{kg} \times 9.8 \frac{\text{m}}{\text{s}^2} = 147 \text{N}$$

(a). We compute the work done by the hiker by focusing on the force that the hiker is exerting on the backpack. We obtain

$$W_H = (F_H \cos\theta)d = F_H(d\cos\theta) = F_H h = 147\text{N} \times 10.0\text{m} = 1470\text{J}$$

(b). Similarly, we compute the work done by gravity by focusing on mg and obtain

$$W_G = -(mg\cos\theta)d = -mg(d\cos\theta) = -mgh = -1470\text{J}$$

(c). The net work done on the backpack is obtained by combining the two contributions

$$W_{net} = W_H + W_G = 1470\text{J} - 1470\text{J} = 0$$

Note that the work done on the backpack is zero since there is no net force acting on it.

Since mechanical energy is often defined as the ability to do work, an object that possesses mechanical energy is able to do work. That is, its mechanical energy enables that object to apply a force to another object in order to cause it to be displaced. In all instances, an object that possesses some form of energy supplies the force to do the work. In some instances, the objects doing the work possess chemical potential energy stored in food or fuel that is transformed into work. For example, a student, a tractor, a pitcher or a motor/chain may be considered as the object doing the work. In the process of doing work, the object (or person) that is doing the work exchanges energy with the object upon which the work is done. When work is done on an object, it gains energy. The energy acquired by the object upon which work is done is known as **mechanical energy**. To ascertain the relationship between work and energy for some mechanical systems, we need to identify work and determine the work done by forces on those systems

Understand-the-Concept Question 5.1:

The Moon revolves around the Earth in a circular orbit, kept there by the gravitational force exerted by the Earth. The gravitational force does

a. positive work on the Moon.
b. negative work on the Moon.
c. no work at all on the Moon.

Explanation:

The definition of work $W = F \cdot d$ states that, in order for an object to do non-zero work, the applied force must have a component in the same direction of motion (i.e., displacement). This means that the work done on the Moon by the gravitational force is zero since it revolves around the Earth in a circular orbit. The directions of the displacement d and gravitational force F, which acts as a centripetal force, are perpendicular to each other (i.e., $\theta = 90^0$).

Answer: c

What is keeping the Moon in a nearly circular orbit around the Earth? The Moon orbits the Earth because its inertia is coupled with the attraction of gravity. From Newton's universal law of gravitation, all orbiting bodies attract each other due to force of gravitation. Also, these objects are "in motion" (i.e.,

WORK AND ENERGY

has inertia). Let us consider an example of inertia to help us learn about the orbit of the Moon. If we see a stone tied to a string is whirled around by someone holding the end of that string, then we see that the weight is moving in a circle about the person. The stone is said to have inertia. Because it is moving, the stone carries a certain "energy" with it. If the person lets go of the string, then the stone continues off on a tangent from the circle it was moving in. Its inertia carried it off when the string was released. If we consider an orbiting body as the stone and gravity as the "string" between the bodies, then we have a good model of the system. Gravity and the inertia of the orbiting body keep things moving the way they were. That is why the Moon orbits the Earth. Since the gravity does no work, the Moon's energy is not changed. Only the direction of motion is changed by gravity.

Figure 5.4: The Moon is the only natural satellite of the Earth, and the fifth largest satellite in the Solar System. It is the largest natural satellite of a planet in the Solar System relative to the size of its primary, having 27% the diameter and 60% the density of Earth, resulting in 1/81 of its mass. The Moon is the second densest satellite after Io, which is a satellite of Jupiter.

It is interesting to note that the radius of Moon's orbit around the Earth is 384,000 km, but it is not constant. The Moon's orbit is getting larger at a rate of about 3.8 centimeters per year. This means that Moon is getting farther from the Earth. The reason for the increase is that the Moon raises tides on the Earth. Because the side of the Earth that faces the Moon is closer, it feels a stronger pull of gravity than the center of the Earth. Similarly, the part of the Earth facing away from the Moon feels less gravity than the center of the Earth. This effect stretches the Earth a bit, making it a little bit oblong. We call the parts that stick out "tidal bulges." The actual solid body of the Earth is distorted a few centimeters, but the most noticeable effect is the tides raised on the ocean. Now, all mass exerts a gravitational force, and the tidal bulges on the Earth exert a gravitational pull on the Moon. Because the Earth rotates faster (once every 24 hours) than the Moon orbits (once every 27.3 days), the bulge tries to "speed up" the Moon and pull it ahead in its orbit. The Moon is also pulling back on the tidal bulge of the Earth, slowing the Earth's rotation. Tidal friction which is caused by the movement of the tidal bulge around the Earth takes energy out of the Earth and puts it into the Moon's orbit, making the Moon's orbit bigger. Paradoxically, the Moon actually moves slower! The Earth's rotation is slowing down due to this. One hundred years from now, the day will be 2 milliseconds longer than it is now. This same process took place billions of years ago. However, the Moon was slowed down by the tides raised on it by the Earth. That is why the Moon always keeps the same face pointed toward the Earth. Since the Earth is so much larger than the Moon, this process (i.e., tidal locking) took place in a few tens of millions of years, which is very quick on astronomical time scales.

5.3 ENERGY CONSERVATION

Although it is not precise, we used the working definition of energy as "the ability to do work" because it can be transformed from one form into another via work as one of the transfer mechanisms. There are many forms of energy, but we focus mainly on two types in this chapter. We combine these two types, (i) kinetic energy and (ii) potential energy, and call them mechanical energy. Kinetic energy is defined as energy of motion, while potential energy is defined as energy associated with forces that depend on the position or configuration of a body (or bodies) and the surroundings. A precise mathematical definition of these energies will be discussed below. Hence, mechanical energy is the energy that is possessed by an object due to its motion or due to its position since it can be either kinetic energy (energy of motion) or potential energy (stored energy of position). In other words, objects have mechanical energy if they are in motion and/or if they are at some position relative to a zero potential energy position. Here, the position of zero potential energy may be chosen arbitrarily. For example, when a brick held at a vertical position above the ground, the zero-height position (i.e., ground) may be chosen as the zero potential position. A moving car possesses mechanical energy due to its motion (i.e., kinetic energy). A moving baseball possesses mechanical energy due to both its high speed (kinetic energy) and its vertical position above the ground (gravitational potential energy).

Conservation of Energy

One important outcome of the concept of energy is the **law of conservation of energy**. This is one of the fundamental laws in physics. The law of conservation of energy was first formulated in the nineteenth century. The principle states that the total amount of energy in an isolated system remains constant over time. This means the total energy is **conserved** over time. For an isolated system, this law indicates that energy can change with its location within the system, and that it can change the form within the system, but that the total energy can be neither created nor destroyed. For instance, chemical energy can become kinetic energy, but the total energy of the system will remain the same. For a mechanical system, a simpler version of this principle states that the total mechanical energy which is the sum of kinetic and potential energy remains the same before and after occurrence of any process:

$$(K.E. + P.E.)_i = (K.E. + P.E.)_f \tag{5.2}$$

As a fundamental concept in physics, the law of conservation of energy provides an explanation for how energy is converted within a system. Generally, one form of energy can be converted into another form of energy. For example, potential energy can be converted to kinetic energy.

5.4 KINETIC ENERGY

Now, let us define the kinetic energy more precisely. The kinetic energy of a particular object is the energy it possesses while in motion. To discuss how kinetic energy is related to a motion of an object, we consider an object of mass m moving in a straight line on a frictionless surface.

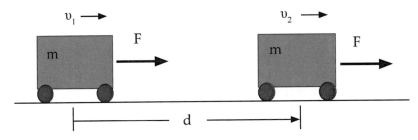

Figure 5.5: When a net force F does work on a rigid body, it causes the body's speed to change over a distance d. The work done by the net force F is the same as the sum of the work done by the action of every force acting the body.

WORK AND ENERGY

By noting that in Fig. 5.5 the displacement and force vectors are in the same direction, we may compute the net work done on the object by the applied force F as

$$W = Fd \tag{5.3}$$

The work done on the cart caused its speed to change from v_1 to v_2. This change is independent of type of force that acted on the body. According to Newton's second law, the relation between the applied force $\mathbf{F}$ and the acceleration a is given by $\mathbf{F} = m\mathbf{a}$. We may now use the velocity-versus-position kinematic equation to express the acceleration a in terms of initial and final velocities as

$$v_2^2 = v_1^2 + 2ad \implies a = \frac{v_2^2 - v_1^2}{2d}. \tag{5.4}$$

This result allows us to write the applied force, which is proportional to the difference between the square of the final and initial velocities, as

$$F = ma = m\frac{v_2^2 - v_1^2}{2d} \tag{5.5}$$

By substituting Eq. (5.5) into Eq. (5.3), we obtain

$$W = Fd = m\frac{v_2^2 - v_1^2}{2} = \frac{1}{2}mv_2^2 - \frac{1}{2}mv_1^2 \tag{5.6}$$

As indicated by Eq. (5.6), the work W done on the object is related to the energy of motion which is equal to one-half of the mass of the object multiplied by the square of the velocity of the object. This is the energy of motion. In general, this energy of motion consists of three types: (i) vibrational, (ii) rotational, and (iii) translational motion. Vibrational energy is due to vibrational motion, and rotational energy is due to rotational motion. Similarly, translational energy is due to motion of the center of mass from one point to another as illustrated in Fig. 5.5.

5.5 WORK-ENERGY PRINCIPLE

As we have seen above, doing work on a body that is initially at rest causes it to be set in motion with some speed v. The work done can be expressed as a function of the body's final speed v and mass m, independent of the type of force that acted on the body. Equation (5.6) indicates the relation between work and energy of motion. This energy of motion which we discussed above is called kinetic energy. This definition can be extended to rigid bodies by defining the work done by the resultant torque (i.e., rotational equivalence of net force) and the energy of rotational motion. We will discuss rotational motion in Chapter 8. The relation between work and kinetic energy is known as the **work-energy principle**. The principle of work and kinetic energy states that the work done by all forces acting on a particle (the work of the resultant force) equals the change in the kinetic energy of the particle. Suppose we define the translational kinetic energy (K.E.) as

$$K.E. = \frac{1}{2}mv^2 \tag{5.7}$$

Then, this definition of kinetic energy allows us to write the work done W of Eq. (5.6) as

$$W = K.E._2 - K.E._1 = \Delta(K.E.) \tag{5.8}$$

Equation (5.8) states that net work done by forces on a particle causes a change in the kinetic energy of the particle. It is noted that the work-energy principle is valid only if W is the net work done on the object (i.e., the work done by all forces acting on the object.) The principle is powerfully simple and gives us a direct relation between the net work done and kinetic energy.

Effects of Work on Kinetic Energy

The effects of work on kinetic energy depend on the sign of work. The work done by the force F can be either positive or negative, depending on the angle between the force vector and displacement vector. For example, suppose we have an object moving with constant velocity. At time $t = 0$ s, a force F is applied. If F is the only force acting on the body, then the object will either increase or decrease its speed depending on whether or not the velocity v and the force **F** are pointing in the same direction. When $\mathbf{F} \cdot \mathbf{v} > 0$, the speed of the object will increase and the work done by the force on the object is positive. When $\mathbf{F} \cdot \mathbf{v} < 0$, the speed of the object will decrease and the work done by the force on the object is negative. When $\mathbf{F} \cdot \mathbf{v} = 0$, we are dealing with rotational motion and the speed of the object remains constant. Note that for the friction force **f**, $\mathbf{f} \cdot \mathbf{v} < 0$ (usually) and the speed of the object for this case is reduced! In short, the net positive work W done on a body causes its kinetic energy to increase by an amount W. Conversely, the net negative work W done on a body causes its kinetic energy to decrease by an amount W.

Understand-the-Concept Question 5.2:

Which object, either a fighter jet aircraft or a Chevy Corvette, can do more work? Why?

Explanation:

The energy associated with the work done by the net force does not disappear after the net force is removed (or becomes zero), it is transformed into the kinetic energy of the body. We account for this transformation by using the work-energy theorem: $W_{net} = \Delta K.E. = K.E._f - K.E._i$. We answer the question by using the theorem. Since both the car and jet aircraft start from rest, they have zero initial kinetic energy. The estimated top speed of F-22 fighter jet is 2,410 km/h and that of 2013 Chevy Corvette is

WORK AND ENERGY

about 240 km/h. Also, the jet aircraft is much more massive than the car, indicating that the final kinetic energy at the top speed will be higher for the jet aircraft. Noting that the change in kinetic energy is due to the work done, we may conclude that the fighter jet aircraft can do more work.

We should remember that the plus or minus sign of the work depends on the perspective. For example, if the body's speed increases then the work done on the body is positive and we may say that its kinetic energy has increased. However, if the body's speed decreases then its kinetic energy decreases and the change in kinetic energy $\Delta K.E.$ is negative. In this case the body does positive work on the system slowing it down. Alternatively speaking the work done on the body is negative.

5.6 POTENTIAL ENERGY

Another important aspect of mechanical energy is potential energy. Potential energy is defined as energy which results from position or configuration. An object may have the capacity for doing work as a result of position in a gravitational field (gravitational potential energy), an electric field (electric potential energy), or a magnetic field (magnetic potential energy). It may have elastic potential energy as a result of a stretched spring or other elastic deformation.

To gain concrete intuition, we first consider gravitational potential energy. Gravitational potential energy is defined as energy that an object possesses because of its position in a gravitational field. The most common use of gravitational potential energy is for an object near the surface of the Earth where the gravitational acceleration can be assumed to be constant $g = 9.80$ m/s². We derive the expression for gravitational potential energy by considering a simple case of lifting an object of mass m vertically near the surface of Earth as shown in Fig. 5.6.

To lift the object, a person must apply an external force F_{ext} upward, against the gravity. The work the person must do to lift the object is

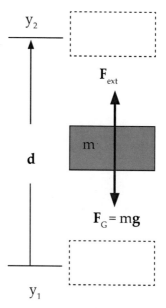

Figure 5.6: Potential energy is the energy of an object or a system due to the position of the body or the arrangement of the particles of the system. In this case, potential energy is associated with the force of gravity. The action of lifting the mass is performed by an external force that works against the force field of the potential. This work is stored in the force field, which is said to be stored as potential energy.

$$W_{ext} = \vec{F}_{ext} \cdot \vec{d} = (F_{ext} \cos 0) \, d = mg(y_2 - y_1). \tag{5.9}$$

Assuming that the applied force moves the object vertically upward very slowly (i.e, with zero acceleration), we apply Newton's second law to determine F_{ext} as

$$\sum_i F_{y,i} = ma_y \Rightarrow 0 = F_{ext} - mg \tag{5.10}$$

Here, we assume that the vertical acceleration a_y is zero since the object is being displaced very slowly. We define the gravitational potential energy of a body at its height y above some reference level as

$$P.E._{grav} = mgy \tag{5.11}$$

Then, we may write the work done on the object by the external force as

$$W_{ext} = mg(y_2 - y_1) = P.E._2 - P.E._1 = \Delta P.E. \tag{5.12}$$

Examples of gravitational potential energy include skydiving and hydroelectricity. Skydiving is the action sport of exiting an aircraft and returning to the surface of Earth with the aid of gravity as shown in Fig. 5.7 while using a parachute to slow down during the final part of the descent. It may or may not involve a certain amount of free-fall, a time during which the parachute has not been deployed and the body gradually accelerates to terminal velocity. Hydropower, or hydroelectricity, is electricity generated by trapping the gravitational potential energy of water as it flows or falls from a high altitude to a low altitude. Is the most widely used form of renewable energy, accounting for 16 percent of global electricity generation (3,427×10^9 kWh of electricity production in 2010), and it is expected to increase about 3.1% each year for the next 25 years. The cost of hydroelectricity is relatively low, making it a competitive source of renewable electricity. The average cost of electricity from a hydroelectric plant with a capacity larger than 10 MW is $0.03 to $0.05 per kWh. This is also a flexible source of electricity since production can be ramped up and down very quickly to adapt to changing energy demands.

Figure 5.7: Gravitational potential energy is energy an object possesses because of its position in a gravitational field. The most common use of gravitational potential energy is for an object near the surface of the Earth where g can be assumed to be constant at about 9.8 m/s². Here, the surface is assumed as the reference point.

We may define the change in potential energy as the work that must be done by an external force to move the object between two points without acceleration. Another type of force that we may apply this approach to compute the work done is the spring force. A restoring force exerted by a spring is written as

$$F_s = -kx \quad (Hooke's\ law) \tag{5.13}$$

where k is the spring constant and x is the amount the spring is stretched or compressed from the natural length of the spring. The work done by the spring is

$$W_s = \overline{F}_s x = \left(\frac{1}{2}kx\right)x = \frac{1}{2}kx^2. \tag{5.14}$$

WORK AND ENERGY

Here, we used the average force because the strength of spring force depends on linearly on x. Equation (5.14) allows us to define elastic potential energy of the spring as

$$P.E._{elastic} = \frac{1}{2}kx^2 \qquad (5.15)$$

Equation (5.15) indicates that the potential energy is stored in the spring when it is deformed (i.e., either stretched or compressed). In general, elastic energy is the potential mechanical energy stored in the configuration of a material or physical system as work is performed to distort its volume or shape. Hence, the elastic potential energy equation of (5.15) may be used to calculate the positions of mechanical equilibrium. Also, this potential energy can be converted into kinetic energy, similar to the gravitational potential energy.

The essence of elastic potential energy and elasticity is reversibility. Forces that are applied to an elastic material transfer energy into the material. When the material yields the energy back to its surroundings, it can recover its original shape. However, all materials have limits to the degree of distortion they can endure without breaking or irreversibly altering their internal structure. Hence, the characterizations of solid materials include specification of its elastic limits. Usually, this specification is expressed in terms of strains. Beyond the elastic limit, a material is no longer stores the energy from mechanical work performed on it in the form of elastic energy.

Examples of elastic potential energy include a toy gun with a spring and trampoline as shown in Fig. 5.8. If we compress the spring in a toy dart gun by exerting a force on it, we know that the state of the compressed spring is different from that of the relaxed spring. By exerting a force on the object through some distance we have changed the energy state of the object. Similarly, if we grab one end of a rubber band and pull on the other, we would realize that the stretched rubber band differs from the same band when it is relaxed. We say that the compressed spring or stretched rubber band stores elastic energy. Here, the energy account is used to describe how an object stores energy when it undergoes a reversible deformation. This energy can be transferred to another object to produce a change. For example, when the spring is released, it can launch a dart. It seems reasonable that the more the spring

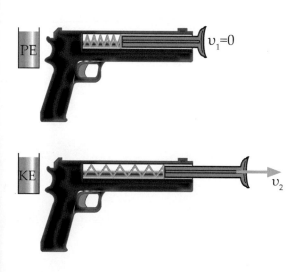

Figure 5.8: Elastic potential energy is potential energy stored as a result of deformation of an elastic object, such as the stretching of a spring or trampoline. It is equal to the work done to stretch the spring or trampoline, which depends upon the spring constant k as well as the distance stretched.

is compressed, the greater the change in speed it can impart to the toy dart. Another example similar to spring is trampoline. A trampoline is a device consisting of a piece of taut, strong fabric stretched over a steel frame using many coiled springs. Because of the elastic property of trampoline, people bounce on trampolines for recreational and competitive purposes. The fabric on which users bounce (commonly known as the 'bounce mat' or 'trampoline bed') is not elastic in itself; the elasticity is provided by the springs that connect it to the frame, which store potential energy.

5.7 CLASSIFICATION OF FORCES

Any forces that we use to describe motion of objects are said to belong one of two types of forces: (i) conservative forces and (ii) non-conservative forces. **Conservative forces** are defined as forces for which the work done does not depend on the path taken but depends only on the initial and final positions. Examples of conservative forces are gravitational force and spring force. Equivalently, if a particle travels in a closed loop, the net work done (the sum of the force acting along the path multiplied by the distance travelled) by a conservative force is zero. If a force is conservative, it is possible to assign a numerical value for the potential energy to any point. When an object moves from one location to another, the force changes the potential energy of the object by an amount that does not depend on the path taken. On the other hand, **non-conservative forces** are defined as forces for which the work done depends on the path taken. Examples of non-conservative force are friction, air resistance, and non-elastic material stress. If the force is not conservative, then defining a scalar potential is not possible, because taking different paths would lead to conflicting potential differences between the start and end points. Non-conservative forces can arise in classical physics due to neglected degrees of freedom or from time-dependent potentials. For instance, friction may be treated without resorting to the use of non-conservative forces by considering the motion of individual molecules. However, this means every molecule›s motion must be considered rather than handling it through statistical methods. For macroscopic systems the non-conservative approximation is far easier to deal with than millions of degrees of freedom.

Although the potential energy can be defined only for conservative forces, both conservative and non-conservative forces can do work. Hence, we may write the work done by the net force as

$$W_{net} = W_C + W_{NC} \tag{5.16}$$

According to the work-energy theorem, the work done by the net force changes the kinetic energy of the object: $W_{net} = \Delta K.E.$ This means we may write Eq. (5.16) as

$$\Delta K.E. = W_C + W_{NC} \tag{5.17}$$

We may rewrite Eq. (5.17) and obtain the work done by the non-conservative forces as

$$W_{NC} = \Delta K.E. - W_C \tag{5.18}$$

The work done by the conservative forces may be written as negative of the change in potential energy which is defined for each force

$$W_C = -\Delta P.E. \tag{5.19}$$

We, therefore, may write the work done by the non-conservative forces as

$$W_{NC} = \Delta K.E. + \Delta P.E. \tag{5.20}$$

WORK AND ENERGY

Equation (5.20) clearly states that the work W_{NC} done by the non-conservative forces acting on an object is equal to the total change in kinetic and potential energy. When there are no non-conservative forces (i.e., $W_{NC} = 0$) acting on the object, we may write Eq. (5.20) as

$$0 = \Delta K.E. + \Delta P.E. \tag{5.21}$$

By writing $\Delta K.E.$ and $\Delta P.E.$, respectively, as the difference in the kinetic and potential energy at points 2 and 1, we obtain

$$0 = (K.E._2 - K.E._1) + (P.E._2 - P.E._1) \tag{5.22}$$

Now, we rewrite Eq. (5.22) by collecting the terms with the subscript 2 on the left hand side of the equation while collecting the terms with the subscript 1 on the right hand side of the equation and obtain

$$K.E._2 + P.E._2 = K.E._1 + P.E._1 \tag{5.23}$$

Equation (5.23) indicates that the sum of kinetic energy and potential energy at point 2 is the same as the sum of these energies at point 1, suggesting that

$$E_2 = K.E._2 + P.E._2 = E_1 = K.E._1 + P.E._1 = \text{constant} \tag{5.24}$$

Observe that the energy conservation of Eq. (5.24) applies for conservative forces only. This allows us to state the principle of conservation of mechanical energy as the following: If only conservative forces are acting on a system, the total mechanical energy of the system neither increases nor decreases in any process. In other words, the total mechanical energy of the system is **conserved.**

Understanding-the-Concept Question 5.3:

Two water slides at a pool are shaped differently but start at the same height h. Two riders, Paul and Kathleen, start from rest at the same time on different slides. (a) Which rider, Paul or Kathleen, is traveling faster at the bottom? (b) Which rider makes it to the bottom first? Ignore friction.

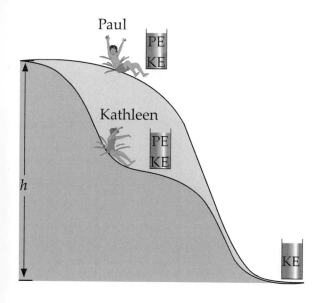

Problem Solving Strategy

What do we know?

Both Paul and Kathleen start at the same height but come down on different slides, which may be considered as frictionless.

What concepts are needed?

We need to apply conservation of mechanical energy. Also, we need to use kinematics equations.

Why?

Only conservative forces are acting on both Paul and Kathleen. The time taken to travel a given distance depends on the speed..

Explanation:

Assuming we can neglect the work done by the non-conservative forces, we may apply energy conservation $E_i = E_f$ and find that the speed of both Paul and Kathleen at the bottom of the slide is the same

$$E_i = E_f \Rightarrow mgh = \frac{1}{2}mv_f^2 \Rightarrow v_f = \sqrt{2gh}$$

indicating that the final speed does not depend on the mass of the person. However, this does not mean that both Paul and Kathleen arrives the bottom at the same time. Noting that the distance d is written as $d = vt$, we may write the time t as $t = d/v$, indicating that a shorter time is required to travel the same distance if a higher speed can be obtained quickly and maintained longer throughout the slide. This may be achieved by converting the gravitational potential energy to kinetic energy faster. According to the schematic diagram above, the shape of the water slide allows Kathleen to achieve this, indicating that she will arrive at the bottom first.

In order to zip down the slide, you need a constant stream of water to reduce friction between you and the fiberglass surface. To maintain this stream, the water park has to get a supply of water to the top of the slide. Most water slides do this with a pump, housed in a building near the base of the slide. We should remember that, when traveling down a water slide, making to the bottom first may not be as important as how fun it is. An excitement factor may increase with increasing acceleration that a person can experience. For this reason, Paul's water slide may be more fun.

Exercise Problem 5.2:

Assuming the height of the hill is 40 m, and the roller-coaster car starts from rest at the top, calculate (a) the speed of the roller-coaster car at the bottom of the hill, and (b) at what height it will have half its speed. Take y = 0 at the bottom of the hill.

Problem Solving Strategy

What do we know?

The roller-coaster car starts from rest at the top and descends the hill.

What concepts are needed?

Assuming that there is no energy loss, we may apply conservation of mechanical energy.

Why?

Since we know the total energy of the car at the top, we may apply the energy conservation principle to compute its speed elsewhere.

WORK AND ENERGY

Solution:

We assume there are no energy losses. Although we may solve this problem by using the kinematic equations, it would be simpler to apply energy conservation. We apply the conservation of energy principle by comparing the total energy at two points: $E_1 = E_2$

$$\frac{1}{2}mv_1^2 + mgy_1 = \frac{1}{2}mv_2^2 + mgy_2 \qquad (2a)$$

Since the initial speed $v_1 = 0$ at the top and $y_2 = 0$ at the bottom of the hill, we may simplify Eq. (2a) and obtain

$$mgy_1 = \frac{1}{2}mv_2^2 \Rightarrow v_2 = \sqrt{2gy_1} = \sqrt{2 \times 9.8 \frac{m}{s^2} \times 40m} = 28 \frac{m}{s} \qquad (2b)$$

If the speed at y_2 is $v_2 = (28 \text{ m/s})/2 = 14$ m/s, then we may, once again, use the conservation of energy principle and obtain

$$\frac{1}{2}mv_1^2 + mgy_1 = \frac{1}{2}mv_3^2 + mgy_3 \Rightarrow mgy_1 = \frac{1}{2}mv_3^2 + mgy_3 \qquad (2c)$$

Here, point 3 presents a mid-hill position. Using Eq. (2c), we may obtain the position of the roller coaster car as

$$y_1 = y_3 + \frac{v_3^2}{2g} \Rightarrow y_3 = y_1 - \frac{v_3^2}{2g} = 40\,m - \frac{(14\,m/s)^2}{2 \times 9.80\,m/s^2} = 30.0\,m$$

A roller coaster ride is a thrilling experience that involves a wealth of physics. Part of the physics of a roller coaster is the physics of work and energy. The ride often begins as a chain and motor (or other mechanical device) exerts a force on the train of cars to lift the train to the top of a very tall hill. Once the cars are lifted to the top of the hill, gravity takes over and the remainder of the ride is an experience in energy transformation. At the top of the hill, the cars possess a large quantity of potential energy. As the cars descend the first drop they lose much of this potential energy in accord with their loss of height. The cars subsequently gain kinetic energy.

Exercise Problem 5.3:

A small mass m slides without friction along the looped apparatus. If the object is to remain on the track, even at the top of the circle (whole radius is r), from what minimum height h must it be released?

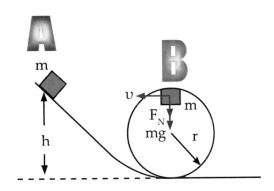

Problem Solving Strategy

What do we know?

A small mass slides down a frictionless track with a loop of radius r.

What concepts are needed?

We need to apply conservation of mechanical energy.

Why?

Since there is no dissipative force, the mechanical energy is conserved. So we may compare the total energy at two points on the looped track.

Solution:

Since the small mass slides without friction, we apply conservation of energy at two points (A and B) and obtain

$$\frac{1}{2}mv_A^2 + mgy_A = \frac{1}{2}mv_B^2 + mgy_B \qquad (3a)$$

Since $v_A = 0$ and $y_A = h$ at point A and $y_B = 2r$ at point B, we simplify Eq. (3a) as

$$mgh = \frac{1}{2}mv_B^2 + mg(2r) \qquad (3b)$$

The speed of a small mass m at point B can be obtained by applying Newton's second law

$$\sum_i F_{i,R} = ma_R \Rightarrow F_N + mg = m\frac{v_B^2}{r} \qquad (3c)$$

Note that derivation of the formula for centripetal force (i.e., mv^2/r) is discussed in Chapter 7. To obtain the minimum speed of the mass at point B, we assume that the mass barely makes it around the loop. This means that the mass almost loses contact with the track. Hence, we set $F_N = 0$ in

Eq. (3c) and solve for the speed at point B to obtain

$$m\frac{v_B^2}{r} = mg \Rightarrow v_B^2 = gr \qquad (3d)$$

We combine the result in Eqs. (3b) and (3d), and then we solve for h. We obtain

$$mgh = \frac{1}{2}m(gr) + 2mgr \Rightarrow h = \frac{5}{2}r \qquad (3e)$$

Remembering Newton's first law which states that an object in motion will continue in motion along a straight line, we realize that passengers going have momentum pushing them straight forward. Similar to Exercise Problem 5.3, on a looping coaster, the general rules of centripetal force (i.e., $Fc = mv^2/r$) are in operation because the train is making a turn at every point during the loop. We should

WORK AND ENERGY

keep in mind Newton's first law which states that an object in motion will stay in motion unless another unbalanced force acts on it. The force that makes the train turn through the loop is centripetal force. The concept of centripetal force will be discussed more extensively in Chapter 7.

Exercise Problem 5.4:

A ball of mass $m = 2.60$ kg, starting from rest, falls a vertical distance $h = 55.0$ cm before striking a vertical coiled spring, which it compresses an amount $y_o = 15.0$ cm. Determine the spring constant of the spring. Assume the spring has negligible mass. Measure all distances from the point where the ball first touches the uncompressed spring ($y = 0$ at this point).

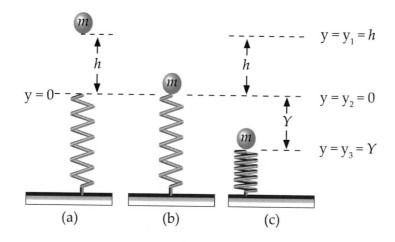

Problem Solving Strategy

What do we know?

When the ball is released, it compresses the spring and stops.

What concepts are needed?

Assuming that there is no air resistance, we may apply conservation of energy.

Why?

Initially, the ball has gravitational potential energy, but this energy is transformed into elastic potential energy when it compresses the spring. It should be note that when the ball stops it has negative potential energy since it is below the reference line.

Solution:

We apply the principle of energy conservation: $E_A = E_B$ and $E = K.E. + P.E.$ Here, we need to note that there are two types of potential energy: gravitational and elastic potential energy. The energy conservation principle for this system may be written as

$$\frac{1}{2}mv_A^2 + mgy_A + \frac{1}{2}kx_A^2 = \frac{1}{2}mv_B^2 + mgy_B + \frac{1}{2}kx_B^2 \qquad (4a)$$

where x represents a distance the spring is compressed. Since $v_A = v_B = 0$ and $x_A = 0$, we may simplify Eq. (4a) and obtain

$$mgh = -mgy_o + \frac{1}{2}ky_o^2 \qquad (4b)$$

We now use Eq. (4b) and solve for the spring constant k. We obtain

$$k = \frac{2mg(h+y_o)}{y_o^2} = \frac{2 \times 2.60\,\text{kg} \times 9.80\,\text{m}/\text{s}^2 \times (55\,\text{cm} + 15\,\text{cm}) \times 1\,\text{m}/100\,\text{cm}}{\left(15\,\text{cm} \times 1\,\text{m}/100\,\text{cm}\right)^2} = 1580\,\frac{\text{N}}{\text{m}} \qquad (4c)$$

The spring constant k is also known as the force constant. It is a measure of the elasticity of the spring. A larger value for a means that more force must be exerted to extend the spring by the same length x. For example, a short spring has a larger spring constant than a long spring if the other aspects of it are the same (thickness, material). If we apply 10N to extend the short spring by 1cm, then to extend the long spring by the same amount, we would need less force, perhaps 5N.

5.8 EFFECT OF DISSIPATIVE FORCES

So far, we have discussed the effects of non-dissipative forces. Let us now focus on the effects of dissipative forces. When dissipative forces are present in a mechanical system, the total mechanical energy (the sum of kinetic and potential energy) decreases upon motion. This energy changes into other forms of energy, such as heat, sound, and light. This process is called the process of dissipation of mechanical energy; it arises because of the presence of various forces of resistance (friction). Examples of dissipative systems include a solid body moving along the surface of another solid in the presence of friction, and a solid body moving through a liquid or gas, among whose particles forces of viscosity (viscous friction) act upon motion. The motion of dissipative systems may be retarded (damped) or accelerated. For example, the oscillations of a weight suspended from a spring will decay because of resistance from the surroundings and because of internal (viscous) resistance arising in the material of the spring itself upon deformation. How does the equation (5.23) which describes energy conservation change if there is a dissipative force F_f (i.e., friction) over a distance d? We may answer this question by computing the work done by the non-conservative force as

$$W_{NC} = -F_f d \qquad (5.25)$$

According to Eq. (5.20), $W_{NC} = \Delta K.E. + \Delta P.E.$, the work done by the non-conservative forces does not remain as mechanical energy but changes the total energy of the system

$$K.E._2 - K.E._1 + P.E._2 - P.E._1 = -F_f d \qquad (5.26)$$

We may compare the total mechanical energy of the system at points 1 and 2 by

$$K.E._1 + P.E._1 = K.E._2 + P.E._2 + F_f d \qquad (5.27)$$

WORK AND ENERGY

Equation (5.27) indicates that the total energy is neither increased nor decreased, but it is transformed. For example, when we use the brakes, it turns the energy of forward motion into heat by using hydraulic pressure to press the brake pads into the brake disc, producing heat. Friction acts between two surfaces in contact. When the surfaces of two bodies are in contact with each other, a force is required to move or slide one of them over the other. This is because there is usually another kind of force, which acts to oppose or prevent the movement of one body over the other. In the process of the surfaces of two bodies rubbing against each other, energy conversion usually takes place as for the case of the brake pads of roller coaster car and track. The friction acting between them changes kinetic energy into thermal or heat energy.

A roller coaster ride as shown in Fig. 5.9 is a thrilling experience which involves the physics of work and energy. The ride often begins as a chain and motor (or other mechanical device) exerts a force on the train of cars to lift the train to the top of a very tall hill. Once the cars are lifted to the top of the hill, gravity takes over and the remainder of the ride is an experience in energy transformation. At the top of the hill, the cars possess a large quantity of potential energy. The car's large quantity of potential energy is due to the fact that they are elevated to a large height above the ground. As the cars descend the first drop they lose much of this potential energy in accord with their loss of height. The cars subsequently gain kinetic energy. The train of coaster cars speeds up as they lose height. Thus, their original potential energy (due to their large height) is transformed into kinetic energy (revealed by their high speeds). As the ride continues, the train of cars is continuously losing and gaining height. Each gain in height corresponds to the loss of speed as kinetic energy (due to speed) is transformed into potential energy (due to height). Each loss in height corresponds to a gain of speed as potential energy (due to height) is transformed into kinetic energy (due to speed). This idealized description implies that a roller coaster could consist of a series of equally tall hills, and the cars would continue forever. That is obviously not the case in the real world.

Based on intuition, we know cars that are moving along a track will eventually slow down and stop, without any additional forces applied to them. As the ride continues, we may notice that the subsequent gain in height followed by loss in height becomes less and less. This decrease in the height is due to dissipative forces acting on the car. This is due to frictional dissipation, in which the cars' motion along the track is converted to mechanically useless forms of energy, specifically heat and sound.

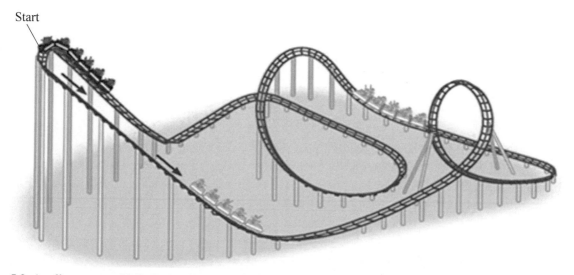

Figure 5.9: A roller coaster ride is designed to give people an entertaining way to experience the principle of conservation of energy, exploiting the interplay between potential and kinetic energy. Eventually, the car slows down and stops due to frictional dissipation.

Also, riders would probably be as familiar with the controlled friction which is demonstrated by the brakes at the end of the ride and is used at the will of the ride's operator. Controlled friction is still a resistance force between two objects. Even though this process represents a loss of energy possessed by the cars, it is not a violation of the principle of conservation of energy. The energy simply moves from the cars to the surrounding environment. Frictional dissipation plays a prominent role in many real world processes, as it does with real roller coasters.

5.9 POWER

As we discussed above, energy can be transferred, dissipated, or transformed in physical processes. For some processes, this may happen quickly, while for other processes, this may happen slowly, even though the same amount of energy may be involved. We distinguish these differences by using the quantity called power. Power is defined as the rate at which energy is transferred, used, or transformed. The unit of power is the joule per second (J/s), known as the watt (i.e., W = J/s) in honor of James Watt (1736-1819) who developed the steam engine in the eighteenth-century. For example, the rate at which a light bulb transforms electrical energy into heat and light is measured in watts. Higher value of wattage means higher power or equivalently larger amount of electrical energy is used per unit time. Also, power is defined as the rate at which work is done (i.e., work done divided by the time to do it) or the rate at which energy is transformed,

$$\bar{P} = \text{average power} = \frac{\text{work}}{\text{time}}. \tag{5.28}$$

Energy transfer can be used to do work, so power is also the rate at which this work is performed. The same amount of work is done when carrying a load up a flight of stairs whether the person carrying it walks or runs, but more power is expended during the running because the work is done in a shorter amount of time. The power expended to move a vehicle is the product of the traction force of the wheels and the velocity of the vehicle. So, we should keep in mind that power is not a measure of the amount of energy, but it is a measure of how quickly energy is transferred.

We may grasp the concept of power by considering the power rating of an automobile engine which is described in terms of horsepower. **Horsepower (hp)** is one of the units of measurement of power, the rate at which work is done. We know that an automobile with a larger horsepower engine has the ability to travel the same distance in shorter time than an identical automobile with a smaller horsepower engine. The most common conversion factor, especially for electrical power, is

$$\text{one horsepower} = 746 W \tag{5.29}$$

Sometimes it is convenient to express power in terms of the net force F applied to an object and average speed $\bar{v}$. This definition of power is readily obtained by using Eq. (5.28) and by noting that the work done over a distance d is $W = Fd$. We may rewrite the expression for average power as

$$\bar{P} = \frac{W}{t} = \frac{Fd}{t} = F\bar{v} \tag{5.30}$$

where $\bar{v} = d/t$. Equation (5.30) indicates that the required power increases with the speed and applied force. It should be noted that the power needed to push an object through a fluid increases as the cube of the velocity. A car cruising on a highway at 50 mph (80 km/h) may require only 10 hp (7.5 kW) to overcome air drag, but that same car at 100 mph (160 km/h) requires 80 hp (60 kW). With a doubling

WORK AND ENERGY

of speed, the drag (force) quadruples. Exerting four times the force over a fixed distance produces four times as much work. At twice the speed the work (resulting in displacement over a fixed distance) is done twice as fast. Since power is the rate of doing work, four times the work done in half the time requires eight times the power.

Understand-the-Concept Question 5.6:

The Nova laser at Lawrence Livermore National Laboratory has ten beams, each of which has power output greater than that of all the power plant in the U.S.A. Where does this power come from?

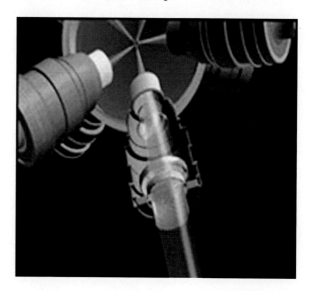

Explanation:

The Nova laser is the largest and most powerful laser in the world. It is located at the Lawrence Livermore National Laboratory in California. The Nova laser is about as long as a football field and is three stories high. The power of the Nova laser is approximately 1.0×10^{14} W (100 terawatts). The main purpose of a laser this powerful is to reproduce the conditions in the center of the sun. The energy produced by the sun results from the process of fusion. The Nova laser is used to fuse hydrogen atoms and produce helium. During this process, enormous amounts of energy are released. For the hydrogen atoms to fuse, they have to be under tremendous pressure and temperature (20 million Kelvin). The amount of energy released from fusion is about 1 million times greater than from any chemical reaction.

We need to remember that the definition of power is the energy transferred per unit time. This means that if the same amount of energy is transferred in a shorter time, then the power will go up. So, to generate the power needed, the 422 MJ of electrical energy is stored in a capacitor bank at a nearby facility and is released in a very short period of time.

The use of the Nova laser to create a fusion reaction is to decrease pollution in generating energy. This is because fusion energy is relatively clean and safe. The Nova laser was a high-power laser built at the Lawrence Livermore National Laboratory in 1984 and was used to conduct advanced inertial confinement fusion (ICF) experiments until its dismantling in 1999. Nova was the first ICF experiment built with the intention of reaching "ignition", a chain reaction of nuclear fusion that releases a large amount of energy. Although Nova failed in this goal, the data it generated clearly defined the problem as being mostly a result of magneto-hydrodynamic instability, leading to the design of the National

Ignition Facility (NIF), Nova's successor. Nova also generated considerable amounts of data on high-density matter physics, regardless of the lack of ignition, which is useful both in fusion power and nuclear weapons research. NIF aims to create a single 5.0×10^{14} peak flash of light that reaches the target from numerous directions at the same time, within a few picoseconds. The design uses 192 beam lines in a parallel system of flashlamp-pumped, neodymium-doped phosphate glass lasers.

Another very powerful laser may be found in the fictional world of the Star Wars movies. In the movie, the power of the first Death Star's super-laser was estimated to have been more than 2.4×10^{32} watts, with an optimum range of 2.0×10^{6} km and a working range of 4.2×10^{8} km. Note that the distance between the Sun and Earth is approximately 1.5×10^{8} km. The laser is powerful enough to destroy a terrestrial planet, and it was the most powerful energy weapon ever built at the time (before the second Death Star shown in Fig. 5.10). The laser is powered by the colossal hyper-matter reactor, and 64 tributary shafts of the Death Star can generate eight beams which can be united to form the primary beam over the central lens of the weapon. These tributary beams were arranged around an invisible, central focusing field, firing in an alternate sequence to build the power necessary to destroy a planet.

In the Star Wars movie, several types of laser, delivering different power, were introduced. A super-laser was a planet-destroying super-weapon that used several separate "super laser beams" combined together as a single beam. This devastating device was designed to destroy planets with one shot at full power. Also, there was smaller-scale super-lasers of lesser power designed for use against capital ships. Yet another small-scale super-laser was a type used by the Geonosians in their droid factories to melt iron ore. The super-laser was composed of several exotic matter beams accelerated and amplified by gigantic focusing magnetic lenses and coils, producing a single powerful beam. Unlike turbo-lasers, it pulled energy from a massive hyper-matter core, converting the energy present in hyperspace into highly unstable particles that were tremendously destructive in normal space. The energy delivered into a target was so great that it could cause the target's atoms to split into matter-antimatter pairs and annihilate themselves, creating hundreds of miniature singularities while generating a powerful surge capable of rupturing the barrier between normal space and hyperspace. As any powered weapon, it drained the amount of energy required to use, resulting in different energy outputs. For example, the Death Star's super-laser could destroy any capital ship with less than 10% power, crack a planet open

Figure 5.10: The power of the Death Star's superlaser was estimated to have been more than 2.4×10^{32} W, with an optimum range of 2.0×10^{6} km and a working range of 4.2×10^{8} km. Powerful enough to destroy a terrestrial planet, it was the most powerful energy weapon ever built at the time.

WORK AND ENERGY

to the core with two 30% shots and blow it to pieces with a full-charged fire. In such cases, the amount of energy involved was so great it would generate minuscule space-time singularities, sending part of the target's mass into hyperspace and generating a massive shockwave from high-energy tachyons from hyperspace. Due to its focusing lenses and beam intensity, the energy wasted was extremely low, leading to a very long shooting range, up to 2.0×10^6 km (over six light-seconds away). At this moment, the lasers in the universe of Star Wars movie are indeed very fictional in comparison to the most powerful laser built by the humans.

Exercise Problem 5.5:

A pump is to lift 8.00 kg of water per minute through a height of 3.50 m. What output rating (Watts) should the pump motor have?

PROBLEM SOLVING STRATEGY

What do we know?

A pump lifts certain amount of water per minute through a height h.

What concepts are needed?

We need to use the definition of power.

Why?

Power is defined as (work) energy per unit time. Work energy is needed to pump water through a height h.

Solution:

We use the definition of power and write

$$\overline{P} = \frac{W}{t} = \frac{Fd}{t} = \frac{8.0 \text{kg} \times 9.80 \frac{\text{m}}{\text{s}^2} \times 3.50 \text{m}}{1 \text{min} \times \frac{60 \text{s}}{\text{min}}} = 4.57 \text{ W}$$

A pump is a device that moves fluids (liquids or gases) by mechanical action. Pumps can be classified into three major groups according to the method they use to move the fluid: direct lift, displacement, and gravity pumps. Pumps operate by some mechanism (typically reciprocating or rotary), and consume energy to perform mechanical work by moving the fluid. Hence, the capacity of a pump is rated in units of horsepower. Also, pumps operate via many energy sources, including manual operation, electricity, engines, or wind power.

Similar to a pump which moves fluids, the human heart moves blood in a person's body. Hence, the human heart is a pump that is made of muscle tissue. It has four chambers: the right atrium and the left atrium, which are located at the top, and the right ventricle and left ventricle, which are located at the bottom. A special group of cells called the sinus node is located in the right atrium.

The sinus node generates electrical stimuli that make the heart contract and pump out blood. Each contraction represents a heartbeat. When the heart contracts it is in a systolic phase and when it rests it is in a diastolic phase. It takes blood about a minute to circulate through the cardiovascular system and pump oxygenated blood throughout the body.

The power of the heart can be calculated by multiplying the pressure by the flow rate. An average person has six liters of blood that circulates every minute, making the flow rate 10^{-4} m³/s (cubic meters per second). The pressure of the heart is about 10^4 N/m², making the heart's power about one watt. This is the power of a typical human heart, but it's different for everyone. The average heart beats about 75 times per minute, which is about five liters of blood per minute. Although this is not much, it enables the heart to complete a tremendous amount of work in a person's lifetime. The human heart beats about 40 million times a year, which adds up to more than 2.5 billion times in a 70-year lifetime. This results in approximately 2 to 3 billion joules of work in a lifetime.

SUMMARY

i. The concepts of work and energy are closely tied to the concept of force. **Energy** is defined as the ability to do work. The "work" done on an object by a constant force (in both magnitude and direction) is defined as

$$W = \vec{F} \cdot \vec{d} = F_{parallel}d = (F \cos\theta)d$$

The unit for describing work is the Joule (J) which equals N·m.

ii. The work-energy principle states that the work done by all forces acting on a particle equals the change in the kinetic energy of the particle. Suppose we define the translational kinetic energy (K.E.) as K.E. = $mv^2/2$. Then the work done, W, as

$$W = K.E._2 - K.E._1 = \Delta(K.E.)$$

The net work done on an object is due to both **conservative and non-conservative** forces.

iii. When only the conservative forces are acting on the system, the sum of kinetic energy and potential energy at point 2 is the same as the sum of these energies at point 1

$$E_2 = K.E._2 + P.E._2 = E_1 = K.E._1 + P.E._1 = \text{constant}$$

This is the principle of **conservation of mechanical energy**. The energy conservation states that if only conservative forces are acting on a system, then the total mechanical energy of the system neither increases nor decreases in any process.

iv. The potential energy is defined only for the conservative forces. The **gravitational potential energy** of a body at its height y above some reference level may be defined as

$$P.E._{elastic} = kx^2/2$$

Also, the elastic **potential energy** (for the spring force $F_s = -kx$) may be defined as

$$P.E._{elastic} = kx^2/2$$

WORK AND ENERGY

The elastic potential energy is the potential mechanical energy stored in the configuration of a material or physical system as work is performed to distort its volume or shape.

v. When the non-conservative forces are also acting on a system, the work done by these non-conservative forces changes in the total mechanical energy of the system. The conservation of energy principle may be rewritten as

$$K.E._1 + P.E._1 = K.E._2 + P.E._2 + F_f d$$

where $F_f d$ represents the work done by the non-conservative forces. The total energy of the system is neither increased nor decreased, but it is transformed in to other forms of energy, such as heat, sound, light, and non-recoverable deformation of the system.

vi. The rate at which work is done (i.e., work done divided by the time to do it) or the rate at which energy is transformed is describe in terms of power which is defined as

$$\bar{P} = \text{average power} = \frac{\text{work}}{\text{time}}$$

The unit for measuring power is watt (W) which is joule per second (i.e., W = J/s).

MORE WORKED PROBLEMS

Problems 5.1:

A pitcher throws a 0.140-kg baseball, and it approaches the bat at a speed of 40.0 m/s. The bat does $W_{NC} = 70.0$ J of work on the ball in hitting it. Ignoring air resistance, determine the speed of the ball after that ball leaves the bat and is 25.0 m above the point of impact.

Solution:

The force exerted by the bat on the ball is the only non-conservative force acting on it. The work done due to this force is

$$W_{NC} = \Delta K.E. + \Delta P.E. = \frac{1}{2}mv_f^2 - \frac{1}{2}mv_i^2 + mg(h_f - h_i)$$

If we take $h_i = 0$ m at the level of the bat, $v_i = 40.0$ m/s just before the bat strikes the ball and v_f to be the speed of the ball and $h_f = 25.0$ m, then we obtain v_f as

$$v_f = \sqrt{\frac{2W_{NC}}{m} + v_i^2 - 2gh_f} = \sqrt{\frac{2 \times 70.0 \text{ J}}{0.140 \text{ kg}} + \left(40.0 \frac{m}{s}\right)^2 - 2 \times 9.80 \frac{m}{s^2} \times 25.0 \text{ m}} = 45.9 \frac{m}{s}$$

Problems 5.2:

A fireman of mass 80 kg slides down a pole. When he reaches the bottom, 4.2 m below his starting point, his speed is 2.2 m/s. By how much has thermal energy increased during his slide?

Solution:

Using ground level as the reference for gravitational potential energy, the conservation of energy equation becomes

$$\Delta K.E. + \Delta P.E. + \Delta E_{thermal} = 0 \Rightarrow \Delta E_{thermal} = -(\Delta K.E. + \Delta P.E.)$$

The change in this gravitational potential energy is

$$\Delta P.E. = \Delta U_g = mgy_f - mgy_i = 0 - 80\,\text{kg} \times 9.80\,\frac{\text{m}}{\text{s}^2} \times 4.2\,\text{m} = 3.3\,\text{kJ}$$

The change in his kinetic energy is

$$\Delta K.E. = K.E._f - K.E._i = 0\,\text{J} - \frac{1}{2} \times 80\,\text{kg} \times \left(2.2\,\frac{\text{m}}{\text{s}}\right)^2 = 0.19\,\text{kJ}$$

So, the increase in the thermal energy is

$$\Delta E_{thermal} = -(190\,\text{J} - 3300\,\text{J}) = 3.1\,\text{kJ}$$

Problems 5.3:

A 3250-kg aircraft takes 12.5 min to achieve its cruising altitude of 10.0 km and cruising speed of 850 km/h. If the plane's engines deliver, on average, 1500 hp of power during this time, what is the efficiency of the engines?

Solution:

Before we get started on this problem, we write the speed in the units of km/h in terms of the units of m/s as

$$850\,\frac{\text{km}}{\text{h}} = 850\,\frac{\text{km}}{\text{h}} \times \frac{1000\,\text{m}}{1\,\text{km}} \times \frac{1\,\text{h}}{3600\,\text{s}} = 236.1\,\frac{\text{m}}{\text{s}}$$

The work required is

$$W = \Delta E = \Delta K.E. + \Delta P.E. = \frac{1}{2}mv^2 + mgh$$

$$= \frac{1}{2} \times 3.25 \times 10^3\,\text{kg} \times \left(236.1\,\frac{\text{m}}{\text{s}}\right)^2 + 3.25 \times 10^3\,\text{kg} \times 9.80\,\frac{\text{m}}{\text{s}^2} \times 10.0 \times 10^3\,\text{m} = 4.091 \times 10^8\,\text{J}$$

WORK AND ENERGY

Noting that the definition of power is $P = E/t$, the engine's output energy is

$$E = Pt = \left(1500\,\text{hp} \times 746\,\frac{\text{W}}{\text{hp}}\right) \times \left(12.5\,\text{min} \times \frac{60\,\text{s}}{1\,\text{min}}\right) = 8.393 \times 10^8\,\text{J}$$

According to the definition of efficiency e, we write

$$e = \frac{E_{output}}{E_{input}} = \frac{W}{E} = \frac{4.091 \times 10^8\,\text{J}}{8.393 \times 10^8\,\text{J}} = 0.487 \Rightarrow e = 48.7\%$$

Problems 5.4:

A 1.50×10^3-kg car starts from rest and accelerates uniformly to 18.0 m/s in 12.0 s. Assume that air resistance remains constant at 400 N during this time. Find (a) the average power developed by this time. Find (a) the average power developed by this time. (b) the instantaneous power output of the engine at $t = 12.0$ s, just before the car stops accelerating.

Solution:

a. The acceleration of the car may be computed as

$$v = v_o + at \Rightarrow a = \frac{v - v_o}{t} = \frac{18.0\,\text{m}/\text{s} - 0}{12.0\,\text{s}} = 1.50\,\frac{\text{m}}{\text{s}^2}$$

Thus, the constant forward force due to the engine is found from

$$\sum_i F_i = ma \Rightarrow F_{engine} - F_{air} = ma$$

$$F_{engine} = F_{air} + ma = 400\,\text{N} + 1.50 \times 10^3\,\text{kg} \times 1.5\,\frac{\text{m}}{\text{s}^2} = 2.65 \times 10^3\,\text{N}$$

The average velocity of the car during this interval is

$$v_{av} = \frac{v + v_o}{2} = 9.0\,\frac{\text{m}}{\text{s}}$$

So, the average power input from the engine during this time is

$$P_{av} = F_{engine} v_{av} = 2.65 \times 10^3\,\text{N} \times 9.0\,\frac{\text{m}}{\text{s}} = 39 \times 10^4\,\text{W} \times \frac{1\,\text{hp}}{746\,\text{W}} = 32.0\,\text{hp}$$

b. At $t = 12.0$ s, the instantaneous velocity of the car is $v = 18.0$ m/s and the instantaneous power input from the engine during this time is

$$P = F_{engine} v = 2.65 \times 10^3\,\text{N} \times 18.0\,\frac{\text{m}}{\text{s}} = 4.77 \times 10^4\,\text{W} \times \frac{1\,\text{hp}}{746\,\text{W}} = 63.9\,\text{hp}$$

Problems

PROBLEMS

1. A person pulls a toboggan for a distance of 35.0 m along the snow with a rope directed 25.0° above the snow. The tension in the rope is 94.0 N. (a) How much work is done on the toboggan by the tension force? (b) How much work is done if the same tension is directed parallel to the snow?

2. A 50-kg sprinter, starting from rest, runs 50 m in 7.0s at constant acceleration. (a) What is the magnitude of the horizontal force acting on the sprinter? (b) What is the sprinter's average power output during the first 2.0 s of his run? (c) What is the sprinter's average power output during the final 2.0 s?

3. A 7.8-g bullet moving at 575 m/s penetrates a tree trunk to a depth of 5.50 cm. (a) Use work and energy considerations to find the average frictional force that stops the bullet. (b) Assuming the fractional force is constant, determine how much time elapses between the moment the bullet enters the tree and the moment it stops moving.

4. A husband and wife take turns pulling their child in a wagon along a horizontal sidewalk. Each exerts a constant force and pulls the wagon through the same displacement. They do the same amount of work, but the husband's pulling force is directed 58° above the horizontal, and the wife's pulling force is directed 38° above the horizontal. The husband pulls with a force whose magnitude is 67 N. What is the magnitude of the pulling force exerted by his wife?

5. As a sailboat sails 52 m due north, a breeze exerts a constant force $\mathbf{F}_1$ on the boat's sails. This force is directed at an angle west of due north. A force $\mathbf{F}_2$ of the same magnitude directed due north would do the same amount of work on the sailboat over a distance of just 47 m. What is the angle between the direction of the force $\mathbf{F}_1$ and due north?

6. A 1200-kg car is being driven up a 5.0° hill. The frictional force is directed opposite to the motion of the car and has a magnitude of $f = 524$ N. A force $\mathbf{F}$ is applied to the car by the road and propels the car forward. In addition to these two forces, two other forces act on the car; its weight $\mathbf{W}$ and the normal force $\mathbf{F}_N$ directed perpendicular to the road surface. The length of the road up the hill is 290 m. What should be the magnitude of $\mathbf{F}$, so that the net work done by all the forces acting on the car is +150 kJ?

7. The force acting on an object is given by $F_x = (8x - 16)$ N, where x is in meters. (a) Make a plot of this force versus x from $x = 0$ to $x = 3.00$ m. (b) From your graph, find the net work done by the force as the object moves from $x = 0$ to $x = 3.00$ m.

8. (a) At the airport, you ride a "moving sidewalk" that carries you horizontally for 25 m at 0.70 m/s. Assuming that you were moving at 0.70 m/s before stepping onto the moving sidewalk and continue at 0.70 m/s afterward, how much work does the moving sidewalk do on you? Your mass is 60 kg. (b) An escalator carries you from one level to the next in airport terminal. The upper level is 4.5 m above the lower level, and the length of the escalator is 7.0 m. How much work does the up escalator do on you when you ride it from the lower level to the upper level? (c) How much work does the down escalator do on you when you ride it from the upper level to the lower level?

9. A college student earning some summer money pushes a lawn mower on a level lawn with a constant force of 250 N at an angle of 30° downward from the horizontal. How far does the student push the mower in doing 1.44×10^3 J of work?

10. The hammer throw is a track-and-field event in which a 7.3 kg ball (the "hammer"), starting from rest, is whirled around in a circle several times and released. It then moves upward in a familiar curving path of projectile motion. In one throw, the hammer is given a speed of 29 m/s. For comparison, a .22 caliber bullet has a mass of 2.6 g and, starting from rest, exits the barrel of

WORK AND ENERGY

a gun with a speed of 410 m/s. Determine the work done to launch the motion of (a) hammer and (b) bullet.

11. An extreme skier, starting from rest, coasts down a mountain that makes an angle of 25.0° with the horizontal. The coefficient of kinetic friction between her skis and the snow is 0.200. She coasts for a distance of 10.4 m before coming to the edge of a cliff. Without slowing down, she skis off the cliff and lands downhill at a point whose vertical distance is 3.50 m below the edge. How fast is she going before she lands?

12. "Rocket man" has a propulsion unit strapped to his back. He starts from rest on the ground, fires the unit, and is propelled straight upward. At a height of 16 m, his speed is 5.0 m/s. His mass, including the propulsion unit, has the approximately constant value of 136 kg. Find the work done by the force generated by the propulsion unit.

13. A student could either pull or push, at an angle of 30° from the horizontal, a 50-kg crate on a horizontal surface, where the coefficient of kinetic friction between the crate and surface is 0.20. The crate is to be moved a horizontal distance of 15 m. (a) Compared with pushing, pulling requires the student to do (1) less, (2) the same, or (3) more works. (b) Calculate the minimum work required for both pulling and pushing.

14. The chin-up is one exercise that can be used to strengthen the biceps muscle. This muscle can exert a force of approximately 800 N as it contracts a distance of 7.5 cm in a 75-kg male. How much work can the biceps muscles (one in each arm) perform in a single contraction? Compare this amount of work with the energy required to lift a 75-kg person 40 cm in performing a chin-up. Do you think the biceps muscle is the only muscle involved in performing a chin-up?

15. A student has six textbooks, each with a thickness of 4.0 cm and a weight of 30 N. What is the minimum work student would have to do place all the books in a single vertical stack, starting with all the books on the surface of the table?

16. A running 62-kg cheetah has a top speed of 32 m/s. (a) What is the cheetah's maximum kinetic energy? (b) Find the cheetah's speed when its kinetic energy is one half of the value found in part (a).

17. The masses of the javelin, discuss, and shot are 0.80 kg, 2.0 kg, and 7.2 kg, respectively, and record throws in the corresponding track events are about 98 m, 74 m, and 23 m, respectively. Neglecting air resistance, (a) Calculate the minimum initial kinetic energies that would produce these throws, and (b) estimate the average force exerted on each object during the throw, assuming the force acts over a distance of 2.0 m. (c) Do your results suggest that air resistance is an important factor?

18. A cyclist approaches the bottom of a gradual hill at a speed of 11 m/s. The hill is 5.0 m high, and the cyclist estimates that she is going fast enough to coast up and over it without pedaling. Ignoring air resistance and friction, find the speed at which the cyclist crests the hill.

19. A skateboarder moving at 5.4 m/s along a horizontal section of a track that is slanted upward by 48° above the horizontal at its end, which is 0.40 m above the ground. When she leaves the track, she follows the characteristic path of projectile motion. Ignoring friction and air resistance, find the maximum height H to which she rises above the end of the track.

20. At a carnival, you can try to ring a bell by striking a target with a 9.00 kg hammer. In response, a 0.400 kg metal piece is sent upward toward the bell, which is 5.00 m above. Suppose that 25.0% of the hammer's kinetic energy is used to do the work of sending the metal piece upward. How fast must the hammer be moving when it strikes the target so that the bell just barely rings?

21. A particular spring has a force constant of 2.5×10^3 N/m. (a) How much work is done in stretching the relaxed spring by 6.0 cm? (b) How much more work is done in stretching the spring an additional 2.0 cm?

22. The elastic energy stored in your tendons can contribute to 35% of your energy needs when running. Sports scientists have studied the change in length of the knee exterior tendon in sprinters and non-athletes. They find (on average) that the sprinters' tendons stretch 41 mm, while non-athletes' stretch only 33 mm. The spring constant for the tendon is the same for both groups, 33 N/mm. What is the difference in maximum stored energy between the sprinters and non-athletes?

23. In a physics lab experiment, a spring clamped to the table shoots a 20 g ball horizontally. When the spring is compressed 20 cm, the ball travels horizontally 5.0 m and lands on the floor 1.5 m below the point at which it left the spring. What is the spring constant?

24. A daredevil wishes to bungee-jump from a hot-air balloon 65.0 m above a carnival midway. He will use a piece of uniform elastic cord tied to a harness around his body to stop his fall at a point 10.0 m above the ground. Model this body as a particle and the cord as having negligible mass and a tension force described by Hooke's force law. In a preliminary test, hanging at rest from a 5.00-m length of the cord, the jumper finds that his body weight stretches it by 1.50 m. He will drop from rest at the point where the top end of a longer section of the cord is attached to the stationary balloon. (a) What length of cord should he use? (b) What maximum acceleration will he experience?

25. In a physics lab experiment, a spring clamped to the table shoots a 20 g ball horizontally. When the spring is compressed 20 cm, the ball travels horizontally 5.0 m and lands on the floor 1.5 m below the point at which it left the spring. What is the spring constant?

26. A 5.00×10^2 kg hot-air balloon takes off from rest at the surface of the earth. The non-conservative wind and lift forces take the balloon up, doing $+9.70\times10^4$ J of work on the balloon in the process. At what height above the surface of the earth does the balloon have a speed of 8.00 m/s?

27. A skier coasts down a very smooth, 10-m-high slope. If the speed of the skier on the top of the slope is 5.0 m/s, what is his speed at the bottom of the slope?

28. The stopping distance of a vehicle is an important safety factor. Assuming that an applied breaking force is constant, show that a vehicle's stopping distance proportional to the square of its initial speed by using the work-energy theorem. If an automobile traveling at 45 km/h is brought to a stop in 50 m, what would be the stopping distance for an initial speed of 90 km/h?

29. If the work required to speed a car up from 10 km/h to 20 km/h is 5.0×10^3 J, what would be the work needed to increase the car's speed from 20 km/h to 30 km/h?

30. A sled is being pulled across a horizontal patch of snow. Friction is negligible. Both the force of pulling and the sled's displacement are in the same direction as, which is along the +x-axis. As a result, the kinetic energy of the sled increases by 38%. By what percentage would the sled's kinetic energy have increased if this force had pointed 62° above the +x axis?

31. Sam's job at the amusement park is to slow down and bring to a stop the boats in the log ride. If a boat and its riders have a mass of 1200 kg and the boat drifts in at 1.2 m/s, how much work does Sam do to stop it?

32. In Exercise 27, the skier has a mass of 60 kg and the force of friction retards his motion by doing 2500 J of work, what is his speed at the bottom of the slope?

33. In the winter sport of curling, players give a 20 kg stone a push across a sheet of ice. A curler accelerates a stone to a speed of 3.0 m/s over a time of 2.0 s. (a) How much force does the curler exert on the stone? (b) What average power does the curler use to bring the stone up to speed?

WORK AND ENERGY

34. In 2.0 minutes, a ski lift raises four skiers at constant speed to a height of 140 m. The average mass of each skier is 65 kg. What is the average power provided by the tension in the cable pulling the lift?

35. A 2.0 hp (horsepower) electric motor on a water-well pumps water from 10 below the surface. The density of water is 1.0 kg per liter (L). How many liters of water can the motor pump in 1 hour?

36. A 3250-kg aircraft takes 12.5min to achieve its cruising altitude of 10.0 km and cruising speed of 850 km/h. If the plane's engines deliver, on average, 1500 hp of power during this time, what is the efficiency of the engines?

37. A 1900-kg car experiences a combined force of air resistance and friction that has the same magnitude whether the car goes up or down a hill at 27 m/s. Going up a hill, the car's engine produces 47 hp more power to sustain the constant velocity than it does going down the same hill. At what angle is the hill inclined above the horizontal?

38. A 650-kg elevator starts from rest and moves upward for 3.00 s with constant acceleration until it reaches its cruising speed, 1.75 m/s. (a) What is the average power of the elevator motor during this period? (b) How does this amount of power compare with its power during an upward trip with constant speed?

39. The elastic energy stored in your tendons can contribute to 35% of your energy needs when running. Sports scientists have studied the change in length of the knee exterior tendon in sprinters and non-athletes. They find (on average) that the sprinters' tendons stretch 41 mm, while non-athletes' stretch only 33 mm. The spring constant for the tendon is the same for both groups, 33 N/mm. What is the difference in maximum stored energy between the sprinters and non-athletes?

40. A 710 kg car drives at a constant speed of 23 m/s. It is subject to a drag force of 500 N. What power is required from the car's engine to drive the car (a) on level ground and (b) up a hill with a slope of 2.0°?

41. In the winter sport of curling, players give a 20 kg stone a push across a sheet of ice. A curler accelerates a stone to a speed of 3.0 m/s over a time of 2.0 s. (a) How much force does the curler exert on the stone? (b) What average power does the curler use to bring the stone up to speed?

MOMENTUM AND COLLISIONS

CHAPTER 6

In a baseball game, when a batter hits a ball using a baseball bat, he exerts a force on the ball. The bat and the baseball make contact for about two milliseconds. During this time, both the bat and baseball experience a force due to the collision. In this collision, as in all collisions, there is a force exerted on both objects. If the contact with the ball can be made longer, then the speed of the ball may be increased. In order to increase the velocity of the ball, the hitter should follow through when hitting the ball. The follow-through of the swing increases the time of collision, which contributes to an increase in the velocity of the ball. The forces experienced by the bat and baseball are equal in magnitude and are in opposite directions. Newton's third law of motion applies to this collision between the bat and the ball. As we may expect from the picture, the force of impact from the bat will change the direction of motion for the baseball.

Although the force causes an acceleration of both objects, the acceleration of the objects is not necessarily equal in magnitude. The baseball, which is the less massive object, receives the greater acceleration. Hence, the force causes the ball to speed up and the baseball bat to slow down. Since the mass of the bat is much greater, the change in velocity is much smaller for the bat than it is for the ball. When they see the baseball flies off from the bat, most observers may have difficulty with this concept because they think the drastic change in the speed of the ball is due to the collision. They may not observe imbalanced forces between the ball and baseball bat. When we account for the difference in the mass of the two objects we know that they are observing different accelerations when the baseball flies off to the outfield at a much faster rate.

We may examine the change in the speed of both the bat and baseball in terms of change in momentum. This change in momentum is called impulse. The impulse experienced by the baseball equals the change in its momentum. Hence, the force changes the momentum of the baseball, when the baseball (i.e., its mass) speeds up. The impulse of the baseball increases its momentum when the baseball bat makes contact with it. The bat and the ball, both experience an equal change in momentum though from the impact. So, obviously, this contact interaction revolves around the conservation of momentum. In this collision, the bat's momentum is reduced and it is transferred to the momentum of the baseball. This conservation law says that there is the same amount of momentum before the collision as there is after the collision.

To see if this is true, we would add up the bat's momentum and the ball's momentum before the contact. Then we would have to check and see if that equals the sum of the momentum of bat and the momentum of the ball after the contact. Verification of the fact that there is the same amount of momentum before the collision as there is after the collision is a proof of one of fundamental principle in physics: conservation of linear momentum. This principle is central to all problems involving collision. We need to remember that both contact and non-contact collision processes may be understood by using the conservation of momentum. Hence, we will discuss how this fundamental principle is used in solving collision problems.

6.1 LINEAR MOMENTUM

Before we discuss the conservation principle, we need to have a better grasp of the concept of momentum, which was introduced above. Momentum can be defined as inertia in motion. This means that something must be moving to have momentum. Momentum is how hard it is to get something to stop or to change directions. For example, a moving train has a great deal of momentum, while a moving ping pong ball does not. In other words, we can easily stop a ping pong ball, even at high speeds. However, it is difficult to stop a train even at low speeds. Our reason for this careful consideration of linear momentum is that along with the law of conservation of energy, the law of conservation of linear momentum is a *very useful principle* for analyzing collisions.

Mathematically, momentum is defined as the product of mass and its velocity. Hence, the **linear momentum p** is a vector and is written as

$$\vec{p} = m\vec{v}. \tag{6.1}$$

MOMENTUM AND COLLISION

Equation (6.1) indicates that, when an object is heavier and/or moving faster, it has more momentum. When the object has more momentum, greater force is needed to change the velocity. Also, the object will exert greater force when it hits something. According to Newton's second law of motion, when it is stated in terms of momentum, the rate of change of momentum of a body is equal to the net force applied to it

$$\sum_i \vec{F}_i = \frac{\Delta \vec{p}}{\Delta t}. \tag{6.2}$$

We already know that mass is a measure of the degree of inertia that an object possesses. Also, we intuitively understand that force is a type of push or pull. However, we do not really have independent definitions of them or procedures for measuring them. The second law of Eq. (6.2) relates them to each other. In so doing, the second law acts as a joint operational definition of them. Mathematically, Newton's second law says that force, F, is equal to the change in momentum, Δp, with the corresponding change in time, Δt. Hence, the net force is

$$\sum_i \vec{F}_i = \frac{\Delta \vec{p}}{\Delta t} = \frac{m\vec{v}_2 - m\vec{v}_1}{\Delta t} = m\frac{\Delta \vec{v}}{\Delta t} = m\vec{a}. \tag{6.3}$$

This brief mathematical statement contains quite a bit of information. As Eq. (6.3) shows that as long as the mass part of the momentum does not change, the change in momentum can be expressed as $\Delta p = \Delta(mv) = m(\Delta v)$. Because we define acceleration, a, as the change in velocity with time $a = \Delta v/\Delta t$, Newton's second law can also be written as $F = ma$. This is a more familiar expression to us and to anyone who had a physics course before. It is interesting to note that since the acceleration a is related to Δv, and Δv is not zero if the direction of v changes even when its size stays the same. An object accelerates if its direction of travel changes while it is moving at constant speed. Newton's second law can also be understood to describe the action of many forces acting on a single object, or many forces acting on many objects. Under these circumstances, the force F is understood to represent the overall force (i.e., net force), and the momentum change Δp to be over all of the objects involved (i.e., net change).

Understanding-the-Concept Question 6.1:

If you pitch a baseball with twice the kinetic energy you gave it in previous pitch, the magnitude of its momentum is

a. 4 times as much.   b. $2(2)^{1/2}$ times as much.   c. doubled.   d. $(2)^{1/2}$ times as much

Explanation:

We may rewrite the kinetic energy by using the definition of momentum and obtain

$$K.E. = mv^2 / 2 = (mv)^2 / 2m = p^2 / 2m.$$

This means that the kinetic energy is proportional to the square of momentum, indicating that the magnitude of momentum must increase $(2)^{1/2}$ times as much when K.E. is twice.

Answer: d

Exercise Problem 6.1:

Water leaves a hose at a rate of 1.5 kg/s with a speed of 20 m/s and is aimed at the side of a car, which stops it. (That is, we ignore any splashing back.) What is the force exerted by the water on the car?

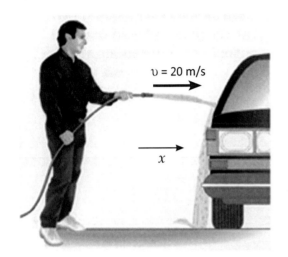

PROBLEM SOLVING STRATEGY

What do we know?
Water leaves a hose and the car stops it.

What concepts are needed?
We need to use Eq. (6.2) and the definition of momentum.

Why?
To stop water, force is needed. We need to compute the force to stop water by using the information about rate of water flow and its speed.

Solution:

We start with the definition of force F as the rate of change in momentum

$$F = \frac{\Delta p}{\Delta t} = \frac{p_2 - p_1}{\Delta t} = \frac{0 - mv_1}{\Delta t} = -\frac{m}{\Delta t} v_1 = -1.5 \frac{\text{kg}}{\text{s}} \times 20 \frac{\text{m}}{\text{s}} = -30 \, \text{N}.$$

This is the force that the car exerts to stop the water. Therefore, the water exerts a force of 30 N on the car.

Exercise Problem 6.2:

Calculate the force exerted on a rocket, given that the propelling gases are expelled at a rate of 1300 kg/s with a speed of 40,000 m/s (at the moment of takeoff.)

PROBLEM SOLVING STRATEGY

What do we know?
The force is exerted on the rocket by the propelling gases.

What concepts are needed?
We need to use Eq. (6.2) and the definition of momentum. We also need to apply Newton's third law

Why?
We may consider the rocket-fuel system as an isolated system. This means the magnitude of the rate of momentum change due to the propelling gases from the fuel equals the force exerted on the rocket.

MOMENTUM AND COLLISION

> **Solution:**
>
> The force (i.e., thrust) exerted on a rocket by the propelling gas is
>
> $$F = \frac{\Delta p}{\Delta t} = \frac{\Delta(mv)}{\Delta t} = v\frac{\Delta m}{\Delta t} = 40,000\frac{m}{s} \times 1,300\frac{kg}{s} = 5.2 \times 10^7 \, N.$$

Thrust is the force that propels a rocket or spacecraft and is measured in pound-force, kilogram, or Newton. Physically, thrust is the result of pressure which is exerted on the wall of the combustion chamber. Higher pressure can be achieved by high speed gas molecules. To create high speed exhaust gases, the high temperatures are necessary. Also, high pressures of combustion are obtained by using a very energetic fuel and by having the molecular weight of the exhaust gases as low as possible. It is also necessary to reduce the pressure of the gas as much as possible inside the nozzle by creating a large section ratio. Rocket thrust results from the high speed ejection of material and does not require any medium to "push against".

To obtain thrust necessary for a launch, the orbiter (i.e., space shuttle) uses three main engines located in the **aft (back) fuselage** (i.e., body of the spacecraft). Each engine is 14 feet (4.3 m) long, 7.5 feet (2. 3 m) in diameter at its widest point (the nozzle) and weighs about 6,700 pounds (3039 kg). The main engines provide the remainder of the thrust (29 percent) to lift the shuttle off the launch pad and into orbit. These engines burn a mixture of liquid hydrogen and oxygen. These fuels are stored in the external fuel tank at a 6:1 ratio. The engines draw liquid hydrogen and oxygen from the fuel tank at an amazing rate, equivalent to emptying a family swimming pool every 10 seconds! The fuel is partially burned in a pre-chamber to produce highly pressured hot gases to drive the turbo-pumps (i.e., fuel pumps). The fuel is then fully burned in the main combustion chamber. The exhaust gases, which are water vapors, leave the nozzle at approximately 6,000 mph (10,000 km/h). Each engine can generate between 375,000 and 470,000 pound (1,668,083 to 2,090,664 N) of thrust. The rate of thrust can be controlled from 65% to 109% maximum thrust. The engines are mounted on gimbals (round bearings) so that it is easier to control the direction of the exhaust, which in turn controls the forward direction of the rocket. Since we can consider the fuel and orbiter as an isolated system, according to Newton's third law, the momentum of the system is conserved. Conservation of momentum dictates that if material is ejected backward, the forward momentum of the remaining rocket must increase since an isolated system cannot change its net momentum.

6.2 LAW OF CONSERVATION OF MOMENTUM

Now, let us see how the conservation of linear momentum for an isolated system can be derived from Newton's third law. This principle states that the total momentum of an isolated system of bodies remains the same, indicating that total momentum before collision is equal to momentum after collision. Here, **system** is defined as a set of objects that interact with each other. However, an **isolated system** is defined as a system in which only forces present are those between the objects of the system. If the net force $\mathbf{F}_{net}$ acting on the system is zero, then the total momentum $\mathbf{P}$ of the system does not change.

If a system is isolated, then the system is subjected to **internal forces**. Here, **internal forces** are defined as forces that act only between particles within the system as shown in Fig. 6.1. Here, each

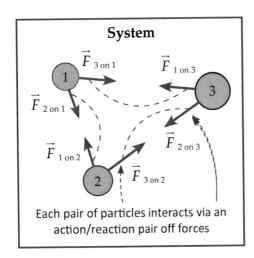
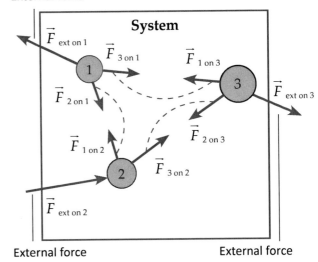

Figure 6.1: A schematic illustration of a system containing three objects interacting with each other internally. In this case, the sum of the forces is zero since there are no external forces acting on the system.

Figure 6.2: A schematic illustration of a system of three objects interacting with each other via internal forces as well as external forces acting on the objects. In this case, the sum of the external forces is non-zero.

pair of particles interacts via an action/reaction pair of forces. In addition to internal forces, other forces can also act on the system as shown in Fig. 6.2. Here, these other forces arise as a result of interactions between the system and surroundings. Hence, these force are called **external forces** and are referred as the forces that act on a system but from outside of the system. The net force F_{net} is due to **external forces**, and an isolated system is one for which $F_{net} = 0$, by definition.

Let us consider an isolated system (i.e., $F_{net} = 0$). For an isolated system (one that does not exchange any matter with the outside and is not acted on by outside forces), we will show that the total momentum of the system remains constant under any process. This important fact, known as the **law of conservation of momentum**, is implied by Newton's third laws of motion. For a system of two particles, the conservation of momentum may be written as

$$m_1\vec{v}_1 + m_2\vec{v}_2 = m_1\vec{v}_1' + m_2\vec{v}_2'. \tag{6.4}$$

Equation (6.4) states that total momentum of the system before collision $P_i = m_1 v_1 + m_2 v_2$ is equal to the total momentum of the system after collision $P_f = m_1 v'_1 + m_2 v'_2$. Here, v_1 and v_2 are the velocity of particle 1 and 2 before the collision, respectively. Also, v'_1 and v'_2 are the velocity of particle 1 and 2 after the collision, respectively. This law (i.e., $P_i = P_f$) holds no matter how complicated the force is between the particles. Similarly, if there are several particles, the momentum exchanged between each pair of particles adds up to zero, so the total change in momentum is zero. This conservation law applies to all interactions, including collisions and separations caused by explosive forces.

Let us derive Eq. (6.4) by considering a collision between two objects as shown in Fig. 6.3. Here, a collision is a short-duration interaction between two bodies simultaneously causing change in motion of bodies involved due to internal forces acted between them. Collisions involve forces and lead to a change in velocity. The magnitude of the velocity difference at impact is called the closing speed. However, we will not be concerned with what type of interaction occurs between the two particles. We will simply assume that there is a change in velocity of the two particles as a result of a collision between them. We will consider the collision as a single isolated event.

MOMENTUM AND COLLISION

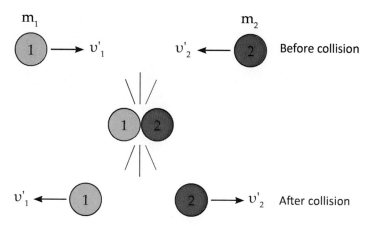

Figure 6.3: A schematic illustration of a collision between two objects. The total momentum of the system before collision is the same as the total momentum of the system after collision, indicating conservation of momentum.

A **collision** is defined as an isolated event in which two or more moving bodies (colliding bodies) exert forces on each other for a relatively short time. The change in momentum (i.e., impulse) due to the collision is written as

$$\vec{F} = \frac{\Delta \vec{p}}{\Delta t} \Rightarrow \Delta \vec{p} = \vec{F} \Delta t. \tag{6.5}$$

A fast-acting force F produces an acceleration, and the greater the force acting on an object, the greater its change in velocity and, hence, the greater its change in momentum. However, changing momentum is also related to how long a time Δt the force acts. This type of impulse is often *idealized* so that the change in momentum produced by the force happens with almost no change in time. This sort of change is a step change and is not physically possible, but this is a useful model for computing the effects of ideal collisions. For particle 1, the change in momentum due to the collision is

$$\Delta \vec{p}_1 = m_1 \vec{v}_1' - m_1 \vec{v}_1 = \vec{F}_{12} \Delta t,$$

where F_{12} is the force exerted on particle 1 by particle 2. Similarly, for particle 2, the change in momentum is

$$\Delta \vec{p}_2 = m_2 \vec{v}_2' - m_2 \vec{v}_2 = \vec{F}_{21} \Delta t,$$

where F_{21} is the force exerted on particle 2 by particle 1. As a consequence of the forces being equal and opposite at each moment, we know that the impulses of the two forces were always equal in magnitude but opposite in direction. This observation, along with the impulse-momentum theorem, is the basis for the derivation of the conservation of momentum law. The impulse-momentum theorem is really equivalent to Newton's second law since it can be derived mathematically from the second law. According to Newton's third law, $F_{12} = -F_{21}$, the force F_{12} is the reaction of the force F_{21}, yielding

$$m_1 \vec{v}_1' - m_1 \vec{v}_1 = \vec{F}_{12} \Delta t = -\vec{F}_{21} \Delta t = -\left(m_2 \vec{v}_2' - m_2 \vec{v}_2 \right).$$

We now collect the terms which contribute to the total momentum before and after the collision, respectively, on the right and left hand side of the equation to obtain

$$m_1 \vec{v}_1' + m_2 \vec{v}_2' = m_1 \vec{v}_1 + m_2 \vec{v}_2. \tag{6.6}$$

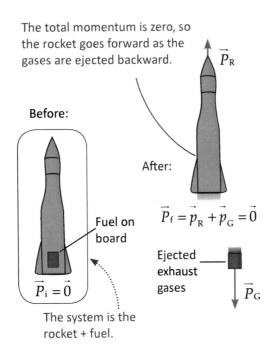

Figure 6.4: A schematic illustration of the law of momentum conservation applied to a fuel-rocket system. Since the total momentum of the system before the launch is zero, the sum of the momentum of the rocket and that of the ejected exhaust gases must also be zero.

Equation (6.6) indicates that the total momentum of the system is unchanged. Note that **if there are external forces, then the total moment is not conserved.** Hence, the total momentum of the system is conserved for collisions occurring only in isolated systems. As an example of an isolated system, we consider a "rocket plus fuel" system shown in Fig. 6.4. For simplicity, suppose that the rocket is at rest, somewhere in space, where no outside forces will exert impulses on our system. Therefore, "rocket plus fuel" is an isolated system and the law of conservation of momentum is applicable. Since both the rocket and the fuel are at rest, their individual momenta are zero, and so the total momentum of the system is zero. When the rocket engine is ignited, the engine ejects the fuel from the rocket. Suddenly, part of the fuel has momentum, mv, toward the bottom as shown in Fig. 6.4. Since the total momentum of the system was zero before the rocket was fired, it must still be zero after it was fired, according to the conservation of momentum principle. This means that the rocket (R) goes forward as the gasses (G) are ejected backward. The conservation of momentum principle yields

$$\vec{p}_R + \vec{p}_G = 0 \Rightarrow \vec{p}_R = -\vec{p}_G.$$

The negative sign means that these two momentum vectors are in opposite directions. Therefore, they cancel. Even though the momentum of the rocket and the momentum of the fuel are the same size, their velocities are not the same size. Since the fuel has a small mass m_G, it gets a larger velocity v_G. Similarly, since the rocket has a large mass M_R, it gets a smaller velocity v_R. These dependences on masses are described by the relation

$$M_R \vec{v}_R = -m_G \vec{v}_G$$

This is the thrust obtained from a rocket engine which exhaust is formed entirely from propellants carried within the rocket before use.

MOMENTUM AND COLLISION

Understand-the-Concept Question 6.2:

A Ping-Pong ball moving East at a speed of 4 m/s collides with a stationary bowling ball. The Ping-Pong ball bounces back to the West, and the bowling ball moves very slowly to the East. Which object experiences the greater magnitude impulse during the collision?

- a. neither, both experienced the same magnitude impulse
- b. it is impossible to tell since the velocities after the collision are unknown
- c. the bowling ball
- d. the Ping-Pong ball

Explanation:

Noting that impulse is defined as the change in momentum, we need to determine which object, Ping-Pong ball or bowling ball, experiences greater change in momentum in the collision. Since momentum is conserved in any collision, the total change in momentum must be zero. This means that the sum of the change in momentum of the Ping-Pong ball and bowling ball is zero, indicating that the magnitude of the impulse for both objects is the same but the direction is opposite to each other.

Answer: a

In summary, because of its inertia, any object with momentum is going to be hard to stop. To stop such an object, it is necessary to apply a force *against* its motion for a given period of time. An object with a large momentum is the harder to stop than one with a small momentum. Thus, the object would require a greater amount of force or a longer amount of time, or both, to bring such an object to a halt. As the force acts upon the object for a given amount of time, the object›s velocity is changed; and hence, the object›s momentum is changed. A force acting on the object for a given amount of time (an action known as impulse). will change the object›s momentum. In other words, an unbalanced force always accelerates an object. This force can either speed up the object or slow it down. If the force is exerted in the opposite direction to the object›s motion, it slows the object down. If a force is exerted in the same direction as the object›s motion, then the force speeds the object up. Either way, a force will change the velocity of an object. Also, if the velocity of the object is changed, then the momentum of the object is changed.

6.3 COLLISION BETWEEN TWO OBJECTS

Now with the understanding of the momentum conservation principle, we are ready to discuss any problems involving collision. Since collision is the most common way that objects interact with each other, this conservation principle will help us solve a wealth of interesting physics problems. A **collision** is defined as an isolated event in which two or more moving bodies (colliding bodies) exert forces on each other for a relatively short time. Although the most common colloquial use of the word "collision" refers to accidents in which two or more objects collide, our use of the word "collision" will not deal with the magnitude of the forces.

Since a collision is defined as short-duration interaction between two or more bodies simultaneously causing change in motion of bodies, it involves forces during a short time, and there is a change in velocity. All collisions conserve momentum but may not conserve energy, we may describe a change in the speed of an object in a collision using the coefficient of restitution. **The coefficient of restitution (CoR)** for two colliding objects is a fractional value representing the ratio of speeds after and before an impact, taken along the line of the impact. For CoR = 1, collision between pairs of objects is elastic, while for CoR < 1 collision between objects is inelastic. For a CoR = 0, the objects effectively "stop" at the collision, not bouncing at all. However, collisions are separated into two different types, depending on whether the energy of the system is conserved or not. These two types of collision are schematically illustrated in Fig. 6.5.

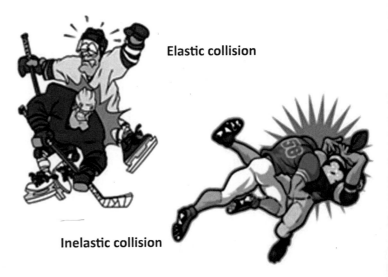

Figure 6.5: A schematic illustration of two types of collision in sports: elastic collision in an ice hockey game and inelastic collision in a football game.

If we compare the system immediately before the collision with that immediately after, the gravitational potential is not expected to change. So, what distinguishes different types of collisions is whether the system also conserves kinetic energy. Specifically, collisions can either be *elastic*, meaning they conserve both momentum and kinetic energy, or *inelastic*, meaning they conserve momentum but not kinetic energy. An inelastic collision is sometimes also called a *plastic collision*. A "perfectly-inelastic" collision is a limiting case of inelastic collision in which the two bodies stick together after impact.

A perfectly elastic collision is defined as one in which there is no loss of kinetic energy in the collision. In reality, any macroscopic collision between objects will convert some kinetic energy to internal energy and other forms of energy, so no large scale impacts are perfectly elastic. However, some problems are sufficiently close to perfectly elastic that they can be approximated as such. In this case, the coefficient of restitution equals to one. **Elastic collision** is defined as a collision in which two objects bounce apart. In this collision, **the total kinetic energy** is conserved

$$\frac{1}{2}m_1 v_1^2 + \frac{1}{2}m_2 v_2^2 = \frac{1}{2}m_1 v_1'^2 + \frac{1}{2}m_2 v_2'^2. \tag{6.7}$$

We do not include the potential energy in Eq. (6.7) because we compare the energy of the system at immediately before and after the collision and at the same position. So, the change in the gravitation potential energy is zero. Collisions in ideal gases approach perfectly elastic collisions, as do scattering interactions of sub-atomic particles which are deflected by the electromagnetic force. Some large-scale interactions like the slingshot type gravitational interactions between satellites and planets are perfectly elastic. Collisions between hard spheres may be nearly elastic, so it is useful to calculate the limiting case of an elastic collision. The assumption of conservation of momentum as well as the conservation of kinetic energy makes possible the calculation of the final velocities in two-body collisions.

On the other hand, in an inelastic collision, part of the kinetic energy is changed to some other form of energy in the collision. In this case, coefficient of restitution does not equal to one. **Inelastic collision** is defined as a collision in which two objects do not bounce apart. In this collision, **the total kinetic energy is not conserved.** However, when we account for the kinetic energy lost in the collision which transformed into other forms of energy, we would find the energy of the system is still conserved

$$\text{K.E.}_1 + \text{K.E.}_2 = \text{K.E.}'_1 + \text{K.E.}'_2 + \text{thermal and other forms of energy.} \tag{6.8}$$

However, keeping a track of other forms of energy may not be easy. Let us think about what happens to the kinetic energy of a car involved in a collision which is dissipated in 10 to 20 milliseconds. A lot of the kinetic energy goes into deforming parts of the vehicle structure and whatever it hits. Also, heat is generated, especially in the brakes if they were applied prior to the impact. Typically, we would hear a loud sound in the collision, but sound energy is quite a small part of the overall energy. Some of the kinetic energy may also have been dissipated in wear of the tires as the result of skidding.

Understanding-the-Concept Question 6.3:

Can this happen?

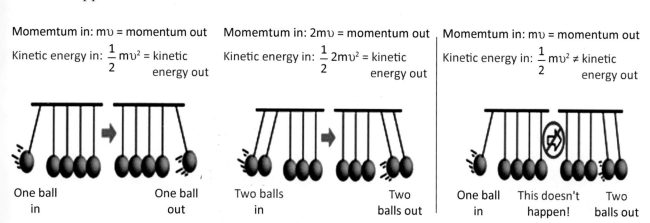

Explanation:

If one ball is pulled away from a Newton's cradle 5-ball system and is allowed to fall, then it strikes the first ball in the series and comes to nearly a dead stop. The ball on the opposite side acquires most of the velocity and almost instantly swings in an arc almost as high as the release height of the last ball. This shows that the final ball receives most of the energy and momentum that was in the first ball. Similarly, if two balls are pulled away and are allowed to fall; they strike the first ball in the series and come to nearly a dead stop.

Again, the two balls on the opposite side acquire most of the velocity and almost instantly swing in an arc almost as high as the release height of the first two balls. This shows that the final two balls receive most of the energy and momentum that was in the first two balls. The impact produces a shock wave that propagates through the intermediate balls. Any efficiently elastic material such as steel will do this as long as the kinetic energy is temporarily stored as potential energy in the compression of the material rather than being lost as heat. In short, with two balls, exactly two balls on the opposite side swing out and back. For a 3-ball swing, the central ball swings without any apparent interruption.

Can more than half the balls be set in motion? The simple equations for the conservation of kinetic energy and conservation of momentum can be used to show this is *possible*, but they cannot be used to predict the final velocities when there are three or more balls in a cradle. The conservation principles only indicate that if one ball is pulled away and is allowed to fall then the two balls on the opposite side would not swing out because these balls cannot acquire the velocity to satisfy both kinetic energy and momentum conservation.

Newton's cradle is named after Sir Isaac Newton. This device demonstrates conservation of momentum and energy via a series of swinging spheres. When one on the end is lifted and released, the resulting force travels through the line and pushes the last one upward. The device is also known as **Newton's balls** or "Executive Ball Clicker".

Exercise Problem 6.3:

A softball of mass 0.220 kg that is moving with a speed of 5.5 m/s collides head-on and elastically with another ball initially at rest. Afterward it is found that the incoming ball has bounced backward with speed 3.7 m/s. Calculate (a) the velocity of the target ball after the collision, and (b) the mass of the target ball.

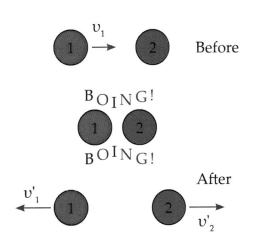

PROBLEM SOLVING STRATEGY

What do we know?

Two objects collide head-on and elastically.

What concepts are needed?

We need to apply the principle of conservation of momentum and kinetic energy.

Why?

When the objects collide elastically, both momentum and kinetic energy of the system are conserved.

Solution:

We need to apply the principle of conservation of momentum and kinetic energy to solve the problem. According to conservation of momentum, we have

$$m_1 v_1 = m_2 v_2' - m_1 v_1' \Rightarrow v_2' = \frac{m_1}{m_2}(v_1 + v_1'). \tag{3a}$$

According to conservation of energy, we have

$$\frac{1}{2}m_1 v_1^2 = \frac{1}{2}m_1 v_1'^2 + \frac{1}{2}m_2 v_2'^2 \Rightarrow v_2'^2 = \frac{m_1}{m_2}(v_1^2 - v_1'^2) \Rightarrow \frac{m_1}{m_2} = \frac{v_2'^2}{v_1^2 - v_1'^2}. \tag{3b}$$

First, we substitute Eq. (3b) into Eq. (3a) and solve for v_2' and obtain

$$\therefore v_2' = \frac{v_2'^2}{v_1^2 - v_1'^2}(v_1 + v_1') = \frac{v_2'^2}{v_1 - v_1'} \Rightarrow v_2' = v_1 - v_1' = 5.5\frac{m}{s} - 3.7\frac{m}{s} = 1.8\frac{m}{s}.$$

Using Eq. (3b), we now solve for mass m_2 and obtain

$$m_2 = m_1 \frac{v_1^2 - v_1'^2}{v_2'^2} = 0.220\,\text{kg} \times \frac{\left(5.5\,\text{m/s}\right)^2 - \left(3.7\,\text{m/s}\right)^2}{\left(1.8\,\text{m/s}\right)^2} = 1.12\,\text{kg}.$$

MOMENTUM AND COLLISION

Particle collisions or scattering (we will use the two terms interchangeably) is one of the most fundamental tools we use to study nature. For example, many elementary particles that do not occur under normal circumstances in nature can be created and detected during energetic collisions of other particles, as is done in high-energy particle accelerators. This is a branch of physics called **particle physics** and studies the elementary constituents of matter and radiation, and the interactions between them. Even when we are looking at an object, we are conducting a scattering experiment, observing how visible light photons are scattered off the object. Essentially all particle physics experiments involve studying the scattering of particles energized by accelerators or natural processes.

WHAT TO DO IN COLLISION PROBLEMS...

In solving any collision problems, first, we need to determine whether we have elastic or inelastic scattering. Also we need to apply the relevant conservation principle. In solving elastic collision problems in one dimension, we need to use both conservation of momentum and conservation of energy. In contrast to an elastic collision, an inelastic collision is a collision in which kinetic energy is not conserved. Hence, conservation of energy is not valid in inelastic collision, but conservation of momentum is still valid.

One extreme form of inelastic collision is completely inelastic collision. In **completely inelastic collision** (i.e., a zero coefficient of restitution) as shown in Fig. 6.6, objects stick together as a result of collision and move with a common final velocity. A perfectly inelastic collision occurs when the maximum amount of kinetic energy of a system is lost. In such a collision, kinetic energy is lost by bonding the two bodies together (i.e., they stick together). This bonding energy usually results in a maximum kinetic energy loss of the system. Note that the energy associated with two objects sticking together is not binding energy.

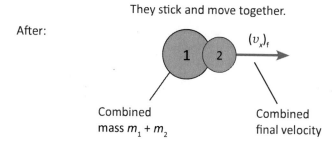

Figure 6.6: A schematic illustration of inelastic collision between two objects colliding head-on. When the collision is completely inelastic, two objects stick together and move as a single object after the collision.

Some examples of inelastic collision

Partially inelastic collisions are the most common form of collisions in the real world. In this type of collision, the objects involved in the collisions do not stick, but some kinetic energy is still lost. Friction, sound and heat are some ways the kinetic energy can be lost through inelastic collisions. For example, a head-on collision between two cars and a mid-air collision between two bald eagles as shown in Fig. 6.7 is inelastic collision. Note that the collision between the two cars is not completely inelastic because parts of cars become detached due to the collision.

Figure 6.7: Two examples of inelastic collision are shown. Both a head-on collision of two cars (left panel) and two tangled bold eagles in mid-air can be considered as inelastic collision.

Unlike completely inelastic collision, **partially inelastic** collisions involve objects which separate after they collide, but which are still deformed in some way by the interaction. Kinetic energy is not conserved. However, it is not easy to figure out what happens afterwards, because there are many possible solutions which satisfy conservation of momentum.

Inelastic collision is commonly used in physics to probe subatomic structure. For example, using a form of inelastic collisions called deep inelastic scattering, particle physicists have determined that the proton and neutron are made of more fundamental particles called quarks. The importance of this discovery, which was made at Stanford Linear Accelerator Center, was recognized with the 1990 Nobel prize, which was awarded to Friedman, Kendall, and Taylor for their investigations of deep inelastic scattering of electrons on protons and bound neutron. This work has played essential role for the development of the quark model. One form of inelastic collision between light and matter is x-ray scattering. The use of x-rays for medical purpose started almost immediately after Röntgen's discovery of x-rays in 1895. Since its discovery, the use of x-rays has been the most important diagnostic imaging method and x-rays still play a central role in radiotherapy.

Exercise Problem 6.4:

The ballistic pendulum is a device used to measure the speed of a projectile, such as bullet. The projectile, of mass m, is first fired into a large block (of wood or other material) of mass M, which is suspended like a pendulum. (Usually, M is somewhat greater than m). As a result of the collision, the pendulum-projectile system swings up to a maximum height. Determine the relationship between the initial speed of the projectile, v, and the height h.

MOMENTUM AND COLLISION

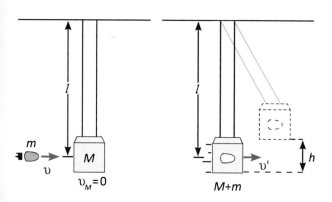

PROBLEM SOLVING STRATEGY

What do we know?

A projectile is embedded in a block after the collision.

What concepts are needed?

We need to apply conservation of momentum first, and then conservation of mechanical energy

Why?

Since the collision is inelastic, only the momentum is conserved. However, if we consider the motion of the bullet-block system after the collision, then the total energy of the system is conserved since the energy is not dissipated. But, we need to apply these principles separately.

Solution:

We separate the problem into two parts. First, we apply conservation of momentum to inelastic collision between the bullet and the pendulum and obtain

$$\vec{P}_i = \vec{P}_f \quad \Rightarrow \quad mv = (m+M)v'. \tag{4a}$$

Second, we apply conservation of energy to the bullet-pendulum system and write down

$$K.E._A + P.E._A = K.E._B + P.E._B.$$

Noting that the energy of the system at the bottom (A) of the pendulum is all kinetic and at the maximum height (B) is all potential, we obtain

$$\frac{1}{2}(m+M)v'^2 = (m+M)gh \quad \Rightarrow \quad \therefore v' = \sqrt{2gh}. \tag{4b}$$

Using the momentum conservation equation of (4a), we obtain

$$mv = (m+M)v' \quad \Rightarrow \quad v = \frac{m+M}{m}v'. \tag{4c}$$

Finally, we substitute Eq. (4b) into Eq. (4c) and obtain

$$v = \frac{m+M}{m}\sqrt{2gh}.$$

A **ballistic pendulum** is a device for measuring a bullet's momentum. From the momentum measurement, it is possible to calculate the velocity and kinetic energy. The ballistic pendulum was invented by Benjamin Robins (1707-1751) in 1742, and was published in his book *New Principles of Gunnery*. This device revolutionized the science of ballistics since it provided the first way to measure the velocity of a bullet accurately. Robins used the ballistic pendulum to measure projectile velocity in two ways. First, he attached the gun to the pendulum and measured the recoil. Since the momentum of the gun is equal to the momentum of the ejecta and the projectile was (in those experiments) the large majority of the mass of the ejecta, the velocity of the bullet could be approximated. Second, he measured the bullet momentum directly by firing it into the pendulum. Robins experimented with musket balls of around one ounce in mass (30 g), while other contemporaries used his methods with cannon shot of one to three pounds (0.5 to 1.4 kg). Ballistic pendulums have been superseded by modern chronographs, which allow direct measurement of the projectile velocity. Although the ballistic pendulum is considered obsolete, it remained in use for a significant length of time and led to great advances in the science of ballistics. The ballistic pendulum is still found in physics classrooms today because of its simplicity and usefulness in demonstrating properties of momentum and energy.

6.4 ELASTIC COLLISION IN MULTI-DIMENSIONAL SPACE

As in the game of billiards shown in Fig. 6.8, collisions in the real world happen in multi-dimensional space. Suppose we consider an elastic collision. In an elastic collision in multi-dimensional space, we need to write conservation of momentum for each spatial direction since momentum is a vector. A two-dimensional collision is a collision in which the two objects are not originally moving along the same line of motion. They could be initially moving at right angles to one another or at least at some angle (other than 0 degrees and 180 degrees) relative to one another. In such cases, vector principles must be combined with momentum conservation principles in order to analyze the collision. The underlying principle of such collisions is that both the "x" and the "y" momentum are conserved, separately, in the collision. The analysis involves determining pre-collision momentum for both the x- and the y-directions. If collision is completely inelastic, then the total momentum of system before the collision (and after) can be determined by using the Pythagorean equation. Since the two colliding objects travel together in the same direction after the collision, the total momentum is simply the total mass of the objects multiplied by their velocity.

The result of a collision between two objects in a plane cannot be predicted only from the momentum and kinetic energy of the objects before the collision. However, as we discussed above, the result of the collision is constrained to obey conservation of momentum, which is a vector relation. This mean that if x and y coordinates are used in the plane, the x and y components of momentum as well as its total magnitude must be the same, before and after the collision if the collision were perfectly elastic. However, ordinary macroscopic collisions usually have significantly less kinetic energy after the collision because some of the kinetic energy transforms into

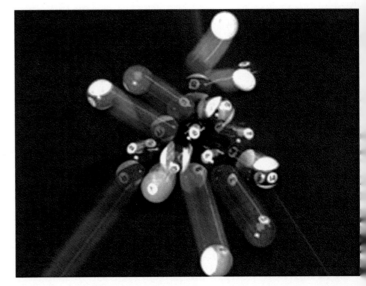

Figure 6.8: The game of billiards, in large part, involves collisions between billiard balls in two dimensions. When two billiard balls collide the collision is nearly elastic, indicating that momentum and kinetic energy are conserved.

MOMENTUM AND COLLISION

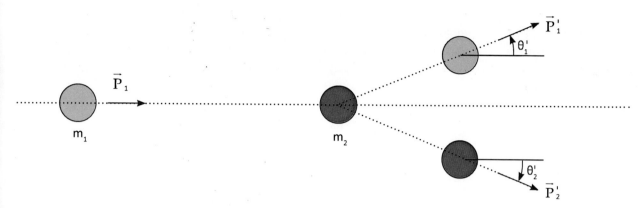

Figure 6.9: A schematic illustration of a collision between two objects in two dimensions. In this case, the total momentum of the system, before and after the collision, in each direction is the same.

other forms. In some inelastic collision, the system has more kinetic energy after the collision. The only way we can have more kinetic energy after the collision would be in the case of some release of energy in the collision such as in a chemical explosion. The chemical explosion may be considered as a perfect inelastic collision occurring in reverse. In the discussion below, the masses, velocities, and angles before and after collision are specified. If one of the velocities (magnitude and direction) is specified after collision, then conservation of momentum determines the other exactly.

Let us see how we may apply both the momentum and energy conservation principle in a two dimensional collision as shown in Fig. 6.9. In two dimensions, the vector nature of momentum becomes especially important when dealing with collision problems in two dimensions. Conservation of momentum is expressed as

$$\vec{p}_1 + \vec{p}_2 = \vec{p}_1' + \vec{p}_2'. \tag{6.9}$$

We may decompose this vector equation in terms of x and y-component equation as follows. For **x-component,** conservation of momentum is written as

$$p_{1x} + p_{2x} = p_{1x}' + p_{2x}' \Rightarrow m_1 v_1 = m_1 v_1' \cos\theta_1' + m_2 v_2' \cos\theta_2'. \tag{6.10}$$

Similarly, for y-component, conservation of momentum is expressed as

$$p_{1y} + p_{2y} = p_{1y}' + p_{2y}' \Rightarrow 0 = m_1 v_1' \sin\theta_2' - m_2 v_2' \cos\theta_2'. \tag{6.11}$$

Since this collision is elastic, the process must also obey conservation of energy

$$\frac{1}{2} m_1 v_1^2 + 0 = \frac{1}{2} m_1 v_1'^2 + \frac{1}{2} m_2 v_2'^2$$

Examination of the kinetic energy after the collision helps to determine whether or not an outcome is reasonable. It is noted that these equations are independent. This means we can solve for three unknowns.

In the general case of a two-dimensional collision between two masses, we cannot anticipate how much kinetic energy will be lost in the collision. Therefore, the velocities of the two masses after the collision are not completely determined by their velocities and directions before the collision. However, conservation of momentum must be satisfies, so that if the velocity of one of the particles after the collision is specified, then the other is determined. In general, to predict the result of a two-dimensional (or three-dimensional) collision, we need to know the initial velocity vectors of both objects, the coefficient of restitution, and the vector along the line of contact. Given this information we can solve collision problems in multi-dimensional space. However, this discussion is beyond the scope of this textbook.

6.5 CENTER OF MASS

So far we have been treating objects as point particles. The motion of real-world objects often appears more complex than this. The reason for this apparent complexity is that most objects are not point particles, but rather they are extended bodies with finite sizes. When dealing with an extended body (instead of single particles), we may describe motion of this body in terms of translational and rotational motion as shown in Fig. 6.6. However, there is one point that moves in the same path that a particle would if it is subject to the same net force. This point is called the **"center of mass"**. This is the reason that we may treat any object as a point particle located at its center of mass. This approach is an excellent way to account for the translational motion of an object. However, in doing so, we neglect the object's rotational motion about the center of mass.

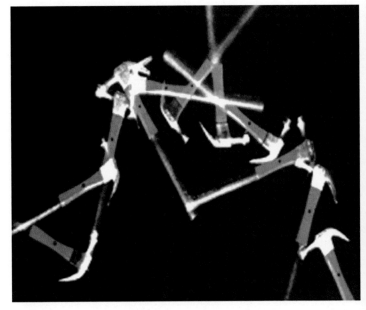

Figure 6.10: Motion of a hammer thrown in air may be described as combination of rotational motion about the center of mass (solid circle) and translational motion of the center of mass. The trajectory of the center of mass motion indicated by the solid circles is a parabolic path.

Before proceeding further, let us first grasp the concept of center of gravity better. This concept is also discussed Chapter 8. The concept of "center of mass" in the form of the "center of gravity" was first introduced by the ancient Greek physicist, mathematician, and engineer Archimedes (287 BC -212 BC) of Syracuse. He worked with simplified assumptions about gravity (i.e., uniform field) and arrived at the mathematical properties of what we now call the center of mass. Archimedes showed that the torque (i.e., rotational analogue of force) exerted on a lever by weights resting at various points along the lever is the same as what it would be if all of the weights were moved to a single point which is their center of mass. The **center of mass** of a distribution of mass in space is defined as the unique point where the weighted relative position of the distributed mass sums to zero. The distribution of mass is balanced around the center of mass and the average of the weighted position coordinates of the distributed mass defines its coordinates. The center of mass is a useful reference point for calculations in mechanics that involve masses distributed in space, such as the linear and angular momentum (i.e., rotational analogue of linear momentum) of planetary bodies and rigid body dynamics. In orbital mechanics, the equations of motion of planets are formulated as point masses located at the centers of mass. The center of mass frame is an inertial frame in which the center of mass of a system is at rest at with respect the origin of the coordinate system.

MOMENTUM AND COLLISION

The general motion of an extended body can be considered as the sum of the translational motion of the center of mass (CM), plus rotational, vibrational, or other types of motion about the CM. Let us see how we can compute the center of mass for a system of massive particles. We may define the position of the center of mass with respect to some point of reference as

$$\vec{r}_{CM} = \frac{\sum_i m_i \vec{r}_i}{\sum_i m_i}, \tag{6.12}$$

where m_i and $\mathbf{r}_i$ denote the mass of particle i in the system and its position vector with respect to a reference point, respectively. We may decompose Eq. (6.12) into the components and write the definition for the center of mass as

$$x_{CM} = \frac{\sum_i m_i x_i}{\sum_i m_i} \quad \text{and} \quad y_{CM} = \frac{\sum_i m_i y_i}{\sum_i m_i}. \tag{6.13}$$

The center of mass is closely related to the center of gravity. The center of gravity is a point at which the force of gravity can be considered to act. The terms "center of mass" and "center of gravity" are used synonymously in a uniform gravitational field to represent the unique point in an object or system which can be used to describe the system's response to external forces and torques. For example, the gravitational force on Earth attracts small objects as if it were pulling them from their center of mass (in this case, the center of mass is also called the **center of gravity**). Two stars in orbit around each other revolve around their collective center of mass.

Why do we care about center of mass? Well, one of the routine but important tasks of many real engineers is to find the center of mass of a complex machine. This routine work is often done with computer aided design software. Just knowing the location of the center of mass of a car, for example, is enough to estimate whether it can be tipped over by maneuvers on level ground. The center of mass of a boat must be low enough for the boat to be stable. Any propulsive force on a space craft must be directed towards the center of mass in order to not induce rotations. Tracking the trajectory of the center of mass of an exploding plane can determine whether or not it was hit by a massive object. Any rotating piece of machinery must have its center of mass on the axis of rotation if it is not to cause much vibration. Also, many calculations in mechanics are greatly simplified by making use of a system's center of mass. In particular, the whole complicated distribution of gravity forces through out a body is equivalent to a single force at the body's center of mass. Many important quantities in dynamics are similarly simplified using the center of mass.

As shown in Fig. 6.7, the center of mass as well as the center of gravity of a system of objects can change as the components of the system move. The term "center of gravity" has implications for all things related to posture, including postural issues such as swayback and designing posture exercise programs and more. The center of gravity is a theoretical place in our body where our mass is considered to concentrate. On Earth, weight and mass are often times used to refer as pretty much the same thing. Recalling our discussion of Chapter 4, we can think of mass as how much resistance our body has when moving - or the force of inertia as it applies to our weight. In outer space, we become weightless, but our mass stays the same. This is due to the fact that the force of gravity does not act on our body in outer space.

Figure 6.11: Individuals in the two pictures are trying to find the balance point under a difficult situation. The center of mass is this balance point. The center of mass can also be considered as the center of gravity, or the point where gravity acts on an object.

Our body on Earth has a center of gravity. However, this position is not fixed. One way of looking at the center of gravity (mass) is the point at which the body's mass is equally balanced. The changes in the center of gravity depend on the position of a person's body (arms up/down, leaning, etc.) Dancers, gymnasts and tight-rope walkers are examples of how the human body compensates for changes in the center of gravity to maintain balance. Usually the center of gravity is located in front of your sacrum bone, at about the second sacral level. The sacrum is made up of five bones fused together vertically. So when we are on planet Earth our weight, or mass, is thought to be concentrated at this point in front of our sacrum. The downward pull of gravity (line of gravity) passes through this point, as well.

To understand the difference between theory and practical application of this concept, let's compare the human body to a baseball for a minute. From a point in the exact center, the baseball is of equal weight and shape all the way around, is it not? So, we can easily see that with any movement of the ball, this center point moves right along with it. However, when we consider center of gravity in the human body, things get more complicated. As we mentioned before, because the body has moving parts (arms, legs, head, various areas of the trunk), every time we do anything, the shape of our overall form changes. And if we carry something like a suitcase or grocery bag or if we wear a backpack, this adds weight, which changes the center of gravity, too. So, we can say that the center of gravity is a constantly changing point in the body that represents where the weight (mass) of the rest of our body is equally balanced in every direction. This point can and does change based on what we are carrying and how we are carrying it, as well as the position we take and the movements we make.

As discussed above, the center of gravity occurs in the body at a point where weight is equally distributed on all sides. From this point, a body can pivot in any direction and remain balanced. When standing evenly over our center of gravity, we are in a state of equilibrium. The line of gravity is an imaginary line that crosses through our center of gravity dividing the mass of the body into two equal halves. This line changes depending on the body's weight distribution. It is a vertical line running from the top of the head, usually around the ear, down to the ground. To keep our body in balance, our posture must correspond with our line of gravity. Gravity affects many parts of our

MOMENTUM AND COLLISION

Center of Mass of Parts of Typical Human Body (full height and mass = 100 units)

Distance of Hinge Points (%)	Hinge Points (•) (Joints)		Center of Mass (x) (% Height Above Floor)	Percent Mass
91.2	Base of skull on spine	Head	93.5	6.9
81.2	Shoulder jont	Trunk and neck	71.1	46.1
	elbow 62.2	Lower arms	55.3	4.2
	wrist 46.2			
52.1	Hip	Upper legs (thighs)	42.5	21.5
28.5	Knee	Lower legs	18.2	9.6
4.0	Ankle	Feet	1.8	3.4
			58.0	100.0

Figure 6.12: A schematic illustration of center of mass for parts of typical human body. The center of the human body changes depending on the relative location of the body parts. When running the center of mass is located in front of the spine below the navel.

body as we age. It compresses the spine, contributes to poor blood circulation and can decrease our flexibility. The gravitational pull also affects our organs, causing them to shift downward, away from their proper position. Gravity is often blamed for the way excess weight accumulates around the midsection.

Exercise Problem 6.5:

A square uniform raft, 18 m by 18 m of mass 6200 kg, is used as a ferry boat. If three cars, each of mass 1200 kg, occupy the NE, SE, and SW corners, determine the center of mass of the loaded ferry boat. (Assume that the ferry boat is a distributed object while the cars can be treated as point masses.)

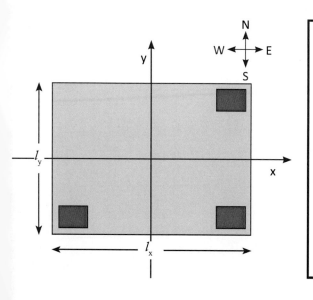

PROBLEM SOLVING STRATEGY

What do we know?

Location of three cars relative to the ferry boat is known.

What concepts are needed?

We need to apply the definition of center of mass for both x- and y-direction.

Why?

According to Eq. (6.13), the center of mass may be computed once the position of objects and their masses are known.

Solution:

We choose the center of the ferry boat as the reference point and compute the center of mass with respect to his point. Along the x-axis, the center of mass is located at

$$x_{cm} = \frac{\sum_i m_i x_i}{\sum_i m_i}$$

$$= \frac{6200\,kg \times 0\,m + 1200\,kg \times 9\,m + 1200\,kg \times 9\,m + 1200\,kg \times (-9\,m)}{6200\,kg + 1200\,kg + 1200\,kg + 1200\,kg}$$

Here, the center of mass is located at the east of the center. Along the y-axis, the center of mass is located at

$$y_{cm} = \frac{\sum_i m_i y_i}{\sum_i m_i}$$

$$= \frac{6200\,kg \times 0\,m + 1200\,kg \times 9\,m + 1200\,kg \times (-9\,m) + 1200\,kg \times (-9\,m)}{6200\,kg + 1200\,kg + 1200\,kg + 1200\,kg}$$

Here, the negative sign indicates that the center of mass is located at the south of the center. The center of mass, however, shifts to a different position as the mass of the cars change. For example, if the mass of the car at the southeastern corner of the ferry increases from 1200 kg to 3600 kg, then the center of mass is located at $(x_{cm}, y_{cm}) = (2.66\,m, -2.66\,m)$.

In a vehicle which relies on center of mass in some way, weight distribution directly affects a variety of vehicle characteristics, including handling, acceleration, traction, and component life. For this reason weight distribution varies with the vehicle's intended usage. For example, a drag car maximizes traction at the rear axle while countering the reactionary pitch-up torque. It generates this counter-torque by placing a small amount of counterweight at a great distance forward of the rear axle. In the airline industry, load balancing is used to evenly distribute the weight of passengers, cargo, and fuel throughout an aircraft, so as to keep the aircraft's center of gravity close to its center of pressure to avoid losing pitch control. In military transport aircraft, it is common to have a loadmaster as a part of the crew; their responsibilities include calculating accurate load information for center of gravity calculations, and ensuring cargo is properly secured to prevent its shifting. In large aircraft and ships, multiple fuel tanks and pumps are often used, so that as fuel is consumed, the remaining fuel can be positioned to keep the vehicle balanced, and to reduce stability problems caused by liquid fuel sloshing around.

6.6 MOTION OF CENTER OF MASS

As shown in Fig. 6.6, when a hammer is thrown with a spin, it appears to execute a complex motion. However, if we follow the movement of the little circle which represents the hammer's center of mass, then we may notice that the center of mass follows a parabolic path. The motion of the center of mass for a system of particles (or an extended body) is directly related to the net force acting on the system as whole. Suppose the three particles lie on the x-axis and have masses m_1, m_2, m_3 and positions x_1, x_2, x_3 respectively. By using the definition of center of mass x_{cm} of Eq. (6.13) we may write

MOMENTUM AND COLLISION

$$(m_1 + m_2 + m_3)x_{cm} = m_1 x_1 + m_2 x_2 + m_3 x_3. \quad (6.14)$$

As the time progresses, these three particles will move and will be located at different positions. This means that the system has a new center of mass position, x'_{cm}. Once again, by using Eq. (6.13), we may write

$$(m_1 + m_2 + m_3)x'_{cm} = m_1 x'_1 + m_2 x'_2 + m_3 x'_3. \quad (6.15)$$

We now subtract Eq. (6.14) from Eq. (6.15) and write the difference equation as

$$(m_1 + m_2 + m_3)(x'_{cm} - x_{cm}) = m_1(x'_1 - x_1) + m_2(x'_2 - x_2) + m_3(x'_3 - x_3). \quad (6.16)$$

Suppose we denote the total mass of the system as $M = m_1 + m_2 + m_3$ and represent the change in the position of the center of mass as $\Delta x_{cm} = x'_{cm} - x_{cm} = v_{cm}\Delta t$. We may rewrite Eq. (6.16) as

$$M v_{cm} \Delta t = m_1 v_1 \Delta t + m_2 v_2 \Delta t + m_3 v_3 \Delta t. \quad (6.17)$$

If we write the change in the position of each particle in the system as $x'_i - x_i = v_i \Delta t$, it is straightforward to see that Δt appearing on the both sides of Eq. (6.17) may be cancelled to obtain

$$M v_{cm} = m_1 v_1 + m_2 v_2 + m_3 v_3. \quad (6.18)$$

It is interesting to note that the quantity $M v_{cm}$ represents the total (linear) momentum of a system of particles. The concept of momentum will be discussed further in Chapter 7. We use the definition of acceleration (i.e., $a = \Delta v/\Delta t$) to write the speed v_i of the i-th particle as $v_i = a_i \Delta t$ and rewrite Eq. (6.18) as

$$M a_{cm} \Delta t = m_1 a_1 \Delta t + m_2 a_2 \Delta t + m_3 a_3 \Delta t. \quad (6.19)$$

Once again, we can easily see that the factor Δt appearing on both sides of Eq. (6.19) may be cancelled to obtain the result that we wanted

$$M a_{cm} = m_1 a_1 + m_2 a_2 + m_3 a_3 = F_1 + F_2 + F_3 = F_{net}. \quad (6.20)$$

Equation (6.20) indicates that the sum of all the forces acting on the system is equal to the total mass of the system times the acceleration of its center of mass (**Newton's second law**). In other words, the center of mass of a system of particles moves as if the total mass of the particles is concentrated there and acted upon by the resultant force. Also, equation (6.20) shows that the motion of the center of mass is only determined by the external forces. Forces exerted by one part of the system on other parts of the system are called internal forces. According to Newton's third law, the sum of all internal forces cancels out (for each interaction there are two forces acting on two parts: they are equal in magnitude but pointing in an opposite direction and cancel if we take the vector sum of all internal forces).

Suppose we consider a simple system of two objects with mass m and M. When the two objects are moving with speed v_m and v_M, it is difficult to tell where their center of motion is located. Most often, the motion of two objects in space is illustrated as with respect to a stationary center of mass. This is the ideal case. For example, when we observe two objects on orbit around their center of mass, such as with twin star, we can usually see them orbit the fixed center of mass between them. Our view is relative to that center of mass, such that $v_{cm} = 0$. The general velocity equation becomes

$$0 = \frac{m\vec{v}_m + M\vec{v}_M}{m + M} \Rightarrow m\vec{v}_m = -M\vec{v}_M$$

This means that the velocity vectors are in opposite directions, when viewed from the center of mass. In this ideal case, it can be said that the motions of the objects mirror each other, according to their respective masses. This "mirroring" can be seen when both objects move apart, move toward each other, go into orbit and fly off into space.

The motion of two objects in space is relative to some fixed reference point. In many cases, we may see the motion with respect to our viewpoint. Ideally, the viewpoint is relative to a stationary center of mass between the two objects.

Understanding-the-Concept Question 6.4:

A rocket is shot into the air. At the moment it reaches its highest point, a horizontal distance d from its starting point. A prearranged explosion separates it into two parts of equal mass. Part I is stopped in mid-air and fall vertically to Earth. Where does part II land? Assume g is constant.

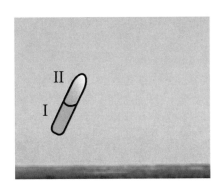

> **PROBLEM SOLVING STRATEGY**
>
> **What do we know?**
>
> The rocket separates into two equal parts at the maximum height.
>
> **What concepts are needed?**
>
> We need to apply the concept that center of mass trajectory of two rocket parts will remain the same after the explosion.
>
> **Why?**
>
> Since the center of mass motion is only determined by the external forces and only internal forces act on the objects in explosion, the center of motion will not be affected.

Explanation:

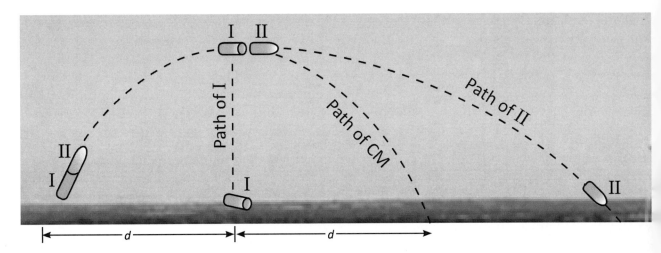

Since the center of mass trajectory is not affected by the internal explosion, the final position of the center of mass for the rocket will be at a distance 2d from the launch point. So, if the half of the rocket

MOMENTUM AND COLLISION

is at a distance d from the launch point, then the other half must be located at a distance 3d from the launch point. This suggests that the rocket can have a longer range if part of the rocket is jettisoned during the flight.

Aside from the consideration of center of mass trajectory, multi-stage rockets have other advantages. The main reason for multi-stage rockets and boosters is that once the fuel is exhausted, the space and structure which contained it and the motors themselves are useless and only add weight to the vehicle which slows down its future acceleration. By dropping stages which are no longer useful, the rocket lightens itself. The thrust of future stages is able to provide more acceleration than if the earlier stage were still attached, or a single, large rocket would be capable of.

When a stage drops off, the rest of the rocket is still traveling near the speed that the whole assembly reached at burn-out time. This means that it needs less total fuel to reach a given velocity and/or altitude. A further advantage is that each stage can use a different type of rocket motor each tuned for its particular operating conditions. Thus the lower stage motors are designed for use at atmospheric pressure, while the upper stages can use motors suited to near vacuum conditions. Lower stages tend to require more structure than upper as they need to bear their own weight plus that of the stages above them, So optimizing the structure of each stage decreases the weight of the total vehicle and provides further advantage.

SUMMARY

i. Mathematically, linear momentum is defined as mass times velocity,

$$\vec{p} = m\vec{v},$$

stating that the more massive something is and/or the faster it's moving the more momentum it has. According to Newton's Second law of motion (in terms of momentum), the rate of change of momentum of a body is equal to the net force applied to it

$$\sum_i \vec{F}_i = \frac{\Delta \vec{p}}{\Delta t}.$$

This definition of the force accounts for the dynamics of a system which changes inertia.

ii. In an isolated system (one that does not exchange any matter with the outside and is not acted on by outside forces), the total momentum is constant. This fact, known as the law of conservation of momentum, is implied by Newton's laws of motion. For a system of two particles undergoing collision, the law of conservation of momentum is written as

$$m_1 \vec{v}_1 + m_2 \vec{v}_2 = m_1 \vec{v}_1' + m_2 \vec{v}_2'.$$

This law holds no matter how complicated the force is between particles. Similarly, if there are several particles, the momentum exchanged between each pair of particles adds up to zero, so the total change in momentum is zero.

iii. **Elastic collision** is defined as a collision in which two objects bounce apart. In this collision, **the total kinetic energy is conserved:**

$$\frac{1}{2}m_1 v_1^2 + \frac{1}{2}m_2 v_2^2 = \frac{1}{2}m_1 v_1'^2 + \frac{1}{2}m_2 v_2'^2.$$

Inelastic collision is defined as a collision in which, **the total kinetic energy is not conserved.** In this collision two objects do not bounce apart and part of the kinetic energy is changed to some other form of energy in the collision. However, when all of energy is accounted for

$$K.E._1 + K.E._2 = K.E.'_1 + K.E.'_2 + \text{thermal and other forms of energy,}$$

the total energy is conserved.

iv. The general motion of an extended body can be considered as the sum of the translational motion of the center of mass, plus rotational, vibrational, or other types of motion about the center of mass (cm) which is defined as

$$\vec{r}_{CM} = \frac{\sum_i m_i \vec{r}_i}{\sum_i m_i} \Rightarrow x_{CM} = \frac{\sum_i m_i x_i}{\sum_i m_i} \text{ and } y_{CM} = \frac{\sum_i m_i y_i}{\sum_i m_i}$$

for a system of particles with the i-th particle with mass m_i is located at r_i. The center of mass is similar to the center of gravity. Center of gravity is a point at which the force of gravity can be considered to act.

MORE WORK PROBLEMS

Problem 6.1 :

At a basketball game, a 120-lb cheerleader is tossed vertically upward with a speed of 4.5 m/s by a male cheerleader. (a) What is the cheerleader's change in momentum from the time she is released to just before being caught if she is caught at the height at which she was released? (b) Would there be any difference if she were caught 0.25 m below the point of release? If so, what is the change then?

Solution:

Noting that the final velocity is equal in magnitude but opposite direction to the initial velocity (i.e., $v_f = -v_i$), we obtain the change in momentum as

$$\Delta p = m \Delta v = m(v_f - v_i) = m(-v_i - v_i) = -2mv_i = -2 \times 120 \text{ lb} \times \frac{1 \text{ kg}}{2.2 \text{ lb}} \times 4.50 \frac{m}{s} = -491 \frac{\text{kg} \times m}{s}.$$

Here, the sign indicates that the direction is downward. b) If she was caught 0.25 m below the point of release, we would expect that there will be difference in the final velocity. We may compute the final velocity as

$$v_f^2 = v_i^2 - 2gy \Rightarrow v_f = \sqrt{v_i^2 - 2gy} = \sqrt{\left(4.50 \frac{m}{s}\right)^2 - 2 \times 9.80 \frac{m}{s^2} \times (-0.25 \text{ m})} = 5.01 \frac{m}{s}.$$

Since the direction is downward, the final velocity is v_f = -5.01 m/s. We now compute the change in momentum as

MOMENTUM AND COLLISION

$$\Delta p = m\left(v_f - v_i\right) = 120\,\text{lb} \times \frac{1\,\text{kg}}{2.2\,\text{lb}} \times \left(-5.01\,\frac{\text{m}}{\text{s}} - 4.50\,\frac{\text{m}}{\text{s}}\right) = -519\,\frac{\text{kg} \times \text{m}}{\text{s}}.$$

We note that downward direction is indicated by the negative sign.

Problem 6.2:

A 45.0-kg girl is standing on a 150-kg plank. The plank, originally at rest, is free to slide on a frozen lake, which is a flat, frictionless surface. The girl begins to walk along the plank at a constant velocity of 1.50 m/s to the right relative to the plank. (a) What is her velocity relative to the surface of the ice? (b) What is the velocity of the plank relative to the surface of the ice?

Solution:

The velocity of the girl relative to the ice v_{gi} is

$$v_{gi} = v_{gp} + v_{pi}.$$

where v_{gp} is the velocity of the girl relative to the plank and v_{pi} is the velocity of the plank relative to ice. Since we are given that $v_{gp} = 1.5$ m/s, we may write v_{gi} as

$$v_{gi} = 1.5\,\frac{\text{m}}{\text{s}} + v_{pi}.$$

We now apply conservation of momentum to this problem and write

$$m_g v_{gi} + m_p v_{pi} = 0 \Rightarrow v_{pi} = -\frac{m_g}{m_p} v_{gi}.$$

This allows us to solve for v_{gi} and obtain

$$v_{gi} = v_{gp} + v_{pi} = v_{gp} - \frac{m_g}{m_p} v_{gi} \Rightarrow v_{gi} = \frac{v_{gp}}{1+\left(\frac{m_g}{m_p}\right)} = \frac{1.50\,\text{m}/\text{s}}{1+\left(\frac{45.0\,\text{kg}}{150\,\text{kg}}\right)} = 1.15\,\frac{\text{m}}{\text{s}}.$$

We, then, solve for v_{pi} and obtain

$$v_{pi} = -\frac{m_g}{m_p} v_{gi} = -\frac{45.0\,\text{kg}}{150\,\text{kg}} \times 1.15\,\frac{\text{m}}{\text{s}} = -0.345\,\frac{\text{m}}{\text{s}}.$$

Here, the negative sign means that the direction is opposite to the girl's motion.

Problem 6.3:

A 60.0-kg person, running horizontally with a velocity of 3.80 m/s, jumps onto a 12.0-kg sled that is initially at rest. a) Ignoring the effects of friction during the collision, find the velocity of the sled and person as they move away. b) The sled and person coast 30.0 m level snow before coming to rest. What is the coefficient of kinetic friction between the sled and the snow?

Solution:

First, we use the conservation of momentum principle to determine v_f by writing

$$mv_i + Mv_{si} = (m+M)v_f \Rightarrow v_f = \frac{mv_i}{m+M} = \frac{60.0\,\text{kg} \times 3.80\,\text{m/s}}{60.0\,\text{kg} + 12.0\,\text{kg}} = 3.17\,\frac{\text{m}}{\text{s}}.$$

Here, the initial velocity of the sled is $v_{si} = 0$ since it is initially at rest. We now use the work-energy principle to write

$$W = \Delta K.E. \Rightarrow F_f d = \frac{1}{2}(m+M)v_f^2.$$

We note that

$$F_f = \mu_k F_N = \mu_k (m+M)g$$

since

$$\sum_i F_{y,i} = ma_y \Rightarrow F_N - (m+M)g = 0.$$

Using this result, we may rewrite the work-energy principle as

$$\mu_k (m+M)gd = \frac{1}{2}(m+M)v_f^2.$$

Now, we solve for the coefficient of kinetic friction μ_k as

$$\mu_k = \frac{1}{2}\frac{v_f^2}{gd} = \frac{(3.17\,\text{m/s})^2}{2 \times 9.80\,\text{m/s}^2 \times 30.0\,\text{m}} = 0.017.$$

Problem 6.4:

The Earth and Moon are separated by a center-to-center distance of 3.85×10^8 m. The mass of the Earth is 5.98×10^{24} kg and that of the moon is 7.35×10^{22} kg. How far does the center of mass lie from the center of the Earth?

Solution:

The center of mass, measured from the center of Earth, is located at

$$x_{cm} = \frac{\sum_i m_i x_i}{\sum_i x_i} = \frac{M_e \cdot 0 + M_m \cdot d}{M_e + M_m} = \frac{0 + 7.35 \times 10^{22}\,\text{kg} \times 3.85 \times 10^8\,\text{m}}{5.98 \times 10^{24}\,\text{kg} + 7.35 \times 10^{22}\,\text{kg}} = 4.67 \times 10^6\,\text{m}.$$

MOMENTUM AND COLLISION

PROBLEMS

1. A golfer, driving a golf ball off the tee, gives the ball a velocity of +38 m/s. The mass of the ball is 0.045 kg, and the duration of the impact with the golf club is 3.0×10^{-3} s. (a) What is the change in momentum of the ball? (b) Determine the average force applied to the ball by the club.

2. A dump truck is being filled with sand. The sand falls straight downward from rest from a height of 2.00 m above the truck bed, and the mass of sand that hits the truck per second is 55.0 kg/s. The truck is parked on the platform of a weight scale. By how much does the scale reading exceed the weight of the truck and sand?

3. A ball of mass 0.150 kg is dropped from rest from a height of 1.25 m. It rebounds from the floor to reach a height of 0.960 m. What impulse was given to the ball by the floor?

4. A tennis player receives a shot with the ball (0.0600 kg) traveling horizontally at 50.0 m/s and returns the shot with the ball traveling horizontally at 40.0 m/s in the opposite direction. (a) What is the impulse delivered to the ball by the racket? (b) What work does the racket do on the ball?

5. A 0.500 kg ball is dropped from rest at a point 1.20 m above the floor. The ball rebounds straight upward to a height of 0.700 m. What are the magnitude and direction of the impulse of the net force applied to the ball during the collision with the floor?

6. A 15.0-g rubber bullet hits a wall with a speed of 150 m/s. If the bullet bounces straight back with a speed of 120 m/s, what is the change in momentum of the bullet?

7. In a simulated head-on crash test, a car impacts a wall at 25 mi/h (40 km/h) and comes abruptly to rest. A 120-lb passenger dummy (with a mass of 55 kg), without a seat belt, is stopped by an air bag, which exerts a force on the dummy of 2400 lb. How long was the dummy in contact with air bag while coming to stop?

8. A 46 kg skater is standing still in front of a wall. By pushing against the wall she propels herself backward with a velocity of -1.2 m/s. Her hands are in contact with the wall for 0.80 s. Ignore friction and wind resistance. Find the magnitude and direction of the average force she exerts on the wall (which has the same magnitude, but opposite direction, as the force that the wall applies to her).

9. A boy catches - with bare hands and his arms rigidly extended - a 0.16-kg baseball coming directly toward him at a speed of 25 m/s. He emits an audible "ouch!", because the ball stings his hands. He learns quickly to move his hands with the ball as he catches it. If the contact time of the collision is increased from 3.5 ms to 8.5 ms in this way, how do the magnitudes of the average impulse force compare?

10. A wagon is coasting at a speed v_A along a straight and level road. When ten percent of the wagon's mass is thrown off the wagon, parallel to the ground and in the forward direction, the wagon is brought to a halt. If the direction in which this mass relative to the wagon remains the same, the wagon accelerates to a new speed v_B. Calculate the ratio v_A/v_B.

11. A small, 100 g cart is moving at 1.2 m/s on an air track when it collides with a larger, 1.00 kg cart at rest. After the collision, the small cart recoils at 0.850 m/s. What is the speed of the large cart after the collision?

12. A proton of mass m moving with a speed of 3.0×10^6 m/s undergoes a head-on elastic collision with an alpha particle of mass $4m$, which is initially at rest. What are the velocities of the two particles after the collision?

13. A neutron in a reactor makes an elastic head-on collision with a carbon atom that is initially at rest. (The mass of the carbon nucleus is about 12 times that of the neutron.) (a) What fraction of

the neutron's kinetic energy is transferred to the carbon nucleus? (b) If the neutron's initial kinetic energy is 1.6×10^{-13} J, find its final kinetic energy and the kinetic energy of the carbon nucleus after the collision.

14. In an elastic head-on collision with a stationary target particle, a moving particle recoils at one third of its incident speed. (a) What is the ration of the particles' masses (m_1/m_2)? (b) What is the speed of the target particle after the collision in terms of the initial speed of the incoming particle?

15. Two blocks of mass m_1 and m_2 approach each other on a horizontal table with the same constant speed, u_o, as measured by a laboratory observer. The blocks undergo a perfectly elastic collision, and it is observed that m_1 stops but m_2 moves opposite its original motion with some constant speed, v. (a) Determine the ratio of the two masses, m_1/m_2. (b) What is the ratio of their speeds, v/u_o?

16. A kid at the junior high cafeteria wants to propel an empty milk carton along a lunch table by hitting it with a 3.0 g spit ball. If he wants the speed of the 20 g carton just after the spit ball hits it to be 0.30 m/s, at what speed should his spit ball hit the carton?

17. In a football game, a receiver is standing still, having just caught a pass. Before he can move, a tackler, running at a velocity of +4.5 m/s, grabs him. The tackler holds onto the receiver, and the two move off together with a velocity of +2.6 m/s. The mass of the tackler is 115 kg. Assuming that momentum is conserved, find the mass of the receiver.

18. By accident, a large plate is dropped and breaks into three pieces. The pieces fly apart parallel to the floor. As the plate falls, its momentum has only a vertical component and no component parallel to the floor. After the collision, the component of the total momentum parallel to the floor must remain zero, since the net external force acting on the plate has no component parallel to the floor. Using the data shown in the drawing, find the masses of pieces 1 and 2.

19. A car of mass m moving at a speed u_1 collides and couples with the back of a truck of mass $2m$ moving initially in the same direction as the car at a lower speed u_2. (a) What is the speed u_f of the two vehicles immediately after the collision? (b) What is the change in kinetic energy of the car-truck system in the collision?

20. A tennis player swings her 1000 g racket with a speed of 10 m/s. She hits a 60 g tennis ball that was approaching her at a speed of 20 m/s. The ball rebounds at 40 m/s. (a) How fast is her racket moving immediately after the impact? You can ignore the interaction of the racket with her hand for the brief duration of the collision. (b) If the tennis ball and racket are in contact for 10 ms, what is the average force that the racket exerts on the ball? (F08, 9.47)

21. Two people are standing on a 2.0-m-long platform, one at each end. The platform floats parallel to the ground on a cushion of air, like a hovercraft. One person throws a 6.0-kg ball to the other, who catches it. The ball travels nearly horizontally. Excluding the ball, the total mass of the platform and people is 118 kg. Due to the throw, the totalmass of 118-kg platform-people recoils. How far does it move before coming to rest again?

22. A 50.0-kg skater is traveling due east at a speed of 3.00 m/s. A 70.0-kg skater is moving due south at a speed of 7.00 m/s. They collide and hold on to each other after the collision, managing to move off at an angle θ south of east, with a speed of u_f. Find (a) the angle θ and (b) the speed u_f, assuming that friction can be ignored.

23. A 100-g bullet is fired horizontally into a 14.9-kg block of wood resting on a horizontal surface, and the bullet becomes embedded in the block. If the muzzle velocity of the bullet is 250 m/s, what is speed of the block immediately after the impact?

MOMENTUM AND COLLISION

24. A ball is dropped from rest from the top of a 6.10-m-tall building, falls straight downward, undergoes inelastic collision with the ground, and bounces back. The ball loses 10.0% of its kinetic energy every time it collides with the ground. How many bounces can the ball make and still reach a windowsill that is 2.44 m above the ground?

25. A 300 g bird flying along at 6.0 m/s sees a 10 g insect heading straight toward it with a speed of 30 m/s. The bird opens its mouth wide and enjoys a nice lunch. What is the bird's speed immediately after swallowing?

26. A 65.0-kg person throws a 0.0450-kg snowball forward with a ground speed of 30.0 m/s. A second person, with a mass of 60.0 kg, catches the snowball. Both people are on skates. The first person is initially moving forward with a speed of 2.50 m/s, and the second person is initially at rest. What are the velocities of the two people after the snowball is exchanged? Disregard friction between the skates and the ice.

27. A car of mass m moving at a speed u_1 collides and couples with the back of a truck of mass $2m$ moving initially in the same direction as the car at a lower speed u_2. (a) What is the speed u_f of the two vehicles immediately after the collision? (b) What is the change in kinetic energy of the car-truck system in the collision?

28. A bullet of mass $m = 8.00$ g is fired into a block of mass $M = 250$ g that is initially at rest at the edge of a table of height $h = 1.00$ m. The bullet remains in the block, and after the impact the block lands $d = 2.00$ m from the bottom of the table. Determine the initial speed of the bullet.

29. A bullet of mass m and speed v passes completely through a pendulum bob of mass M. The bullet emerges with a speed of $v/2$. The pendulum bob is suspended by a stiff rod of length l and negligible mass. What is the minimum value of v such that the bob will barely swing through a complete vertical circle?

30. Gayle runs at a speed of 4.00 m/s and dives on a sled, initially at rest on the top of a frictionless, snow-covered hill. After she has descended a vertical distance of 5.00 m, her brother, who is initially at rest, hops on her back, and they continue down the hill together. What is their speed at the bottom of the hill if the total vertical drop is 15.0 m? Gayle's mass is 50.0 kg, the sled has mass of 5.00 kg, and her brother has a mass of 30.0 kg.

31. A fireworks rocket is moving at a speed of 45.0 m/s. The rocket suddenly breaks into two pieces of equal mass, which fly off with velocities u_1 and u_2, as shown in the drawing. What is the magnitude of (a) u_1 and (b) u_2?

32. A 100-kg astronaut (mass included space gear) on a space-walk is 5.0 m from a 3000-kg space capsule and at the full length of her safety cord. To return to the capsule, she pulls herself along the cord. Where do the astronaut and capsule meet?

33. A 20 g ball of clay traveling east at 2.0 m/s collides with a 30 g ball of clay traveling 30° south of west at 1.0 m/s. What are the speed and direction of the resulting 50 g blob of clay?

34. A spaceship of mass 2.0×10^6 kg is cruising at a speed of 5.0×10^6 m/s when the antimatter reactor fails, blowing the ship into three pieces. One section, having a mass of 5.0×10^5 kg, blows in straight backward with a speed of 2.0×10^6 m/s. A second piece, with mass 8.0×10^5 kg, continues forward at 10×10^6 m/s. What are the direction and speed of the third piece?

35. Two objects of masses m and $3m$ are moving toward each other along the x-axis with the same initial speed u_o. The object with mass m is traveling to the left and the object with mass $3m$ traveling to the right. They undergo an elastic glancing collision such that m is moving downward after the collision at right angles from its initial direction. (a) Find the final speeds of the two objects. (b) What is the angle θ at which the object with mas $3m$ is scattered?

36. The earth and moon are separated by a center-to-center distance of 3.85×10^8 m. The mass of the earth is 5.98×10^{24} kg and that of the moon is 7.35×10^{22} kg. How far does the center of mass lie from the center of the earth?

37. The drawing shows a sulfur dioxide molecule. It consists of two oxygen atoms and a sulfur atom. A sulfur atom is twice as massive as an oxygen atom. Using this information and the data provided in the drawing, find (a) the x coordinate and (b) the y coordinate of the center of mass of the sulfur dioxide molecule. Express your answers in nanometers (1 nm = 10^{-9} m).

38. Three particles, each with a mass of 0.25 kg, are located at (-4.0 m, 0), (2.0 m, 0), and (0, 3.0 m) and are acted on by forces $\mathbf{F}_1 = (-3.0 \text{ N})\,\hat{y}$, $\mathbf{F}_2 = (5.0 \text{ N})\,\hat{y}$, and $\mathbf{F}_3 = (4.0 \text{ N})\,\hat{x}$, respectively. Find the acceleration (magnitude and direction) of the center of mass of the system [Hint: Consider the components of the acceleration.]

CIRCULAR MOTION

CHAPTER 7

We may remember that amusement parks are thrilling places to spend the long days of summer. We may not have thought about these parks as huge physics classrooms, but in fact all of the rides are built with the laws of physics in mind. It is playing with these laws that make these rides so fun and scary. We will take a look at how the rides at amusement parks involve the forces, energy types, and laws of physics. As we discussed in Chapter 5, roller coaster rides involve force, kinetic and potential energy, and energy conservation principle. Another type of amusement park ride which involves physics is a Ferris wheel. Imagine riding on a Ferris wheel. As we know, a Ferris wheel rider moves in a circular path, indicating that rotational motion is involved in this ride. From the laws of motion discussed in Chapter 4, we know Newton's first law states that an object in motion stays in motion and that motion is in a straight path, not a circular path. Circular motion is a little more complicated than linear motion. So, what happens if the rider's motion is circular? Since the rider is traveling in a circular path, an outside force must be acting on her to change the direction of motion from a straight path. This is similar to circular motion of a ball attached to a string. Here, the force is exerted on the ball by the string. The string pulls the ball back toward us, providing the force that pulls towards the center of the circular path. This force is known as centripetal force. The word centripetal means "center-seeking".

The carousel is another amusement park ride in which horses are attached to a rotating platform so that they go round and round in a horizontal circle. Here, the centripetal force again keeps the horses and riders traveling in a circular path just as the string provided the centripetal force for the ball. As long as the ride is moving slowly enough, the centripetal force of the platform can keep everyone and everything on board. In theory, if the carousel starts moving really fast, centrifugal force («center-fearing») takes over and breaks the hold the platform (centripetal force) had on the riders and the riders would fly off.

As we will discuss below, centrifugal force is actually not a real force. It is a fictitious force. If the centripetal force that pulls an object into the center stops working (e.g. the string breaks), then it is the object›s inertia that takes over and sends the object traveling in a straight path. We can test this outside by spinning the ball around us and letting go of the string. If centrifugal force was a real force, the ball would move straight away from the center at the point where the string was let go. But it does not. Instead, the ball follows its path of inertia and moves in a straight path that is tangent to the circular path.

Similar to translational motion, rotational motion of an object may be described in two parts: (i) rotational kinematics and (ii) rotational dynamics. In this chapter, we focus on rotational kinematics and defer the discussion on rotational dynamics to Chapter 8. It turns out that there are a similar set of equations which describe the rotational motion of objects. However, as we shall see below, they involve a different set of variables.

7.1 ROTATIONAL MOTION VERSUS CIRCULAR MOTION

In the real world, there are lots of objects moving in a circular path as well as rotating about some rotational axis. For example, objects moving in a circular path include an artificial satellite orbiting the Earth at constant height, a stone which is tied to a rope and is being swung in circles, a car turning through a curve in a race track, an electron moving perpendicular to a uniform magnetic field, and a gear turning inside a mechanism.

What is the difference between rotational motion and circular motion?

The difference between circular motion and rotational motion may be understood by considering motion of a stone-pole system in which the stone is attached to one end of the pole while other end is pivoted. In this example, circular motion is the stone attached to the pole if we only study the physical behavior of the stone. However, rotational motion refers to a whole body with a pivot point rotating about this point and the physical behavior of the whole body is studied. Let us examine another example. Suppose we consider a communication satellite orbiting around Earth which spins about

CIRCULAR MOTION

an axis once per day as illustrated in Fig. 7.1. A simple question we may ask is "Which object is executing rotational motion and which one is executing circular motion?"

It may be obvious to know that Earth's motion is rotational and the satellite's motion is circular. Then we may ask another question "What allows us easily distinguish the difference between rotational and circular motion?" It appears that motion of the satellite is similar to that of the stone in the previous example. Both the satellite and stone have a center of curvature for their path of motion. However, similar to the pole in the previous example, Earth is spinning about its axis of rotation. It appears that the common feature in the stone-pole and satellite-Earth system is that there is a pivot point or an axis of rotation, respectively.

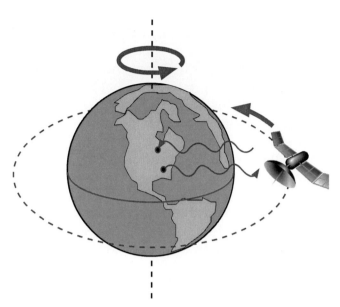

Figure 7.1: A schematic illustration of rotational and circular motion in the satellite-Earth system. Earth spins about its axis, while the satellite moves in a circular path with respect to the center of Earth.

To elaborate the difference between rotational and circular motion, let us look at few more examples. As shown in Fig. 7.2, there are clothes in a dryer drum and travel luggage on a baggage carousel. Let us see how we may tell the difference between rotational motion and circular motion in these two cases.

We may easily see that the clothes in a rotating dryer drum are in circular motion because it is moving in a circular path about a center point. However, the dryer drum's motion is rotational because it is spinning about an axis of rotation, which is the center of the dryer drum. Similarly, the travel luggage is moving in a circular path about a center point, but the baggage carousel is rotating about an axis of rotation. As may see from these examples, whether we consider an object moving in a circular path or consider a point on a rotating body, we may describe the motion by using rotational kinematics.

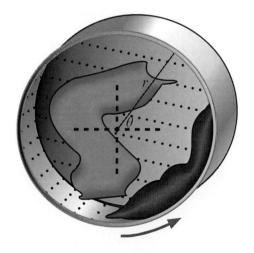 

Figure 7.2: A schematic illustration of both rotational and circular motion. In the left panel, motion of the clothes is circular, but motion of the dryer drum is rotational. Similarly, in the right panel, a travel bag is moving in a circular path, but the baggage carousel is executing rotational motion.

7.2 ROTATIONAL KINEMATICS

Rotational kinematics describes the circular and rotational motion of points, bodies (objects) and systems of bodies (groups of objects) without consideration of the causes of motion. Suppose we consider purely rotational motion of rigid bodies. Here, purely rotational motion is defined as motion in which all points in the body move in circles, and a rigid body is defined as a body with a definite shape such that its constituent particles stay in fixed positions relative to one another. In pure rotational motion, the constituent particles of a rigid body rotate about a fixed axis in circular trajectories. The particles which compose the rigid body are always at a constant perpendicular distance from the axis of rotation since their internal distances within the rigid body are locked. Furthermore, all particles in the rigid body rotate the same way, but they do not have the same speed. A particle which is farther away from the axis of rotation has a greater speed as it rotates about the rotational axis. This clearly indicates that rotation of a rigid body comprises circular motion of individual particles.

Rotational motion of an object may be described in the same way as the translational motion of a point particle. When dealing with the rotation of the object, it is simpler to consider the body itself as rigid. As we have discussed above, a body is generally considered rigid when the separations between all the particles remains constant throughout the objects motion, so for example parts of its mass are not flying off. In a realistic sense, all things are deformable. However, this effect on the motion is minimal and negligible. Thus the rotation of a rigid body over a fixed axis is referred to as rotational motion. Noting this we may recognize that, similar to translational kinematics, we need to define the following quantities to describe rotational motion of an object: (i) angular displacement, (ii) angular velocity, and (iii) angular acceleration. These quantities are similar to displacement, velocity, and acceleration in translational motion. As we will see below, the set of equations for describing rotational kinematics is identical to that for translational kinematics. First, let us discuss about angular displacement by asking the following question:

How can we describe an angular displacement?

Angular displacement of a body is defined as the angle in radians (also, in either degrees or revolutions) through which a point or line has been rotated in a specified sense, about a specified axis. This specified axis is usually the axis of rotation and the angle is measured with respect to a reference line as shown in Fig. 7.3. When an object such as a compact disc rotates about its axis, the motion cannot simply be analyzed as a particle, since circular motion involves a changing velocity and acceleration at any time t. It should be noted that both velocity and acceleration are vectors. This indicates that either magnitude or direction, or both, may change when the velocity and acceleration change.

We now define the angular coordinate by noting that the circumference of a circle of radius r is

$$\text{Circumference} = 2\pi r$$

We may think of the circumference of a circle as the arc length of an arc with angle 2π. In this context, the arc length ℓ for an angle θ (i.e., angular displacement) may be written as

$$\ell = \theta\, r \Rightarrow \theta = \frac{\ell}{r} \tag{7.1}$$

where θ is the angle measured in radians. The relation

$$360^0 = 2\pi \text{ rad}$$

CIRCULAR MOTION

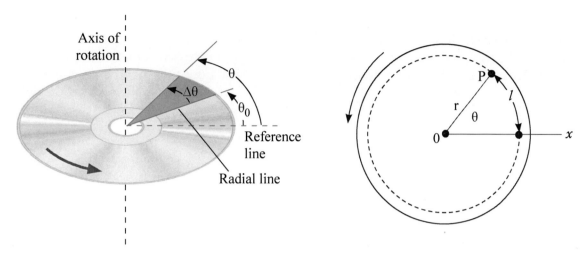

Figure 7.3: A schematic illustration of how to determine the angular displacement of a rotating object. The angular displacement is the angle with respect to a reference line, usually a horizontal line. The angular displacement of point P (right panel) may be converted to the distance traveled using the radius r.

indicates that 1 rad = $360°/2\pi$ = $57.3°$. Angle subtended by a circular arc, if quoted in rodians, is equal to the arc length divided by the arc radius. A measurement of an angle in radians is numerically equal to the length of a corresponding arc of a unit circle. The radian is the standard unit of angular measure, used in many areas of mathematics. The unit was formerly an SI supplementary unit, but this category was abolished in 1995 and the radian is now considered an SI derived unit. The SI unit of solid angle measurement is the steradian.

Exercise Problem 7.1

A child rolls a ball on a level floor. The ball rolls 4.5 m to another child. If the ball makes 15.0 revolutions, what is its diameter?

Solution:

The distance traveled by the ball in one revolution is

$$d_{1rev} = 2\pi r,$$

indicating that the radius r of the ball may be computed from the total distance traveled. The total distance d_{tot} traveled is

$$d_{tot} = n_{rev} d_{1\,rev} = n_{rev}(2\pi r) \Rightarrow r = \frac{d_{tot}}{2\pi n_{rev}}$$

where n_{rev} is the number of revolutions. Since the diameter is twice the radius, we obtain

$$2r = \frac{d_{tot}}{\pi n_{tot}} = \frac{4.5\,\text{m}}{\pi \times 15} = 0.095\,\text{m} = 9.5\,\text{cm}.$$

As we can see from Example Problem 7.1, the rotational motion of an object is closely related to its translational motion. Similar to translational kinematics, rotational kinematics describe how objects rotate. To describe rotational kinematics, in addition to angular displacement, we need to know both angular velocity and angular acceleration. These two physical quantities have their counterparts in translational motion. **Angular velocity** ω is analogous to linear velocity v. Remembering the definition of average linear velocity in Chapter 2, we may define the average angular velocity as

$$\bar{\omega} = \frac{\Delta \theta}{\Delta t} = \frac{\theta - \theta_o}{\Delta t}. \tag{7.2}$$

Also, the instantaneous angular velocity may be defined as

$$\alpha = \lim_{\Delta t \to 0} \frac{\Delta \omega}{\Delta t}. \tag{7.3}$$

For rotational motion, it is noted that all points in a rigid body rotate with the same angular velocity. The **angular velocity** is defined as the rate of change of angular displacement. Also, it is a vector quantity (more precisely, a pseudo-vector) which specifies the angular speed (rotational speed) of an object and the axis about which the object is rotating. The SI unit of angular velocity is radians per second (i.e., rad/s). However, it may be measured in other units such as degrees per second, degrees per hour and so forth. The direction of the angular velocity vector is perpendicular to the plane of rotation. This direction is usually specified by the right-hand rule. We discuss this right hand rule in Chapter 8. Similarly, **angular acceleration** α is analogous to linear acceleration a. Remembering the definition of linear acceleration from Chapter 2, we may define the average angular acceleration as

$$\bar{\alpha} = \frac{\Delta \omega}{\Delta t} = \frac{\omega - \omega_o}{\Delta t}. \tag{7.4}$$

Also, the instantaneous angular acceleration may be defined as

$$\alpha = \lim_{\Delta t \to 0} \frac{\Delta \omega}{\Delta t}. \tag{7.5}$$

Angular acceleration is the rate of change of angular velocity. In SI units, it is measured in radians per second squared (i.e., rad/s²). Since all points of a whole rigid body rotate with the same angular velocity and acceleration when the object rotates, it is noted that both ω and α are properties of the rotating body as a whole.

As we can see from the rotating tires of a moving car, there is a relationship between angular and linear speed. Here, the angular speed of a tire is the rate at which the central angle of a spinning wheel is changing and the linear speed of the car is the rate at which the distance traveled by the wheel is changing. The relation between these two velocities may be written as

$$v = \frac{\Delta \ell}{\Delta t} = \frac{r \Delta \theta}{\Delta t} = r\omega. \tag{7.6}$$

Here, the distance traveled is equal to as the arc length $\Delta \ell$ due to the angular displacement $\Delta \theta$. Although ω is the same for every point in the rotating body at any instant, the linear velocity v is greater for points farther from the axis since v is linearly proportional to r.

For an object in a circular motion, there are two contributions to acceleration. These two contributions are (i) transverse or tangential acceleration a_{tan} and (ii) radial or centripetal acceleration a_R. These two accelerations are perpendicular to each other. For non-uniform circular motion in which

CIRCULAR MOTION

the speed along the curved path changes, these two accelerations contribute to yield the net acceleration as shown in Fig. 7.4. First, the transverse acceleration of the object is equal to the rate of change of the angular speed around the circle times the radius of the circle. This means that whenever there is changing angular velocity ω the tangential acceleration a_{tan} is non-zero. For example, if the angular velocity at time t_1 is ω_1 and that at time t_2 is ω_2, the change in angular velocity is $\Delta\omega = \omega_2 - \omega_1$. The time interval during which the angular velocity changes is $\Delta t = t_2 - t_1$. So, according to the definition, the tangential acceleration a_{tan} is given by

$$a_{tan} = \frac{\Delta v}{\Delta t} = \frac{r\Delta\omega}{\Delta t} = r\alpha. \tag{7.7}$$

Equation (7.7) indicates the direct relationship between the angular acceleration α and tangential linear acceleration a_{tan}.

Second, the centripetal acceleration a_R is due to the changing direction of the linear motion of particles in a rotating rigid body. For simplicity, if we assume that the angular speed ω is constant, a particle at point P in Fig. 7.4 changes its direction of motion continuously while keeping the tangential speed $v = r\omega$ the same. This changing direction implies that the acceleration is directed toward the center of the circular motion (i.e., point 0). A detailed derivation of the centripetal acceleration is discussed in Sec. 7.4. This implies that the force acting on a body in uniform circular motion is also directed toward the center of the circle. It is centripetal, which means "center seeking". The uniform angular velocity is related to the centripetal acceleration as

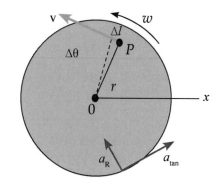

Fig. 7.4: A schematic illustration of a rotating object. When the angular velocity of the object changes with time, each point has both tangential acceleration a_{tan} and radial acceleration a_R, which are perpendicular to each other.

$$a_R = \frac{v^2}{r} = \frac{(r\omega)^2}{r} = r\omega^2. \tag{7.8}$$

In short, with uniform circular motion (i.e., constant speed along a circular path), a body experiences an acceleration resulting in velocity of a constant magnitude but changing direction. The total acceleration which accounts for the contributions from tangential and centripetal accelerations for non-uniform circular motion is

$$\vec{a} = \vec{a}_{tan} + \vec{a}_R. \tag{7.9}$$

Non-uniform circular motion is any case in which an object moving in a circular path has a varying speed. There is still a contribution to the net acceleration from the centripetal acceleration because the direction of motion constantly changes. However, in addition, the tangential acceleration a_{tan} is non-zero since the speed is also changing. Therefor, an object executing non-uniform circular motion experiences both centripetal force and tangential forces.

It is helpful to note that, while the centripetal force is a physical force which is directed towards the center, there may also be the so-called 'centrifugal force' appearing to act outward on the body. As we discuss below in Sec. 7.4, this force is really a pseudo force experienced in the frame of reference of the body in circular motion, due to the body's linear momentum at a tangent to the circle.

7.3 ROLLING MOTION

As an example of relation between linear translational motion and rotational motion, we consider rolling motion. **Rolling motion** is a type of motion that combines rotation (usually, of an axially symmetric object) and translation of an object with respect to a surface. Under ideal conditions the object and surface are in contact with each other without sliding. In this case, the rotational speed at the line or point of contact is equal to the translational speed, and the motion is "pure rolling". In practice, due to small deformations at the contact area, some sliding does occur. Ignoring this negligible deformation, we may write the relation between linear speed v and angular speed ω as

$$v = r\omega \tag{7.10}$$

Also, we write the relation between linear acceleration a and angular acceleration α as

$$a = r\alpha \tag{7.11}$$

Rolling without slipping involves both rotation and translation. Whether or not this condition can be achieved depends on static friction between the rolling object and the ground.

In real situations, the friction force between a rotating object and surface which facilitates rolling motion dissipated part of energy of the system. For example, when an automobile coasts along the road as shown in Fig. 7.5, the resistive force of rolling friction acting on tires slows down the motion. The rolling friction of the tire is slightly affected by the static friction of the rubber on the pavement. The adhesion effect of the rubber adds a little more to the rolling friction. However, the major contribution to the rolling friction is the deformation of the tires while rolling. The coefficient of friction for the automobile tires can be determined experimentally, but it only applies to the specific configuration. Nevertheless, rolling resistance is much lower than sliding friction. Hence, rolling objects, typically, require much less energy to be moved than sliding ones. As a result, such objects will move more easily if they experience a force with a component along the surface, for instance gravity on a tilted surface, wind pushing, and pulling by an engine.

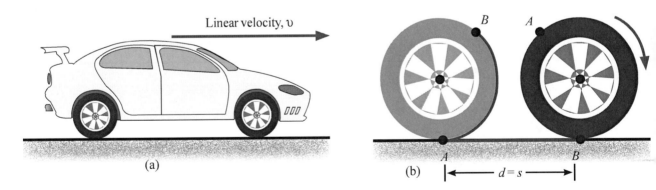

Figure 7.5: A schematic illustration of rolling motion of tires on a car. Here, the friction between the tire and the ground yields the relation between rotational motion of the tire and translational motion of the car.

Angular velocity and frequency of rotation

As we can see from an object in circular motion, any rotational motion is repetitive since the position of a particle appears to be at the same location after the 2π angular displacement. For any objects exhibiting a repetitive motion, we may use either angular or linear frequency to describe the motion. **Angular frequency** ω (also referred to by the terms angular speed, radial frequency,

CIRCULAR MOTION

circular frequency, orbital frequency, and radian frequency) is a scalar measure of rotation rate. Hence, angular frequency (or angular speed) is the magnitude of the vector quantity, angular velocity. Hence, there is relationship between the angular velocity ω and the frequency of rotation f. This relation is written as

$$f = \frac{\omega}{2\pi} \Rightarrow \omega = 2\pi f. \tag{7.12}$$

The unit for frequency f is "Hertz" (Hz) which is revolution per second (i.e., rev/s). Similarly, period T is defined as the time required for one complete revolution

$$T = \frac{1}{f}. \tag{7.13}$$

This means that an object which rotates with uniform angular velocity $\omega = 2\pi f$ turns through f revolutions in 1 second. Hence, the object turns through 2π radians (i.e., it executes a complete circle) in T seconds.

Exercise Problem 7.2

A 70 cm-diameter wheel accelerates uniformly from 160 rpm to 280 rpm in 4.0 s. Determine its (a) angular acceleration. Also, compute (b) the radial and tangential components of the linear acceleration of a point on the edge of the wheel 2.0 s after it has started accelerating.

Solution:

Using the definition, we compute the angular acceleration α and obtain

$$\alpha = \frac{\Delta \omega}{\Delta t} = \frac{\omega_f - \omega_i}{\Delta t} = \frac{(280-160)\,\text{rev}/\text{min} \times 2\pi/\text{rev} \times 1\,\text{min}/60\text{s}}{4.0\text{s}} = 3.14\,\frac{\text{rad}}{\text{s}^2}.$$

Similarly, we compute the tangential acceleration a_{tan} and obtain

$$a_{tan} = \frac{\Delta v}{\Delta t} = \frac{r\Delta\omega}{\Delta t} = r\alpha = 0.35\,\text{m} \times 3.13\,\frac{\text{rad}}{\text{s}^2} = 1.1\,\frac{\text{m}}{\text{s}^2}.$$

For an object with angular velocity ω, the radial acceleration a_R is

$$a_{rad} = \frac{v^2}{r} = \frac{(r\omega)^2}{r} = r\omega^2 = r(\omega_i + \alpha t)^2$$

$$= 0.35\,\text{m} \times \left(160 \times \frac{2\pi}{60}\,\frac{\text{rad}}{\text{s}} + 3.14\,\frac{\text{rad}}{\text{s}^2} \times 2.0\text{s}\right)^2 = 1.9 \times 10^2\,\frac{\text{m}}{\text{s}^2}$$

Here, the wheel has angular acceleration α, indicating that its angular velocity ω increases with time. So, we used the relation $\omega = \omega_i + \alpha t$ to compute the angular velocity at time t. This equation is similar to one of the kinematic equations that we used to describe translational motion of an object.

Kinematics Equations

Exercise Problem 7.2 suggests that there is much similarity between rotational and translational kinematics. In describing circular motion of an object, it is helpful to recognize that circular kinematic equations are very similar to linear kinematic equations. The main difference is that in circular kinematics the bodies are always rotating, like a weight at the end of a string we are swinging around, instead of moving in a straight path as in translational kinematics. We do not derive the equations for circular kinematics here. However, for the purposes of comparing circular kinematics with linear kinematics, we consider the equations for uniform angular acceleration and tabulated them below:

	Linear	Circular
Speed versus time	$v = v_0 + at$	$\omega = \omega_0 + \alpha t$
Position versus time	$x = x_0 + v_0 t + \frac{1}{2}at^2$	$\theta = \theta_0 + \omega_0 t + \frac{1}{2}\alpha t^2$
Speed versus position	$v^2 = v_0^2 + 2a(x - x_0)$	$\omega^2 = \omega_0^2 + 2\alpha(\theta - \theta_0)$
Average speed	$\bar{v} = \dfrac{v + v_0}{2}$	$\bar{\omega} = \dfrac{\omega + \omega_0}{2}$

Because the equations describing rotational and translational motion are mathematically equivalent, we may simply substitute the rotational variables, such as angular displacement θ, angular velocity ω, and angular acceleration α, into the kinematic equations which we derived in Chapter 2 and replace the corresponding translational variables. We do not need to derive these equations since the procedures for deriving the circular kinematic equations are the same as those derived in one-dimensional translational kinematic equations. Thus we can simply state the equations, alongside their translational analogues.

Exercise Problem 7.3

A bicycle slows down uniformly from $v_0 = 8.40$ m/s to rest over a distance of 115 m. Each wheel and tire has an overall diameter of 68.0 cm. Determine (a) the angular velocity of the wheels at the initial instant, (b) the total number of revolution each wheel rotates in coming to rest, (c) the angular acceleration of the wheel, and (d) the time it took to come to a stop.

(a) Bike as seen from the ground ($t = 0$).

(b) From rider's reference frame the ground is moving to the rear ar an initial speed of 8.40 m/s ($t = 0$).

CIRCULAR MOTION

> **PROBLEM SOLVING STRATEGY**
>
> **What do we know?**
>
> The bicycle slows down over a given distance.
>
> **What concepts are needed?**
>
> We need to apply the rotational kinematic equations. Also, we need to use the relation between translational and rotational motion for rolling motion of the wheels.
>
> **Why?**
>
> The bicycle wheels roll on the surface. Also, the initial speed of the bicycle is reduced to the final speed of zero over the distance of 115m, indicating that the speed-versus-position equation is needed.

Solution:

a. Using the relation between the linear velocity and angular velocity (i.e., $v_0 = r\omega_0$), we obtain the initial angular velocity ω_0 as

$$\omega_0 = \frac{v_0}{r} = \frac{v_0}{d/2} = \frac{8.40\,\text{m/s}}{0.68\,\text{m}/2} = 24.7\,\text{rad/s}.$$

b. The number of revolution n_{rev} for the bicycle wheel may be obtained from the distance x traveled divided by the circumference of the wheel as

$$n_{rev} = \frac{x}{2\pi \cdot d/2} = \frac{115\,\text{m}}{2\pi \times 0.68\,\text{m}/2} = 53.8\,\text{rev}.$$

c. The angular acceleration α of the wheel may be determined by using the speed-versus-position equation

$$\omega^2 = \omega_0^2 + 2\alpha(\theta - \theta_0)$$

We solve for the angular acceleration α and obtain

$$\alpha = \frac{\omega^2 - \omega_0^2}{2(\theta - \theta_0)} = \frac{-(24.7\,\text{rad/s})^2}{2 \times 2\pi \times 53.8\,\text{rev}} = -0.902\,\frac{\text{rad}}{\text{s}^2}.$$

Note that 2π is in radian so that the final unit for α is in rad/s².

d. The time that the bicycle takes to come to rest is determined by using the kinematic equation, $\omega = \omega_0 + \alpha t$. We solve for the time t and obtain

$$t = \frac{\omega - \omega_0}{\alpha} = \frac{-24.7\,\text{rad/s}}{-0.902\,\text{rad/s}^2} = 27.4\,\text{s}.$$

As we can see from motion of the bicycle wheels in Exercise Problem 7.3, **rolling motion** is a combination motion in which an object rotates about an axis as it moves along a straight-line trajectory. For the bicycle wheel rolling down the road, it is rolling without slipping. Similarly, for a disk or sphere rolling along a horizontal surface, the motion can be considered in two ways: (i) combination of rotational and translational motion, and (ii) pure rotational motion. If we consider the combination of rotational and translational motion, then the center of mass is in translational motion and the rest of the body is rotating around it. However, if we consider purely rotational motion, then the whole object is revolving around a point on the object in contact with the surface. This point of contact changes with time.

Let us consider combination of rotational and translational motion as shown in Fig. 7.6. When the object makes one complete revolution, it has moved a distance equal to the circumference, and each point on the exterior has touched the ground once. When the object rotates through an angle θ, the distance that the center of mass has moved is $\Delta x_{cm} = \theta R$ where R is the radius. As the object rolls one revolution without slipping, a point on the rim indicated by the black dot follows a cycloid path, representing the combined motion, as shown in Fig. 7.6.

As indicated by the horizontal distance traveled by the black dot, this means that if the object rotates about its axis one complete revolution then the center of mass moves forward one circumference $\Delta x_{cm} = 2\pi R$ because the object does not slip. Another interesting point to note is that the velocity of two different points with the same radial distance from the axis of rotation is different for a rolling object. **The point on the bottom of a rolling object is instantaneously at rest**, and the point on the top as shown in Fig. 7.7 is twice the center of mass velocity.

As illustrated in Fig. 7.7, the velocity of any point on the disk as seen by an observer on the ground is the vector sum of the velocity with respect to the center of mass ($v_{rel,cm}$) and the velocity of the center of mass with respect to the ground (v_{cm})

$$\vec{v}_{ground} = \vec{v}_{rel,cm} + \vec{v}_{cm}. \tag{7.14}$$

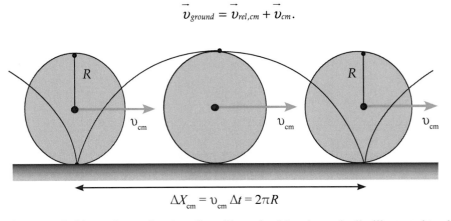

Figure 7.6: The trajectory of a blue point on the rim of a rolling wheel is schematically illustrated to show the combined motion of translation and rotation lead to a cycloid path.

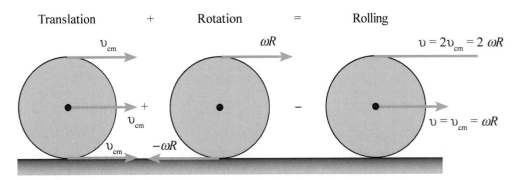

Figure 7.7: Due to the combination of rotation and translation motion in a rolling object, the speed of a particle on the wheel depends on its position relative to the ground, which is the point of contact with the surface.

CIRCULAR MOTION

This means that the point at the bottom and that at the top of a wheel has the minimum and maximum speed, respectively. The point on the top of the wheel has a speed (relative to the ground) that is twice the velocity of the center of mass. If our car is traveling down the highway at 70 mph, the tops of the wheels are moving at 140 mph while the bottoms of the wheels are moving at 0 mph.

Exercise Problem 7.4

A bicycle with 0.80-m-diameter tires is coasting on a level road at 5.6 m/s. A small red dot has been painted on the tread of the rear tire. (a) What is the angular speed of the tires? (b) What is the speed of the red dot when it is 0.80 m above the road? (c) What is the speed of the red dot when it is 0.40 m above the road?

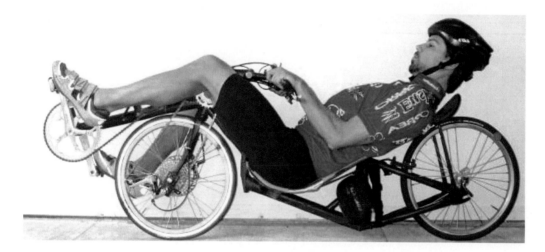

Solution:

a. Since the tires are rolling without slipping, the angular velocity ω of the tires is

$$\omega = \frac{v_{cm}}{R} = \frac{5.6 \text{ m/s}}{0.40 \text{ m}} = 14 \text{ rad/s}.$$

b. The black dot on the rear tire is undergoing translation and rotation. At the top of the tire, it has a translational velocity equals to the speed of the bike plus an additional velocity, which equals ωR, due to the rotation of the tire. Therefore, the speed of the red dot at the top is

$$v = v_{cm} + \omega R = 2v_{cm} = 2 \times 5.6 \text{ m/s} \cong 11 \text{ m/s}$$

c. The dot has a translational velocity equal to the velocity of the center of mass of the tire in the horizontal direction. The tangential velocity of the dot is in the vertical direction and its magnitude is equal to ωR. The velocity of the dot is equal to the sum of these two vectors. Since the tire is rolling without slipping $v_{cm} = \omega R$. The speed of the dot is equal to

$$v = \sqrt{v_{cm}^2 + v_{cm}^2} = 7.9 \, \text{m}/\text{s}$$

This is the reason why pebbles fly out of tire tread when a car is moving fast. For example, consider a 1.3 g pebble stuck in a tread of a 0.83 m-diameter (i.e., $2r$) automobile tire, held in place by static friction of at most 3.3 N. When the car starts from rest and gradually accelerates on a straight road, a greater centripetal force is needed to keep the pebble stuck in the tread. Finaly, when the car moves at a speed larger than $v_{max} = (mgr)^{1/2} = 16.2$ m/s, the pebble flies out of the tire tread. Here, μ is the coefficient of friction.

7.4 DYNAMICS OF UNIFORM CIRCULAR MOTION

In Sec 7.2, we stated that any object moving in a circular path always has a component of acceleration directed toward the center of the path. To prove this point, we consider an object experiencing "uniform circular motion". Uniform circular motion can be described as the motion of an object in a circle at a constant speed. As shown in Fig. 7.8, the magnitude of the velocity remains constant but the direction of the velocity changes continuously as the object moves around the circle.

When the object revolves in a circle continuously, it is accelerating even when the speed remains constant (i.e., $v_1 = v_2 = v$). At all instances, the object is moving tangent to the circle since the direction of the velocity vector is the same as the direction of the object's motion. The velocity vector is directed tangent to the circle. Hence, an object moving in a circle is accelerating.

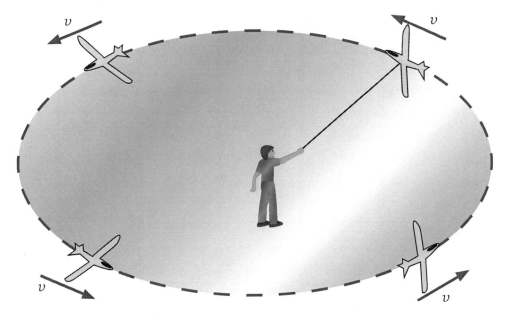

Figure 7.8: A schematic illustration of uniform circular motion. When an object moves in a circular path with a constant speed, it changes the direction of motion. Hence, it accelerates. This is called centripetal acceleration.

CIRCULAR MOTION

Why?

Uniform circular motion implies that the direction of motion is being changed constantly. Accelerating objects are objects which change their velocity with time. This means that either the speed (i.e., magnitude of the velocity vector) or the direction must be changed. An object in uniform circular motion is moving with a constant speed. Nonetheless, it is accelerating due to its change in direction. The direction of the acceleration is inwards. Figure 7.9 illustrates the changes in the direction of velocity by means of a vector arrow.

The final motion v_2 characterizes the direction of the net force acting on an object undergoing uniform circular motion. According to Newton's second law, the net force acting upon such an object is directed towards the center of the circle since the tangential acceleration is zero. The net force is said to be an inward or **centripetal force**. Without such an inward force, an object would continue in a straight line, never deviating from its direction. Yet, with the inward net force directed perpendicular to the velocity vector, the object is always changing its direction and undergoing an inward acceleration.

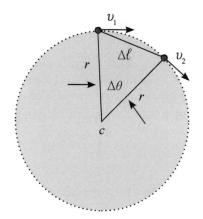

Figure 7.9: A schematic illustration of a wheel rotating with respect to the axis of rotation at the center (i.e., point C). Due to angular displacement $\Delta\theta$, the red point on the rim changes the direction from v_1 to v_2.

Centripetal Acceleration

Now, let us see why that the centripetal force is always directed inward. If an object is traveling in a circle then it is accelerating, even if its speed is constant. This is because velocity is a vector, and so a change in the velocity means a change in the direction. Acceleration, according to its definition, is the change in velocity per change in time. This is a sideways acceleration because its direction is perpendicular to the direction of velocity. It causes the object to "turn". When the object is moving along a circle, this means that the acceleration is along the radius, pointing toward the center of the circle. This may be easily seen from the definition **acceleration** written as

$$\vec{a} = \frac{\vec{v}_2 - \vec{v}_1}{\Delta t} = \frac{\Delta \vec{v}}{\Delta t}. \tag{7.15}$$

If the time interval Δt is very small, then the arc length $\Delta \ell$ and the angular displacement $\Delta \theta$ are also very small. If v_1 is the initial velocity vector, then the velocity vector v_2 at Δt later will be almost parallel to v_1, as shown in Fig. 7.10. This means that $\Delta v = v_2 - v_1$ will be essentially perpendicular to both v_1 and v_2.

The acceleration vector a of Eq. (7.15) is in the same direction as Δv and points towards the center of the circle. So, this is called **centripetal acceleration** (i.e., "center seeking" acceleration) or radial

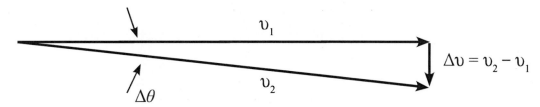

Figure 7.10: A schematic illustration of a small change in the direction of velocity vector. Although the vectors v_1 and v_2 do not differ much, the direction of the vector $\Delta v = v_2 - v_1$ is perpendicular to both v_1 and v_2.

acceleration (since it is directed along the radius, toward the center of the circle). We may compute the magnitude of the centripetal acceleration by noting that

$$\frac{\Delta v}{v} = \Delta\theta = \frac{\Delta\ell}{r} \Rightarrow \Delta v = \frac{v}{r}\Delta\ell. \qquad (7.16)$$

Here, we may consider Δv as the arc length of a circle with radius $v = |v_1| = |v_2|$. This is similar to the arc length $\Delta\ell$ for part of circle with an angular displacement $\Delta\theta$ and radius r. Using Eq. (7.16), we may write the magnitude of the centripetal acceleration a_R as

$$a_R = \frac{\Delta v}{\Delta t} = \frac{v}{r}\frac{\Delta\ell}{\Delta t} = \frac{v^2}{r}. \qquad (7.17)$$

Equation (7.17) indicates that an object moving in a circle of radius r with constant speed v has acceleration whose direction is toward the center of the circle and whose magnitude is

$$a_R = \frac{v^2}{r}. \qquad (7.18)$$

As indicated above, circular motion is often described in terms of the **frequency** f (i.e., revolutions per second) or the **period** T (i.e., the time required for one complete revolution) as

$$T = \frac{1}{f} \quad \text{and} \quad v = \frac{\Delta x}{\Delta t} = \frac{2\pi r}{T}. \qquad (7.19)$$

because it is a repetitive motion. So, for a particle under uniform circular motion, it covers a constant distance in completing circular trajectory in one revolution. This distance is equal to the perimeter of the circle $s = 2\pi r$. Further the particle covers the perimeter with constant speed. It means that the particle travels the circular trajectory in a constant time given by its time period T as $T = 2\pi r/v$. This implies that the frequency f and speed v is related as $1/f = 2\pi r/v$.

Exercise Problem 7.5

The Moon's nearly circular orbit about the Earth has a radius of about 384,000 km and a period T of 27.3 days. Determine the acceleration of the Moon toward the Earth.

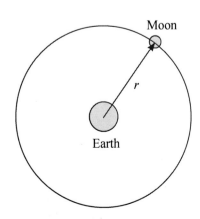

PROBLEM SOLVING STRATEGY

What do we know?

The Moon's orbit around the Earth is nearly circular.

What concepts are needed?

The acceleration of the Moon toward the Earth is due to centripetal acceleration.

Why?

The Moon orbits around the Earth with a constant speed, indicating that there is no tangential acceleration.

CIRCULAR MOTION

Solution:

Using Eqs. (7.17) and (7.18), respectively, for the centripetal acceleration and the speed

$$a_R = \frac{v^2}{r} \text{ and } v = \frac{2\pi r}{T},$$

we obtain the expression for the radial acceleration (or centripetal acceleration)

$$a_R = \frac{(2\pi r)^2}{T^2 r} = \frac{4\pi^2 r}{T^2}.$$

We now substitute the values for r and T to compute the magnitude of a_R and obtain

$$a_R = \frac{4\pi^2 \times 3.84 \times 10^5 \text{ km} \times \frac{1000 \text{ m}}{1 \text{ km}}}{\left(27.3 \text{d} \times \frac{24 \text{h}}{1 \text{d}} \times \frac{3600 \text{s}}{1 \text{h}}\right)^2} = 2.72 \times 10^{-3} \frac{\text{m}}{\text{s}^2}$$

The acceleration a_R, which reflects the force ma_R maintaining the moon's orbit around Earth, is much smaller than the gravitational acceleration $g = 9.8$ m/s².

It is interesting that a tiny second moon may once have orbited Earth before catastrophically slamming into the other one. As a new study suggests, this titanic clash could explain why the two sides of the surviving lunar satellite are so different from each other. The second moon around Earth would have been about 750 miles (1,200 km) wide. Scientists suspect it could have been formed from the same collision between the planet and a Mars-sized object and helped in creating the moon we see in the sky today. This type of collisions and the gravitational interactions (discussed in Sec. 7.5) have played the central role in the formation of the current Earth-Moon system we see today. The gravitational tug of war between the Earth and Moon slowed the rate at which it whirls, such it now always shows just one side to Earth. The far side of the Moon remained a mystery for centuries until 1959, when the Soviet Luna 3 spacecraft first snapped photos of it. The far side is sometimes erroneously called the dark side, even though it has days and nights just like the near side.

Centripetal force versus centrifugal force

An object in circular motion at constant speed experience either centripetal force or centrifugal force. **Centripetal force** is a "real" force that prevents the object from "flying out" and keeping it moving with a uniform speed along a circular path. The centripetal force counteracts the centrifugal force. What is centrifugal force? The word centrifugal is from Latin word *centrum* which means "center" and *fugere* which means "to flee". Hence, the **centrifugal force** is the apparent outward force that draws a rotating body away from the center of rotation. It is caused by the inertia of the body as its path is continually redirected. In Newtonian mechanics, the term centrifugal force is used to refer to one of two distinct concepts: an inertial force (also called a "fictitious" force) observed in a non-inertial reference frame, and a reaction force corresponding to a centripetal force as shown in Fig. 7.11. This **centrifugal force** describes the tendency of an object following a curved path to fly outwards, away from the center of the curve. It is not really a force because it results from inertia.

Centrifugal force arises as a result of the tendency of an object to resist any change in its state of rest or motion. The conception of this force has evolved since the time of Huygens, Newton, Leibniz, and Hooke. Its modern conception as a fictitious force arising in a rotating reference frame evolved in the eighteenth and nineteenth centuries.

Centrifugal force is a virtual force (i.e., not a real force). However, its effect is as real as a real force. There are some situations we can be in that the centrifugal force can have us accelerate without speeding up. One of these situations is a carnival ride which spins around in a circle at constant speed. In this ride, we feel pressed against the wall very tightly, and then the floor drops out. Most people would believe they were moving steadily, with their bodies being pressed tightly against the wall (outward, in a centrifugal direction). This is centripetal force. However, we know that this is not really what happens. When moving fast, a great deal of force is required to make us change direction. Our body "wants" to continue in a straight line. The curved wall gets in the way. The wall pushes in against our body. The "outward force" is just our body trying to move in a straight line. It is not a force at all. It is inertia, our body resisting the effects of the forces it feels. Virtual forces exist when our body is accelerating. When we see objects moving at a steady speed with steady change in direction, they appear to accelerate.

A ball in a moving car is also a good example, demonstrating centrifugal force. Suppose we place a ball on a car seat while moving at a steady speed and have the driver slam on the brakes. We will observe the ball appears to be pushed forward and off the seat. Of course, there is no force acting on the ball, and it just continues to move forward. However, it appears as if the ball is experiencing a force. In reality, it is the car that felt the backward force of the brakes. Another example is a heavy ball hanging from a spring in an elevator. As the elevator begins to rise, the ball begins to move, as if someone pushed down on it. It is the ball just "trying" to stay still as the elevator accelerates upward. While rising, we can stop the motion. It will start again when the elevator stops. The faster the acceleration (the more we can "feel" it in our body), the stronger the virtual forces appear to be.

Another example is how a centrifuge separates different components in a fluid. A centrifuge is a piece of equipment, generally driven by an electric motor (some older models were spun by hand), that puts an object in rotation around a fixed axis, applying a force perpendicular to the axis. A centrifuge is used to separate the components of blood in blood banks. The centrifuge works using the sedimentation principle in which the centrifugal acceleration causes denser substances to separate out along the radial direction (the bottom of the tube). By the same token lighter objects will tend to move to the top (of the tube; in the rotating picture, move to the center).

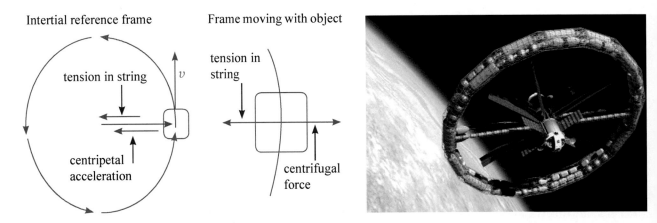

Figure 7.11: A schematic illustration of the difference between the centripetal force and centrifugal force (left panel). When a person is on an accelerating reference frame such as a space station slowly rotating about an axis (right panel), the person will experience a fictitious force known as centrifugal force (e.g., artificial gravity).

CIRCULAR MOTION

Artificial gravity is another example of centrifugal acceleration. Artificial gravity is the variation (increase or decrease) of apparent gravity (g-force) via artificial means. This variation can be achieved in space as well as on the Earth. Artificial gravity can be achieved by the use of the centrifugal force and some fictitious forces due to linear acceleration. The creation of artificial gravity is considered desirable for long-term space travel or habitation, for ease of mobility, for in-space fluid management, and to avoid the adverse long-term health effects of weightlessness in a space-station shown in Fig. 7.11. Without g-force, space adaptation syndrome occurs in some humans and animals. Many adaptations occur over a few days, but over a long period of time bone density decreases, and some of this decrease may be permanent. The minimum g-force required to avoid bone loss is not known because nearly all current experience is with g-forces of $1g$ (on the surface of the Earth) or $0g$ in orbit. Also, there has been insufficient time spent on the Moon to determine whether or not lunar gravity is sufficient to prevent the bone loss. A limited amount of experimentation has been performed with chickens since they are bipeds.

Similar experiments had been performed on mice. Mice can experience high g-force over long periods in large centrifuges on the Earth. Rats have been exposed to continuous artificial gravity of $1g$ during Russian biosatellite missions lasting two weeks. The muscle and bone loss in these animals was found to be less than rats in zero gravity. Also, sensory perception may change in the artificial gravity environment. As a test, astronauts were exposed to artificial gravity levels ranging from 0.2 to $1g$ for a few minutes during several spaceflight missions, using linear sleds or rotating chairs. When the g level was lower than $0.5g$, they did not perceive any changes in their spatial orientation at the inner ear level, where the sensory receptors for gravity perception are located.

Exercise Problem 7.6

A 0.150 kg ball on the end of a 1.10 m long cord (negligible mass) is swinging in a vertical circle. Determine the minimum speed the ball must have at the top of its arc so that it continues moving in a circle. Calculate the tension in the cord at the bottom of the arc assuming the ball is moving at twice the speed at the top.

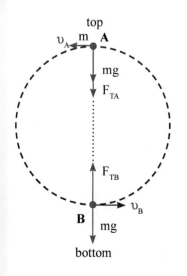

Problem Solving Strategy

What do we know?

The ball is attached to a string and is swinging in a vertical circle.

What concepts are needed?

We need to apply Newton's second law to the ball in circular motion. Also, we need to draw a free-body diagram to determine the net force directed towards the center of the motion.

Why?

The radial forces acting on the ball are keeping it on the vertical circular path.

Solution:

We start with Newton's second law applied to circular motion and write

$$\sum_i F_{Ri} = ma_R.$$

At the top of the loop (at point A), we analyze the free-body diagram and write

$$mg + F_{TA} = ma_R = m\frac{v_A^2}{r}.$$

Since the minimum speed is obtained when the ball barely makes the loop, the tension on the string should vanish. So, we set $F_{TA} = 0$ and solve for the speed at the top. We obtain

$$mg = m\frac{v_A^2}{r} \Rightarrow v_A = \sqrt{gr} = \sqrt{9.8 \text{ m/s}^2 \times 1.10 \text{m}} = 3.28 \text{ m/s}.$$

At the bottom of the loop, again, we analyze a free-body diagram and write

$$F_{TB} - mg = ma_R = m\frac{v_B^2}{r}$$

We solve for the tension F_{TB} at point B and obtain

$$F_{TB} = mg + m\frac{v_B^2}{r} = 0.15\text{kg} \times 9.8 \text{ m/s}^2 + 0.15\text{kg} \times \frac{(2 \times 3.28 \text{ m/s})^2}{1.10\text{m}} = 7.34 \text{ N}$$

Exercise Problem 7.6 indicates that vertical circular motion near the surface of Earth can change the acceleration from the usual gravitational acceleration of g. For example, when an acrobatic airplane is pulling up in a $+g$ maneuver, the pilot is experiencing several times the Earth's gravity as inertial acceleration in addition to the force of gravity. The cumulative vertical axis forces acting upon his body make him momentarily 'weigh' many times more than normal. Positive, or "upward" g, drives blood downward to the feet of a seated or standing person. More naturally, the feet and body may be seen as being driven by the upward force of the floor and seat, upward with respect to the blood. Resistance to positive g varies. A typical person can handle about $5g$ (49 m/s²) before losing consciousness, but through the combination of a special flight suit (i.e., anti-g suit) and efforts to strain muscles the modern fighter pilots can typically handle a sustained $9g$ (88 m/s²). Both the flight suit and muscles act to force blood back into the brain

Exercise Problem 7.7

A flat puck (mass M) is rotated in a circle on a frictionless air hockey tabletop, and is held in this orbit by a light cord which is connected to a dangling mass m through the central hole. Show that the speed of the puck is given by $v = (mgR/M)^{1/2}$.

CIRCULAR MOTION

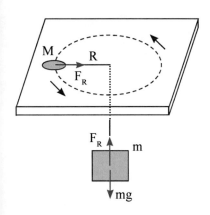

> **PROBLEM SOLVING STRATEGY**
>
> **What do we know?**
> The flat puck is moving in a circular path, while the mass which is connected to the puck via string is not moving.
>
> **What concepts are needed?**
> We need to apply Newton's second law to the flat puck due to its circular motion. However, we apply Newton's first law to the dangling mass.
>
> **Why?**
> For the flat puck, a radial force (i.e., tension) is keeping it on the circular path. For the dangling mass, it is not moving, indication that its vertical acceleration is zero.

Solution:

First, we draw a free-body diagram for each mass. For the dangling mass m, we apply Newton's first law and obtain

$$\sum_i F_{y,i} = ma \Rightarrow F_R - mg = 0 \Rightarrow F_R = mg. \tag{7a}$$

For the flat hockey puck, we apply Newton's second law for circular motion and obtain

$$\sum_i F_{Ri} = Ma_R \Rightarrow F_R = M\frac{v^2}{R} \Rightarrow mg = M\frac{v^2}{R}. \tag{7b}$$

We solve Eq. (7b) for the speed v of the hockey puck and obtain

$$v = \sqrt{\frac{mgR}{M}}. \tag{7c}$$

Equation (7c) indicates that, for a fixed m and M, a larger speed v leads to a larger radius R of the circular path because the amount of the radial force exerted on the flat puck remains the same. This is the reason why one must make a wider turn at a street corner if one's car is moving faster.

Exercise Problem 7.8

A 1200 kg car rounds a curve of radius 70 m banked at an angle of 12°. If the car is traveling at 90 km/h, will a friction force be required? If so how much friction force is needed and what is its direction?

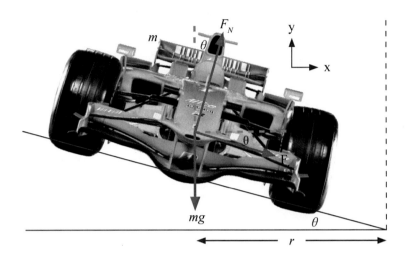

PROBLEM SOLVING STRATEGY

What do we know?

The car is going around a banked track.

What concepts are needed?

We need to decompose the motion of the car into the vertical and horizontal components. We need to apply Newton's law of motion.

Why?

The horizontal motion (i.e., along the x-direction) is circular while the vertical motion is translational (i.e., along the y-direction).

Solution:

First, we check to see if a friction force is required. We assume that friction is not required and compute the bank angle. Analyzing the free-body diagram shown above, if the friction force is not required, we obtain the net force acting on the car in the y- direction as

$$\sum_i F_{yi} = ma_y = F_N \cos\theta - mg = 0 \Rightarrow F_N = \frac{mg}{\cos\theta}, \tag{8a}$$

and the net force in the x-direction as

$$\sum_i F_{xi} = ma_R = F_N \sin\theta = m\frac{v^2}{r}. \tag{8b}$$

CIRCULAR MOTION

Now, we substitute Eq. (8a) into Eq. (8b) and obtain

$$\frac{mg}{\cos\theta}\sin\theta = m\frac{v^2}{r} \Rightarrow \tan\theta = \frac{v^2}{rg}. \tag{8c}$$

We substitute the numerical values into Eq. (8c) and find that

$$\tan\theta = \tan(12.0°) = 0.21 \neq \frac{\left(90\,\text{km}/\text{h} \times 1000\,\text{m}/1\text{km} \times 1\text{h}/3600\text{s}\right)^2}{70\,\text{m} \times 9.8\,\text{m}/\text{s}^2} = 0.91.$$

This result indicates that friction is needed to keep the car on the track because the bank angle of 12° is not sufficient. So, we now include the friction force and carry out the same calculation as above. Newton's second law applied along the y- and x-direction yields

$$\sum_i F_{yi} = ma_y \Rightarrow F_N\cos\theta - mg - F_f\sin\theta = 0 \Rightarrow F_N = \frac{1}{\cos\theta}\left(mg + F_f\sin\theta\right) \tag{8d}$$

and

$$\sum_i F_{xi} = ma_R \Rightarrow m\frac{v^2}{r} = F_N\sin\theta + F_f\cos\theta, \tag{8e}$$

respectively. We substitute Eq. (8d) into Eq. (8e) and rewrite Eq. (8e) as

$$m\frac{v^2}{r} = \frac{\sin\theta}{\cos\theta}\left(mg + F_f\sin\theta\right) + F_f\cos\theta \tag{8f}$$

We solve for the friction force F_f from Eq. (8f) and obtain

$$F_f = \frac{m}{\tan\theta\sin\theta + \cos\theta}\left(\frac{v^2}{r} - g\tan\theta\right)$$

$$= \frac{1200\,\text{kg}}{\tan 12° \sin 12° + \cos 12°}\left[\frac{\left(90000\,\frac{\text{m}}{\text{h}} \times 1\text{h}/3600\text{s}\right)^2}{70\,\text{m}} - 9.8\frac{\text{m}}{\text{s}^2} \times \tan 12°\right] = 8.0 \times 10^3\,\text{N}$$

When a car is making a turn at high speed, a high bank angle is needed to keep it on the track. This is most visible in Daytona. Daytona is a superspeedway that can seat 168,000 race fans. Here, a superspeedway is any track over two miles (3.2 km) in length (excluding road courses) and featuring high-banked turns. At 2.5 miles (4.02 km) long, Daytona is second in length only to Talladega which is 2.66 miles (4.3 km) long. The banking is 18 degrees at the start/finish line, 31 degrees in turns. Thirty-one degrees of banking is pretty steep. To get an idea of this bank angle, the top of the track near the outside wall is more than 35 feet (10.7 m) above the infield. This height is higher than the roof of a typical two-story house. As a result of the extreme banking in the corners and the length of the track, stock cars can travel at speeds in excess of 200 mph (322 km/h). While those speeds are relatively safe for a car with an extreme amount of down force such as a Grand-Am sports car or an Indy car, stock cars have a tendency to lift off the ground when they reverse direction at high speeds, as in the case of a spin.

7.5 GRAVITY AND CIRCULAR MOTION

In planetary motion, the most important force which keeps the planets in orbit is the gravitational force. This force in most cases involving planetary motion acts as a centripetal force. Also, as we discuss in Chapter 4, the gravity is derived from the gravitational force. Let us examine a description of gravity from the point of view of universal law of gravitation. This implies that gravity should behave in similar ways regardless of where we are in the universe.

This important principle was discovered by Newton. According to a popular story, Newton was sitting under an apple tree, an apple fell on his head, and he suddenly thought of the universal law of gravitation. As in all such legends, this is almost certainly not a true story in its details even though it may contain some elements of what actually happened. Most likely, the correct version of the story is that, when Newton observed an apple falling from a tree, he may have thought along the following lines. The apple is accelerated since its initial velocity was zero when it was hanging on the tree but it has changed to move toward the ground. According to Newton's second law, there must be a force that acts on the apple to cause this acceleration. This force is called as "gravity", and the acceleration associated with this force is the "acceleration due to gravity". This line of thought may be continued to imagine the apple tree which is twice as high. Again, in this case, we would expect the apple to be accelerated toward the ground. So, this conclusion suggests that this force of gravity reaches to the top of the tallest apple tree. This line of reasoning may have led Newton to find one of the most important principles. Application of this universal law of gravitation allows us to understand not only the planetary motion but also the formation of our universe.

We recall, from chapter 4, **Newton's law of universal gravitation** which states that every point mass in the universe attracts every other point mass with a force that is directly proportional to the product of their masses and inversely proportional to the square of the distance between them. A simple version of this law is that every particle in the universe exerts an attractive force on every other particle as shown in Fig. 7.12. The force that each exerts on the other is directed along the line joining the particle. The two forces shown in Fig. 7.12 may be considered as action-reaction forces. The magnitude of the force is given by

$$F = G \frac{m_1 m_2}{r^2}, \tag{7.20}$$

where G is the universal gravitational constant $G = 6.673 \times 10^{-11}$ N·m²/kg². **Basic property of Newton's law of gravity** is the following. First, the force is inversely proportional to the square of the distance between the objects (i.e., inverse-square law). Second, the force is directly proportional to the product of the masses of the two objects.

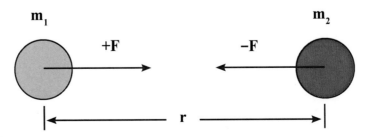

Figure 7.12: A schematic illustration of the gravitational force between two massive objects. According to universal law of gravitation, two objects exert equal force on each other, but in opposite directions.

CIRCULAR MOTION

Universal law of gravitation may be used to understand the concept of weight and gravitational acceleration. We recall, from Chapter 4, that we may use universal law of gravitation to determine the weight of an object on the surface of Earth. The weight of the object with mass m is

$$W = G\frac{M_E m}{r_E^2} = mg, \qquad (7.21)$$

where $M_E = 5.98 \times 10^{24}$ kg and $r_E = 6.38 \times 10^6$ m. A simple calculation yields the value of gravitational acceleration as $g = 9.80$ m/s². In Chapter 3, we used this value of g to describe projectile motion of an object on the surface of Earth. Another useful aspect of Newton's law of universal gravitation is the **universality** of gravity. It should be noted that Newton's place in the "*Physics Hall of Fame*" is not due to his discovery of gravity but rather due to his discovery that gravitation is universal. This means that **all** objects attract each other with a force of gravitational attraction.

Gravity is universal. Because of gravity, when we drop something, it falls down, instead of up. Well, everybody knows that this is what happens on the surface of Earth. But, what does this really mean? What is gravity and why is it universal? As we noted earlier, **gravity** is the natural phenomenon by which physical bodies appear to attract each other with a force proportional to their masses. It is most commonly experienced as the agent that gives weight to objects with mass and causes them to fall to the ground when dropped. As it turned out, the phenomenon of gravitation itself, however, is a byproduct of a more fundamental phenomenon described by general relativity, which suggests that space-time is curved according to the presence of matter through a yet-to-be discovered mechanism. The notion of universal implies that what happens on the surface of Earth can also happen on the surface other planets. For example, the free-fall acceleration g on the surface of a planet is determined from the law is

$$g_{planet} = G\frac{M_{planet}}{R_{planet}^2}. \qquad (7.22)$$

This picture of gravity indicates that our weight on the moon is a function of the moon's gravity. The principle that works on the surface of Earth works also on the surface of Moon. Hence, the weight of an object on the moon is

$$W = F_{moon\ on\ m} = mg_{moon} = G\frac{M_{moon} m}{R_{moon}^2}. \qquad (7.23)$$

The gravitational acceleration on the moon is

$$g_{moon} = G\frac{M_{moon}}{R_{moon}^2}. \qquad (7.24)$$

Because gravity is universal, we know that Eq. (7.24) is useful for estimating the value of g_{moon} without being on the surface of Moon and testing its validity. However, there are few things difficult to understand from the universal law of gravitation. First, gravity is a force that attracts all physical objects towards each other but it is unknown why this happens. Second, it is unclear why if the mass of an object is greater, then the force of gravity is stronger. The moon is 1/4 the size of Earth, so the moon's gravity is much less than the earth's gravity, 83.3% (or 5/6) less to be exact. Based on these aspects of gravity, "weight" is a measure of the gravitational pull between two objects. So, we would weigh much less on the moon. These difficulties questions can be answered by general relativity, in terms of the curved space-time concept.

Gravity has played an important part in making the universe the way it is. Astronomers theorize that the universe began billions of years ago with an enormous explosion of matter called the **big bang**. Since the big bang, the universe has been expanding rapidly. Gravity slowly pulled the gas and dust together. As the cloud of gas and dust shrank, it spun faster and faster and flattened, forming a rotating disk. Gravity pulled most of the gas into the center of the disk, where the sun eventually formed. About 200 million years after the big bang, the force of gravity caused matter to come together to form the first stars. Also, gravity is what makes pieces of matter clump together into planets, moons, and stars. Gravity caused the planets, their moons, and other bodies in the solar system to form. About 5 billion years ago, a giant cloud of gas and dust began to pull together to form our solar system. Gravity is what makes the planets orbit the stars. Gravity is the force that keeps the planets and other bodies in the solar system in orbits around the sun. For example, Earth orbits the Sun. Gravity also caused stars to gather together in galaxies, huge collections of stars. Gravity is what makes the stars clump together in huge, swirling galaxies. Gravity is a force that pulls objects toward each other, and the force of gravity acts between all objects in the universe.

Understanding-the-Concept Question 7.1

Two planets have the same surface gravity, but planet B has twice the mass of planet A. If planet A has radius r, what is the radius of planet B?

  a. $0.707r$ b. r c. $1.41r$ d. $4r$

Explanation:

In answering the question, we apply Newton's universal law of gravitation, $F_G = GmM/R^2$. According to this law, the magnitude of surface gravity may be computed from $g = GM/R^2$. If planet B has twice the mass of planet A (i.e., $M_B = 2M_A$), while having the same surface gravitational acceleration, we may write

$$g = G\frac{2M_A}{R_B^2} = G\frac{M_A}{R_A^2} \Rightarrow R_B^2 = 2R_A^2 = 2r^2$$

We solve for R_B and obtain that $R_B = 1.414r$.

Answer: c

The universal law of gravitation can help us understand how small and large objects orbit around Earth. The skies above Earth are teeming with large and small man-made objects. The U.S. space surveillance network uses radar to track more than 13,000 such items that are larger than four inches (ten centimeters). This celestial clutter includes everything from the International Space Station (ISS) and the Hubble Space Telescope to defunct satellites, rocket stages, or nuts and bolts left behind by astronauts. Also, there are millions of smaller, harder-to-track objects such as flecks of paint and bits of plastic. This is the result of human activities in space. For half a century, humans have been putting satellites into orbit around Earth to serve a variety of functions. The Soviets launched the first, Sputnik 1, in October of 1957 just to prove they could. Four months later, the U.S. responded with Explorer 1. Since then, some 2,500 satellites have been sent aloft. These include Hubble and the ISS, the Russian Mir space station, the 27-satellite Global Positioning System, as well as hundreds of others. Some of the purposes of these satellites include providing communication relays, broadcasting television and radio signals, and helping scientists predict weather. These man-made objects circle

CIRCULAR MOTION

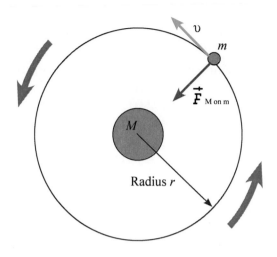

Figure 7.13: A schematic illustration of a satellite's orbit around the Earth. According to the universal law of gravitation, the speed of the satellite is determined by the orbital radius r (i.e., $v = (GM/r)^{1/2}$). The right panel illustrates a large number of objects in near Earth orbits.

Earth in orbits that range from as near as 150 miles (240 km) to 22,500 miles (36,200 km) away. Satellites in **low-Earth orbit** (LEO) stay within 500 miles (800 km) and travel extremely fast, 17,000 miles an hour (27,400 km/h) or more. The reason for this fast speed is to keep them from being drawn back into Earth's atmosphere. Most satellites around Earth are found in the LEO range. Other objects are sent much farther into space and placed in what is called **geosynchronous orbit**. This orbit allows the satellite to match the Earth's rotation and "hover" over the same spot at all times. Weather and television satellites are generally in this category.

Let us consider the orbital motion of a satellite around Earth. When a lighter body of mass m (i.e., satellite) is orbiting a massive body of mass M, the speed of satellite, according to Newton's second law, is

$$F_{M \text{ on } m} = G\frac{Mm}{r^2} = ma_r = m\frac{v^2}{r}. \tag{7.25}$$

In Eq. (7.25), we assume that the orbit is circular. Here, the gravitational force keeps the satellite in a circular motion. It acts as a centripetal force, $F_c = ma_R$. From Eq. (7.25), we derive the speed of the satellite v as

$$v = \sqrt{\frac{GM}{r}} = \frac{2\pi r}{T}. \tag{7.26}$$

Since the speed of the satellite v, according to Eq. (7.26), is related to the period T of the circular orbit, we may use Eqs. (7.25) and (7.26) to derive the expression for T as

$$T^2 = \left(\frac{4\pi^2}{GM}\right)r^3. \tag{7.27}$$

Equation (7.27) states the square of the period (i.e., T^2) of orbit is proportional to the cube of the radius (i.e., r^3) of the orbit. This is one of three Kepler's laws. In astronomy, **Kepler's laws of planetary motion** are three scientific laws describing orbital motion, originally formulated to describe the motion of planets around the Sun. Kepler's laws are: (i) The orbit of every planet is an ellipse with the Sun at one of the two foci; (ii) A line joining a planet and the Sun sweeps out equal areas during equal intervals of time; and (iii) The square of the orbital period of a planet is directly proportional to the cube of the semi-major axis of its orbit. In the early 1600s, Kepler (1571-1630) proposed three laws of planetary motion. Kepler was able to summarize the carefully collected data of his mentor, Brahe (1546-1601).

Kepler's three statements accurately described the motion of planets in a sun-centered solar system. Kepler's explanations for the underlying reason for such motion are no longer accepted because they are not based on the fundamental principles but are based on phenomenological observations. However, the actual three laws themselves are still considered an accurate description of the motion of any planet and any satellite. It is worth noting that Kepler's three laws of planetary motion can be derived from Newton's laws of motion and his universal law of gravitation, suggesting that even the heavenly bodies obey the laws of physics.

Exercise Problems 7.9

The *asteroid belt* circles the sun between the orbits of Mars and Jupiter. One asteroid has a period of 5.0 earth years. What are the asteroid's orbital radius and its speed?

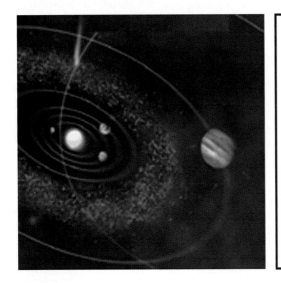

PROBLEM SOLVING STRATEGY

What do we know?

The asteroid circles the sun. The period of one asteroid is 5.0 years.

What concepts are needed?

We need to apply the universal law of gravitation between two objects (i.e., sun and the asteroid). The speed of the asteroid is the circumference of the orbit divided by the period.

Why?

The gravitational force between the sun and asteroid is keeping the asteroid in a circular orbit around the sun.

Solution:

The gravitational force between the Sun (mass = M_s) and asteroid provides the centripetal acceleration required for circular motion

$$F = G\frac{M_s m_a}{r_a^2} = m_a \frac{v_a^2}{r_a} \Rightarrow \frac{GM_s}{r_a} = \left(\frac{2\pi r_a}{T_a}\right)^2 \Rightarrow r_a = \left(\frac{GM_s T_a^2}{4\pi^2}\right)^{1/3}$$

Substituting the universal gravitational constant $G = 6.67 \times 10^{-11}$ N·m²/kg², mass of the Sun $M_s = 1.99 \times 10^{30}$ kg, and the time period T_a of asteroid, we obtain

$$r_a = 4.37 \times 10^{11} \text{ m}$$

Noting that $T_a = 5.0$ Earth years $= 1.5779 \times 10^8$ s, we computed the velocity v_a of the asteroid in its orbit as

$$v_a = \frac{2\pi r_a}{T_a} = \frac{2\pi \times 4.37 \times 10^{11} \text{ m}}{1.5779 \times 10^8 \text{ s}} = 1.7 \times 10^4 \frac{\text{m}}{\text{s}}.$$

CIRCULAR MOTION

There are lots of asteroids moving within our solar system. However, most of these asteroids are located in the asteroid belt. The **asteroid belt** is the region of the solar system which is located roughly between the orbits of the planets Mars and Jupiter. It is occupied by numerous irregularly shaped bodies called asteroids or minor planets. The asteroid belt formed from the primordial solar nebula as a group of planetesimals. Planetesimals are the smaller precursors of the planets, which in turn formed protoplanets. Protoplanets are large planetary embryos that originate within protoplanetary discs and have undergone internal melting to produce differentiated interiors. They are believed to form out of kilometer-sized planetesimals that attract each other gravitationally and collide. Between Mars and Jupiter, however, gravitational perturbations from the giant planet imbued the protoplanets with too much orbital energy for them to accrete into a planet. Collisions became too violent, and instead of fusing together, the planetesimals and most of the protoplanets shattered. As a result, most of the asteroid belt›s mass has been lost since the formation of the solar system. Some fragments can eventually find their way into the inner Solar System, leading to meteorite impacts with the inner planets. Asteroid orbits continue to be appreciably perturbed whenever their period of revolution about the Sun forms an orbital resonance with Jupiter. At these orbital distances, a Kirkwood gap occurs as they are swept into other orbits.

Exercise Problem 7.10

A geosynchronous satellite is one that stays above the same point on the equator of the Earth. Such satellites are used for such purposes as cable TV transmission, for weather forecasting, and as communication relays. Determine (a) the height above the Earth's surface such a satellite must orbit and (b) such satellite's speed.

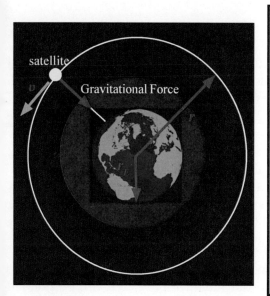

Problem Solving Strategy

What do we know?

We have a geosynchronous satellite. This means that the period of the orbit is one day. The orbit of the satellite is circular

What concepts are needed?

We need to apply the universal law of gravitation and Newton's second law for a circular motion.

Why?

The satellite is moving in a circular orbit at constant speed, but it accelerates because its direction changes constantly. Also, the gravitational force due to the Earth is keeping the satellite in this orbit.

Solution:

According to Newton's second law, the net radial force acting on the satellite is

$$\sum_i F_i = m_{sat} a_R.$$

The force applied to the satellite is the Newton's universal gravitational force

$$G\frac{m_{sat}m_E}{r^2} = m_{sat}\frac{v^2}{r}.$$

This is the force which keeps that satellite in a circular orbit around the Earth. Since the period of the orbit T is 1 day, we may express the speed of the satellite as $v = 2\pi/T$ and compute the radius r of orbit as

$$G\frac{m_E}{r^2} = \frac{(2\pi r)^2}{rT^2} \Rightarrow r^3 = G\frac{m_E T^2}{4\pi^2}$$

The cube of the radius of orbit measured from the center of Earth is

$$r^3 = \frac{6.67\times 10^{-11}\,\text{N}\cdot\text{m}^2/\text{kg}^2 \times 5.98\times 10^{24}\,\text{kg} \times (24\times 3600\,\text{s})^2}{4\pi^2} = 7.54\times 10^{22}\,\text{m}^3.$$

This yields the radius r as $r = 4.23\times 10^7$ m. Now, we compute the height above Earth's surface as

$$height = r - r_E = 4.23\times 10^4\,\text{km} - 6.38\times 10^3\,\text{km} \cong 3.6\times 10^4\,\text{km}.$$

The satellite's speed v for maintaining a geosynchronous orbit is

$$v = \sqrt{\frac{Gm_E}{r}} = \sqrt{\frac{6.67\times 10^{-11}\,\text{N}\times\text{m}^2/\text{kg}^2 \times 5.98\times 10^{24}\,\text{kg}}{4.23\times 10^7\,\text{m}}} = 3.07\times 10^3\,\frac{\text{m}}{\text{s}}.$$

A **geosynchronous orbit (GSO)** is an orbit around Earth with an orbital period of one sidereal orbit (approximately 23 hours 56 minutes and 4 seconds), matching the Earth's sidereal rotation period. The synchronization of rotation and orbital period means that, for an observer on the surface of the Earth, an object in geosynchronous orbit returns to the exactly same position in the sky after a period of one sidereal day. Over the course of a day, the object›s position in the sky traces out a path which is typically in the form of an analemma, whose precise characteristics depend on the inclination and eccentricity of the orbit. Here, an analemma is a curve representing the changing angular offset of the Sun from its mean position on the celestial sphere as viewed from the Earth. A special case of this orbit is **the geostationary orbit**. Popularly, the term "geosynchronous" may sometimes be used to mean, specifically, geostationary. This is a circular geosynchronous orbit at zero inclination, that is, directly above the equator. A satellite in a geostationary orbit appears stationary, always at the same point in the sky, to ground observers.

Communications satellites are often given geostationary orbits, or close to geostationary, so that the satellite antennas that communicate with them do not have to move, but can be pointed permanently at the fixed location in the sky where the satellite appears. A **geosynchronous satellite** is a satellite in geosynchronous orbit, with an orbital period the same as the Earth's rotation period, and over the course of a day traces out a path in the sky that is typically some form of analemma. A special case of geosynchronous satellite is the **geostationary satellite**, which has a geostationary orbit which is a circular geosynchronous orbit directly above the Earth's equator. Another type of geosynchronous orbit used by satellites is the Tundra elliptical orbit. Geosynchronous satellites have the advantage of remaining permanently in the same area of the sky, as viewed from a particular location on Earth, and so permanently within view of a given ground station. Such satellites are

CIRCULAR MOTION

often used form a **geosynchronous network** which is a network based on communication with or through geosynchronous satellites. These satellites have revolutionized global communications, television broadcasting and weather forecasting. Also, they have a number of important defense and intelligence applications. However, one disadvantage of geostationary satellites is that radio signals take approximately 0.25 of a second to reach and return from the satellite, resulting in a small but significant signal delay. This delay is a result of high altitude and increases the difficulty of telephone conversation. Also, this delay reduces the performance of common network protocols such as TCP/IP, but it does not present a problem with non-interactive systems such as television broadcasts.

SUMMARY

i. To describe either circular or rotational motion of an object, we define the following quantities: (i) angular displacement, (ii) angular velocity, and (iii) angular acceleration. **Angular displacement** θ is defined as the angle through which a rigid object rotates about a fixed axis. The **average angular velocity** is define as

$$\bar{\omega} = \frac{\Delta\theta}{\Delta t} = \frac{\theta - \theta_o}{\Delta t}.$$

This is analogous to linear average velocity. We may define angular acceleration as

$$\bar{\alpha} = \frac{\Delta\omega}{\Delta t} = \frac{\omega - \omega_o}{\Delta t}.$$

Angular acceleration α is the rate of change of angular velocity ω. This is analogous to linear acceleration is called either tangential or transverse acceleration. For an object in uniform circular motion, the acceleration due to changing direction is *toward* the center of the circle. This is known as the centripetal acceleration. The angular velocity is related to the centripetal acceleration as

$$a_R = \frac{v^2}{r} = \frac{(r\omega)^2}{r} = r\omega^2.$$

ii. When no sliding takes place, the rolling motion is referred to as "pure rolling". In this case, the relation between linear speed v and angular speed ω is given by

$$v = r\omega.$$

Also, the relation between linear acceleration a and angular acceleration α is written as

$$a = r\alpha.$$

iii. Circular motion is often described in terms of the **frequency** f (i.e., revolutions per second) or the **period** T (i.e., the time required for one complete revolution). The frequency f is inversely related to the period T, $T = 1/f$.

iv. **Newton's law of universal gravitation** states that every point mass in the universe attracts every other point mass with a force that is directly proportional to the product of their masses and inversely proportional to the square of the distance between them. The law indicates that every particle in the universe exerts an attractive force on every other particle. The force that each exerts on the other is directed along the line joining the particle. The magnitude is given by

$$F = G\frac{m_1 m_2}{r^2}.$$

where G is the universal gravitational constant $G = 6.673 \times 10^{-11}$ N·m^2/kg^2.

MORE WORKED PROBLEMS

Problem 7.1

If 1.2 g pebble is stuck in a tread of a 0.75-m-diameter automobile tire, held in place by static friction that can be at most 3.6 N. The car starts from rest and gradually accelerates on a straight road. How fast is the car moving when the pebble flies out of the tire thread?

Solution:

At the point just before release, the force acting on the pebble is

$$F = ma = m\frac{v^2}{r}.$$

We solve for v (the speed of the top of the tire) and obtain

$$v = \sqrt{\frac{Fr}{m}} = \sqrt{\frac{3.6\text{N} \times 0.375\text{m}}{0.0012\text{kg}}} = 33.5\frac{\text{m}}{\text{s}}.$$

Since $v = 2v_{cm}$, we may write $v_{cm} = v/2$, indicating that the car is going at half the speed of the pebble. The speed of the center of mass is

$$v_{cm} = 16.8\frac{\text{m}}{\text{s}}$$

Problem 7.2

A 40.0-kg child swings in a swing supported by two chains, each 3.00 m long. The tension in each chain at the lowest point is 350 N. Find (a) the child's speed at the lowest point and (b) the force exerted by the seat on the child at the lowest point. (Ignore the mass of the seat.)

Solution:

If T is the tension in each of the two support chains, then the net force acting on the child at the lowest point on the circular path is obtained from

$$\sum_i F_{y,i} = ma_c \Rightarrow 2T - mg = m\frac{v^2}{r}.$$

So, the speed v at this point is

$$v = \sqrt{r\left(\frac{2T}{m} - g\right)} = \sqrt{3.00\text{m}\left(\frac{2 \times 350\text{N}}{40\text{kg}} - 9.80\frac{\text{m}}{\text{s}^2}\right)} = 4.81\frac{\text{m}}{\text{s}}.$$

The upward force that the seat exerts on the child at this lowest point is

$$F_{seat} = 2T = 2 \times 350\text{N} = 700\text{N}.$$

CIRCULAR MOTION

Problem 7.3

Two objects attract each other with a gravitational force of magnitude 1.00×10^{-8} N when separated by 20.0 cm. If the total mass of the object is 5.00 kg, what is the mass of each?

Solution:

We know that $m_1 + m_2 = 5.0$ kg or $m_2 = 5.0$ kg $- m_1 = M - m_1$. Here, $M = m_1 + m_2$ is the total mass. Using this, we may write

$$F = G\frac{m_1 m_2}{r^2} = G\frac{m_1(M-m_1)}{r^2} \Rightarrow \frac{Fr^2}{G} = m_1 M - m_1^2.$$

We may rewrite this equation as

$$m_1^2 - m_1 M + \frac{Fr^2}{G} = 0.$$

Noting that $M = 5.00$ kg and $Fr^2/G = 1.0 \times 10^{-8}$ N$\times(0.200$ m$)^2/6.67 \times 10^{-11}$ N·m^2/kg$^2 = 6.00$ kg^2, we rewrite the quadratic equation in a simpler form

$$m_1^2 - 5.00\,\text{kg}\,m_1 + 6.00\,\text{kg}^2 = 0 \Rightarrow (m_1 - 3.00\,\text{kg})(m_1 - 2.00\,\text{kg}) = 0.$$

This factorization yields the answer $m_1 = 3.00$ kg, so $m_2 = 2.00$ kg. The answer of $m_1 = 2.00$ kg and $m_2 = 3.00$ kg is physically equivalent to the previous answer.

Problem 7.4

For a scene in a movie, a stunt driver drives a 1.5×10^3-kg pickup truck with a length of 4.25 m around a circular curve with a radius of curvature of 0.333 km. The truck is to curve off the road, jump across a gully 10.0 m wide, and land on the other side 2.96 m below the initial side. What is the minimum centripetal acceleration the truck must have in going around the circular curve to clear the gully and land on the other side?

Solution:

To cleat the gully, the truck must travel a minimum horizontal distance of 10.0 m + 4.25 m = 14.25 m. First, we use kinematics to compute the velocity of the truck by considering it as a horizontal projectile

$$y = y_o + v_{oy}t - \frac{1}{2}gt^2 \Rightarrow y = -\frac{1}{2}gt^2.$$

This allows us to determine the time of flight

$$t = \sqrt{-\frac{2y}{g}} = \sqrt{-\frac{2(-2.96\,\text{m})}{9.80\,\frac{\text{m}}{\text{s}^2}}} = 0.777\,\text{s}.$$

So, by using another kinematics equation, we may determine the x-component of velocity

$$x = x_o + v_{ox}t \Rightarrow v_{ox} = \frac{x}{t} = \frac{14.25\,\text{m}}{0.777\,\text{s}} = 18.3\,\frac{\text{m}}{\text{s}}.$$

Therefore, the minimum centripetal acceleration of the truck must be

$$a_R = \frac{v_{ox}^2}{r} = \frac{\left(18.3\,\text{m}/\text{s}\right)^2}{333\,\text{m}} = 1.01\,\frac{\text{m}}{\text{s}}.$$

PROBLEMS

1. A turntable rotates counterclockwise at 78 rpm. A speck of dust on the turntable is at $\theta = 0.45$ rad at $t = 0$ s. What is the angle of the speck at $t = 8.0$ s? Your answer should be between 0 and 2π rad.

2. In degrees and radians, what is the angular width (a) of a full Moon as viewed from the Earth and (b) of a full Earth as viewed from the Moon? [Hint: Use data from Appendix.]

3. A turntable rotates counterclockwise at 78 rpm. A speck of dust on the turntable is at $\theta = 0.45$ rad at $t = 0$ s. What is the angle of the speck at $t = 8.0$ s? Your answer should be between 0 and 2π rad.

4. Electrical wire with a diameter of 0.75 cm is wound on a spool with a radius of 30 cm and a height of 24 cm. (a) Through how many radians must the spool be turned to wrap one even layer of wire? (b) What is the length of his wound wire?

5. A ball of radius 0.200 m rolls along a horizontal table top with a constant linear speed of 3.60 m/s. The ball rolls off the edge and falls a vertical distance of 2.10 m before hitting the floor. What is the angular displacement of the ball while the ball is in the air?

6. The driver of a car sets the cruise control and ties the steering wheel so that the car travels at a uniform speed of 15 m/s in a circle with a diameter of 120 m. (a) Through what angular distance does the car move in 4.00 min? (b) What linear distance does it travel in this time?

7. A bicycle is rolling down a circular hill that has radius of 9.00 m. The angular displacement of the bicycle is 0.900 rad. The radius of each wheel is 0.400 m. What is the angle (in radians) through which each tire rotates?

8. Your roommate is working on his bicycle and has the bike upside down. He spins the 60-cm-diameter wheel, and your notice that a pebble stuck in the tread goes by three times every second. What are the pebble's speed and acceleration?

9. A rotating wheel requires 3.00 s to rotate 37.0 revolutions. Its angular velocity at the end of the 3.00-s interval is 98.0 rad/s. What is the constant angular acceleration (in rad/s²) of the wheel?

10. In 1.5 min, a car moving at constant speed travels halfway around a circular track that has a diameter of 1.0 km. What are the car's (a) angular speed and (b) tangential speed?

11. A baton twirler throws a spinning baton directly upward. As it goes up and returns to the twirler's hand, the baton turns through four revolutions. Ignoring air resistance and assuming that the average angular speed of the baton as 4.80 rad/s, determine the height to which the center of the baton travels above the point of release.

CIRCULAR MOTION

12. A Ferris wheel with a diameter of 35.0 m starts from rest and achieves its maximum operational tangential speed of 2.2 m/s in a time of 15.0 s. (a) What is the magnitude of the wheel's angular acceleration? (b) What is the magnitude of the tangential acceleration after the maximum operational speed is reached?

13. A ball with a radius of 15 cm rolls on a level surface, and the translational speed of the center of mass is 0.25 m/s. What is the angular speed about the center of mass if the ball rolls without slipping?

14. A cylinder with a diameter of 20 cm rolls with an angular speed of 0.50 rad/s on a level surface. If the cylinder experiences a uniform tangential acceleration of 0.018 m/s² without slipping until its angular speed is 1.25 rad/s, through how many complete revolutions does the cylinder rotate during the time it accelerates?

15. An adventurous archeologist ($m = 85.0$ kg) tries to cross a river by swinging from a vine. The vine is 10.0 m long, and his speed at the bottom of the swing is 8.00 m/s. The archeologist does not know that the vine has a breaking strength of 1000 N. Does he make it across the river without falling in?

16. (a) What is the tangential acceleration of a bug on the rim of a 10.0-in.-diameter disk if the disk accelerates uniformly from rest to an angular speed of 420.0 rad/min in 3.00 s? (b) When the disk is at its final speed, what is the tangential velocity of the bug? One second after the bug starts from rest, what are its (c) tangential acceleration, (d) centripetal acceleration, and (e) total acceleration?

17. A satellite moves on a circular earth orbit that has a radius of 6.7×10^6 m. A model airplane is flying on a 15-m guideline in a horizontal circle. The guideline is parallel to the ground. Find the speed of the plane such that the plane and the satellite have the same centripetal acceleration.

18. Consider a baggage carousel at an airport. Your suitcase has not slid all the way down the slope and is going around at a constant speed on a circle ($r = 11.0$ m) as the carousel turns. The coefficient of static friction between the suitcase and the carousel is 0.760, and the angle θ in the drawing is 36.0°. How much time is required for your suitcase to go around once?

19. A bicycle chain is wrapped around a rear sprocket ($r = 0.039$ m) and a front sprocket ($r = 0.10$ m). The chain moves with a speed of 1.4 m/s around the sprockets, while the bike moves at a constant velocity. Find the magnitude of the acceleration of a chain link that is in contact with (a) rear sprocket, (b) neither sprocket, and (c) the front sprocket.

20. The large blade of a helicopter is rotating in horizontal circle. The length of the blade is 6.7 m, measured from its tip to the center of the circle. Find the ratio of the centripetal acceleration at the end of the blade to that which exists at a point located 3.0 m from the center of the circle.

21. A jet flying at 123 m/s banks to make a horizontal circular turn. The radius of the turn is 3810 m, and the mass of the jet is 2.00×10^5 kg. Calculate the magnitude of the necessary lifting force.

22. A typical laboratory centrifuge rotates at 4000 rpm. Test tubes have to be placed into a centrifuge very carefully because of the very large accelerations. (a) What is the acceleration at the end of a test tube that is 10 cm from the axis of rotation? (b) For comparison, what is the magnitude of the acceleration a test tube would experience if dropped from a height of 1.0 m and stopped in a 1.0-ms-long encounter with a hard floor?

23. A peregrine falcon in a tight, circular turn can attain a centripetal acceleration 1.5 times the free-fall acceleration. If the falcon is flying at 20 m/s, what is the radius of the turn?

24. A computer is reading data from a rotating CD-ROM. At a point that is 0.030 m from the center of the disc, the centripetal acceleration is 120 m/s². What is the centripetal acceleration at a point

that is 0.050 m from the center of the disc?

25. A motorcycle has a constant speed of 25.0 m/s as it passes over the top of a hill whose radius of curvature is 126 m. The mass of the motorcycle and driver is 342 kg. Find the magnitude of (a) the centripetal force and (b) the normal force that acts on the cycle.

26. A rigid massless rod is rotated about one end in a horizontal circle. There is a mass m_1 attached to the center of the rod and a mass m_2 attached to the outer end of the rod. The inner section of the rod sustains three times as much tension as the outer section. Find the ratio m_2/m_1.

27. A car with a constant speed of 83.0 km/h enters a circular flat curve with a radius of curvature of 0.400 km. If the friction between the road and the car's tires can supply a centripetal acceleration of 1.25 m/s², does the car negotiate the curve safely? Justify your answer.

28. A rectangular plate is rotating with a constant angular acceleration about an axis that passes perpendicularly through one corner, as the drawing shows. The tangential acceleration measured at corner A has twice the magnitude of that measured at corner B. What is the ratio L_1/L_2 of the lengths of the sides of the rectangle?

29. A top is a toy that is made of spin on its pointed end by pulling on a string wrapped around the body of the top. The string has a length of 64 cm and is wrapped around the top at a place where its radius is 2.0 cm. The thickness of the string is negligible. The top is initially at rest. Someone pulls the free end of the string, thereby unwinding it and giving the top an angular acceleration of +12 rad/s². What is the final angular velocity of the top when the string is completely unwound?

30. A child, hunting for favorite wooden horse, is running on the ground around the edge of a stationary merry-go-around. The angular speed of the child has a constant value of 0.250 rad/s. At the instant the child spots the horse, one-quarter of a turn away, the merry-go-around begins to move (in the direction the child is running) with a constant angular acceleration of 0.0100 rad/s². What is the shortest time it takes for the child to catch up with the horse?

31. In an automatic clothes dryer, a hollow cylinder moves the clothes on a vertical circle (radius r = 0.32 m), as the drawing shows. The appliance is designed so that the clothes tumble gently as they dry. This means that when a piece of clothing reaches an angle of θ above the horizontal, it loses contact with the wall of the cylinder and falls onto the clothes below. How many revolutions per second should the cylinder make in order that the clothes lose contact with the wall when θ = 70.0°?

32. One type of slingshot can be made from a length of rope and a leather pocket for holding the stone. The stone can be thrown by whirling it rapidly in a horizontal circle and releasing it at the right moment. Such a slingshot is used to throw a stone from the edge of a cliff, the point of release being 20.0 m above the base of the cliff. The stone lands on the ground below the cliff at a point X. The horizontal distance of point X from the base of the cliff (directly beneath the point of release) is thirty times the radius of the circle on which the stone is whirled. Determine the angular speed of the stone at the moment of release.

33. A race car travels with a constant tangential speed of 75.0 m/s around a circular track of radius 625 m. Find (a) the magnitude of the car's total acceleration and (b) the direction of its total acceleration relative to the radial direction.

34. A student investigating circular motion places a dime 10 cm from the center of a 33.333-rpm record on a turntable. The record player can accelerate at 1.42 rad/s². The student notes that the dime slides outward 2.25 s after she has switched on the turntable. (a) Why does the dime slide outward? (b) What is the coefficient of the static friction between the dime and the record?

35. A car rounds a banked curve where the radius of curvature of the road is R, the banking angle

CIRCULAR MOTION

is θ, and the coefficient of static friction is μ. (a) Determine the range of speeds the car can have without slipping up or down the road. (b) What is the range of speeds possible if $R = 100$ m, $\theta = 10°$, and $\mu = 0.10$ (slippery conditions)?

36. To create artificial gravity, the space station shown in the drawing is rotating at a rate of 1.00 rpm. The radii of the cylindrically shaped chambers have the ratio $r_A/r_B = 4.00$. Each chamber A simulates gravity with an acceleration of 10.0 m/s². Find values for (a) r_A, (b) r_B, and (c) the acceleration due to gravity that is simulated in chamber B.

37. A block is hung by a string from the inside roof of a van. When the van goes straight ahead at a speed of 28 m/s, the block hangs vertically down. But when the van maintains this same speed around an unbanked curve (radius = 150 m), the block swings toward the outside of the curve. Then the string makes an angle θ with the vertical. Find θ. (F09, 5.21)

38. A motorcycle is traveling up one side of a hill and down the other side. The crest of the hill is circular arc with a radius of 45.0 m. Determine the maximum speed that the cycle can have moving over the crest without losing contact with the road.

39. A curve of radius 120 m is banked at an angle of 18°. At what speed can it be negotiated under icy conditions where friction is negligible?

40. Satellite A orbits a planet with a speed of 10,000 m/s. Satellite B is twice as massive as satellite A and orbits at twice the distance from the center of the planet. What is the speed of satellite B?

41. A projectile is fired straight upward from the Earth's surface at the South Pole with an initial speed equal to one third the escape speed. (a) Ignoring air resistance, determine how far from the center of the Earth the projectile travels before stopping momentarily. (b) What is the altitude of the projectile at this instant?

42. In an old-fashioned amusement park ride, passengers stand inside a 3.0-m-tall, 5.0-m-diameter hollow steel cylinder with their backs against the wall. The cylinder begins to rotate about a vertical axis. Then the floor on which the passengers are standing suddenly drops away! If all goes well, the passengers will "stick" to the wall and not slide. Clothing has static coefficient of friction against steel in the range 0.60 to 1.0 and a kinetic coefficient in the range 0.40 to 0.70. What is the minimum rotational frequency, in rpm, for which the ride is safe?

43. A roller coaster car crosses the top of a circular loop-the-loop at twice the critical speed. What is the ratio of the car's apparent weight to its true weight?

44. A satellite has a mass of 5850 kg and is in a circular orbit $4.1×10^5$ m above the surface of planet. The period of orbit is two hours. The radius of the planet is $4.15×10^6$ m. What is the true weight of the satellite when it is rest on the planet?

45. A starship is circling a distant planet of radius R. The astronauts find that the free-fall acceleration at their altitude is half the value at the planet's surface. How far above the surface are they orbiting? Your answer will be a multiple of R.

46. A 20 kg sphere is at the origin and a 10 kg sphere is at $(x, y) = (20$ cm, 0 cm$)$. At what point or points could you place a small mass such that the net gravitational force on it due to the sphere is zero?

47. A satellite orbiting the moon very near the surface has a period of 110 min. Use this information, together with the radius of the moon from the table on the inside of the back cover, to calculate the free-fall acceleration on the moon's surface.

48. An artificial satellite circling the Earth completes each orbit in 110 minutes. (a) Find the altitude

of the satellite. (b) What is the value of g at the location of the satellite?

49. Show that the escape speed from the surface of a planet of uniform density is directly proportional to the radius of the planet.

50. The space shuttle orbits 300 km above the surface of the earth. (a) What is the force of gravity on a 1.0 kg sphere inside the space shuttle? (b) The sphere floats around inside the space shuttle, apparently "weightless." How is this possible?

51. A 75-kg weighs 735 N on the Earth's surface. How far above the surface for the Earth would he have to go to "lose" 10% of his body weight?

52. A satellite is placed in orbit 6.00×10^5 m above the surface of Jupiter. Jupiter has a mass of 1.90×10^{27} kg and a radius of 7.14×10^7 m. Find the orbital speed of satellite.

EQUILIBRIUM AND ROTATIONAL DYNAMICS

CHAPTER 8

This is a picture from a motion picture called "Indiana Jones and the Raiders of the Lost Ark" (1981) in which Dr. Jones, the leading character, is running for his life with a giant spherical stone ball rolling down after him. Of course, in the movie, he barely makes it out of an ancient temple with an ancient artifact. If the stone ball had been sliding on a frictionless surface (as a simplifying assumption), Would Indiana Jones have been able to escape from the temple? As we discussed in Chapter 7, rolling motion is a combination of translational and rotational motion. Does rolling motion lead, to a different outcome than sliding motion? This is answered in this chapter. Another aspect of rotational motion is its dynamics, which accounts for the effects of torque acting on the system. Torque is the application of force in a way that tends to produce rotation. The most obvious example of torque in action is the operation of a crescent wrench loosening a lug nut, and a close second is a playground seesaw. Both provide an easy means of illustrating the two ingredients of torque, force and moment arm, which we discuss in this chapter. In any object experiencing torque, there is a pivot point, which on the seesaw is the balance-point, and which in the wrench-and-lug nut combination is the lug nut itself. This is the area around which all the forces are directed. In each case, there is also a place where force is being applied. On the seesaw, it is the seats, each holding a child of differing weight (which is a downward force). Torque is also crucial to the operation of gyroscopes for navigation, and of various motors, both internal-combustion and electrical. As for what torque is and how it works, it is best to discuss its relationship to rotational motion of actual objects in the physical world. We now examine below these concepts associated with rotational dynamics in a more concrete theoretical framework.

8.1 ROTATIONAL DYNAMICS

As we discussed before, a complex motion of an object may be decomposed into translational and rotational motion. In this chapter, we focus on rotational dynamics. As we have seen in Chapter 7, rotational motion may be defined as the motion of a rigid body which takes place in such a way that all of its particles move in circles about an axis with a common angular velocity and angular acceleration. Rotation of a particle about a fixed point in space is also considered as rotational motion. Beside this simple example, there are lots of other examples of rotational motion that we can see around us. For instance, rotational motion is illustrated by (1) the fixed speed of rotation of the Earth about its axis, (2) the varying speed of rotation of the flywheel of a sewing machine, (3) the rotation of a satellite about a planet, (4) the motion of an ion in a cyclotron, and (5) the motion of a Foucault pendulum which demonstrates the rotation of the Earth. In considering rotational dynamics of an object, it is useful to ask the following question.

What makes an object start rotating about an axis?

The answer is "Force". Similar to the reason why an object moves in a certain direction, an applied force acting on a particular point on the object with respect to an axis of rotation leads to rotational motion. However, the effect of this force, as defined by both the magnitude and direction, may not be the same even though the magnitude of the force is the same. This may compel us to ask another question.

What is the effect of this force?

The effect of force depends on its magnitude and direction as well as a place where the force is applied. Both the direction of the force and the place is described in terms of the perpendicular distance from the axis of rotation to the line of action. This distance is called the "lever arm" or "moment arm". Here, the line of action is a straight line extension along which the force is applied.

EQUILIBRIUM AND ROTATIONAL DYNAMICS

Torque

To cause a rotational motion, torque is needed. Torque is a measure of how much force acting on an object causes that object to rotate. Hence, torque is the rotational equivalence of force, which leads to translational motion of an object. Just as a force is a push or a pull, a torque can be thought of as a twist to an object. A torque tends to rotate an object about an axis of rotation, fulcrum, or pivot. Since torque is a rotational equivalence of force, it is a vector. It has both magnitude and direction. It may be defined more precisely as a cross product of two vectors. Mathematically, torque τ is defined as a

$$\vec{\tau} = \vec{r} \times \vec{F} = rF \sin\theta \; (\hat{r} \times \hat{F}), \tag{8.1}$$

where r is a displacement vector representing the separation distance between the axis of rotation and the point where a force F is applied, and θ is the angle between the two vectors F and r, as shown in Fig. 8.1. Here, the direction of the torque is denoted by $\hat{r} \times \hat{F}$. The magnitude of the torque is defined as the product of the lever-arm distance and force. In Eq. (8.1), the distance $r \sin\theta$ is the lever-arm distance. The effects of non-zero torque on an object are to produce an angular acceleration. The direction of torque is in the direction perpendicular to the plane of two vectors, F and r. We may determine the direction of τ by applying the right-hand rule, as shown in Fig. 8.3 below.

Loosely speaking, torque is a measure of the turning force on an object such as a bolt or a flywheel. For example, pushing or pulling the handle of a wrench connected to a nut or bolt produces a torque (turning force) that loosens or tightens the nut or bolt. The ability of a torque to cause a rotation or a twisting motion depends on the following three quantities shown in Fig. 8.1: i) magnitude F of the force, ii) distance r from the pivot to the point at which the force is applied, and iii) angle θ at which the force is applied. In other words, the magnitude of a torque depends on the above three quantities.

Wrenches are used to provide leverage when applying rotating force to do jobs like turn bolts and tighten fittings as shown in Fig. 8.2. So, we may use a torque wrench, which is a tool for "converting" force into torque. In other words, when turning the wrench, an applied force of a specific magnitude yields an proportionally large applied torque.

The magnitude of the torque with respect to a given axis of rotation is equal to the force times the lever arm with respect to that axis. The lever arm is the perpendicular distance from the axis of rotation to the line of action of the force as shown in Fig. 8.2. The line of action extends in the direction of the force vector, and passes through the point at which the force acts. The moment arm, $r \sin\theta$, extends from the pivot to the line of action and is perpendicular to the line of action. The direction of the torque, defined according to Eq. (8.1), is determined by using the right-hand rule.

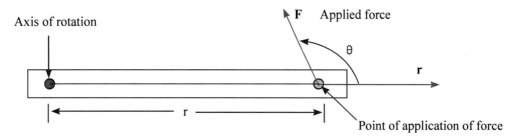

Figure 8.1: A schematic illustration of the physical quantities in defining torque, which is the rotational analog of force. Torque due to force F tends to cause an object to rotate.

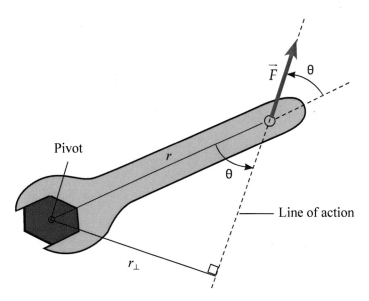

Figure 8.2: The magnitude of torque is determined by the magnitude of the force F and the moment arm, $r_\perp = r \sin \theta$, which is the perpendicular distance from the pivot and the line of action.

Right-hand rule

The right-hand rule is a common mnemonic for understanding notation conventions for vectors in three dimensions. It was invented for use in electromagnetism by John Fleming (1849-1945) in the late nineteenth century. When choosing three vectors that must be at right angles to each other, there are two distinct solutions, so when expressing this idea in mathematics, one must remove the ambiguity of which solution is meant. The direction is determined by the right-hand rule, but there are several different ways to apply this rule as shown in Fig. 8.3.

One way to apply the right-hand rule is by curling the fingers of our right hand around a screw and by holding it with our thumb pointing up. Turn the screw counter-clockwise, as viewed looking down upon the tip of our extended thumb. The direction of this turning motion can also be described as turning in the direction from the base of our fingers toward the tips of our fingers. If it moves in the direction of our thumb, which is up in this case, the screw is called right-handed. If it moves into our hand, which is the opposite direction that our thumb is pointing, the screw would be left-handed.

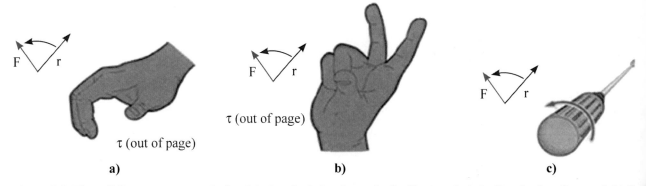

Figure 8.3: Three different ways to apply the right-hand rule is schematically illustrated. a) Curling the four fingers, initially pointing in the direction of **r**, towards the direction of **F** leads the thumb to point o page, indicating the direction of **r**×**F**. b) Index finger, middle finger and thumb of the right hand are used to determine the direction. c) Curling the fingers from **r** to **F** yields the thumb to point out of the page as the direction.

EQUILIBRIUM AND ROTATIONAL DYNAMICS

Most screws are right-handed. Another way to apply the right-hand rule to determine the direction of a cross product **r**×**F** between two vectors **r** and **F** is by aligning four index fingers along the vector **r** in the plane of these two vectors and curl the fingers toward the vector **F**. Here, we need to remember the following two points: i) initially, the thumb must be perpendicular to the four index fingers, and ii) the palm of the right-hand must be oriented in such a way to be able to curl the four index fingers toward the vector **F**. The direction of the cross product **r**×**F** is indicated by the thumb.

Exercise Problem 8.1:

The biceps muscle exerts a vertical force on the lower arm. Calculate the torque about the axis of rotation through the elbow joint, assuming the muscle is attached 5.0 cm from the elbow.

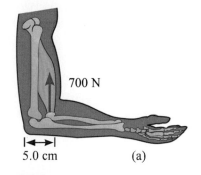
(a)

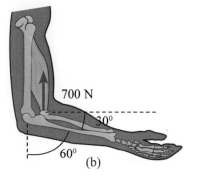
(b)

PROBLEM SOLVING STRATEGY

What do we know?

The pivot is the elbow joint. The force is applied 5.0 cm form the pivot.

What concepts are needed?

Since we are asked to compute the torque, we need to apply the definition of torque and the right-hand rule.

Why?

Since the torque is a vector, we need to compute both its magnitude and direction.

Solution:

a. Applying the definition of torque in Eq. (8.1), we compute the torque as

$$\vec{\tau} = \vec{r} \times \vec{F} = rF\sin\theta \ (\odot\square)$$
$$= 0.05\text{m} \times 700\text{N} \times \sin 90^\circ \ (\odot\square) = 35\text{m·N} \ (\odot\square)$$

b. In this case, the torque is computed in the same way as in part a), but $\theta = 120^\circ$,

$$\vec{\tau} = \vec{r} \times \vec{F} = rF\sin\theta \ (\odot\square)$$
$$= 0.05\text{m} \times 700\text{N} \times \sin(90^\circ + 30^\circ) \ (\odot\square)$$
$$= 0.05\text{m} \times 700\text{N} \times \sin 60^\circ \ (\odot\square) = 30\text{m·N} \ (\odot\square)$$

Note the symbol ⊙□ means the direction is out of the page. Similarly, the symbol ⊗ means the direction is into the page. It is interesting to note that both torque τ and energy E have the same dimensions (i.e., N•m).

There are many levers within the skeletal structure of the human body. The forearm is a great example of a lever. The biceps muscles in the forearm flex it while the tendons attach this muscle close to the elbow. The fulcrum, load and effort can be easily identified. Levers work in the human body by a means of coordination between muscles. They work together by use of a combination between the fulcrum, which rotates the lever, a load, which is the applied force by the lever, and effort as the force is applied by the person. As shown in the figure of Exercise Problem 8.1, different muscles (including the triceps not shown in the above illustration) of our arm are used to flex and extend the arm because muscles work only by contraction. There are three forces acting on the forearm: the force from the biceps, the force at the elbow joint, and the force from the load being lifted. Because the elbow joint is motionless, it is natural to define our torques by using the joint as the axis. Now, the situation becomes quite simple because the force exerted by the upper arm at the elbow has $r = 0$, and therefore it creates no torque. We can ignore this force completely. In general, we would call this the fulcrum of the lever. In general, a lever may be used either to increase or to reduce a force. Why did our arms evolve so as to reduce force? In general, our body is built for compactness and maximum speed of motion rather than maximum force. This is the main anatomical difference between us and the Neanderthals (their brains covered the same range of sizes as those of modern humans), and it seems to have worked for us. As with all machines, the lever is incapable of changing the amount of mechanical work we can do. A lever that increases force will always reduce motion, and vice versa, leaving the amount of work unchanged.

Net torque

We have discussed above the effect of a single torque on the rotational motion of an object. What happens if there are multiple torques acting on the object? We may answer this question by making an analogy with the translational motion of an object with a multiple number of forces acting on it. When considering translational motion, there may be more than one force acting on an object, and each of these forces may act on different point on the object, these forces were added to obtain the net force. When forces $F_1, F_2, F_3, F_4 \ldots$ exert torques $\tau_1, \tau_2, \tau_3, \tau_4, \ldots$, the net torque about a pivot point is the sum of the torques due to the applied forces:

$$\vec{\tau}_{net} = \vec{\tau}_1 + \vec{\tau}_2 + \vec{\tau}_3 + \cdots = \sum_i \vec{\tau}_i. \tag{8.2}$$

For example, there are five forces acting on different parts of the system shown in Fig 8.4. Here, the axle acts as the axis of rotation. Also, the axle exerts a force on the object to keep $F_{net} = 0$, indicating that the translational motion is uniform. However, these forces may yield non-zero net torque. While the other four forces yield either positive or negative torque about the axle, the force F_{axle} does not exert torque because it is exerted on the axle (i.e., no moment arm).

In general, when there is more than one force acting on different points with respect to an axis of rotation, each force will contribute to the net torque. The rotational motion of the object is determined by the **net torque** which is the sum of the individual torques. When the net force on the system is zero, the torque measured from any point in space is the same.

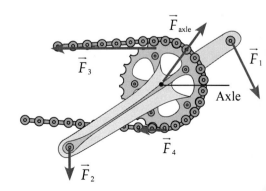

Figure 8.4: The net torque about an axis of rotation is the sum of the torque contributions due to all of the forces acting on the system. Any force that goes through the axis of rotation does not contribute to the net torque.

EQUILIBRIUM AND ROTATIONAL DYNAMICS

Understanding-the-Concept Question 8.1:

If a constant net torque is applied to an object, the object will

    a.   rotate with constant angular velocity.

    b.   rotate with constant angular acceleration.

    c.   having an increasing moment of inertia.

    d.   having a decreasing moment of inertia.

Explanation:

Similar to Newton's second law which states that a constant net force applied to an object makes it accelerate, if a constant net torque is applied, then the object will rotate with constant angular acceleration. It should be noted that net torque is the angular equivalence of net force.

Answer: b

8.2 RIGID BODY IN EQUILIBRIUM

We have seen that whenever there is either unbalanced force or torque acting on a rigid body, it will move with either linear acceleration or angular acceleration, respectively. A question we may ask is "Under what condition is the rigid body in equilibrium?" We recall that a rigid body is an idealization of a solid body in which deformation is neglected. In other words, the distance between any two given points of a rigid body remains constant in time regardless of external forces exerted on it. Even though such an object cannot physically exist due to relativity, normally objects can be assumed to be perfectly rigid if they do not move near the speed of light. As a way to examine the conditions for equilibrium, we consider a special case of motion in which the net force and the net torque on a system are both zero. This means that there is no linear acceleration and no angular acceleration. In the absence of any types of acceleration, there are two types of equilibrium: i) system is at rest, and ii) center of motion of system is moving at constant velocity. When an object is at rest, it is said to be in static equilibrium. More specifically, a standard definition of static equilibrium is that a system of particles in which all the particles are at rest and the total force acting on each particle is permanently zero. This is a strict definition, but often the term "static equilibrium" is used in a more relaxed manner, interchangeably with "mechanical equilibrium". A rigid body in mechanical equilibrium is undergoing neither linear nor rotational acceleration. However, it could be translating or rotating at a constant velocity. When an object is moving at constant speed, it is in dynamic equilibrium. Here, in both cases, we define equilibrium as a situation in which two or more forces (or torques) actingon an object orsystem are balanced.

Conditions for Equilibrium

When a rigid body is in equilibrium, the sum of forces acting on particles of the system is zero, and also the sum of torques is zero. This means that the vector sum of all *external forces* is zero, implying that the object is either at rest or in uniform translational motion. Another condition for equilibrium is that the sum of the moments of all *external forces* about any line (axis of rotation) is zero. This condition means that the net torque about an axis of rotation must be zero, implying that the object has either no rotational motion. For a rigid body to be in mechanical equilibrium, the sum of the forces and torques acting on it must add up to zero. The two conditions are written formally as the following:

i. The vector sum of all "external forces" must be zero (First condition for equilibrium):

$$\sum_i F_{x,i} = 0, \quad \sum_i F_{y,i} = 0, \quad \sum_i F_{z,i} = 0. \tag{8.3}$$

ii. The sum of all external torques (i.e., moment of external forces) must be zero (Second condition for equilibrium):

$$\sum_i \tau_i = 0 \tag{8.4}$$

If the first condition is satisfied (i.e., the total force acting on the system is zero), it can be shown that the total torque on the system is independent of the choice of origin. Therefore, we can show the second condition is valid by choosing the origin in a way that simplifies the calculation. Typically one chooses to calculate torques about the point at which the most number of forces act on.

Understanding-the-Concept Question 8.2:

Why do tightrope walkers carry a long, narrow beam?

Explanation:

A wire-walker may use a pole for balance or may stretch out his or her arms perpendicular to his/her trunk in the manner of a pole. This technique provides several advantages. It distributes mass away from the pivot point and changes the moment of inertia (see section 8.4). This reduces angular velocity. It takes longer to sweep out the same angle because the increase in the moment of inertia of the pole produces a smaller angular acceleration due to stray torque. The result is less tipping. In addition, this will create an equal and opposite torque on his/her body. By doing this, the wire-walker may maintain static rotational equilibrium. Sometimes the pole is weighted and has a dip at the ends. This provides additional stability by lowering the center of mass.

EQUILIBRIUM AND ROTATIONAL DYNAMICS

Understanding-the-Concept Question 8.3:

The bar is being used as a lever to pry up a large rock. The small rock acts as a fulcrum. The force F_p required at the long end of the bar can be quite a bit smaller than the rock's weight mg, since the torques balance in the rotation about fulcrum. If, however, the leverage is not quite good enough, and the rock does not budge, what are two ways to increase the leverage?

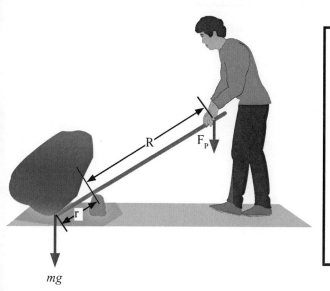

PROBLEM SOLVING STRATEGY

What do we know?

The small rock acts as a fulcrum. The bar is used as a lever. The force F_p is required.

What concepts are needed?

The magnitude of torque is computed from its definition in Eq. (8.1).

Why?

We need know what variables need to be changes to increase the magnitude of torque.

Explanation:

In prying up the large rock by using a long pry bar and by using a small rock as a fulcrum, we need to maximize the applied torque. However, if the leverage is not quite good enough, then we may increase the leverage in the following two ways: i) we move the small rock closer to the large rock so that the distance r is decreased and ii) we increase the length R by using a longer pry bar. Another way to increase the applied torque without applying stronger F_p is to change the direction of the applied force so that its direction is perpendicular to the pry bar. So, we need to insert our pry bar between the underside of the rock and over the fulcrum and slowly pry it out by applying a force perpendicular to the pry bar.

As suggested in Understanding-the-Concept Question 8.3, we may use a pry bar and a small rock to move a much larger rock by utilizing torque. There is another mechanical device which serves the similar function. Levers utilize torque to assist us in lifting or moving objects. Examples of levers include can openers, car jacks, and millions of other everyday devices. A lever is used to change the direction of a force, and every lever has a **fulcrum** point, about which the lever pivots. The two arms of a lever are not equal. A lever allows us to move a heavy weight with a small force by taking advantage of this inequality, as illustrated in the figure of Understanding-the-Concept Question 8.3. Also, by applying half the force on the longer side of the lever, but using a pry bar with twice the length, we can move the same weight on the short side of the lever. For example, we can move a 1 kilogram weight half a meter by putting a half kilogram weight on the long arm of the lever, and letting it push down one meter. The ratio of the long arm to the short arm tells us the **mechanical advantage** of the lever, which is the ratio of work put into the system to work done. If the long arm is twice as long as the short arm, the ratio is 2:1, and the mechanical advantage is 2. So we could move the weight with half the force needed if we are willing to move twice as far. This comes in handy when we have got a motor that can apply only so much force.

Exercise Problem 8.2:

A 5.0 m long ladder leans against a wall at a point 4.0 m above the ground. The ladder is uniform and has mass 12.0 kg. Assuming the wall is frictionless (but the ground is not), determine the forces exerted on the ladder by the ground and the wall.

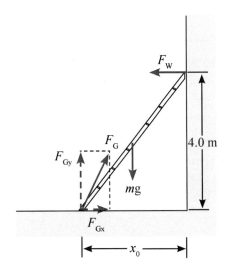

PROBLEM SOLVING STRATEGY

What do we know?
The length and mass of the ladder and the places where the ladder is making contact with the wall are given. The ladder is in equilibrium.

What concepts are needed?
The equilibrium conditions need to be applied.

Why?
There is no translational and rotational motion for the ladder. The ladder is in static equilibrium.

Solution:

First, we determine the length of the ladder by the Pythagorean equation and obtain

$$\ell^2 = h^2 + x_o^2 \Rightarrow x_o = \sqrt{\ell^2 - h^2} = 3.0\,\text{m}.$$

Now, we apply the conditions for equilibrium. Since the ladder has no translational motion either in y- or x-direction, we apply the Newton's first law and write

$$\sum_i F_{yi} = 0 \Rightarrow F_{Gy} - mg = 0 \Rightarrow F_{Gy} = mg$$

for y-direction, and

$$\sum_i F_{xi} = 0 \Rightarrow F_{Gx} - F_w = 0 \Rightarrow F_{Gx} = F_w$$

for x-direction, respectively. Here, F_{Gy} is the normal force acting on the ladder and F_{Gx} is the friction force. To solve the problem, we still need to determine F_{Gx} and F_w. This may be done by using

$$\sum_i \tau_i = 0 \Rightarrow F_w h - mg\frac{x_o}{2} = 0$$

which is the remaining condition for equilibrium, stating that the net torque about the bottom of the ladder is zero. Here, we should note that we may choose an axis of rotation to be anywhere since the ladder is not rotating. We solve for F_w and obtain

$$F_w = \frac{mgx_o}{2h} = \frac{12.0\,\text{kg} \times 9.8\,\text{m}/\text{s}^2 \times 3.0\,\text{m}}{2 \times 4.0\,\text{m}} = 44\,\text{N}.$$

EQUILIBRIUM AND ROTATIONAL DYNAMICS

Combining the result, we compute the magnitude of the force that the ground is exerting on the ladder as

$$F_G = \sqrt{F_{Gx}^2 + F_{Gy}^2} = \sqrt{(44N)^2 + (12kg \times 9.8 \, m/s^2)^2} = 126 \, N$$

Finally, we compute the direction of the force F_G by using the definition of the tangent function. The angle θ between F_G and the ground (along +x-direction) is

$$\theta = \tan^{-1} \frac{mg}{F_w} = \tan^{-1} \frac{12kg \times 9.8 \, m/s^2}{44N} = 69.5°.$$

Exercise Problem 8.3:

Calculate the tension F_T in the wire that supports the 30 kg beam, and the force F_H exerted by the wall on the beam (give magnitude and direction)

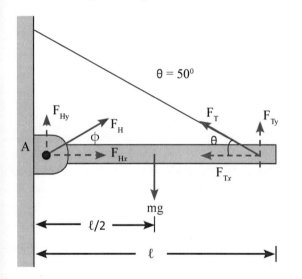

PROBLEM SOLVING STRATEGY

What do we know?

This mass of the beam is given. Also, the horizontal beam is being supported by both the wall and the wire.

What concepts are needed?

The two conditions for equilibrium need to be applied.

Why?

The beam is in static equilibrium.

Solution:

We apply the conditions for equilibrium. Applying Newton's first law along the x-direction, we obtain

$$\sum_i F_{xi} = 0 \Rightarrow F_H \cos\phi - F_T \cos\theta = 0 \Rightarrow F_H \cos\phi = F_T \cos\theta$$

Similarly, applying Newton's first law along the y-direction, we obtain

$$\sum_i F_{yi} = 0 \Rightarrow F_H \sin\phi + F_T \sin\theta - mg = 0$$

Finally, at equilibrium, the net torque about any fixed point is zero. We compute the net torque about point A as

$$\sum_i \tau_i = 0 \Rightarrow -mg \, \ell/2 + F_T \sin\theta \, \ell = 0$$

We solve for F_T and obtain

$$F_T = \frac{mg}{2\sin\theta} = \frac{30\text{kg} \times 9.8\,\text{m}/\text{s}^2}{2 \times \sin 50°} = 192\,\text{N}$$

We rewrite the equilibrium condition for the y-axis and obtain

$$F_H \sin\phi + F_T \sin\theta - mg = 0 \Rightarrow F_H \sin\phi + \frac{mg}{2} - mg = 0 \Rightarrow F_H \sin\phi = \frac{mg}{2}$$

We now use the trigonometry identity, $\sin^2\phi + \cos^2\phi = 1$, and obtain F_H by using the results obtained above and by writing the magnitude of $\mathbf{F}_H$ as

$$F_H^2 = F_H^2 \cos^2\phi + F_H^2 \sin^2\phi = (F_T \cos\theta)^2 + \left(\frac{mg}{2}\right)^2$$

$$F_H = \sqrt{(192\text{N} \times \cos 50°)^2 + \left(30\text{kg} \times 9.8\,\text{m}/\text{s}^2 \times 0.5\right)^2} = 192\,\text{N}$$

Finally, we find the direction of $\mathbf{F}_H$ by applying the definition of tangent function as

$$\tan\phi = \frac{F_H \sin\phi}{F_H \cos\phi} = \frac{mg/2}{F_T \cos\theta} \Rightarrow \phi = \tan^{-1}\frac{mg}{2F_T \cos\theta} = \tan^{-1}\frac{30\text{kg} \times 9.8\,\text{m}/\text{s}^2}{2 \times 192\text{N} \times \cos 50°} = 50.0°.$$

The type of support structure shown in Exercise Problem 8.3 may also be seen in a suspension bridge. A **suspension bridge** is a type of bridge in which the deck (the load-bearing portion) is hung below suspension cables on vertical suspenders. The first examples of this type of bridge were built in the fifteenth century in Tibet and Bhutan. Bridges without vertical suspenders have a long history in many mountainous parts of the world. This type of bridge has cables suspended between towers, plus vertical suspender cables that carry the weight of the deck below, upon which traffic crosses. This arrangement allows the deck to be level or to arc upward for additional clearance. Like other suspension bridge types, this type is often constructed without structure failures. The suspension cables must be anchored at each end of the bridge, since any load applied to the bridge is transformed into a tension in these main cables. The main cables continue beyond the pillars to deck-level supports, and further continue to connections with anchors in the ground. The roadway is supported by vertical suspender cables or rods, called hangers. In some circumstances the towers may sit on a bluff or canyon edge where the road may continue directly to the main span. Otherwise, the bridge will usually have two smaller spans, running between pair of pillars and the highway, which may be supported by suspender cables, or may use a truss bridge to make this connection. In the latter case, there will be very little arc in the outboard main cables. Outside Tibet and Bhutan, construction of this type of bridge dates from the early nineteenth century. Now-a-days, a suspension bridge may be seen quite commonly. For example, Golden Gate Bridge in San Francisco, California is a suspension bridge. The construction of this bridge began on January 1933 and completed on April 1937.

8.3 CENTER OF GRAVITY

As we noted in Chapter 6, the center of gravity and center of mass may be almost considered as synonymous because the same constraints that apply to the center of mass also apply to the center of gravity. In the real world, the center of gravity will not be exactly equal to the center of mass. Why? To answer this question, let us examine the center of gravity little more closely. The **center of gravity** is a geometric property which represent the average location of an objects. Hence, the center of gravity (of a rigid body) is defined as the point at which the torque due to the weight of the object is applied. The similarity between the concept of center of gravity and center of mass suggests that we may describe the motion of any object in terms of the **translational** motion of its center of gravity and the **rotational** motion about its center of gravity, if it is free to rotate.

In flight, both airplanes and rockets rotate about their centers of gravity. A kite, on the other hand, rotates about the bridle point. However, the trim of a kite still depends on the location of the center of gravity relative to the bridle point because the weight for every object always acts through the center of gravity. Determining the center of gravity is very important for an aircraft. Airplane balances at the proper center of gravity location is an absolute requirement for safe flight. For a large commercial aircraft, a large adjustment to the location of the center of gravity can be made by distribution of the fuel load, passengers or cargo. As shown in Fig. 8.5, a shift in the center of gravity may make an aircraft difficult to control during the flight.

In general, determining the center of gravity requires a complicated procedure because the mass (and weight) may not be uniformly distributed throughout the object. Hence, the general case requires the use of calculus which we are not equipped with so we will discuss below a simpler case. For example, if the mass is uniformly distributed, the problem is greatly simplified. If the object has a line (or plane) of **symmetry**, the center of gravity lies on the line of symmetry. For a solid block of uniform material, the center of gravity is simply at the average location of the physical dimensions.

To motivate the definition of a center of gravity more concretely, we consider the center of gravity for a system of following two objects: i) a horizontal uniform board (weight W_1) and ii) a uniform box (weight W_2)

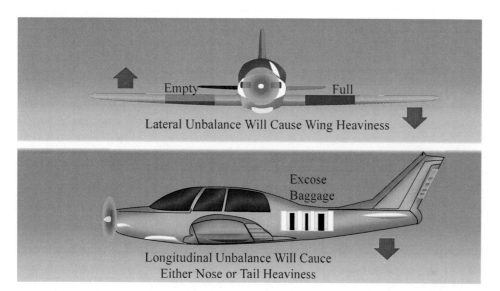

Figure 8.5: Effects a shift in the center of gravity which results in either lateral (top panel) or longitudinal (bottom panel) unbalance of an airplane is schematic illustrated.

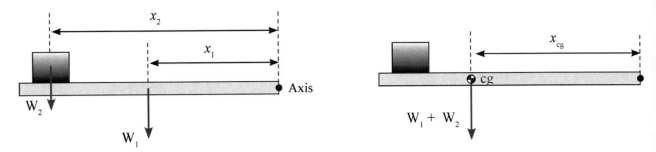

Figure 8.6: The center of gravity for a system of objects is the average location of the weight of the objects. The center of gravity (cg) is schematically illustrated for a system of two objects.

The net torque created about the axis of rotation by these objects may be written as

$$\tau_{net} = \sum_i \tau_i = W_1 x_1 + W_2 x_2 = (W_1 + W_2) x_{cg}. \quad (8.5)$$

We solve for the center of gravity, x_{cg}, and obtain

$$x_{cg} = \frac{W_1 x_1 + W_2 x_2}{W_1 + W_2} = \frac{m_1 g x_1 + m_2 g x_2}{m_1 g + m_2 g} = x_{cm}. \quad (8.6)$$

When the gravitational acceleration experienced by all of the objects in the system is the same, the center of gravity x_{cg} is the same as the center of mass x_{cm}. We may generalize the result of two objects to a group of N objects and compute the center of gravity as

$$x_{cg} = \frac{W_1 x_1 + W_2 x_2 + ... + W_N x_N}{W_1 + W_2 + ... + W_N} = \frac{\Sigma_i W_i x_i}{\Sigma_i W_i}. \quad (8.7)$$

As indicated by Eq. (8.6), the center of mass of a rigid body is also called its center of gravity in circumstances where the gravitational acceleration is uniform. Since gravitational acceleration of Earth may be taken as uniform at all places, the center of mass and the center of gravity are effectively the same. When an accurate value of the gravitational acceleration is needed, we need to account for the non-uniformity of gravitational field. We know from Chapter 2 that gravitational field of the Earth is not uniform, indicating that the center of gravity is not the same as the center of mass. In this case, the center of mass is a fixed property which is the average location of the mass of the body, and it has nothing to do with gravity.

As we discussed above, the center of gravity of an aircraft is important for its stability during it flight. Here, the **center of gravity** is the point at which the aircraft would balance if it were possible to suspend it at that point. It is the theoretical point at which the entire weight of the aircraft is assumed to be concentrated. Its position is calculated after supporting the aircraft on at least two sets of weighing scales or load cells, and noting the weight shown on each set of scales or load cell. The center of gravity affects the stability of the aircraft. To ensure the aircraft is safe to fly, the center of gravity must fall within specified limits established by the aircraft manufacturer.

In preparing for a flight, the pilot (and ground crew) has to assume the responsibility (and is required by law) because he/she has control over both loading cargo and fuel management the two variable factors which can change both the total weight/balance, and center of gravity

EQUILIBRIUM AND ROTATIONAL DYNAMICS

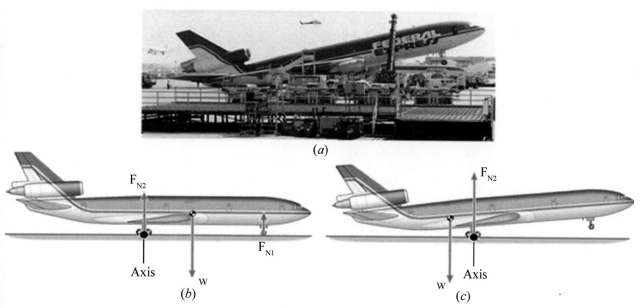

Figure 8.7: A shift in the center of gravity (cg) results in (a) longitudinal unbalance of an airplane. This shift from (b) the stable position to (c) an unstable position is due to too much weight on the rear of the aircraft.

This information is available to the pilot in the form of aircraft records, operating handbooks and placards in baggage compartments and or fuel caps. The owner of the aircraft has the responsibility to make sure that up-to-date information is available to the pilot. While performing the weight/balance calculations it is sometimes necessary to move cargo or passengers about the aircraft to remain within the center of gravity limits. The pilot has a couple of options when the center of gravity is not within limits after calculating the numbers. Defueling the aircraft (may not be viable) or redistributing cargo or passengers are some of the options. As shown in Fig. 8.7, redistribution of cargo can definitely leads to a shift in the location of the center of gravity. Here, when heavier weights are placed on the rear of the airplane, they cause the center of gravity to shift to the opposite side of the rear landing gears, which are acting as the axis of rotation and lead the fuselage to rotate. This type of accident is not so uncommon for a cargo airplane.

How important is the center of gravity for the performance of the airplane in flight? For many aircraft, the center of gravity must be within a small distance and very close to the wing spar and the center of lift as shown in Fig. 8.7. If the center of gravity is too far forward or backward, then the aircraft has to be trimmed to fly level. When it is trimmed, the extra lift (or negative lift) by the tail can reduce the total lift of the airplane or result in decreased airspeed due to the attitude of the airplane. For extreme cases, aircraft have taken off and the cargo load has shifted aft, causing the airplane to pitch its nose up to an extreme angle such that it could not recover or fly level and they have crashed.

Exercise Problem 8.4:

A massless, rigid board is placed across two bathroom scales that are separated by a distance of 2.00 m. A person lies on the board. The scale under his head reads 425 N, and the scale under his feet reads 315 N. (a) Find the weight of a person. (b) Locate the center of gravity of the person relative to the scale beneath his head.

> **PROBLEM SOLVING STRATEGY**
>
> **What do we know?**
>
> The length of the rigid board is given. Also, the force of support for a person at his head and at his feet is also provided. The acceleration of the person is zero since he is not moving.
>
> **What concepts are needed?**
>
> We need to apply Newton's first law and the definition of center of gravity.
>
> **Why?**
>
> The person's weight may be determined from Newton's first law and the center of gravity relative his head may be computed by applying the definition.

Solution:

a. Since the system is in equilibrium, we may apply Newton's first law for the vertical direction, we obtain

$$\sum_i F_i = 0 \Rightarrow F_h + F_f - W = 0$$

We solve for the weight W and obtain

$$W = F_h + F_t = 425\text{N} + 315\text{N} = 740\text{N}.$$

b. We now compute the center of gravity of the person relative to the scale at the person's head by calculating the net torque and dividing it by W

$$x_{cg} = \frac{F_h x_h + F_f x_f}{W} = \frac{425\text{N} \times 0\text{m} + 315\text{N} \times 2.00\text{m}}{740\text{N}} = 0.851\text{m}.$$

This indicates that center of gravity (or center of mass) of the person is not at the middle of the body since the weight distribution with the body is not uniform. This may be seen easily by noting that the person's head is heavier than the feet.

Similar to objects, there is a center of gravity for each person. A human's center of gravity is the location on the body where the weight of the body is concentrated. A person's height and weight both affect the location of the center of gravity. Weight is a measurement of the force of gravity on the body, and height affects how the person balances that gravitational pull during movement. Calculating a person's center of gravity can help him/her balance better. The lowest center of gravity is between the hips. However, the center of gravity shifts with each body movement. The center of gravity for adults is the hips. However, as the person grows older, a stooped posture is common because of the changes from osteoporosis and normal bone degeneration, and the knees, hips, and elbows flex. This stooped posture results in the upper torso being the center of gravity for the elderly person.

8.4 INERTIA FOR A ROTATING BODY

As for the inertia of a body in linear motion, the inertia of a body rotating about an axis may be described by using the moment of inertia. In other words, an object that is rotating at constant angular velocity will remain rotating unless acted upon by an external torque. In this way, the moment of inertia plays the same role in rotational dynamics as mass does in linear dynamics, describing the relationship between angular momentum and angular velocity, torque and angular acceleration. Here, the **moment of inertia** is a property of a distribution of mass in space that measures its resistance to rotational acceleration about an axis. This scalar moment of inertia becomes an element in the inertia matrix when a distribution of mass is measured around three axes in space. This inertia matrix appears in the calculation of the angular momentum, kinetic energy and resultant torque in the dynamics of a rigid body. For simplicity, we will only consider a distribution of mass which is measured with respect to a single axis of rotation. We should note that the rotational analogue of inertia is the mass term in Newton's second law.

Let us see how the moment of inertia of a rotating object derived from Newton's second law. Suppose we consider the angular acceleration α of a rotating body is proportional to the net torque τ applied to it. Then, we may write

$$\vec{F} = m\vec{a} = mr\vec{\alpha}. \tag{8.8}$$

For a single particle, we may write torque τ as

$$\vec{\tau} = \vec{r} \times \vec{F} = mr^2\alpha \, (\hat{r} \times \hat{\alpha}). \tag{8.9}$$

The moment of inertia force on a single particle around an axis multiplies the mass of the particle by the square of its distance to the axis. Of course, a rigid body has a large number of particles. For a rotating body with a large number of particles, we can write the net torque as the sum the torque τ_i contribution from each particle

$$\sum_i \tau_i = \left(\sum_i m_i r_i^2\right)\alpha = I\alpha \tag{8.10}$$

where

$$I = \sum_i m_i r_i^2 = m_1 r_1^2 + m_2 r_2^2 + \cdots \tag{8.11}$$

is called "moment of inertia." Moment of inertia is the name given to rotational inertia which is the rotational analog of mass, inertia for linear motion. Moment of inertia appears in the relationships for dynamics in rotational motion. Equation (8.11) indicates that the moment of inertia must be specified with respect to a chosen axis of rotation and that the moments of inertia of individual particles in a rigid body sum to define the moment of inertia of the body rotating about an axis. In general, the moment of inertia indicates the relative difference in how difficult or easy it will to set any object in motion about an axis of rotation. For rigid bodies moving in a plane, such as a compound pendulum, the moment of inertia is a scalar. Many systems use a mass with a large moment of inertia to maintain a rotational velocity and resist small variations in applied torque. For example, the long pole held by a tight-rope walker maintains a zero angular velocity resisting the small torque applied by the walker to maintain balance. Another example is the rotating mass of a flywheel which maintains a constant angular velocity resisting the torque variations in a machine.

Understanding-the-Concept Question 8.4:

Which case has greater moment of inertia?

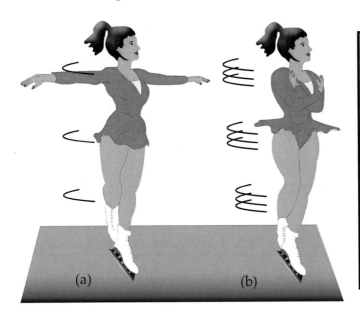

PROBLEM SOLVING STRATEGY

What do we know?

A figure skater is spinning about her body as the axis of rotation.

What concepts are needed?

We need to apply the definition of moment of inertia.

Why?

When the position of her arms changes, she changes the distribution of her mass with respect to her body.

Explanation:

Figure skaters pay unconscious tribute to their intuitive understanding of physics when they draw in their arms as they spin. In general, a more compact body has a smaller moment of inertia than a more extended of the same mass. The moment of inertia for a figure skater (or any other object) is a measure of how easy or difficult it is to change the rate of rotation. As the figure skaters draw their arms close, they decrease their moment of inertia and their rate of rotation increases accordingly. As we will discuss below, this is due to that the angular momentum is conserved.

Exercise Problem 8.5:

Two "weights" of mass 5.0 kg and 7.0 kg are mounted 4.0 m apart on a light rod (whose mass can be ignored). Calculate the moment of inertia of the system (a) for rotation about an axis halfway between the weights, and (b) for rotation about an axis 0.5 m to the left of the 5.0 kg mass.

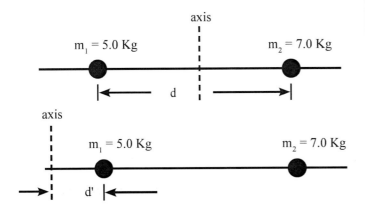

PROBLEM SOLVING STRATEGY

What do we know?

The positions of two masses relative to the axis are given.

What concepts are needed?

We need to apply Eq. (8.11), the definition of moment of inertia.

Why?

We have a system of two point masses.

EQUILIBRIUM AND ROTATIONAL DYNAMICS

Solution:

a. By using the definition, we compute the moment of inertia of the system as

$$I = \sum_i m_i r_i^2 = m_1 \left(\frac{d}{2}\right)^2 + m_2 \left(\frac{d}{2}\right)^2 = \frac{(m_1+m_2)d^2}{4} = \frac{12.0\,\text{kg}}{4} \times (4.0\,\text{m})^2 = 48\,\text{kg}\cdot\text{m}^2.$$

b. When the axis of rotation is shifted, the moment of inertia is changed. The moment of inertia about the new axis of rotation is

$$I = \sum_i m_i r_i^2 = m_1 (d')^2 + m_2 (d+d')^2$$

The numerical value of I is obtained by substituting the values

$$I = 5.0\,\text{kg} \times (0.5\,\text{m})^2 + 7.0\,\text{kg} \times (4.5\,\text{m})^2 = 143\,\text{kg}\cdot\text{m}^2.$$

Since the moment of inertia is defined with respect to a specific rotational axis, the distribution of mass of an object with respect to an axis of rotation is important in determining the magnitude of the inertia. The moment of inertia of point with respect to an axis is defined as the product of the mass times the distance from the axis squared. This indicates that when an object's mass is distributed farther away from the axis, the moment of inertia becomes larger.

Moment of inertia for some useful objects

The moment of inertia of any extended object is built from the basic definition of Eq. (8.11). The continuous mass distributions require an infinite sum of all of the point mass moments which make up the whole. This is accomplished by using integration (i.e., calculus) over all the mass. For this reason,

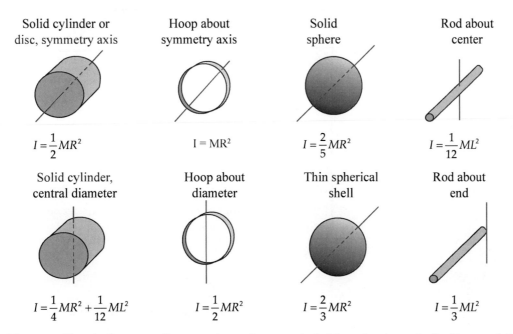

Figure 8.8: Moment of inertia for some of commonly used geometrical objects is schematically illustrated. The position of the axis of rotation is indicated by a red line.

the moment for a different shape of object is shown below in Fig. 8.8 without its derivation. However, it is worth noting that the moment of inertia for an extended object depends on the location of the axis of rotation as well as its shape.

Understanding-the-Concept Question 8.5:

Consider two uniform solid spheres where one has twice the mass and twice the diameter of the other. The ratio of the larger moment of inertia to that of the smaller moment of inertia is

    a. 2   b. 8   c. 4   d. 10   e. 6

Explanation:

The moment of inertia of an extended object shown in Fig. 8.8 is proportional to the mass and to the square of the distance R form the axis of rotation (i.e., $I \propto mR^2$). This indicates that if mass is increased twice and the radius is increased twice, then the new larger moment of inertia is increased by $2\times(2)^2 = 8$ times from the smaller moment of inertia.

Answer: b

Exercise Problem 8.6

Consider a pulley of mass $M = 4.00$ kg and radius $R = 33.0$ cm. Suppose a bucket of weight 15 N (mass $m = 1.53$ kg) is hanging from the cord, which we assume not to stretch or slip on the pulley. (a) Calculate the angular acceleration of the bucket. (b) Determine the angular velocity ω of the pulley and the linear velocity v of the bucket at $t = 3.00$ s if the pulley (and bucket) start from rest at $t = 0$. There is a frictional torque of $\tau_{fr} = 1.10$ N·m at the axle.

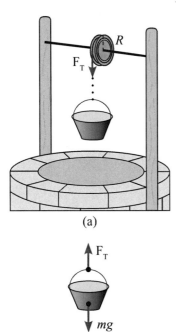

PROBLEM SOLVING STRATEGY

What do we know?

The bucket with mass M accelerates vertically down, while the pulley rotates counterclockwise. The cylindrical shape pulley will have an angular acceleration.

What concepts are needed?

We need to apply the Newton's second law to describe the translational motion of the bucket and rotational motion of the pulley. Also, we need the relationship between these two different types of motion.

Why?

The bucket falls down and accelerates, but both the bucket and pulley are connected by a cord.

EQUILIBRIUM AND ROTATIONAL DYNAMICS

Solution:

For convenience, we consider the pulley and bucket separately. For the rotation of the pulley, we write the rotational equivalence of the Newton's second law. We note that the shape of the pulley is a disk and that the axis of rotation is the center of the disk. With this consideration, we write the Newton's second law as

$$\sum_i \tau_i = I\alpha \Rightarrow F_T R - \tau_{fr} = I\alpha \tag{6a}$$

where $I = MR^2/2$. For the bucket, we apply the Newton's law along the y-direction and obtain

$$\sum_i F_i = ma \Rightarrow mg - F_T = ma. \tag{6b}$$

Equation (6b) allows us to solve for the tension F_T on the string

$$F_T = mg - ma = mg - mR\alpha \tag{6c}$$

by applying the rolling condition for the pulley (i.e., $a = R\alpha$). This allows us to find the expression for the angular acceleration α of the pulley from

$$I\alpha = F_T R - \tau_{fr} = (mg - mR\alpha)R - \tau_{fr} = mgR - mR^2\alpha - \tau_{fr} \tag{6d}$$

which is obtained by substituting F_T into Eq. (6a) for the net torque

$$(I + mR^2)\alpha = mgR - \tau_{fr} \Rightarrow \alpha = \frac{mgR - \tau_{fr}}{I + mR^2}. \tag{6e}$$

The numerical value for the angular acceleration is

$$\alpha = \frac{mgR - \tau_{fr}}{I + mR^2} = \frac{15\text{N} \times 0.33\text{m} - 1.10\text{N}\cdot\text{m}}{\tfrac{1}{2} \times 4.0\text{kg} \times (0.33\text{m})^2 + 1.53\text{kg} \times (0.33\text{m})^2} = 1.0 \times 10^1 \frac{\text{rad}}{\text{s}^2}$$

and the linear acceleration is

$$a = R\alpha = 0.33\text{m} \times 10.0 \frac{\text{rad}}{\text{s}^2} = 3.3 \frac{\text{m}}{\text{s}^2}.$$

Since the angular acceleration is constant, we compute the angular velocity by using the rotational kinematics equation

$$\omega = \omega_0 + \alpha t = 10.0 \frac{\text{rad}}{\text{s}^2} \times 3.00\text{s} = 30.0 \frac{\text{rad}}{\text{s}}.$$

Applying the no slip condition, we compute the linear velocity of the bucket as

$$v = R\omega = 0.33\text{m} \times 30.0 \frac{\text{rad}}{\text{s}} = 9.9 \frac{\text{m}}{\text{s}}$$

Exercise Problem 8.6 indicates that, when the problem is examined in the view of dynamics, we conclude that the bucket accelerates as it falls. This acceleration results in the angular acceleration of the pulley since the bucket is connected to the pulley by the cord. We may consider this problem in another perspective. We may see this problem in terms of energy. As the bucket falls, the potential energy of the bucket is converted into the kinetic energy of the bucket and pulley. Of course, some of the potential energy is dissipated via the frictional force. However, we should keep in mind that the rotating pulley will have rotational kinetic energy whereas the falling bucket will have translational kinetic energy. Here, both translation and rotational motion should be treated equally.

8.5 ENERGIES OF A MOVING RIGID BODY

As we discussed in Chapter 5, the **kinetic energy** of an object is the energy which it possesses due to its motion. When an object is rolling, we may describe its motion as a combination of translational and rotational motion. This means that the rolling rigid body has two types of kinetic energy contributions: i) translational kinetic energy, and ii) rotational kinetic energy. The **translational kinetic energy** is the energy possessed by an object due to its linear motion and is written as

$$K.E._{tran} = \frac{1}{2}mv^2. \tag{8.12}$$

The **rotational energy** or **angular kinetic energy** is the kinetic energy due to rotation of an object. We may derive the expression for this rotational energy by accounting for the kinetic energy of each particle with tangential speed v_i. By summing these kinetic energy contributions, we may write K.E.$_{rot}$ as

$$K.E._{rot} = \sum_i \frac{1}{2}m_i v_i^2 = \frac{1}{2}\left(\sum_i m_i r_i^2\right)\omega^2 = \frac{1}{2}I\omega^2. \tag{8.13}$$

Since both translational energy and rotational energy is part of the total kinetic energy, we may write the total kinetic energy as the sum of both translational and rotational energy

$$K.E. = \frac{1}{2}mv^2 + \frac{1}{2}I\omega^2. \tag{8.14}$$

In describing the motion of rolling objects, we must keep in mind that the kinetic energy is divided between translational kinetic energy and rotational kinetic energy. Another key point to remember from Chapter 7 is that, according to the condition for rolling without slipping, the linear velocity of the center of mass is equal to the angular velocity times the radius (i.e., $v = r\omega$). This relation indicates that, for a rolling object, the ratio of the translational and rotational kinetic energy is determined by its shape.

Exercise Problem 8.7:

What will be the speed of a solid sphere of mass M and radius R when it reaches the bottom of an incline if it starts from rest at a vertical height h and rolls without slipping? Ignore losses due to dissipative forces and compare your result to that for an object sliding down a frictionless incline.

EQUILIBRIUM AND ROTATIONAL DYNAMICS

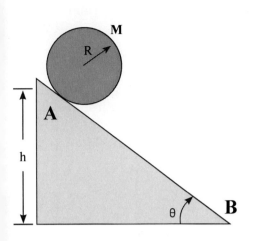

PROBLEM SOLVING STRATEGY

What do we know?

A solid sphere of mass M and radius R either rolls or slides down the incline.

What concepts are needed?

We need to apply the energy conservation law.

Why?

We are asked to ignore losses due to dissipative forces and a frictionless incline. When there is no energy dissipation, the mechanical energy of the system is conserved.

Solution:

Since we assume that there are no dissipative forces, the mechanical energy of the system is conserved. We apply conservation of energy principle to the problem (i.e., $K.E._A + P.E._A = K.E._B + P.E._B$). We note that the initial kinetic energy and final potential energy is zero. Also, we note that the kinetic energy has two contributions: (i) translational kinetic energy and (ii) rotational kinetic energy. Combining this, we write

$$Mgh = \frac{1}{2}Mv^2 + \frac{1}{2}I_{cm}\omega^2 \qquad (7a)$$

where $I_{cm} = (2/5)MR^2$ for the solid sphere. Noting that the rolling motion of the sphere indicates that $v = R\omega$, we rewrite Eq. (7a) as

$$Mgh = \frac{1}{2}Mv^2 + \frac{1}{2}\left(\frac{2}{5}MR^2\right)\left(\frac{v}{R}\right)^2 = \left(\frac{1}{2} + \frac{1}{5}\right)Mv^2.$$

We now solve for v and obtain

$$v = \sqrt{\frac{10}{7}gh}.$$

If the object slides down, then Eq. (7a) simplifies to

$$Mgh = Mv^2/2$$

since the angular speed $\omega = 0$. The speed of the solid sliding the incline changes to

$$v = \sqrt{2gh}$$

This speed is greater! But, why is this speed becomes greater?

If a body rolls down a hill, then its gravitational potential energy converts into two kinetic energies: a translational kinetic energy and a rotational kinetic energy. This means that it must have less translational energy than an object which just slides. So, all other things being equal, rolling objects move more slowly than sliding ones.

Understanding-the-Concept question 8.7: Who is faster?

Several objects roll without slipping down an incline of vertical height H, all starting from rest at the same moment. The objects are a thin hoop (or a plain wedding band), a marble, a solid cylinder (D-cell), an empty soup can, and an unopened soup can. In addition a greased box slides down without friction. In what order do they reach the bottom of the incline?

Explanation:

As indicated in Example Problem 8.7, the speed of an object coming down an incline is reduced as part of potential energy is converted to the rotational kinetic energy. Since the rotational kinetic energy increases with the moment of inertia, a rolling object with the largest moment of inertia will reach the incline slowest. This means that the sliding box will have the largest speed. So, it will reach the bottom of the incline first. However, the hoop has the largest moment of inertia. So, it will have the smallest translational kinetic energy. Hence, it will reach the bottom last as shown below

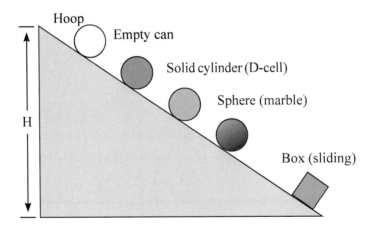

Understanding-the-Concept Question 8.8:

Do you think that Indiana Jones, as shown in the picture at the beginning of this chapter, be happier if the rock was sliding down instead of rolling down?

Explanation:

As shown in the movie, Dr. Jones barely makes out the temple. Assuming that there is no friction and the rock slides, instead of rolling down, the speed of the rock would be greater since the initial potential energy will be converted all to the translational kinetic energy. This means that Indiana would not be able to escape from the temple.

Work (Rotational)

There is a one-to-one correspondence between quantities in pure translation and pure rotation. This correspondence allows us to actually write down the relations for rotational work and kinetic energy by just inspecting corresponding relations for pure translational motion. Work done on a rigid body rotating about a fixed axis is determined in the similar way as that for the translational work

$$W = F\Delta\ell = F(r\Delta\theta) = \tau\Delta\theta. \qquad (8.15)$$

The linear displacement $\Delta\ell$ in Eq. (8.15) is the arc length resulting from the angular displacement. So, the work done on a rotating object is the torque τ time the angular displacement $\Delta\theta$. Here, τ is the rotational analog of F and the angular displacement is the rotational analog of translational displacement.

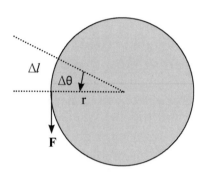

Figure 8.9: The rotational equivalence of work done on an object is due to an angular displacement Dθ due to the applied force F.

To get a more concrete idea, we consider a simple rotational situation in which a rope is wrapped around a solid wheel as shown in Fig. 8.9. The wheel has radius $r = 2.0$ m and mass $m = 10$ kg. If one pulls the rope with force $F = 100$ N for a distance of $\Delta\ell = 1.0$ m, the one does work of $W = F(\Delta\ell) = 100$ N×1.0 m = 100 J. If this was expressed in terms of rotational work, then one needs to compute the torque τ and angular displacement $\Delta\theta$. The force creates a torque of $\tau = r \cdot F = 2.0$ m×100N = 200 N·m and the wheel's rim spins through an angular displacement of $\Delta\theta = \Delta\ell/r = 1.0$ m/2.0 m = 0.5 rad. Thus (the rotational equivalence of) the work done is $W = \tau\Delta\theta = 200$ N·m×0.5 rad = 100 J, which is identical to the result of earlier calculation.

8.6 ANGULAR MOMENTUM

Similar to the role of linear momentum in describing translational dynamics, angular momentum is a useful quantity in rotational dynamics. **Angular momentum** or **rotational momentum** is a vector which represents the product of a body's rotational inertia and rotational velocity about a particular axis. The angular momentum of a system of particles (e.g. a rigid body) is the sum of angular momenta of the individual particles. A freely-rotating disk (like a Frisbee in flight or a tire rolling down a hill) has angular momentum. For a rigid body rotating around an axis of symmetry (e.g. the blades of a ceiling fan), the angular momentum can be expressed as the product of the body's moment of inertia I (i.e., a measure of an object›s resistance to changes in its rotation velocity) and its angular velocity ω with respect to its axis,

$$\vec{L} = I\vec{\omega}. \qquad (8.16)$$

It is noted that Eq. (8.16) defines the angular momentum of a symmetrical body, such as a spinning flywheel. The physical units for the angular momentum L is kg·m²/s. It is noted that angular momentum is a rotational analog of linear momentum. The angular momentum L of a particle about a given origin is defined as

$$\vec{L} = \vec{r} \times \vec{p} \qquad (8.17)$$

where r is the position vector of the particle relative to the origin, p is the linear momentum of the particle, and × denotes the cross product (or vector product).

Let us think about a likely event on a playground involving angular momentum. Suppose we consider a girl, with mass m, is running with a tangential velocity v, to spin a playground merry-go-round as shown in Fig. 8.10 and then jumps on at its edge. Jumping on the merry-go-around, she and the merry-go-round (mass M, radius R, and $I = cMR^2$) then spin together with a constant angular velocity ω_f. If her initial velocity is tangent to the circular merry-go-round, then what is ω_f? What concept should we use to approach this problem? This is an example involving a rotational collision. This is a completely inelastic collision because both the girl and merry-go-round move together after the collision.

Figure 8.10: A girl spins a playground merry-go-round to an angular speed ω by exerting a force at an edge to generate its angular momentum Iω. When she hops on the merry-go-around, the angular speed decreases due to the conservation of angular momentum.

The system clearly has angular momentum after the completely inelastic collision, but where is the angular momentum beforehand? It is with the girl. Her linear momentum can be converted to an angular momentum relative to an axis through the center of the merry-go-around. Here, we can treat her as a point particle with mass m and initial velocity v_i. We may apply conservation of angular momentum in considering the problem. However, before proceeding further, let us see what type of condition will lead to the system to conserve angular momentum. To do this, we ask the following question.

What happens when the net torque acting on the system is zero?

In our discussion on linear momentum, we found out that when no net force acts on an object or a closed system of objects, the linear momentum remains constant or conserved. Similarly, the **law of conservation of angular momentum** states that when no external torque is exerted on an object or a closed system of objects, no change of angular momentum can occur. Let us discuss how this angular momentum conservation principle arises by first considering Newton's second law

$$\vec{F} = \sum_i \vec{F}_i = m\vec{a} = \frac{\Delta \vec{p}}{\Delta t} \Rightarrow F = \frac{\Delta p}{\Delta t}. \tag{8.18}$$

We now use Eq. (8.18) to compute the net torque exerted on the object

$$\tau = \sum_i \tau_i = \sum_i r_i F_i = \sum_i r_i \frac{\Delta p_i}{\Delta t} = \frac{\Delta\left[\sum_i r_i m_i (r_i \omega)\right]}{\Delta t} = I\alpha. \tag{8.19}$$

EQUILIBRIUM AND ROTATIONAL DYNAMICS

Similarly, we may define the torque τ as the rate of change in the angular momentum $L = I\omega$ and write

$$\tau = \frac{\Delta L}{\Delta t} = \frac{\Delta(I\omega)}{\Delta t} = I\frac{\Delta \omega}{\Delta t} = I\alpha. \quad (8.20)$$

Equation (8.20) indicates that if the net torque exerted on the object is zero (i.e., $\tau = 0$), then we may write

$$\tau = I\alpha = 0 \Rightarrow \frac{I\omega_f - I\omega_i}{\Delta t} = 0 \Rightarrow I\omega_f = I\omega_i. \quad (8.21)$$

Hence, the angular momentum before an event involving only internal torques or no torques is equal to the angular momentum after the event.

8.7 CONSERVATION OF ANGULAR MOMENTUM

Angular momentum is conserved in a system where there is no net external torque. This conservation principle helps to explain many diverse phenomena. For example, the increase in rotational speed of a spinning figure skater as the skater's arms are contracted as shown in Fig. 8.11 is a consequence of conservation of angular momentum. The very high rotational rates of neutron stars can also be explained in terms of angular momentum conservation. Moreover, angular momentum conservation has numerous applications in physics and engineering (e.g., gyrocompass). According to the principle, the total angular momentum of rotating body remains constant if the net force (i.e., net torque) acting on it is zero, suggesting that L is conserved in any collisions

$$L_i = L_f \quad (8.21)$$

We may use the definition of angular momentum and write this conservation principle as

$$I_i\omega_i = I_f\omega_f \quad (8.22)$$

This conservation principle is one of the fundamental principles in physics. Also, it is a very useful tool for solving problems involving rotational motion of objects in a system. Similar to the collision problems involving translational motion of objects, angular momentum may be transferred from one object to another within the system.

The conservation of angular momentum explains the angular acceleration of an ice skater as she brings her arms and legs close to the vertical axis of rotation. By bringing part of mass of her body closer to the axis she decreases her body's moment of inertia. Because angular momentum is constant in the absence of external torques, the angular

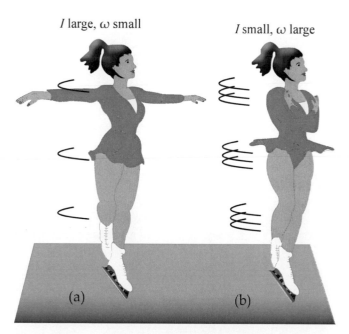

Figure 8.11: According to conservation of angular momentum, (a) when the moment of inertia is large, the angular velocity is small. However, (b) when the moment of inertia is small, the angular velocity is large to keep the angular momentum L the same.

velocity (rotational speed) of the skater has to increase. Also, the conservation of angular momentum is used extensively in analyzing what is called central force motion. If the net force on some body is directed always toward some fixed point, the center, then there is no torque on the body with respect to the center, and so the angular momentum of the body about the center is constant. Conservation of angular momentum is extremely useful when dealing with the orbits of planets and satellites. Also, this principle is useful when analyzing the Bohr model (see Chapter 27) of the atom.

Understanding-the-Concept Question 8.9: Spinning bicycle wheel

A person is holding a spinning bicycle wheel standing on a stationary frictionless turntable. What will happen if the teacher suddenly flips the bicycle wheel over so that it is spinning in the opposite direction?

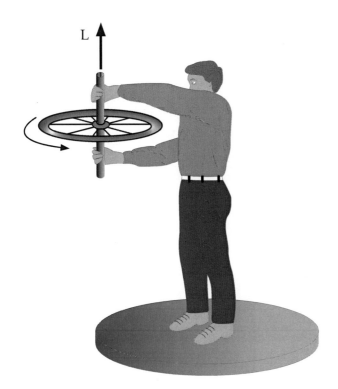

Problem Solving Strategy

What do we know?

A mass m attached to the end of a string is revolving in a circle without any dissipation.

What concepts are needed?

We need to apply the law of angular momentum conservation.

Why?

The torque due to string tension which keeps the mass in a circular motion is zero since the string passes through the hole. When the net torque is zero, the angular momentum of the system is conserved.

Explanation:

Since there are no external forces acting on the system, the total angular momentum of the system is conserved (i.e., $L_i = L_f$). The initial angular momentum of the system is equal to the angular momentum of the bicycle wheel rotating in counterclockwise direction (i.e., $L_i = L$) since the turntable is not rotating initially. When the rotating bicycle wheel is flipped, its angular momentum is now $-L$. However, the total angular momentum of the system which is the sum of angular momentum of the bicycle wheel and turntable is equal to L (i.e., $L_i = L_f = L$). This means that the angular momentum of the turntable L_{turn}, which is given by the relation $L_f = L = L_{turn} - L$, is $L_{turn} = 2L$. Both the turntable and person are now rotating in the counterclockwise direction.

EQUILIBRIUM AND ROTATIONAL DYNAMICS

Exercise Problem 8.8:

A mass m attached to the end of a string revolves in a circle on a frictionless table top. The other end of the string passes through a hole in the table. Initially, the mass revolves with a speed $v_1 = 2.4$ m/s in a circle of radius $r_1 = 0.80$ m. The string is then pulled slowly through the hole so that the radius is reduced to $r_2 = 0.48$ m. What is speed v_2 of the mass now?

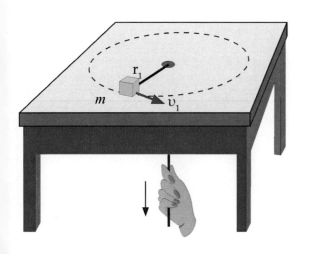

Problem Solving Strategy

What do we know?

A mass is flipping a spinning bicycle wheel while standing on a stationary frictionless turntable.

What concepts are needed?

We need to apply the principle of angular momentum conservation.

Why?

There is no external torque acting on the system. Also, we may treat flipping the bicycle wheel as a collision which transfers momentum.

Solution:

The force is exerted toward the axis so the lever arm is zero. This mean that the net torque acting on the mass is

$$\tau = 0.$$

Using conservation of angular momentum, we write the angular momentum of the mass before (1) and after (2) the string is pulled as

$$I_1 \omega_1 = I_2 \omega_2, \tag{8a}$$

where the moment of inertia I is given by $I = mr^2$ since we may consider the mass as a point-like object located at a distance r from the axis of rotation. Now, we rewrite Eq. (8a) as

$$mr_1^2 \omega_1 = mr_2^2 \omega_2.$$

We solve the final angular speed ω_2 and obtain

$$\omega_2 = \omega_1 \frac{r_1^2}{r_2^2}.$$

Noting that $\omega = v/r$, we compute the final speed of the mass as

$$\therefore \omega_2 = \frac{v_2}{r_2} = \frac{v_1}{r_1} \cdot \frac{r_1^2}{r_2^2} \Rightarrow v_2 = v_1 \frac{r_1}{r_2} = 2.4 \frac{m}{s} \times \frac{0.8\,m}{0.48\,m} = 4.0 \frac{m}{s}.$$

Exercise Problem 8.9:

An asteroid of mass 1.0×10^5 kg, traveling at a speed of 30 km/s relative to Earth, hits the Earth at the equator. It hits the Earth tangentially and in the direction of the Earth's rotation. Use angular momentum to estimate the fractional change in the angular speed of the Earth as a result of the collision.

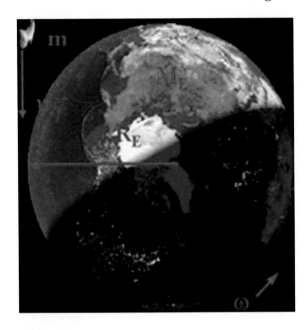

PROBLEM SOLVING STRATEGY

What do we know?

An asteroid hits the Earth tangentially in the direction of the Earth's rotation.

What concepts are needed?

We need to apply the law of angular momentum conservation. Also, we need to know the angular momentum of the asteroid before it hits the Earth.

Why?

We may consider the asteroid-Earth system as an isolated system. The net torque acting on this isolated system is zero (i.e., no external torque).

Solution:

Using conservation of angular momentum, we write

$$I_E\omega + R_E m v = I_E \omega'$$

where the moment of inertia of the earth is $I_E = (2/5)M_E R_E^2$, assuming that a spherical shape for the earth, and the angular moment of the asteroid before the impact is $L = |\vec{r}\times\vec{p}| = R_E m v$. We may solve for the final angular speed ω' and obtain

$$\omega' = \omega + \frac{mv}{\tfrac{2}{5}M_E R_E}$$

The fractional change, according to its definition, is

$$\frac{\omega'-\omega}{\omega} = \frac{5mv}{2M_E R_E \omega} = \frac{5\times 1.0\times 10^5 \text{kg}\times 30\,\text{kg/s}}{2\times 5.97\times 10^{24}\text{kg}\times 6.38\times 10^3\text{km}\times \tfrac{2\pi}{24\text{h}}\times \tfrac{1\text{h}}{3600\text{s}}} = 3\times 10^{-18}.$$

This exercise problem indicates that the change in the angular speed of Earth due to the impact of a small asteroid is negligible. However, we cannot say the same for its effect on the surface of Earth where it collides. The asteroid has the kinetic energy of $E = mv^2/2 = 4.5\times 10^{13}$J. This energy is equal to the energy released from a 10.75 kiloton nuclear device, which is smaller than the energy yield of the first atomic bomb used in World War II which is known as "Little Boy" and had a yield of 13 to 18 kiloton of TNT.

EQUILIBRIUM AND ROTATIONAL DYNAMICS

The rotation of Earth about its axis means that it has angular momentum. The answer to "why does the Earth rotate" leads back to the violent history of Earth. The answer lies within the law of conservation of angular momentum. Some scientists believe that large impacts in the early history of the Solar System may have affected the rotation of some planets. One may have caused Venus to rotate in retrograde and another could account for the 23 degree axial tilt of the Earth. The main reason that the Earth has never stopped rotating since the early periods of the solar nebula is that there is nothing in the vacuum of space to stop it. The amount of angular momentum of an object with respect to a fixed point is proportional to the object's moment of inertia and to the angular velocity relative to the fixed point. Multiplying those three factors gives the angular momentum. The direction of angular momentum is the same as that of the angular velocity. The direction of momentum is the direction of the velocity. In the absence of torque both the amount and the direction of the angular momentum will not change. The only thing that may stop the Earth's rotation is the gravitational forces of the Sun after it begins to expand into its red giant phase. Unfortunately, when this happens, there will be no life on Earth to find out.

In understanding the planetary motion, it is useful to know that, when two objects with a particular angular moment relative to a common fixed point interact, they affect each other's angular momentum and the total angular momentum is conserved. When we consider an orbit, the angular momentum is distributed between the spin of the planet itself and the angular momentum of its orbit, but, if a planet is found to spin more slowly than expected, astronomers suspect that the planet has a satellite, since the total angular momentum is shared between the planet and its satellite for the sake of conservation.

SUMMARY

i. Torque is a measure of how much force is acting on an object causes that object to rotate. Mathematically, torque is defined as the cross product of the lever-arm distance and force, which tends to produce rotation. Torque τ is a vector and is defined as a

$$\vec{\tau} = \vec{r} \times \vec{F} = rF\sin\theta \, (\hat{r} \times \hat{F})$$

where r is the separation distance between the axis of rotation and a point of application of force F, and θ is the angle between the two vectors F and r. The direction of the torque (vector) is determined by using the right-hand rule for a cross product between two vectors.

ii. The **two conditions for mechanical equilibrium** state that, for a rigid body to be at rest, the sum of the forces and torques acting on it must add up to zero:

1. First condition for equilibrium is

$$\sum_i F_{x,i} = 0, \quad \sum_i F_{y,i} = 0, \quad \sum_i F_{z,i} = 0.$$

2. Second condition for equilibrium is

$$\sum_i \tau_i = 0$$

iii. The **center of gravity** is the average location of the weight of an object. For a group of N objects, we may compute the center of gravity as

$$x_{cg} = \frac{W_1 x_1 + W_2 x_2 + \cdots + W_N x_N}{W_1 + W_2 \cdots + W_N} = \frac{\sum_i W_i x_i}{\sum_i W_i}.$$

We note that center of mass for a rigid body may also be called as its center of gravity.

iv For a rotating body with a large number of particles, we may write the **net torque** as

$$\sum_i \tau_i = I\alpha$$

where

$$I = \sum_i m_i r_i^2 = m_1 r_1^2 + m_2 r_2^2 + \cdots$$

is called "**moment of inertia.**" This is the sum of the moment of inertia contribution from each particle in the rigid body. The moment of inertia force on a single particle around an axis is the product of the mass of the particle and the square of its distance to the axis.

v. Any rolling rigid body has two types of kinetic energy: i) translational kinetic energy, and ii) rotational kinetic energy. The **translational kinetic energy** is the energy possessed by an object due to its linear motion and is written as

$$K.E._{tran} = \frac{1}{2} m v^2.$$

The **rotational energy** or **angular kinetic energy** is the kinetic energy due to the rotation of an object and written as

$$K.E._{rot} = \frac{1}{2} I \omega^2.$$

vi. **Angular momentum** or **rotational momentum** is a vector which represents the product of a body's rotational inertia I and rotational velocity ω about a particular axis

$$\vec{L} = I\vec{\omega}.$$

The angular momentum L of a particle about a given origin is defined as

$$\vec{L} = \vec{r} \times \vec{p}.$$

where r is the position vector of the particle relative to the origin, p is the linear momentum of the particle, and × denotes the cross product (or vector product).

vii. The **law of conservation of angular momentum** can arise when no external torque acts on an object or a closed system of objects, indicating no change of angular momentum can occur. This means that the angular momentum is conserved

$$L_i = L_f$$

We may use the definition of angular momentum to rewrite this conservation principle as

$$I_i \omega_i = I_f \omega_f.$$

This conservation principle is one of the fundamental principles in physics.

EQUILIBRIUM AND ROTATIONAL DYNAMICS

MORE WORKED PROBLEMS

Problem 8.1

A platform is rotating at an angular speed of 2.2 rad/s. A block is resting on this platform at a distance of 0.3 m from the axis. The coefficient of static friction between the block and the platform is 0.75. Without any external torque acting on the system, the block is moving away from the axis. Ignore the moment of inertia of the platform and determine the smallest distance from the axis at which the block can be relocated and still remain in place as the platform rotates.

Solution:

The block will just start to move when the centripetal force on the block just exceeds $F_{s,max}$. Thus if r_f is the smallest distance from the axis at which the block stays at rest when the angular speed of the block is ω_f then

$$F_s = \mu_s F_N = \mu_s mg \quad \text{and} \quad F_c = ma_c \Rightarrow F_s = m\frac{(r_f \omega_f)^2}{r_f} = mr_f \omega_f^2.$$

This indicates that

$$\mu_s mg = mr_f \omega_f^2 \Rightarrow \mu_s g = r_f \omega_f^2.$$

Since there are no external torques acting on the system, angular momentum will be conserved

$$I_i \omega_i = I_f \omega_f,$$

where $I_i = mr_i^2$ and $I_f = mr_f^2$. We may substitute the expression for the moment of inertias and obtain

$$\left(mr_i^2\right)\omega_i = \left(mr_f^2\right)\omega_f \Rightarrow r_i^2 \omega_i = r_f^2 \omega_f.$$

We now use this result to obtain

$$\mu_s g = r_f \omega_i^2 \frac{r_i^4}{r_f^4} = \frac{\omega_i^2 r_i^4}{r_f^3}.$$

We solve for r_f and obtain

$$r_f = \left(\frac{\omega_i^2 r_i^4}{\mu_s g}\right)^{1/3} = \left[\frac{(2.2\,\text{rad}/\text{s})^2 \times (0.30\,\text{m})^4}{0.75 \times 9.80\,\text{m}/\text{s}^2}\right]^{1/3} = 0.17.$$

Problem 8.2

An industrial flywheel with a moment of inertia of 4.25×10² kg·m² rotates with a speed of 7500 rpm. (a) How much work is required to bring the flywheel to rest? (b) If this work is done uniformly in 1.5 min, how much power is expanded?

Solution:

First, we convert 7500 rpm into

$$7500 \frac{\text{rev}}{\text{min}} \times \frac{2\pi \text{ rad}}{1 \text{ rev}} \times \frac{1 \text{ min}}{60 \text{ s}} = 785.4 \frac{\text{rad}}{\text{s}}.$$

From the work-energy principle, we write

$$W = \Delta K.E. = \frac{1}{2} I \omega_f^2 - \frac{1}{2} I \omega_i^2.$$

Noting that $\omega_f = 0$, we may simplify

$$W = 0 - \frac{1}{2} I \omega_i^2 = -\frac{1}{2} \times 4.25 \times 10^2 \text{ kg} \cdot \text{m}^2 \times \left(785.4 \frac{\text{rad}}{\text{s}}\right)^2 = -1.31 \times 10^8 \text{ J}.$$

Now, we use the definition of power ($P = W/t$) to compute the power expanded

$$P = \frac{1.31 \times 10^8 \text{ J}}{1.5 \text{ min} \times \frac{60 \text{ s}}{\text{min}}} = 1.46 \times 10^6 \text{ Watt}.$$

Problem 8.3

A uniform 2.0-kg cylinder of radius 0.15 m is suspended by two strings wrapped around it as shown in the figure below with two 1.0-kg hanging masses. As the cylinder descends, the strings unwind from it. What is the acceleration of the center of mass of the cylinder? (Neglect the mass of the string.)

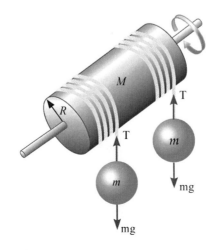

EQUILIBRIUM AND ROTATIONAL DYNAMICS

Solution:

We apply Newton's second law. For the center of mass, we write

$$\sum_i \vec{F}_i = M\vec{a} \Rightarrow \sum_i F_{y,i} = -2T + Mg = Ma.$$

(1)

For rotation about center of mass, we write

$$\sum_i \tau_i = 2TR = I\alpha = \frac{1}{2}MR^2\alpha = \frac{1}{2}MRa.$$

We note that $I = (1/2)MR^2$ for a cylinder and $\alpha = a/R$. The tension T is given by

$$2T = \frac{1}{2}Ma.$$

(2)

By adding equations (1) and (2), we obtain

$$Mg = Ma + \frac{1}{2}Ma \Rightarrow g = \frac{3}{2}a.$$

The acceleration a is given by

$$a = \frac{2}{3}g = \frac{2}{3} \times 9.80 \frac{m}{s^2} = 6.5 \frac{m}{s^2}.$$

Problem 8.4

A man is attempting to raise a 7.5-m-long, 28 kg flagpole that has a hinge at the base by pulling on a rope attached to the top of the pole as shown in the figure below. With what force does the man have to pull on the rope to hold the pole motionless in this position?

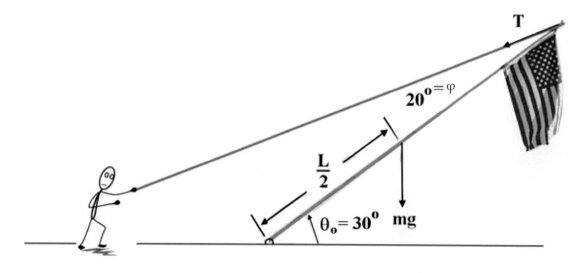

Solution:

The clockwise torque is due to the weight of the flagpole

$$\tau_{cw} = rF\sin\theta = \frac{L}{2}mg\sin(90° - \theta) = \frac{7.5\,m}{2} \times 28\,kg \times 9.80\frac{m}{s^2} \times \sin 60° = 891\,N\cdot m.$$

In the counterclockwise direction, we want to know the force (tension in the rope)

$$\tau_{ccw} = rF\sin\theta = LT\sin\varphi.$$

Since the net torque exerted about the hinge point is zero, we write

$$\tau_{net} = 0 = \sum_i \tau_i = \tau_{ccw} - \tau_{cw} \Rightarrow \tau_{ccw} = \tau_{cw}.$$

We solve for T and obtain

$$T = \frac{\tau_{ccw}}{L\sin\varphi} = \frac{\tau_{cw}}{L\sin\varphi} = \frac{891\,N\cdot m}{7.5\,m \times \sin 20°} = 350\,N.$$

PROBLEMS

1. The steering wheel of a car has a radius of 0.19 m, while the steering wheel of a truck has a radius of 0.25 m. The same force is applied in the same direction to each. What is the ratio of the torque produced by this force in the truck to the torque produced in the car?

2. A rotational axis is directed perpendicular to the plane of square and is located as shown in the drawing. Two forces F_1 and F_2, are applied to diagonally opposite corners, and act along the sides of the square, first as shown in part a and then as shown in part b of the drawing. In each case the net torque produced by the forces is zero. The square is one meter on a side, and the magnitude of F_2 is three times that of F_1. Find the distances a and b that locate the axis.

3. A 10-kg solid disk of radius 0.50 m is rotated about an axis through its center. If the disk accelerates from rest to an angular speed of 3.0 rad/s while rotating 2.0 revolutions, what net torque is required?

4. One end of a meter stick is pinned to a table, so the stick can rotate freely in a plane parallel to the tabletop. Two forces, both parallel to the tabletop, are applied to the stick in such a way that the net torque is zero. One force has a magnitude of 2.00 N and is applied perpendicular to the length of the stick at the free end. The other force has magnitude of 6.00 N and acts at a 30.0° angle with respect to the length of the stick. Where along the stick is the 6.00-N force applied? Express this distance with respect to the end that is pinned.

5. Children with masses m_1 and m_2 sit at opposite ends of seesaw with length L and mass M. What distance D should the pivot point be from the first child if the seesaw is to balance? Your answer will be an expression written in terms of three masses and the length L.

6. A uniform board is leaning against a smooth vertical wall. The board is at an angle θ above the horizontal ground. The coefficient of static friction between the ground and the lower end of the board is 0.650. Find the smallest value for the angle θ, such that the lower end of the board does not slide along the ground.

EQUILIBRIUM AND ROTATIONAL DYNAMICS

7. An inverted "V" is made of uniform boards and weighs 356 N. Each side has the same length and makes a 30.0° angle with the vertical, as the drawing shows. Find the magnitude of the static frictional force that acts on the lower end of each leg of the "V."

8. A 5.0-N uniform meter-stick is pivoted so that it can rotate about a horizontal axis through one end. If a 0.15-kg mass is suspended 75 cm from the pivoted end, what is the tension in the string?

9. Two identical, side-by-side springs with spring constant 240 N/m support a 2.00 kg hanging box. By how much is each spring stretched?

10. A uniform beam of mass m pivoted at its lower end, with a horizontal spring attached between its top end and a vertical wall. The beam makes an angle θ with the horizontal. Find expressions for (a) the distance d the spring is stretched from equilibrium and (b) the components of the force exerted by the pivot on the beam.

11. One end of a uniform 4.0-m-long rod of weight W is supported by a cable at an angle of $\theta = 37°$ with the rod. The other end rests against a wall, where it is held by friction. The coefficient of static friction between the wall and the rod is $m_s = 0.50$. Determine the minimum distance x from point A at which an additional weight w (the same as the weight of the rod) can be hung without causing the rod to slip at point A.

12. For the system shown in Fig. 8.41, $m_1=9.0$ kg, $m_2=3.0$ kg, $\theta=30°$, and the radius and mass of the pulley are 0.10 m and 0.10 kg, respectively. (a) What is the acceleration of the masses? (Neglect friction and the string's mass.) (b) If the pulley has a constant frictional torque of 0.050 m N when the system is in motion, what is the acceleration of the masses? [Hint: Isolate the forces. The tensions in the strings are different. Why?]

13. Two springs attached to a box that can slide on a frictionless surface. In the block's equilibrium position, neither spring is attached. What is the net force on the block if it is moved 15 cm to the right of its equilibrium position? Hint: There is zero net force on the point where the two springs meet. This implies a relationship between the amounts the two springs stretch.

14. A 1220-N uniform beam is attached to a vertical wall at one end and is supported by a cable at the other end. A 1960-N crate hangs from the far end of the beam. Using the data shown in the drawing, find (a) the magnitude of the tension in the wire and (b) the magnitude of the horizontal and vertical components of the force that the wall exerts on the left end of the beam.

15. The model airplane is flying at a speed of 22 m/s on a horizontal circle of radius 16 m. The mass of the plane is 0.90 kg. The person holding the guideline pulls it in until the radius becomes 14 m. The plane speeds up and the tension in the guideline becomes four times greater. What is the net work done on the plane?

16. A horizontal 800-N merry-go-round of radius 1.50 m is started from rest by a constant horizontal force of 50.0 N applied tangentially to the merry-go-round. Find the kinetic energy of the merry-go-round after 3.00 s. (Assume it is s solid cylinder.)

17. The drawing shows an A-shaped stepladder. Both sides of the ladder are equal in length. This ladder is standing on a frictionless horizontal surface, and only the crossbar (which has a negligible mass) of the "A" keeps the ladder from collapsing. The ladder is uniform and has a mass of 20.0 kg. Determine the tension in the crossbar of the ladder.

18. A very light tripod has 70-cm-long legs that can be spread out so that each makes a 20° angle with the vertical. The points where the three legs rest on the ground form an equilateral triangle. If you attach an object to the tripod, such as a camera with a long lens, what is the maximum possible horizontal distance between the center of gravity of the object and the center of the tripod, if the tripod is to remain stable?

19. A thin, rigid, uniform rod has a mass of 2.00 kg and a length of 2.00 m. (a) Find the moment of inertia of the rod relative to an axis that is perpendicular to the rod at one end. (b) Suppose all the mass of the rod were located at a single point. Determine the perpendicular distance of this point from the axis in part (a), such that this point particle has the same moment of inertia as the rod. This distance is called the **radius of gyration** of the rod.

20. Two thin rectangular sheets (0.20 m × 0.40 m) are identical. In the first sheet the axis of rotation lies along the 0.20-m side, and in the second it lies along the 0.40-m side. The same torque is applied to each sheet. The first sheet, starting from rest, reaches its final angular velocity in 8.0 s. How long does it take for the second sheet, the starting from rest, to reach the same angular velocity? (F09, 9.41)

21. A stationary bicycle is raised off the ground, and its front wheel (m = 1.3 kg) is rotating at an angular velocity of 13.1 rad/s (see the drawing). The front break is then applied for 3.0 s, and the wheel slows down to 3.7 rad/s. Assume that all the mass of the wheel is concentrated in the rim, the radius of which is concentrated in the rim, the radius of which is 0.33 m. The coefficient of kinetic friction between each brake pad and the rim is m_k = 0.85. What is the magnitude of the normal force that *each* brake pad applies to the rim?

22. A uniform solid cylinder of mass M and radius R rotates on frictionless horizontal axle. Two objects with equal masses m hang from light cords wrapped around the cylinder. If the system is released from rest, find (a) the tension in each cord and (b) the acceleration of each objects have descended a distance h.

23. A tennis ball, starting from rest, rolls down the hill in the drawing. At the end of the hill the ball becomes airborne, leaving at an angle of 35° with respect to the ground. Treat the ball as a thin-walled spherical shell, and determine the range x.

24. A solid, horizontal cylinder of mass 10.0 kg and radius 1.00 m rotates with an angular speed of 7.00 rad/s about a fixed vertical axis through its center. A 0.250-kg piece of putty is dropped vertically on to the cylinder at a point 0.900 m from the center of rotation and sticks to the cylinder. Determine the final angular speed of the system.

25. A uniform 2.0-kg cylinder of radius 0.15 m is suspended by two strings wrapped around it. As the cylinder descends, the strings unwind from it. What is the acceleration of the center of mass of the cylinder? (Neglect the mass of the string.)

26. A 150-kg merry-go-round in the shape of a uniform, solid, horizontal disk of radius 1.50 m is set in motion by wrapping a rope about the rim of the disk and pulling on the rope. What constant force must be exerted on the rope to bring the merry-go-round from rest to an angular speed of 3.100 rad/s in 2.00 s?

27. A 240-N sphere 0.20m in radius rolls without slipping 6.0 m down a ramp that is inclined at 37° with the horizontal. What is the angular speed of the sphere at the bottom of the slope if it starts from rest?

28. An airliner lands with a speed of 50.0 m/s. Each wheel of the plane has a radius of 1.25 m and moment of inertia of 110 kg·m². At touchdown, the wheels begin to spin under the action of friction. Each wheel supports a weight of 1.40×10⁴ N, and the wheels attain their angular speed in 0.480 s while rolling without slipping. What is the coefficient of kinetic friction between the wheels and the runway? Assume that the speed of the plane is constant.

29. What is the angular momentum of the moon around the earth? The moon's mass is 7.4×10²² kg and it orbits 3.8×10⁸ m from the Earth.

EQUILIBRIUM AND ROTATIONAL DYNAMICS

30. The period of the Moon's rotation is the same as the period of its revolution: 27.3 days (sidereal). What is the angular momentum for each rotation and revolution? (Because the periods are equal, we see only one side of the Moon from Earth.)

31. An ice skater spinning with outstretched arms has an angular speed of 4.0 rad/s. She tucks in her arms, decreasing her moment of inertia by 7.5%. (a) What is the resulting angular speed? (b) By what factor does the skater's kinetic energy change? (Neglect any frictional effects.) (c) Where does the extra kinetic energy come from?

32. A steel ball rolls down an incline into a loop-the-loop of radius R. (a) What minimum speed must the ball have at the top of the loop in order to stay on the track? (b) At what vertical height (h) on the incline, in terms of the radius of the loop, must the ball be released in order for it to have the required minimum speed at the top of the loop? (Neglect frictional losses.) (c) Now, consider the loop-the-loop of a roller coaster. What are the sensations of the riders if the roller coaster has the minimum or a greater speed at the top of the loop? [Hint: In case the speed is below the minimum, seat and shoulder straps hold the riders in.]

33. A student sits on a rotating stool holding two 3.0-kg objects. When his arms are extended horizontally, the objects are 1.0 m from the axis of rotation and the he rotates with an angular speed of 0.75 rad/s. The moment of inertia of the student plus stool is 3.0 kg·m² and is assumed to be constant. The student then pulls in the objects horizontally to 0.30 m from the rotation axis. (a) Find the new angular speed of the constant. (b) Find the kinetic energy of the student before and after the objects are pulled in.

34. A thin, uniform rod is rotating at an angular velocity of 7.0 rad/s about an axis that is perpendicular to the rod at its center. As the drawing indicates, the rod is hinged at two places, one-quarter of the length from each end. Without the aid of external torques, the rod suddenly assumes a "u" shape, with the arms of the "u" parallel to the rotation axis. What is the angular velocity of the rotating "u"?

35. A small 0.500-kg object moves on a frictionless horizontal table in a circular path of radius 1.00 m. The angular speed is 6.28 rad/s. The object is attached to a string of negligible mass that passes through a small hole in the table at the center of the circle. Someone under the table begins to pull the string downward to make the circle smaller. If the string will tolerate a tension of no more than 105 N, what is the radius of the smallest possible circle on which the object can move?

36. A comet approaches the Sun and is deflected by the Sun's gravitational attraction. This event is considered a collision, and b is called the impact parameter. Find the distance of closest approach (d) in terms of the impact parameter and the velocities (u_o at large distances and v at closest approach). Assume that the radius of the Sun is negligible compared to b. (The tail of a comet always "points" away from the Sun.)

SOLIDS AND FLUIDS

CHAPTER 9

As the pinnacle of bridge technology, a suspension bridge is capable of spanning up to 7,000 feet. It manages this feat by successfully dealing with two important forces called compression and tension. Compression is a force that acts to compress or shorten the object when it is acting on. On the other hand, tension is a force that acts to expand or lengthen the object it is acting on. Suspension bridges are made to sway with the wind and to be fairly flexible, but lots of compression and tension are present. Suspension bridges are also better at withstanding earthquakes than bridges with a more rigid construction. However, because of their lightweight nature, they may sway too dangerously in extreme conditions and may become unsafe to use. Due to lots of compressive and tensile stress, suspension bridges that had flaws in the support system, even small flaws, have been known to collapse. The original Tacoma Narrows Bridge, which was built in 1940, collapsed under strong winds only four months after it opened. As the bridge sways in extreme weather, a lot of stress is also put on the supports. This means that any support has to be very heavy and durable. This is a major concern when the supports are installed into softer earth at the bottom of a waterway.

Compression and tension are present in all bridges, and it is the job of the bridge design to handle these forces without buckling or snapping. These mechanical stresses may be understood in terms of a simple, everyday example of compression and tension on a spring. When we press down, or push the two ends of the spring together, we compress it. The force of compression shortens the spring. When we pull up, or pull apart the two ends, we create tension in the spring. The force of tension lengthens the spring. In a bridge, buckling is what happens when the force of compression overcomes an object's ability to handle compression, and snapping is what happens when the force of tension overcomes an object's ability to handle tension. The best way to deal with these forces is to either dissipate them or transfer them. To dissipate force is to spread it out over a greater area, so that no single spot has to bear the brunt of the concentrated force. To transfer force is to move it from an area of weakness to an area of strength, an area designed to handle the force. In a modern suspension bridge with two tall towers through which the cables are strung, the towers are supporting the majority of the roadway's weight. The force of tension on the cables pulls up the suspension bridge's deck, but because it is a suspended roadway, the cables transfer the tension to the towers, which dissipate this force directly into the earth where they are firmly entrenched. The supporting cables which run between the two anchorages are the recipients of the tension forces. The cables are literally stretched from the weight of the bridge and its traffic as they run from anchorage to anchorage. The anchorages are also under tension, but since they are held firmly to the earth, the tension they experience is dissipated, similar to the towers. Almost all suspension bridges have, in addition to the cables, a supporting truss system beneath the bridge deck (a deck truss). This helps to stiffen the deck and reduce the tendency of the roadway to sway and ripple.

Another major topic of discussion for this chapter is fluids. More specifically, we will discuss the forces acting on an object in fluids. One aspect of this discussion is to understanding why a steel plate on an aircraft carrier floats but the same steel plate by itself sinks. The standard definition of floating was first recorded by Archimedes. What he found out was that an object in fluids experiences an upward force equal to the weight of the fluid displaced by the object. So if a boat weighs 1,000 pounds (or kg), it will sink into the water until it has displaced 1,000 pounds (or kg) of water. Provided that the boat displaces 1,000 pounds of water before the whole thing is submerged, the boat floats. It is not very hard to shape a boat in such a way that the weight of the boat has been displaced before the boat is completely underwater. The reason is that a good portion of the interior of any boat is air (unlike a cube of steel, which is solid steel throughout). The average density of a boat which is the combination of the steel and air is very light compared to the average density of water. So very little of the boat actually has to submerge into the water before it has displaced the weight of the boat. Then, the next question is "How does the water molecules know when 1,000 pounds of them have gotten out of the way?" As we shall see below, it turns out that the actual act of floating has to do with **pressure** rather than weight. If we take a column of water 1 inch square and 1 foot tall, it weighs about 0.44 pounds depending on the temperature of the water (if we take a column of water 1 cm square by 1 meter tall, its weight is about 100 grams). That means that a 1-foot-high column of water exerts 0.44 pounds per square inch (PSI). Similarly, a 1-meter-high column of water exerts 9,800 pascals (Pa). If we were to submerge a box with a pressure gauge attached into water, then the pressure gauge would measure the pressure of the water at the submerged depth. In other words, if we were to submerge the box 1 foot into the water, the gauge would read 0.44 PSI (if we submerged it 1 meter, it would read 9,800 Pa). This means that the bottom of the box has an

SOLIDS AND FLUIDS

upward force being applied to it by that pressure. So if the box is 1 foot square and it is submerged 1 foot, then the bottom of the box is being pushed up by a water pressure of (12 inches ×12 inches × 0.44 PSI) 62 pounds (if the box is 1 meter square and submerged 1 meter deep, the upward force is 9,800 Newton). This just happens to exactly equal the weight of the cubic foot or cubic meter of water that is displaced! This upward water pressure pushing on the bottom of the boat is causing the boat to float. Each square inch (or square centimeter) of the boat that is underwater has water pressure pushing it upward, and this combined pressure floats the boat. This example indicates the property of object in fluids may also be understood by employing the concept of force and energy which we studied in previous chapters.

9.1 EFFECTS OF AN APPLIED FORCE

Let us first examine the effect of applied forces on solid objects. Applied forces can change the shape of objects. If a small force is applied, like in Fig 9.1, then the object stretches. The length ΔL of stretch increases with increasing the strength of the force. However, if the forces are great enough, then the object will either break or fracture. The fracture of a solid almost always occurs due to the development of certain displacement discontinuity surfaces within the material. If a displacement developed in this case is perpendicular to the surface of displacement, then it is called a normal tensile crack or simply a crack. However, if a displacement developed is tangent to the surface of displacement, then it is called a shear crack, slip band, or dislocation. The word "fracture" is often applied to bones of living creatures (i.e., a bone fracture) and applied to crystals or crystalline materials such as gemstones or metal. Sometimes, in crystalline materials, individual crystals fracture without the body actually separating into two or more pieces. However, depending on the substance which is fractured, a fracture reduces strength in most substances or inhibits transmission of light in optical crystals. On the other hand, if the forces are not great enough, then the object will change its length.

Let us consider the effects of a force exerted on an object such as vertically suspended metal rod, as shown in Fig. 9.1. We may describe the effects of an applied force by treating the rod as a spring and using Hooke's law

$$F = k(\Delta L), \tag{9.1}$$

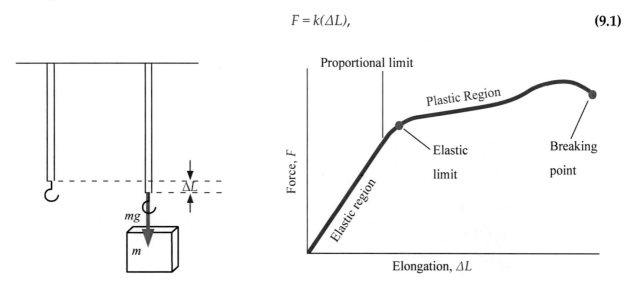

Figure 9.1: A schematic illustration of the effects of a load on a long vertical rod. The load leads to the elongation of the rod (left panel). The elastic property of a material on the right panel shows that the amount of applied force is directly proportional to elongation in the elastic region.

which describes a force exerted on an object attached to a spring with the spring constant k. The Hooke's law states that the extension produced in the spring ΔL is directly proportional to the load. This law is valid for only small deformation of the spring within the elastic regime. We may ask "Why are we using Hooke's law to describe the effects of an applied force?" As we discussed in Chapter 4, a toy model of solids consists of an array of balls interconnected by springs. For real materials, the balls represent molecules and the springs represent the bonds between them. In this model, the chemical bonds between the atoms either stretch or compress when a force is applied. The amount that the bonds stretch or compress depends on the magnitude of the applied force, similar to that in a real spring.

The initial straight-line section in the F versus ΔL graph of Fig. 9.1 represents the modulus of elasticity (or Young's modulus or stiffness) of the material. If the stress is removed during this elastic strain period, the material will revert to its original shape. However, if the load or stress goes beyond this linear section, then we enter the area of plastic strain and the material will now never revert to its original shape. It would be reasonable to think that as long as we keep the stress within this linear elastic region, we would be quite safe since the strength of material would not fail. However, this is not always the case. As we will see below, there are three ways in which materials, such as solder joints, can fail under low applied stresses. This indicates that every solid object has a breaking point. For example, a concrete column which can support lots of weight appears quite strong, but it too has a breaking point. This breaking point occurs at a lower force when the force is applied along the sides of the concrete column as in Fig. 9.2.

When the applied force is small enough so that the length change is within the elastic limit, the amount of elongation of an object depends linearly on the force and on its dimension. The change in the length ΔL due to the applied force F is given by

$$F = k\Delta L \Rightarrow \Delta L = \frac{A}{kL_0}\frac{F}{A}L_0 \Rightarrow \Delta L = \frac{1}{E}\frac{F}{A}L_0, \tag{9.2}$$

Figure 9.2: The 1989 San Francisco earthquake illustrates that all materials have breaking points. When large forces were applied, the concrete columns supporting a freeway could not withstand the effects of these forces. When the applied forces are beyond the breaking point, the concrete columns will fracture and break.

SOLIDS AND FLUIDS

where L_o is the original length of the object, A is the cross-sectional area where the force is applied, and E is the elastic modulus (or Young's modulus). Note that the constant kL_o/A in Eq. (9.2) has the dimensions of elastic modulus, indicating we may describe the elastic property of a material using Hooke's law and its elastic modulus.

The elastic properties of a material are characterized by the elastic modulus E. The elastic modulus relates two physical quantities: stress and strain. The definitions of stress and strain are given as follows: **stress** is defined as force per unit area (i.e., F/A with units of N/m^2); and **strain** is defined as the ratio of change in length to original length $\Delta L/L_o$. Using the definition of stress and strain, we may rewrite Eq. (9.2) as

$$\Delta L = \frac{1}{E}\frac{F}{A}L_o \Rightarrow \frac{F}{A} = E\frac{\Delta L}{L_o}. \qquad (9.3)$$

This means the elastic modulus E is the ratio of stress and strain, which may be written as

$$E = \frac{F/A}{\Delta L/L_o} = \frac{\text{stress}}{\text{strain}}. \qquad (9.4)$$

The **elastic modulus** E (or Young's modulus Y) describes a substance's tendency to be deformed elastically (i.e., non-permanently) when a force is applied to it. The elastic modulus is defined as the slope of a stress versus strain curve in the elastic deformation region. This means that a stiffer material will have a higher elastic modulus. It may be easier to think about the difference between stress and strain by using an analogy illustrated in Fig 9.3.

Since stress is the restoring force arising from the deformation divided by the area to which the force is applied, the person who is about to be executed is experiencing lots of (emotional) stress. However, as strain is the ratio of the change in length caused by the stress to the original length of the object, the executioner feels lots of strain on his hands since he may have been on his job for a long time.

Figure 9.3: The difference between stress and strain is schematically illustrated. The person who is about to be executed feels (psychological) stress, while the executioner feels stain on the ligaments of his hands.

Understanding-the-Concept Question 9.1

A mass of 50 kg is suspended from a steel wire of diameter 1.0 mm and length 11.2 m. How much will the wire stretch? The Young's modulus for steel is 20×10^{10} N/m^2.

    a. 1.5 cm   **b.** 2.5 cm   **c.** 3.5 cm   **d.** 4.5 cm

Explanation:

In this question, we are given the elastic modulus E of steel, the cross-sectional area $A = \pi r^2$, the length L_o, and the load $F = mg$ on the wire. Since we are asked to calculate the amount of elongation due to the load, we use Eq. (9.2) and solve for ΔL to obtain

$$\Delta L = \frac{1}{E}\frac{F}{A}L_o = \frac{1}{E}\frac{mg}{pr^2}L_o = \frac{1}{20\times 10^{10} \text{ N}/\text{m}^2} \frac{50\text{ kg} \times 9.80 \text{ m}/\text{s}^2}{p(0.5\times 10^{-3} \text{ m})^2} \times 11.2 \text{ m} = 0.0349 \text{ m}.$$

The calculation indicates that the steel wire will stretch by 3.5 cm.

It is interesting to know that spider's silk is tougher than steel. Spider silk is a polymer material made of thin crystalline sheets of proteins bound together by amorphous layers of amino acids. Recently, physicists in Germany found that it can be made as much as eight times stronger by adding small quantities of metal.

Answer: c

It is noted that normal stress arises from the component of a force vector which is perpendicular (or antiparallel) to the material's cross section on which it acts. Our calculation shows that exerting a force on a wire by pulling on the both ends leads to elongation of the wire, We may achieve a similar effect by pressing on the both ends of a rigid rod. This suggests that there are a number of different ways in which stress (applied force per unit area) may lead to deformation of a rigid object. In this case, the direction of the applied forces is useful for distinguishing the difference.

Types of Stress

Excluding psychological stress, there are three types of physical stress that account for the effects of applied forces on an object. Here, we need to pay attention to the direction of applied forces. The three types of stress are (i) tensile stress, (ii) compressive stress, and (iii) shear stress, as illustrated in Fig. 9.4. **Tensile stress** (or tension) is the stress leading to expansion due to a force pulling down. In this case, the length of an object tends to increase in the tensile direction when two opposite forces with the equal magnitude are applied on the object. The volume of the material stays constant. **Compressive stress** is the stress on materials that leads to a smaller volume due to a force of compression. Compressive stress to rigid objects such as bars and columns leads to shortening since the material is under compression. Compressive stress is the opposite of tensile stress. A **shear stress** is defined as the component of stress coplanar with a material cross-section.

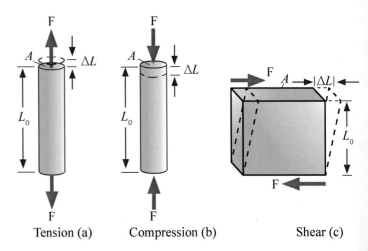

Figure 9.4: A schematic illustration of the direction of applied forces for the three types of stress. a) Tensile stress and b) compressive stress change the length, but c) shear stress deforms the object without changing its volume.

SOLIDS AND FLUIDS

Shear stress arises from equal and opposite forces which are applied across the opposite faces of an object. Here, applied forces are parallel to the cross-section.

In materials science, the strength of a material is its ability to withstand an applied stress without failure. In accounting for the strength of materials in the real world, we need to deal with loads, deformations and the forces acting on them. A load which is applied to a mechanical part of a machine will induce internal forces within the member called stresses. The stresses acting on the material cause deformation of the material. It is noted that when applied stress is greater than the strength the material, it will lead to a fracture as shown in Fig. 9.5. Tensile stress

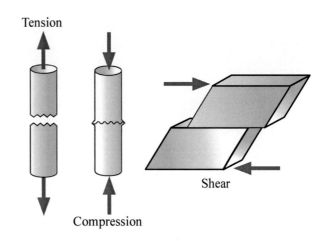

Figure 9.5: An illustration of the effects of too much stress on objects. When applied stress is larger than the strength of a material, the object will either break or fracture.

may be increased until it reaches the tensile strength, which is the *limit state* of stress. One can increase the compressive stress until compressive strength is reached. When this limit state is reached, the materials will react with ductile behavior (as in most metals, some soils and plastics) or with fracture in case of brittle materials (as in geo-materials, cast iron, glass, etc.). In engineering, **shear strength** is the strength of a material or component against shear. This type of structural failure occurs when the material or component fails due to shear stress. A shear load is a force that tends to produce a sliding failure on a material along a plane that is parallel to the direction of the force. For example, when a paper is cut with scissors, the paper fails in shear.

A material's strength depends on its microstructure. There are ways to change the strength of a material because the conditions of the engineering processes can alter this microstructure. The variety of mechanisms that alter the strength of a material includes work hardening, solid solution, precipitation hardening and grain boundary strengthening. The effects of these mechanisms can be explained both quantitatively and qualitatively. Sometimes, it is useful to be aware that, in attempts to make the material stronger, strengthening mechanisms may be accompanied by degeneration of some other mechanical properties of the material. For example, although the yield strength is maximized with decreasing grain size in the grain boundary strengthening process, very small grain sizes ultimately make the material brittle. In general, the yield strength of a material is a useful indicator of the material›s mechanical strength.

Table. 9.1: Ultimate strength of materials (Force/Area)

Material	Tensile strength (N/m^2)	Compressive strength (N/m^2)	Shear strength (N/m^2)
Iron, cast	170×10^6	550×10^6	170×10^6
Steel	500×10^6	500×10^6	250×10^6
Brass	250×10^6	250×10^6	200×10^6
Aluminum	200×10^6	200×10^6	200×10^6
Concrete	2×10^6	20×10^6	2×10^6
Brick		35×10^6	
Marble		80×10^6	

	Tensile strength	Compressive strength	Shear strength
Granite		170×10^6	
Wood (pine)			
(parallel to grain)	40×10^6	35×10^6	5×10^6
(perpendicular to grain)		10×10^6	
Nylon	500×10^6		
Bone (limb)	130×10^6	170×10^6	

The effects of dynamic loading may be the most important practical consideration of the strength of materials, especially the problem of fatigue. Repeated loading often initiates brittle cracks, which grow until failure occurs. The cracks always start at places where stress is concentrated, especially at the locations of changing cross-section area of the material, near holes, and corners.

As indicated on Table 9.1, the shear strength of materials tends to be the weakest. As noted earlier, **shear stress** is the force which tends to cause deformation of a material by slippage along a plane or planes parallel to the imposed stress. When shear stress is applied, dimension of the object does not change significantly, but the shape does. In geology, the resultant shear is related to the downslope movement of earth materials and to earthquakes. This shear stress may occur in solids, but it is related to fluid viscosity in liquids. The **shear strain** is computed from

$$\Delta L = \frac{1}{G} \frac{F}{A} L_0, \qquad (9.5)$$

where G is the shear modulus, and the cross-sectional area A is parallel to a force F as in Fig. 9.4c, yielding the sheared face to move a distance ΔL. The **shear modulus** is also known as the rigidity. It is a numerical constant that describes the elastic properties of a solid under the application of transverse internal forces. These forces lead to torsion and in twisting a metal pipe about its lengthwise axis. When a shear stress is applied on a small cube, the cubic volume is slightly distorted in such a way that two of its faces slide parallel to each other over a small distance and two other faces change from squares to diamond shapes. Hence, the shear modulus is a measure of the ability of a material to resist transverse deformation. It is a valid index of elastic behavior only for small deformations, after which the material is able to return to its original configuration. Large shearing forces lead to flow and permanent deformation or fracture.

There are lots of ways that shear stresses act on our body. Research has shown that pressure and soft tissue shear stress are leading sources of pain and discomfort in the human body. Increased shear force leads to occlusion (blockage) of blood flow, which is one of the most important factors behind pressure sores and discomfort. Luckily, our body has a way to reduce this stress. For example, a bursa is a fluid-filled sac that decreases shear forces between tissues of the body. Excess stress on a bursa causes Trochanteric Bursitis, as shown in Fig. 9.6. Trochanteric bursitis (inflammation of a bursa) which is caused by excessive stress on the bursa between the iliotibial band and the greater trochanter is characterized by painful inflammation of the bursa located just superficial to the greater trochanter of the femur. So, the signs and symptoms include pain over the outer aspect of the hipbone, which often is exacerbated when lying on the affected side, standing on the affected leg, or excessive walking. Treatment which often includes rest, ice, and compression is helpful. Also, physical therapy including stretching and progressive strengthening, and steroid injection may be helpful.

Also, an excess shear stress can cause a bone fracture. Bone fractures are determined by the mode of the applied loads and their orientation. Bones are strongest in compression, less strong in tension and weakest in shear. Under some loading conditions, there are either tensile and shear or compressive

SOLIDS AND FLUIDS

and tensile loads at a given position, depending on how forces are applied. Bones usually break by shear (twisting) stresses or under tension, but not under compression. A shearing force will produce a fracture parallel to the direction of the applied force as shown in Fig. 9.6. Since a bone is weakest under shear stress, fractures of this nature are common even with minimal trauma. If the fractured bone is left untreated, it can lead to angular limb deformity.

Exercise Problem 9.1:

PROBLEM SOLVING STRATEGY

What do we know?

A compressive force is exerted on a bone of certain cross sectional area.

What concepts are needed?

We need to apply the definition of stress and compute the change in the length resulting from this stress.

Why?

We are asked to see if the bone can withstand certain applied stress.

If a compressive force of 3.6×10^4 N is exerted on the end of 20 cm long bone of cross sectional area 3.6 cm². (a) Will the bone break, and (b) if not, by how much does it shorten? (The Young's modulus of bone is $E_{bone} = 15 \times 10^9$ N/m².)

Solution:

The maximum compressive strength of a bone is roughly 170×10^6 N/m². The stress exerted on the bone is

$$P = \frac{F}{A} = \frac{3.6 \times 10^4 \, \text{N}}{3.6 \, \text{cm}^2 \left(1 \, \text{m}/100 \, \text{cm}\right)^2} = 100 \times 10^6 \, \frac{\text{N}}{\text{m}^2}.$$

Therefore, there is not enough stress to break the bone, but it will lead to the length change. The change in the length is

$$\Delta L = \frac{1}{E} \frac{F}{A} L_o = \frac{100 \times 10^6 \, \text{N}/\text{m}^2}{15 \times 10^9 \, \text{N}/\text{m}^2} \times 20 \, \text{cm} = 0.133 \, \text{cm} = 1.33 \, \text{mm}.$$

Exercise Problem 9.1 raises an interesting question. Is it safe for kids to lift weights? According to Mayo clinic, strength training is okay for kids because it offers kids many benefits, but there are important caveats to keep in mind. For kids, light resistance and controlled movements with a special emphasis on proper technique and safety are best. A child can do many strength training exercises with his or her own body weight or inexpensive resistance tubing. Free weights and machine weights are other options. However, we should not confuse strength training with weightlifting, bodybuilding

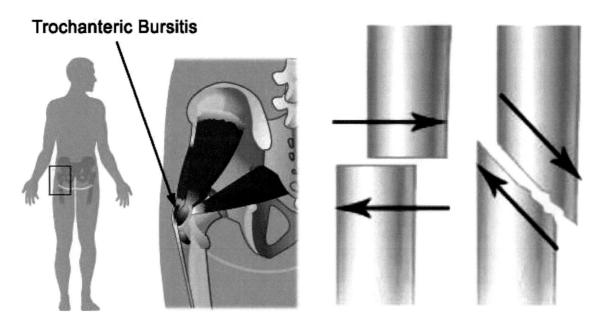

Figure 9.6: The effects of shear stresses on our body are illustrated. When a body is in motion, shear force between tissues of the body may be applied. A bursa is a fluid-filled sac that decreases this shear force. The right panel illustrates the effects of too much shear stress on bones.

or powerlifting. These activities are largely driven by competition, with participants vying to lift heavier weights or build bigger muscles than those of other athletes. This can put too much strain on young muscles, tendons and areas of cartilage that have not yet turned to bone (growth plates). This is problematic especially when proper technique is sacrificed in favor of lifting larger amounts of weight.

9.2 VOLUME DEFORMATION:

As we discussed above, when compressive stress is applied on a material, its length becomes reduced. Similarly, if an object is subject to inward forces from all sides, its volume will decrease. When an object is placed in a fluid as shown in Fig. 9.7, the inward forces are exerted on the object from all sides. This applied pressure (i.e., force per area) reduces the volume of the object. However, the object returns to its original volume when the pressure is removed. The bulk modulus measures the ability of a substance to withstand changes in volume when under compression on all sides. It is equal to the applied pressure divided by the relative deformation. The bulk modulus is sometimes referred to as the inverse compressibility.

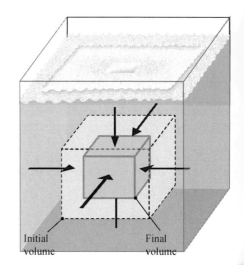

Figure 9.7: When an object is submerged in a fluid, pressure (i.e., force per area) is applied on the object from all sides. The volume of this object changes due to changes in the applied pressure.

In this case, the relative deformation is the change in volume divided by the original volume. This relative deformation is commonly called strain. Thus, if the original volume V_o of a material is reduced by an applied pressure P to a new volume V, the strain may be expressed as the change in volume $\Delta V = V_o - V$ divided by the original volume (or $(V_o - V)/V_o$). The fractional amount of change in the volume $\Delta V/V_o$ may be described by

SOLIDS AND FLUIDS

$$\frac{\Delta V}{V_0} = -\frac{1}{B}\Delta P. \tag{9.6}$$

where ΔP is the change in pressure, B is the bulk modulus, ΔV is the change in volume, and V_0 is the original volume. This suggests that we may define the bulk modulus B of a material as

$$B = -\frac{\Delta P}{(\Delta V/V_0)}. \tag{9.7}$$

where **bulk modulus** is a numerical constant that describes the elastic properties of a solid or fluid when it is under pressure on all surfaces. Because the denominator (i.e., strain) is a ratio without dimensions, the dimensions of the bulk modulus are the same as those of pressure, force per unit area. In the English system, the bulk modulus may be expressed in units of pounds per square inch (usually abbreviated to psi). In the metric system, on the other hand, the unit for B is pascal which is Newton per square meter (N/m^2).

Table 9.2: Elastic moduli of materials at normal atmospheric pressure (i.e., pressure at the sea level) are shown. There is no variation in temperature during processes.

Material	Elastic modulus E (N/m^2)	Shear modulus G (N/m^2)	Bulk modulus B (N/m^2)
Solids			
Iron, cast	100×10^9	40×10^9	90×10^9
Steel	200×10^9	80×10^9	140×10^9
Brass	100×10^9	35×10^9	80×10^9
Aluminum	70×10^9	25×10^9	70×10^9
Concrete	20×10^9		
Brick	14×10^9		
Marble	50×10^9		70×10^9
Granite	45×10^9		45×10^9
Wood (pine)			
(parallel)	10×10^9		
(perpendicular)	1×10^9		
Nylon	5×10^9		
Bone	15×10^9	80×10^9	
Liquids			
Water			2.0×10^9
Alcohol (ethyl)			1.0×10^9
Mercury			2.5×10^9
Gases			
Air, H_2, He, CO_2			1.01×10^9

Understanding-the-Concept Question 9.2

At a depth of about 1030 m in the sea the pressure has increased by 100 atmospheres (to about 10^7 N/m²). By how much has 1.0 m³ of water has been compressed by this pressure? The bulk modulus of water is 2.3×10^9 N/m².

    a. 2.3×10^{-3} m³   **b.** 3.3×10^{-3} m³   **c.** 4.3×10^{-3} m³   **d.** 5.3×10^{-3} m³

Explanation:

We are given information about the pressure change ΔP, the initial volume V_o, and the bulk modulus B. Using Eq. (9.6), we compute the volume compression ΔV as

$$\frac{\Delta V}{V_o} = -\frac{1}{B}\Delta P \Rightarrow \Delta V = -\frac{V_o}{B}\Delta P = -\frac{1.0\,\text{m}^3}{2.3 \times 10^9\,\text{N}/\text{m}^2} \times 10^7\,\frac{\text{N}}{\text{m}^2} = -4.35 \times 10^{-3}\,\text{m}^3.$$

Here, the minus sign indicates that the final volume is smaller than the initial volume.

Answer: c

    Even though we are living under constant 14.7 pounds per square inch (PSI), or 1 kg per square cm, of pressure pushing down on our bodies at sea level, we do not feel it. Our body compensates for this weight by pushing out with the same force. However, the situation is different under water. Since water is much heavier than air, this pressure increases as we venture into the water. For every 33 feet down we travel, one more atmosphere (14.7 PSI) pushes down on us. For example, at 66 feet, the pressure equals 44.1 PSI, and at 99 feet, the pressure equals 58.8 PSI. To travel into this high-pressure environment, we have to make some adjustments. Humans can travel to three to four-atmosphere environment without much difficulty. To go farther, submarines are needed. Animals that live in this watery environment can undergo large pressure changes in short amounts of time. Sperm whales make hour-long dives 7,380 feet (2,250 m) down. This dive results in a pressure change of more than 223 atmospheres!

Exercise Problem 9.2

One liter of alcohol (1000 cm³) in a flexible container is carried to the bottom of the sea, where the pressure is 2.6×10^6 N/m². What will be its volume there?

PROBLEM SOLVING STRATEGY

What do we know?

One liter of alcohol is at the bottom of the sea.

What concepts are needed?

We need to apply the relationship between the volume change and the pressure change.

Why?

We are asked to compute the volume change of the alcohol resulting from the increase in the pressure.

SOLIDS AND FLUIDS

Solution:

From Eq. (9.6), the change in the volume is given by

$$\frac{\Delta V}{V_o} = \frac{V - V_o}{V_o} = -\frac{1}{B}(P - P_o)$$

Here, the bulk modulus for alcohol is $B = 1.0 \times 10^9$ N/m² and $P_o = 1$ atm $= 1.013 \times 10^5$ N/m². So, we may use these values to compute the final volume V and obtain

$$V = V_o - \frac{1}{B} V_o (P - P_o) = V_o \left(1 - \frac{P - P_o}{B}\right)$$

$$= 1000 \, \text{cm}^3 \left(1 - \frac{2.6 \times 10^6 \, \text{N/m}^2 - 1.013 \times 10^5 \, \text{N/m}^2}{1.0 \times 10^9 \, \text{N/m}^2}\right) = 997.5 \, \text{cm}^3$$

All real fluids are compressible, and almost all fluids expand when heated. **Compressibility** is a measure of the relative volume change of a fluid or solid as a response to a pressure (i.e., stress) change. Usually a fluid may be considered incompressible when the velocity of fluid is greater than one-third of speed of sound in the fluid, or if the fluid is a liquid. The incompressible fluids are treated is simply by assuming the density is constant. This assumption of constant density yields a simple relationship for the state of substances. Hence, the variation of density of the fluid with pressure is a primary factor in deciding whether a fluid is incompressible.

9.3 FLUIDS

The term "fluid" is used to refer particular states of matter. Matter can exist in four phases (or states). Three commonly known states of matter are (i) solid, (ii) liquid, and (iii) gas. We may define these states in terms of material's ability to form different shapes. A **solid** is matter in which the molecules are very close together and cannot move around. Examples of solids include rocks, wood, and ice (frozen water). A solid maintains a fixed shape and fixed size even if a large force is applied. A **liquid** is matter in which the molecules are close together and move around slowly. Examples of liquids include drinking water, mercury at room temperature, and lava (molten rock). Liquids do not maintain a fixed shape but they are not readily compressible. A **gas** is matter in which the molecules are widely separated, move around freely, and move at high speeds. Examples of gasses include the gasses we breathe (nitrogen, oxygen, and others), the helium in balloons, and steam (water vapor). Gasses can maintain neither a fixed shape nor a fixed volume. We refer liquids and gasses as "fluids" since they do not maintain a fixed shape and they have ability to flow.

The fourth state of matter, as shown in Fig. 9.8, is plasma. **Plasma** is a gas composed of free-floating ions (atoms stripped of some electrons become positively charged) and free electrons (negatively charged). Plasma conducts electrical currents. Plasma was discovered by William Crookes in 1879. The word "plasma" was first applied to ionized gas by Irving Langmuir in 1929. Plasma is by far the most common form of matter. Plasma in the stars and in the tenuous space between them makes up over 99% of the visible universe. Perhaps, most of that space is not visible. There are many

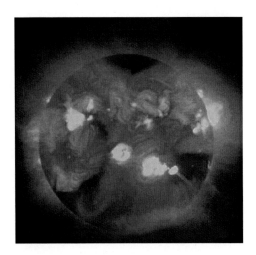 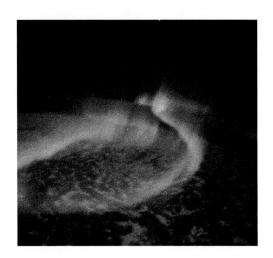

Figure 9.8: A collection of ionized atoms, known as plasma, is the fourth state of matter. Both Sun (left panel) and an aurora (right panel) are examples of this state.

different types of plasmas. There is plasma in stars (including our Sun). The solar wind in our Solar System is made of plasma. This state occurs at very high temperature. It is noted that temperature and density range of plasma is very broad, from relatively cool and tenuous (like aurora) to very hot and dense (like the central core of a star). Ordinary solids, liquids, and gases are both electrically neutral and too cool or dense to be in a plasma state.

As shown in Fig. 9.8, aurorae are an example of plasma. An **aurora** (from the Latin word *aurora*, "sunrise") is a natural light display in the sky particularly in high-latitude (Arctic and Antarctic) regions, caused by the collision of energetic charged particles with atoms in the high altitude atmosphere (thermosphere). The charged particles originate in the magnetosphere and solar wind and, on Earth, are directed by the Earth's magnetic field into the atmosphere.

Some useful terminology

Before discussing the fundamental principles governing fluids, first, we need to define some useful terms such as density, specific gravity, and pressure. The **Density** ρ of a fluid is defined as its mass per unit volume and is written as

$$\rho = \frac{m}{V} \Rightarrow m = \rho V. \tag{9.8}$$

In some cases (for instance, in the United States oil and gas industry), density is defined as its weight per unit volume. This quantity is more properly called specific weight. The **Specific gravity (S.G.)** is the ratio of the density of a substance to the density (mass of the same unit volume) of a reference substance. Apparent specific gravity is the ratio of the weight of a substance with a specified volume to the weight of the reference substance with the equal volume. The reference substance is nearly always water at 4 °C for liquids or air for gases. It is noted that the density of water ρ_{water} is 1.00 g/cm³ or 1.00×10³ kg/m³. **Pressure** is the ratio of force to the area over which that force is distributed. Here, the force F is understood to be acting perpendicular to the surface area A. This definition of pressure is identical to that for stress. Hence, the two words, pressure and stress, may be used interchangeably. Pressure is expressed as

SOLIDS AND FLUIDS

$$P = \frac{F}{A} \tag{9.9}$$

and the units for pressure is pascal (Pa) which is N/m². Sometimes pound-per-square-inch (PSI) (i.e., lbf/in²) is used to measure pressure. Another unit for the pressure is atmosphere (atm). One atm is the pressure of the air at sea level. Some useful conversion factors for pressure are

$$1 \text{ atm} = 1.013 \times 10^5 \text{ N/m}^2 = 101.3 \text{ kPa}$$

$$1 \text{ bar} = 1.00 \times 10^5 \text{ N/m}^2 = 100 \text{ kPa}$$

Automobile tire pressures are typically measured in PSI. The recommended tire pressure for a vehicle will vary depending on the kind of tire, the type of vehicle, and the use for which the tire is applied. If we are considering a car, light truck, SUV or other passenger vehicle, there is usually a placard on the door, in the glove compartment or the fuel compartment cover which specifies the best tire pressure for the vehicle. This may be as low as 24 PSI or as high as 35 PSI in most cases. The safest range is within 1 to 2 pounds of the level on the door sticker, or the owner's manual specifications. We should never operate using the maximum pressure which is stamped on the side of the tire as normal conditions. This maximum is the greatest pressure which the tire is designed to resist. This is not the recommended operating pressure. If a tire is used with less than the optimum pressure, then we may notice that the outside edges of the tread will be wearing faster than the center. This is a clear sign that we need to use more pressure. If the tire is wearing only in the center of the tread faster than the outside edges, then we need to reduce the pressure. The vehicle's weight and the load carried also play a part. The heavier the load the more pressure required in the tires.

There are three types of pressure: (i) absolute pressure, (ii) gauge pressure, and (iii) atmospheric pressure. Tire pressure, which is measured using a pressure gauge is gauge pressure. Everyday pressure measurements, such as for tire pressure, are usually made relative to ambient air pressure. In other cases, measurements are made relative to a vacuum or to some other ad hoc reference. **Absolute pressure** is zero-referenced against a perfect vacuum, so it is equal to gauge pressure plus atmospheric pressure. **Gauge pressure** is zero-referenced against ambient air pressure, so it is equal to absolute pressure minus atmospheric pressure. Negative signs are usually omitted. To distinguish a negative pressure, the value may be appended with the word "vacuum" or the gauge may be labeled a "vacuum gauge." Hence, gauge pressure is the pressure over and above atmospheric pressure, and the absolute pressure is written as

$$P = P_A + P_G. \tag{9.10}$$

where P_A is the atmospheric pressure and P_G is the gauge pressure. Sometimes differential pressure which is defined as the difference in pressure between two points is used for convenience. This includes absolute pressure, gauge pressure, and differential pressure. Absolute pressure is zero referenced against a perfect vacuum. This is the method of choice when measuring quantities where absolute values must be determined.

To measure pressure, we often used a pressure gauge similar to that shown in Fig. 9.9. A well designed and properly sized pressure gauge should provide accurate service for years. We may use the pressure gauges to judge the performance of spray nozzles, pumps and other fluid components. The two most popular types of pressure gauges are shown in the right panel of Fig. 9.9. These types are regular and (more reliable) filled. The two most common causes for gauge failure are pipe vibration and water condensation, which can lead to freezing in colder environments. The delicate connecting links, pivots, and pinions of a regular gauge are sensitive to both condensation and vibration. Filled

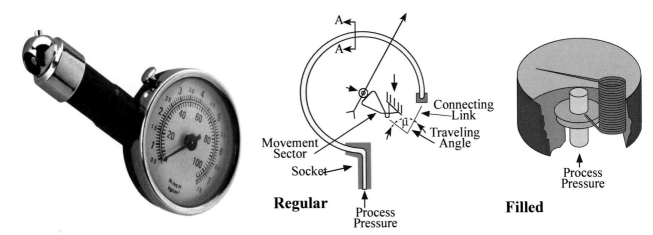

Figure 9.9: A pressure gauge measures gauge pressure. A tire gauge (left panel) measures gauge pressure of tires. The right panel illustrates some of the designs for measuring gauge pressure accurately.

gauges last longer because they have fewer moving parts and the housing is filled with a viscous glycol or silicon fluid. The fill in a gauge helps dampen pointer vibration and eliminates corrosion due to condensed water in humid air

Understanding-the-Concept Question 9.3:

You insert a straw of length L into a tall glass of your favorite beverage. You place your finger over the top of the straw so that no air can get in or out, and such that the distance from the bottom of your finger to the top of the liquid is h. Does the air in the space between your finger and the top of the liquid have a pressure P that is

- a. greater than
- b. equal to
- c. less than

the atmospheric pressure P_A outside the straw?

Explanation:

As we expect, the liquids tend to fall downward in the presence of gravity. However, the air outside tends to fill up the empty space that would be created by the drainage of liquid. In this process the air, instead of penetrating through the liquid column, just pushes the whole column of liquid upward and keeps it intact inside the straw. If the straw had a wide opening, this will not happen. If we were to try this experiment on the moon, it would not work. The level of the liquid in the straw would not lift up when we lift the straw (with the end covered by a finger) because there is no atmosphere. There is no air pressure P_A to support the weight of the liquid.

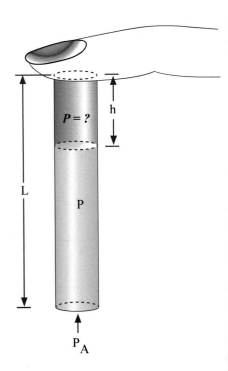

Answer: c

SOLIDS AND FLUIDS

Similar to the straw in Understanding-the-Concept Question 9.3, there is a pressure difference between inside and outside of a suction cup. This pressure difference holds it on the vertical glass wall. A guy suggested this idea to NASA. Can this idea work in space? A story about a guy who suggested this idea to NASA goes as follows. A guy walks in the main office of NASA and said "I've got an idea. When astronauts have to step outside for a spacewalk, why not have them carry suction cups? They could inch their way along the outside of the capsule and never have to worry." Well? What should we think? It sounds like it might work since some very adventurous people have used big suction cups to climb up the fronts of glass buildings before. However, it may not be a good idea to climb too high because a suction cup will be less effective as the person go up higher. This means the suction cups in space are no good at all! Why? The answer is in how a suction cup works in the first place. It works because of air pressure. Air pressure is the weight of the Earth's atmosphere pressing on things. We do not feel the air pressure on us because it is evenly balanced, inside and out. In fact, there is an atmospheric pressure of 15 pounds per square inch on everything around us, including on us. When we set the suction cup down on a surface, the air pressure is balanced on all sides. Force the air out of the bowl, however, and now the air pressure is all concentrated on one side. The weight of Earth's atmosphere is actually what holds the cup against a smooth surface! That is neat in itself, but we may see the problem. Up on top of a high mountain there is less air pressing down, so suction cups will not work as well. Similarly, out in space there is no air pressure at all. So, a suction cup would not hold for an instant.

9.4 OBJECT IN FLUID

When an object is immersed in fluid, the surrounding fluid exerts pressure in all directions. An object in air is also considered as being immersed in fluid, but the effects of pressure exerted in all directions may not be easily noticed because of its low density. So, we consider a liquid which has much higher density than that of air. In a liquid, the pressure exerted on the object is much greater that in the air. Hence, the effects of pressure exerted on the object by the molecules in the fluid are more noticeable. It should be noted that at any point (i.e., depth) in a fluid at rest, the same pressure is exerted in all directions. The force due to fluid pressure always acts perpendicular to any surface on which the fluid is in contact with, as shown in Fig. 9.10.

The reason that the pressure at a given depth is the same in all directions is due to the fact the molecules or atoms in a fluid are constantly colliding on the surface of a submerged object. On the average, only the normal component (to the surface) of the net force is non-zero. This is another statement of the fact that pressure is not a vector and thus has no direction associated with it when it is not in contact with some surface. Hence, the pressure on a submerged object is **always perpendicular to the surface at each point on the surface.**

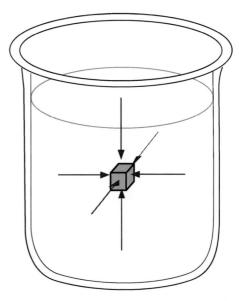

Figure 9.10: An object submerged in a fluid, pressure is always is exerted perpendicular to the surface at each point on the surface which is in contact with the fluid.

Pressure

If a fluid is within a container and the depth of an object placed in that fluid can be measured, then the deeper the object is placed in the fluid, the more pressure it experiences. This is due to the weight of the fluid above it. The denser the fluid above it, the more pressure is exerted on the object that is submerged, due to the weight of the fluid. The pressure due to the weight of liquid at a depth h below the surface (i.e., due to the weight of the column of liquid) is given by

$$P = \frac{F}{A} = \frac{\rho A h g}{A} = \rho g h. \tag{9.11}$$

Equation (9.11) is derived by considering the fact that when an object is submerged in a liquid the force exerted on the object is the weight of fluid directly above the object. This means that if any object is submerged in a liquid at the same depth, then the same amount of force is exerted on the object. Hence, we may conclude that **the pressure at equal depths within a uniform liquid is the same.** The fluid pressure is proportional to the specific gravity at any point and to the height of the fluid above the point.

It is noted that Eq. (9.11) for pressure $P = \rho g h$ is valid only for fluids whose density is constant and does not change with depth (i.e., if the fluid is incompressible)

$$\Delta P = \rho g (\Delta h). \tag{9.12}$$

The compressibility of water is a function of pressure and temperature. At 0 °C, the compressibility is 5.1×10^{-10} Pa^{-1} at the limit of zero pressure. At the zero-pressure limit, the compressibility reaches a minimum of 4.4×10^{-10} Pa^{-1} around 45 °C before increasing again with increasing temperature. As the pressure is increased, the compressibility decreases, being 3.9×10^{-10} Pa^{-1} at 0 °C and 100 MPa (mega-pascal). Due to its low compressibility, water, is often assumed to be incompressible. The low compressibility of water means that even in the deep oceans at 4 km depth, where pressures are 40 MPa, there is only a 1.8% decrease in volume, which is a small volume change.

The pressure depends on the depth, rather than the volume of fluid. This means no matter how much water is in a dam, as long as objects are submerged at equal depth from the surface, they experience the same pressure. Hence, as illustrated in the schematic design of the Hoover Dam in Fig. 9.11, the deeper we go under the water, the greater the pressure of the water pushing down on us. For every 33 feet (10 meters) we go down, the pressure increases by 14.7 PSI (1 bar). In the deepest ocean, the pressure is equivalent to the weight of an elephant

(a)

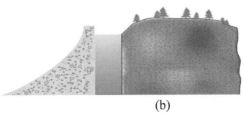

(b)

Figure 9.11: (a) Hoover Dam, which was constructed between 1931 and 1936, provides irrigation water and produce hydroelectric power. (b) A side view of the Hoover Dam shows that the design is an ach-gravity type to withstand the water pressure at the bottom of the dam.

SOLIDS AND FLUIDS

balanced on a postage stamp, or the equivalent of one person trying to support 50 jumbo jets! As we know, many animals that live in the sea have no trouble at all with high pressure. Whales, for instance, can withstand dramatic pressure changes because their bodies are more flexible. Their ribs are bound by loose, bendable cartilage, which allows the rib cage to collapse at pressures that would easily snap our bones.

Understanding-the-Concept question 9.4:

You are originally 1.0 m beneath the surface of a pool. If you dive to 2.0 m beneath the surface, what happens to the absolute pressure on you?

- a. It quadruples.
- b. It more than doubles.
- c. It doubles.
- d. It less than doubles.

Explanation:

As indicated by Eq. (9.10), the absolute pressure is the sum of atmospheric pressure and gauge pressure. When we dive beneath the surface of a pool, the absolute pressure on us increases because the gauge pressure on us increases with increasing depth. However, the atmospheric pressure on us remains the same. This means that if we dive 2.0 m beneath the surface, the gauge pressure is twice large than 1.0 m beneath the surface, while the atmospheric pressure does not change. Hence, the absolute pressure at 2.0 m beneath the surface is less than twice of that at 1.0 m beneath the surface.

Answer: d

Exercise Problem 9.3

Determine the water gauge pressure at a house at the position at the bottom of a hill fed by a full tank of water 5.0 m deep and connected to the house by a pipe that is 100 m long at an angle of 60° from the horizontal. Neglect turbulence, and friction and viscous effects. How high would the water shoot if it came vertically out of a broken pipe in front of the house?

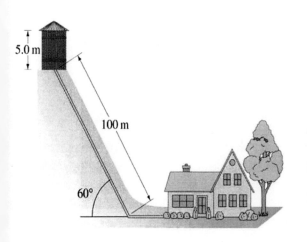

PROBLEM SOLVING STRATEGY

What do we know?

We are given information about the vertical position of the broken pipe relative to the surface of water.

What concepts are needed?

We need to use the relationship between the changes in the pressure and depth below the water surface.

Why?

We need to compute the pressure difference will for the water from the broken pipe to shoot up.

Solution:

The gauge pressure of the water at the house is

$$\Delta P = \rho g(\Delta h) = \rho g(\ell \sin\theta + h')$$

$$= 1.0\times10^3 \text{ kg/m}^3 \times 9.8 \text{ m/s}^2 \times (100\text{m}\times \sin 60° + 5 \text{ m}) = 9.0\times10^5 \text{ N/m}^2.$$

Using this value of ΔP, we compute the height the water will shoot and obtain

$$\Delta P = \rho g h \Rightarrow \Delta h = \frac{\Delta P}{\rho g} = \frac{9.0\times10^5 \text{ N/m}^2}{1.00\times10^3 \text{ kg/m}^3 \times 9.8 \text{ m/s}^2} = 92\text{ m}.$$

It is interesting to note that $\Delta h = 92$ m is the original height of the water level in the tank, measured from the ground.

The gauge pressure is what allows water to shoot up from a broken water main. The higher gauge pressure leads water to shoot up higher. One of most dramatic examples of broken water mains happened in Australia. On October 29, 2012, a newspaper in Victoria, Australia reported that up to 2 million litres of water were lost in the rupture. As a result of this rupture, water shot up to 50 meters into the air for about an hour in the residential suburb, as shown below in Fig. 9.12. In determining how high water will shoot up depends on the pressure difference as well as on a number of other factors such as viscosity and turbulence, which we discuss below. In the real world, due to these other contributing factors, Eq. (9.12) cannot be used to estimate the water height accurately.

Figure 9.12: A jet of water is expelled out vertically from a broken water main. According to Pascal's principle, the height reached by the water depends on a number of factors including the water pressure at the bottom and the gravitational acceleration.

In our discussion below, we focus on the four governing principles of fluid. These governing principles are (i) Pascal's principle, (ii) Archimedes' principle, (iii) continuity principle, and (iv) Bernoulli's principle.

9.5 PASCAL'S PRINCIPLE

Pascal's law is an important principle in fluid mechanics which reveals how pressure is transmitted anywhere in a confined incompressible fluid, **Pascal's principle** states that pressure applied to a confined fluid increases the pressure throughout by the same amount. This means that any externally applied pressure is transmitted to all parts of the enclosed fluid. For example, the pressure at the bottom of a jug is equal to the externally applied pressure on the top of the fluid plus the static fluid pressure from the weight of the liquid.

SOLIDS AND FLUIDS

Pascal's principle may be used to obtain mechanical advantage. Here, **mechanical advantage** is a measure of the force amplification achieved by using a tool which includesa mechanical device or machine system. Ideally, the device preserves the input power and trades off forces against movement, to obtain a desired amplification in the output force. This mechanical advantage is obtained by using a hydraulic system. Hydraulic systems use an incompressible fluid, such as oil or water, to transmit forces from one location to another within the fluid. Most aircraft use hydraulics in the braking systems and landing gear. Sometimes, a compressible fluid is used to serve asimilar functions. For example, pneumatic systems use compressible fluid, such as air, in their operation. Some aircraft utilize pneumatic systems for their brakes, landing gear, and movement of flaps.

Let us discuss some examples in where Pascal's principle is applied. Suppose we consider a hydraulic lift as shown below in Fig 9.13. According to Pascal's principle, when there is an increase in pressure at any point in a confined fluid, there is an equal increase at every other point in the container such that

$$P_{in} = P_{out}. \tag{9.13}$$

We may use the definition pressure (i.e., $P = F/A$) and write

$$\frac{F_{in}}{A_{in}} = \frac{F_{in}}{A_{in}} \Rightarrow F_{out} = F_{in} \frac{A_{out}}{A_{in}}. \tag{9.14}$$

Since the mechanical advantage is defined as the ratio of output and input forces, we write

$$\frac{F_{out}}{F_{in}} = \text{mechanical advantage} = \frac{A_{out}}{A_{in}} = \frac{\pi r_{out}^2}{\pi r_{in}^2} = \left(\frac{r_{out}}{r_{in}}\right)^2. \tag{9.15}$$

assuming that the area of the piston in Fig. 9.13 is circular. This means that if the cylinder on the left has a cross-section area of 1 square inch, while the cylinder on the right has a cross-section area of 10 square inches. The cylinder on the left has a weight (force) of 1 pound acting downward on the piston, which lowers the fluid 10 inches. As a result of this force, the piston on the right lifts a 10 pound weight a distance of 1 inch. The 1 pound load on the 1 square inch area causes an increase in pressure on the fluid in the system. This pressure is distributed equally throughout and acts on every square inch of

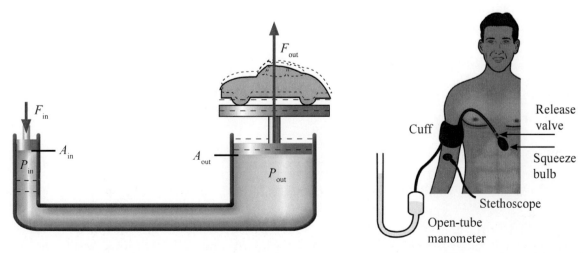

Figure 9.13: Pascal's principle may be applied in situations involving pressure at two different points. This principle is useful in measuring the mechanical advantage (left panel) and blood pressure of a person (right panel).

the 10 square inch area of the large piston. As a result, the larger piston lifts up a 10 pound weight. The larger the cross-section area of the second piston, the larger the mechanical advantage, and the more weight it lifts. This large multiplication of force (hydraulic press principle) is possible due to transmission of externally applied pressure to other part a of the enclosed fluid.

For monitoring blood pressure, an inflatable cuff is placed on the upper arm of a person at the same level as the heart. Blood flow is detected just below the cuff, and corresponding pressure is transmitted to a mercury-filled manometer as shown in Fig. 9.13. Atmospheric pressure adds to blood pressure in every part of the circulatory system (As noted in Pascal's principle, the total pressure in a fluid is the sum of the pressures from different sources. Here, it is the pressure from the heart and the atmosphere.) However, atmospheric pressure has no net effect on blood flow since it adds to the pressure coming out of the heart and going back into it, too. What is important is how much *greater* the blood pressure is than the atmospheric pressure. Blood pressure measurements, like tire pressures, are thus made relative to atmospheric pressure (i.e., gauge pressure).

How can we measure pressure?

There are a number of ways we can measure pressure. The piece of lab equipment specifically designed to measure the pressure of gasses is known as the barometer. A **barometer** uses the height of a column of mercury to measure gas pressure in millimeters of mercury. The mercury in the tube is pushed up from the dish until the pressure at the bottom of the tube (due to the mass of the mercury) is balanced by the atmospheric pressure. Air pressure is measured by using a mercury barometer as shown in Fig. 9.14

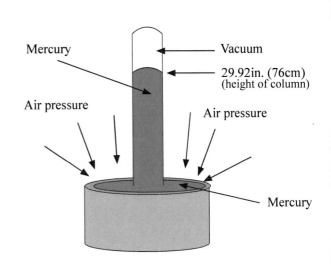

Another way to measure pressure is by using a device called "open-tube manometer". The open-tube manometer measures pressure by using a liquid in a container as shown in Fig. 9.15. A manometer relies on Pascal's principle to measure pressure in gasses by measuring the difference in height h of two levels of liquid. Sometimes "mercury" is used. It is noted that "mm-Hg" is equivalent to a pressure of 133 N/m² since 1 mm = 10^{-3} m and ρ_{Hg} =13.6×10³ kg/m³. We may see this easily by using Eq. (9.12) for pressure which we derived above and by computing

Figure 9.14: A schematic drawing of a simple mercury barometer with vertical mercury column and reservoir base. The height of the mercury column is related to the atmospheric pressure.

$$\Delta P = \rho g h = 13.6 \times 10^3 \text{kg/m}^3 \times 9.8 \text{m/s}^2 \times 1.0 \times 10^{-3} \text{m} = 1.33 \times 10^2 \text{N/m}^2.$$

mm-Hg is called "torr" to honor Torricelli (1608-1647), who invented barometer. Sometimes, a closed-tube manometer is used to measure pressure. Closed-tube manometers look similar to regular manometers except that the end that is open to the atmospheric pressure in a regular manometer is sealed and contains a vacuum.

SOLIDS AND FLUIDS

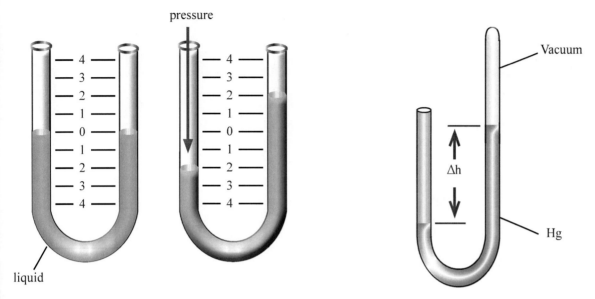

Figure 9.15: The U-Tube which was invented by Huygens in 1661 is one of many instruments used to measure pressure. Here, the column will rise or fall until its weight is in equilibrium with the pressure differential between the two ends of the tube.

9.6 ARCHIMEDES' PRINCIPLE

Archimedes' principle is a physical law stating that the upward buoyant force exerted on a body immersed in a fluid is equal to the weight of the fluid the body displaces. In other words, an object immersed in a fluid is buoyed up by a force equal to the weight of the fluid it actually displaces. Archimedes' principle is an important and underlying concept in the field of fluid mechanics and is helpful for understanding why a very heavy ship such as an aircraft carrier can float. This principle is named after its discoverer, Archimedes of Syracuse (287 BC - 212 BC).

The buoyant force occurs because the pressure in a fluid increases with depth. In other words, the upward pressure from the bottom of a submerged object is greater than the downward pressure on its top surface, as shown in Fig. 9.16.

We now discuss the derivation of buoyant force. Buoyancy arises from the fact that fluid pressure increases with depth, and from the fact that the increased pressure is exerted in all directions (Pascal's principle), so that there is an unbalanced upward force on the bottom of a submerged object. Since the immersed object in Fig. 9.15 is exactly supported by the difference in pressure, it follows that the buoyant force on the solid object is equal to the weight of the fluid displaced (Archimedes' principle). The direction of the force is always toward the surface of the fluid. The buoyant force F_B is given by

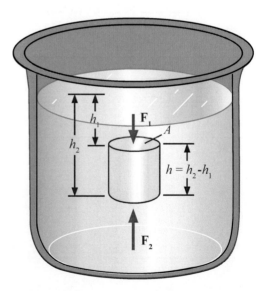

Figure 9.16: A schematic illustration of the physical quantities that are involved in the derivation of the buoyant force acting on a submerged object, based on Archimedes' principle.

$$F_B = F_2 - F_1 = \rho_F g A(h_2 - h_1) = \rho_F g V = m_F g. \tag{9.16}$$

This means that objects with equal volumes feel equal amount of buoyant forces. Suppose we have equal sized balls of cork, aluminum, and lead, with respective specific gravities of 0.2, 2.7, and 11.3. If the volume of each is 10 cubic centimeters then their masses are 2, 27, and 113 gm. Each would displace 10 grams of water, yielding apparent masses of -8 (the cork would accelerate upward), 17 and 103 grams respectively. The behavior of the three balls would certainly be different upon release from rest in the water. The cork would bob up, the aluminum would sink, and the lead would sink more rapidly. However, the buoyant force on each is the same because of identical pressure environments and equal water displacement. The difference in behavior comes from the comparison of that buoyant force with the weight of the object.

For an immersed object...

When an object is immersed in water, its apparent weight is less than its true weight. This is due to the buoyant force on a submerged object is equal to the weight of the liquid displaced by the object and the direction of this force is in the opposite direction to the weight. For water, with a density of one gram per cubic centimeter, this provides a convenient way to determine the volume of an irregularly shaped object as well as its density. A free-body diagram for an object immersed in fluid, as shown in Fig. 9.17, is useful in determining the apparent weight of the object.

An object floats on a fluid if its density is less than that of the fluid. It is wrong to think that light things float on water while heavy things sink. For example, a heavy log will float down a river, but a light grain of sand will sink. However, whether an object will float or sink still has something to do with how heavy the object is. To find out if an object will sink or float in water, we have to compare the weight of the object to the weight of an equal amount of water. Take a grain of sand as an example. Since a grain of sand sinks in water, do we think that the grain of sand weighs more or less than a tiny amount of water, or is the same size as the grain of sand? The sinking of a grain of sand indicates that its density is larger than that of water.

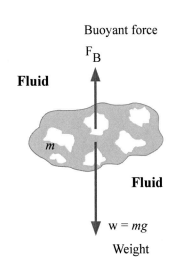

Figure 9.17: Two forces, buoyant force and weight, are acting on a submerged object in fluid. When the buoyant force acting on a person is greater than weight, the person floats (right panel).

SOLIDS AND FLUIDS

A ship (or a submarine) can float because the weight of water it displaces is equal to the weight of the ship. This displacement of water creates an upward force called the **buoyant force** and acts opposite to gravity, which would pull the ship down. Unlike a ship, a submarine can control its buoyancy, thus allowing it to sink and surface at will. To control its buoyancy, the submarine has **ballast tanks** and auxiliary tanks (or trim tanks) that can be alternately filled with water or air. When the submarine is on the surface, the ballast tanks are filled with air and the submarine's overall density is less than that of the surrounding water.

As the submarine dives, the ballast tanks are flooded with water and the air in the ballast tanks is vented from the submarine until its overall density is greater than the surrounding water and the submarine begins to sink (**negative buoyancy**). A supply of compressed air is maintained aboard the submarine in air flasks, for life support, and for use with the ballast tanks. In addition, the submarine has movable sets of short "wings," called **hydroplanes** on the stern (back), that help to control the angle of the dive. The hydroplanes are angled so that water moves over the stern, which forces the stern upward; therefore, the submarine is angled downward.

Understanding-the-Concept Question 9.5

A steel ball sinks in water but floats in a pool of mercury. Where is the buoyant force on the ball greater?

- a. Floating on the mercury.
- b. Submerged in the water.
- c. It is the same in both cases.
- d. Cannot be determined from the information given.

Explanation:

According to Eq. (9.16), the buoyant force on the steel ball is equal to the weight of the fluid displace by the ball. Also, the ball will float when the buoyant force acting on the completely submerged ball is greater than its weight. On the other hand, the ball will sink when the buoyant force is less than its weight. Hence, the buoyant force on the ball is greater floating on the mercury.

Answer: a

There is an interesting story about how the great mathematician Archimedes discovered buoyancy started streaking. When Archimedes solved a puzzling problem given to him by his king, he ran through the streets naked, shouting Eureka! Prior to Archimedes' mad dash, King Hiero II had asked him for a favor. The king had asked his goldsmith to make a crown out of pure gold, and he suspected the goldsmith had been dishonest and threw some silver into the mix. Archimedes had to determine whether the goldsmith had been dishonest, without damaging the crown. Archimedes had to solve a problem very similar to Exercise Problem 9.5.

Exercise Problem 9.5:

When a crown of mass 14.7 kg is submerged in water, an accurate scale reads only 13.4 kg. Is the crown made of gold?

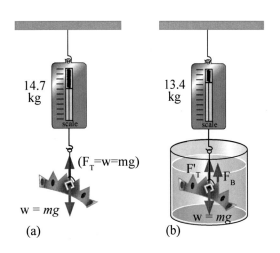

PROBLEM SOLVING STRATEGY

What do we know?

We are given the mass of the crown and its apparent mass in water.

What concepts are needed?

We need to apply the Archimedes' principle.

Why?

In water, the buoyant force reduces the tension on the string. In air, the string tension equals the weight of the crown.

Solution:

The apparent weight of the submerged object is

$$W' = F_T = W - F_B = \rho_o g V - \rho_F g V$$

We now obtain the specific gravity of the object by taking the ratio of the object's weight and buoyant force

$$\frac{W}{W - W'} = \frac{\rho_o g V}{\rho_F g V} = \frac{\rho_o}{\rho_F} = \frac{14.7\,\text{kg}}{14.7\,\text{kg} - 13.4\,\text{kg}} = 11.3.$$

Therefore, the density of the unknown object is

$$\rho_o = 11.3 \rho_F = 11,300 \frac{\text{kg}}{\text{m}^3}.$$

Noting that the density of lead and gold is given, respectively, as

$$\rho_{Pb} = 11.3 \times 10^3 \frac{\text{kg}}{\text{m}^3} \quad \text{and} \quad \rho_{Au} = 19.3 \times 10^3 \frac{\text{kg}}{\text{m}^3}$$

We conclude that the crown is not made of gold. It is made out of lead.

As we can see, this was not a trivial problem! Archimedes knew that silver was less dense than gold, but he did not know any way of determining the relative the density (mass/volume) of an irregularly shaped crown. The weight could be determined using a balance or scale, but the only way known to determine volume, using the geometry of the day, was to beat the crown into a solid sphere or cube. Since Hiero had specified that damage to the crown would be viewed with less than enthusiasm, Archimedes did not wish to risk the king's wrath by pounding the crown into a cube and hoping that post-analysis it could be made all better again.

While in the public baths, Archimedes observed that the level of water rose in the tub when he entered the bath. He realized this was the solution to his problem and supposedly, in his excitement, he leaped up and ran naked through the streets back to his laboratory, shouting "Eureka, Eureka!" (I've got it!). Later, he demonstrated to Hiero and his court how the amount of water overflowing

SOLIDS AND FLUIDS

a tub could be used to measure a volume. His calculations indicated the goldsmith was, indeed, an embezzler. History does not record the fate of the unscrupulous artisan. Archimedes' observation has been formalized into Archimedes' principle.

9.7 FLUID DYNAMICS

Fluid dynamics is a branch of fluid mechanics that deals with **fluid flow.** This discipline includes aerodynamics (the study of air and other gases in motion) and **hydrodynamics** (the study of liquids in motion). Fluid dynamics has a wide range of applications, including calculating forces and moments on aircraft, determining the flow rate of petroleum through pipelines and blood through veins, predicting weather patterns, understanding nebulae in interstellar space and modeling fission weapon detonation. Some of its principles are even used in traffic engineering, where traffic is treated as a continuous fluid.

As a simple starting point for studying fluid dynamics, we consider a motion of fluid around an object. Fluid flow can be very smooth, calm, and regular. This type of fluid flow is obtained at low speed and is easy to characterize. However, in general, the fluid flow is not so smooth at high speed. It can become erratic and starts flowing in random patterns and leads turbulent flow. This type of fluid flow is difficult to characterize. We may distinguish these two types of flow as (i) streamline or laminar flow and (ii) turbulent flow.

Streamlined or laminar flow is defined as a smooth flow such that neighboring layers of the fluid slide by each other smoothly. The flow of a fluid moving with a moderate speed has fluid layers moving past other layers as if some sheets are moving over other layers. In laminar flow, viscous shear stresses act between these layers of the fluid which defines the velocity distribution among these layers of flow. In this flow, the shear stresses are defined by Newton's equation for shear stress. On the other hand, **turbulent flow** is defined as a rough erratic flow. As the flow speed of the otherwise calm layers increases, these smoothly moving layers start moving randomly. With further increase in flow velocity, the flow of fluid particles becomes completely random and no such laminar layers exist anymore. Shear stresses in the turbulent flow are more than those in laminar flow. These two flow types are shown in Fig. 9.18. Also, in characterizing the flow type, a dimensionless parameter called "Reynolds number" is used. The Reynolds number is defined as the ratio of inertial and viscous force to characterize these two types of flow patterns. With increase in flow velocity and the initial forces, the Reynolds number increases. For moderate flows (i.e., laminar flow), the Reynolds number is below 2000. However, for turbulent flows, it is well above 2300. For the transition region between the two types of flow, the Reynolds number varies between 2000 and 4000.

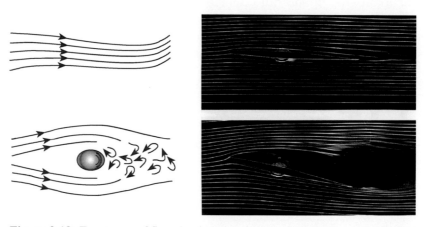

Figure 9.18: Two types of flow, laminar (two upper panels) flow and turbulent (two lower panels) flows, are illustrated. These flow types may be characterized by a dimensionless parameter known as the Reynolds number.

The process of a laminar flow becoming turbulent is known as the **laminar-turbulent transition**. This is an extraordinarily complicated process which is not fully understood at present. However, we do understand the process which proceeds through a series of stages. A boundary layer can make transition to turbulence through a number of paths. Which path is realized depends physically on the initial conditions such as initial disturbance amplitude and surface roughness.

Continuity equation: Flow through an enclosed pipe

Let us examine the flow of liquid through a pipe. In general, the flow of liquid along a pipe can be determined by the use of the **Bernoulli equation** and the **continuity equation**. The former represents conservation of energy, which in Newtonian fluids has either potential or kinetic energy, and the latter ensures that what goes into one end of a pipe must come out at the other end! It must be stressed that only **incompressible liquids** are being considered. We consider the steady laminar flow of a fluid through an enclosed tube or pipe and ask the following question:

What happens to the speed of the fluid when the size of the tube changes?

Any fluid moving through a pipe obeys the law of continuity. This law states that the product of average velocity v, pipe cross-sectional area A, and fluid density ρ for a given flow stream must remain constant as illustrated in Fig. 9.19. Fluid continuity is an expression of a more fundamental law of physics: the conservation of mass. Here, **incompressible flow** refers to a flow in which the material density is constant within a fluid parcel. This means that an infinitesimal volume moves with the velocity of the fluid.

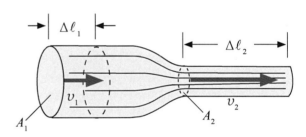

Figure 9.19: A schematic illustration of incompressible flow of a fluid through a close pipe. When the fluid density does not change, the volume flow rate in two regions with a different cross-sectional area A remains the same.

First, we consider the mass flow rate, $\Delta m / \Delta t$, of a fluid with density ρ_1 entering the region with cross-sectional area A_1 at speed v_1 as shown in Fig. 9.18 by writing

$$\frac{\Delta m_1}{\Delta t} = \frac{\rho_1 \Delta V_1}{\Delta t} = \frac{\rho_1 A_1 \Delta \ell_1}{\Delta t} = \rho_1 A_1 v_1. \tag{9.17}$$

Later, the same fluid will flow through the region with cross-sectional area A_2 at speed v_2 since no fluid flows in or out through the sides. This means the mass flow rate in region 1 is the same as that in region 2. So we may write

$$\frac{\Delta m_1}{\Delta t} = \frac{\Delta m_2}{\Delta t} \Rightarrow \rho_1 A_1 v_1 = \rho_2 A_2 v_2. \tag{9.18}$$

Equation (9.18) is called the "continuity equation". Since we are considering incompressible fluid, the density of the fluid will not change from region 1 to region 2. So, we set $\rho_1 = \rho_2$ and simplify the continuity equation of Eq. (9.18) and write it as

$$A_1 v_1 = A_2 v_2. \tag{9.19}$$

SOLIDS AND FLUIDS

Eq. (9.19) states that the volume rate of flow for the fluid in two different regions remain the same. Here, the quantity Av is the volume rate of flow

$$Av = A\frac{\Delta \ell}{\Delta t} = \frac{\Delta V}{\Delta t}. \qquad (9.20)$$

That the **volumetric flow rate**, (aka asrate of fluid flow) measures the volume of fluid that passes through a given cross-sectional area of a pipe per unit time. Hence, according to Eq. (9.19), the fluid flows faster through a narrow channel and flows slower through a wider channel.

Understanding-the-Concept Question 9.6

Water flows through a pipe. The diameter of the pipe at point B is larger than at point A. Where is the speed of the water greater?

- a. Point A
- b. Point B
- c. Same at both A and B
- d. Cannot be determined from the information given.

Explanation:

According to the continuity equation of (9.19), the flow speed is greater in a narrower channel (i.e., in the region where the cross-sectional area is smaller) when water flows through a pipe. This indicates that the water speed is greater at point A.

Answer: a

Exercise Problem 9.6

A 17 cm – radius air duct is used to replenish the air of room 9.2 m × 5.0 m × 4.5 m every 10 minutes. How fast does the air flow in the duct?

Solution:

We need to use the expression for the volume flow rate Av and write

$$Av = \frac{\Delta V}{\Delta t} \Rightarrow v = \frac{1}{A}\frac{\Delta V}{\Delta t}.$$

The speed of the air flow is computed as

$$v = \frac{1}{\pi r^2}\frac{\Delta V}{\Delta t} = \frac{1}{\pi(0.17\text{m})^2}\frac{9.2\text{m}\times 5.0\text{m}\times 4.5\text{m}}{10\text{min}\times(60\text{ s}/\text{min})} = 3.8\frac{\text{m}}{\text{s}}.$$

The volumetric flow rate is important in ventilation. **Ventilating** (the V in HVAC) is the process of "changing" or replacing air in any space to provide high indoor air quality (i.e., to control temperature, replenish oxygen, or remove moisture, odors, smoke, heat, dust, airborne bacteria, and

> **PROBLEM SOLVING STRATEGY**
>
> **What do we know?**
> We are given the dimensions of the room and the air duct. Also, we know the time needed to replenish the air in the room.
>
> **What concepts are needed?**
> We need to apply the continuity principle and the definition of volume flow rate.
>
> **Why?**
> We are asked to compute the speed of air flow through the air duct.

carbon dioxide). Ventilation is used to remove unpleasant smells and excessive moisture, introduce outside air, to keep interior building air circulating, and to prevent stagnation of the interior air. There are two types of measurements for the flow rate. For compliance sampling for industrial hygiene/occupational health, the volumetric air flow rates are measured. It is the volume of air at the existing pressure and temperature at the sampling site. The US Environmental Protection Agency (EPA) also specifies this type of measurement for fine particle ($PM_{2.5}$). Here, $PM_{2.5}$ is particulate matter less than 2.5 microns in diameter composed of very small bits of ash, wood tars, soot and other substances created by combustion. Electronic meters such as orifice plate type flow detector is used to "read out" the volumetric flow rate (or Q_a). EPA also uses Q_s which is known as standard air flow rate for reporting PM_{10}. This means that the flow rate is reported to standard conditions. For the US EPA, these conditions are 25 °C and 1 atmosphere pressure. In other words, this is the mass of air flowing which is universally referred to as mass flow rate in US. However, throughout most of the world Q_s is referred to as mass flow. The standard conditions outside of the U.S. are 0 °C and 1013.25 millibars (i.e., 1 atm).

Fluid flow is also important for understanding a number of interesting phenomena. For example, a fluid flowing past the surface of a body exerts a force on the surface. **Lift** force is the component of this force that is perpendicular to the oncoming flow direction. It contrasts with the drag force, which is the component of the surface force parallel to the flow direction. If the fluid is air, the force is called an aerodynamic force.

9.8 BERNOULLI'S PRINCIPLE

Many interesting phenomena associated with fluid flow may also be understood in terms of energy conservation. This energy conservation as described by the Bernoulli's principle provides the relationship between the velocity and pressure exerted by a moving fluid: with increasing velocity of a fluid, the pressure exerted by that fluid decreases. **Bernoulli's principle** states that for the steady laminar flow through an enclosed tube or pipe, a place where the velocity of a fluid is high the pressure is low, and a place where the velocity is low, the pressure is high. Airplanes get a part of their lift by taking advantage of Bernoulli's principle. Race cars employ Bernoulli's principle to keep their rear wheels on the ground while traveling at high speeds.

SOLIDS AND FLUIDS

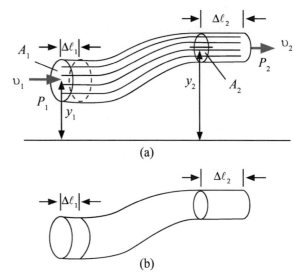

Figure 9.20: Bernoulli's principle states that energy of the moving fluid at two different points is the same. This principle allows us to understand how wings produce lift for airplanes and helicopters to fly.

Let us see how the energy conservation equation for describing the Bernoulli's principle is derived. Suppose we consider a fluid flowing through an enclosed pipe as shown in Fig. 9.20. The work done on the fluid at point 1 is given by

$$W_1 = F_1 \Delta \ell_1 = P_1 A_1 \Delta \ell_1. \qquad (9.21)$$

Similarly, the work done on the fluid at point 2 is given by

$$W_2 = P_2 A_2 \Delta \ell_2. \qquad (9.22)$$

Here, the minus sign denotes that the force exerted on the fluid is opposite to the motion. Also, since the fluid changes its elevation as it flows through the pipe, the work done by gravity has to be accounted for. The work done by gravity is given by

$$W_3 = -mg(y_2 - y_1). \qquad (9.23)$$

The net work W done on the fluid is computed by adding above three work contributions

$$W = W_1 + W_2 + W_3 = P_1 A_1 \Delta \ell_1 - P_2 A_2 \Delta \ell_2 - mg(y_2 - y_1). \qquad (9.24)$$

We may now relate this work done on the fluid to the change in its kinetic energy by using work-energy principle. Using the work-energy principle (i.e., the net work done on a system is equal to its change in kinetic energy), we may write Eq. (9.24) as

$$\frac{1}{2}mv_2^2 - \frac{1}{2}mv_1^2 = P_1 A_1 \Delta \ell_1 - P_2 A_2 \Delta \ell_2 - mg(y_2 - y_1) \qquad (9.25)$$

We may rewrite Eq. (9.25) into a more convenient form by dividing each term by the volume of the fluid and obtain

$$\frac{1}{2}\frac{m}{A_2 \Delta \ell_2}v_2^2 - \frac{1}{2}\frac{m}{A_1 \Delta \ell_1}v_1^2 = P_1 - P_2 - \frac{m}{A_2 \Delta \ell_2}gy_2 + \frac{m}{A_1 \Delta \ell_1}gy_1. \qquad (9.26)$$

Since the volume of the fluid in two regions remains the same, $A_1 \Delta \ell_1 = A_2 \Delta \ell_2$, we obtain

$$\frac{1}{2}\rho v_2^2 - \frac{1}{2}\rho v_1^2 = P_1 - P_2 - \rho g y_2 + \rho g y_1. \quad (9.27)$$

We now make further simplification by collecting the terms with the identical indices on the same side of equation and obtain

$$P_1 + \frac{1}{2}\rho v_1^2 + \rho g y_1 = P_2 + \frac{1}{2}\rho v_2^2 + \rho g y_2. \quad (9.28)$$

Equation (9.28) indicates that the energy density of the fluid which is written as

$$P + \frac{1}{2}\rho v^2 + \rho g y = constant. \quad (9.29)$$

remains the same at every point in the fluid. Equation (9.28) is a form of the law of energy conservation. This conservation of energy, which is known as Bernoulli's principle, is useful for solving problems. For a non-viscous, incompressible fluid in steady flow, the sum of pressure, potential and kinetic energies per unit volume is constant at any point, indicating that increase in kinetic energy of a fluid is offset by a reduction of the "static energy" associated with pressure. The fluid is assumed incompressible and inviscid (i.e., the fluid does not generate drag).

Exercise Problem 9.7

Water at a pressure of 3.8 atm at a street level flows into an office building at a speed of 0.60 m/s through a pipe of 5.0 cm in diameter. The pipe tapers down to 2.6 cm in diameter by the top floor, 20 m above. Calculate the flow velocity and the pressure in such a pipe on the top floor. Ignore viscosity. Pressures are gauge pressures.

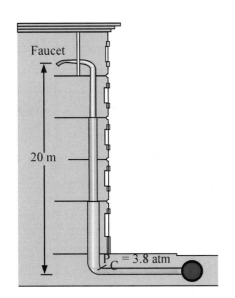

PROBLEM SOLVING STRATEGY

What do we know?

We know the water gauge pressure at the street level and the dimensions of the pipe at both the street level and at the top floor.

What concepts are needed?

We need to apply both the continuity principle and Bernoulli principle.

Why?

To compute the water pressure at the top floor using Bernoulli principle, we need to calculate the water speed by using the continuity equation.

SOLIDS AND FLUIDS

Solution:

First, we use the continuity equation to determine the speed v_2 at point 2:

$$v_2 = v_1 \frac{A_1}{A_2} = v_1 \frac{\pi(d_1/2)^2}{\pi(d_2/2)^2} = v_1 \frac{d_1^2}{d_2^2} = 0.60 \frac{m}{s} \times \left(\frac{0.050m}{0.026m}\right)^2 = 2.22 \frac{m}{s}. \qquad (7a)$$

We now use Bernoulli's equation to determine the pressure P_2 at point 2

$$P_1 + \frac{1}{2}\rho v_1^2 + \rho g y_1 = P_2 + \frac{1}{2}\rho v_2^2 + \rho g y_2. \qquad (7.b)$$

With some algebraic calculation, we solve for the pressure difference $P_1 - P_2$ to obtain

$$P_1 - P_2 = \rho\left[\frac{1}{2}(v_2^2 - v_1^2) + g(y_2 - y_1)\right]$$

$$= 1000 \frac{kg}{m^3}\left[\frac{1}{2}\left[\left(2.22\frac{m}{s}\right)^2 - \left(0.60\frac{m}{s}\right)^2\right] + 9.8\frac{m}{s^2} \times 20m\right]$$

$$= 1.983 \times 10^5 \frac{kg}{ms^2} \cdot \frac{1 atm}{1.013 \times 10^5 \, N/m^2} = 1.957 \, atm$$

The gauge pressure at the top floor is

$$P_2 = P_1 - 1.957 \, atm = 3.8 \, atm - 1.957 \, atm = 1.84 \, atm \text{ (gauge pressure)}.$$

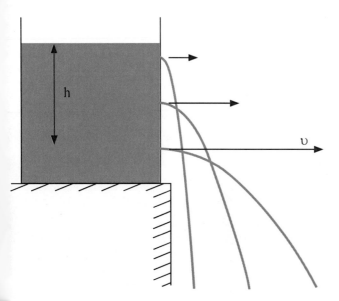

Figures 9.21: The speed of a liquid flowing out of an open tank under the force of gravity, according to Torricelli's theorem is directly proportional to the vertical distance between the liquid surface and the center of the opening.

Let us examine a few special cases of Bernoulli's equation to see how we may apply this principle in solving problems. One special application of the Bernoulli's equation is the derivation of Torricelli's law. **Torricelli's law** is a theorem in fluid dynamics relating the speed of fluid flowing out of an opening to the height of fluid above the opening as shown in Fig. 9.21. Torricelli's law states that the speed of efflux, v, of a fluid through a sharp-edged hole at the bottom of a tank filled to a depth h is the same as the speed that a body (in this case a drop of water) would acquire in falling freely from a height h.

We start the derivation by noting that both at the top and at the small opening where the fluid comes out is at equal atmospheric pressure (i.e., $P_1 = P_2 = P_A$). This assumption allows us to simplify the Bernoulli's equation of Eq. (9.28) and obtain

$$\frac{1}{2}\rho v_1^2 + \rho g y_1 = \frac{1}{2}\rho v_2^2 + \rho g y_2. \tag{9.30}$$

We now solve for the speed of the fluid at the small opening which is at the depth h from the top and obtain

$$v_1^2 = 2g(y_2 - y_1) \Rightarrow v_1 = \sqrt{2g(y_2 - y_1)}. \tag{9.31}$$

Equation (9.31) states that v_1 is the same speed as that of a freely falling object after fall the same height. This law in a different form was discovered by Torricelli in 1643. As we can see, the falling fluid is a particular example of Bernoulli's principle. Another interesting example of Bernoulli's principle is the lift force exerted on the wings of an airplane as air flows around it.

One popular explanation of the lift force acting on an aircraft involves Bernoulli's equation. As the aircraft moves forward into a stream of air, the wing deflects the air. Some of the air moves to flow above the wing while some of the air moves to flow below the wing. As shown in Fig. 9.22 the wing is curved to help the air that flows above the wing move more quickly than the air that was able to flow below the non-curved bottom of the wing. We may assume that there is no appreciable change in height due to small thickness of the wing. So, we set $y_1 = y_2$ in Eq. (9.28) and obtain

$$P_1 + \frac{1}{2}\rho v_1^2 = P_2 + \frac{1}{2}\rho v_2^2 \Rightarrow P_u + \frac{1}{2}\rho v_b^2 = P_d + \frac{1}{2}\rho v_t^2. \tag{9.32}$$

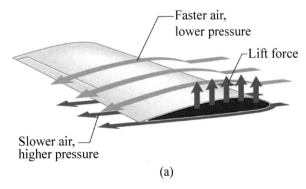

Figure 9.22: Lift is commonly associated with the wing of a fixed-wing aircraft, even though it is generated by propellers, helicopter rotors, and rudders. a) The lift is generated as the air is turned both above and below the wing so both the upper and lower surface contribute to the flow turning. b) A fixed-wing aircraft with a large surface area for the wing tends to experience a stronger lift force.

SOLIDS AND FLUIDS

A net upward force (dynamic lift) is determined as the difference in the upper (P_u) and lower (P_d) pressure times the area A of the wing

$$F_{net} = (P_u - P_d)A = \frac{1}{2}\rho(v_t^2 - v_b^2)A. \tag{9.33}$$

Equation (9.33) indicates the lift force is due to the air which is moving more quickly above the curved wing starts to put less pressure on the wing while it adjusts to its new stream. Meanwhile, the air that is moving at a consistent speed below the wing maintains its rate of pressure. This quick differential produces lift. The higher air pressure can push the wing upward into the space where the air pressure is lower. This difference in air pressures is a result of Bernoulli's principle which states that as the speed of a moving fluid (like air) increases, the pressure within that fluid decreases. Note that the above explanation is based on the assumption of "equal transit times." This maybe over simplified. The true theory of the lift force is on an aircraft wing is much more complicated.

9.9 REAL FLUIDS

There is one important difference between ideal fluids that we considered above and real fluids. Real fluids have viscosity. **Viscosity** is a frictional force between adjacent layers of fluid as the layers move past one another. In liquids, viscosity is mainly due to the cohesive forces between the molecules. However, in gases, viscosity arises from collisions between the molecules.

what is the effect of viscosity?

Because of viscosity, in order to have steady flow of any real fluids, a pressure difference between the ends of the tube is necessary. This means that for a real fluid such as blood will not flow unless there is a pressure difference between any two points. Also, the direction of flow is determined by the value of pressure since the fluid will flow from a region of high pressure to a region of low pressure.

Flow in tubes: Poiseuille's equation

Real fluids have some resistance between their molecules, particles or layers. In other words, the real fluids have some viscosity. A pressure difference, as shown in Fig. 9.23, is needed to overcome the resistance due to viscosity and induce flow. In the case of smooth flow (laminar flow), the volume flow rate is given by the pressure difference divided by the viscous resistance. This resistance depends linearly upon the viscosity and the length, but it depends on the

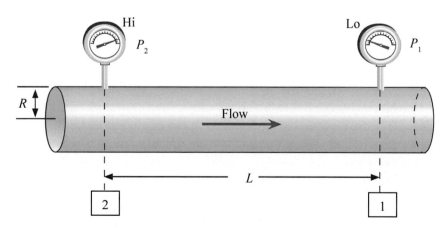

Figure 9.23: In the case of smooth flow of real fluid, a pressure differential is needed between two points because the viscous resistance of the fluid depends on the viscosity, the length, and the radius of the pipe.

fourth power of the radius (which is quite dramatic). Poiseuille's law is found to be in reasonable agreement with experimental data for uniform liquids (called Newtonian fluids) in the cases where there is no appreciable turbulence. Poiseuille's equation, which describes the volume rate of flow Q for a fluid, can be used to describe the flow of real fluids such as water and blood.

A more precise form of Poiseuille's equation, applicable to laminar flow of a fluid in a round tube, is

$$Q = \frac{\Delta V}{\Delta t} = \frac{\pi R^4 (P_2 - P_1)}{8\eta L}. \tag{9.34}$$

where R is the inside radius of the tube, L is the length of the tube, $P_2 - P_1$ is the pressure difference between the ends, and η is the coefficient of viscosity. Equation (9.34) may be useful for understanding some treatments for high blood pressure. **High blood pressure**, sometimes called **arterial hypertension**, is a chronic medical condition in which the blood pressure in the arteries is elevated. This requires the heart to work harder than normal to circulate blood through the blood vessels. Blood pressure is summarized by two measurements, systolic and diastolic. These measurements depend on whether the heart muscle is contracting (systole) or relaxed between beats (diastole). Normal blood pressure at rest is within the range of 100 to140 mm-Hg systolic (top reading) and 60 to 90 mm-Hg diastolic (bottom reading). High blood pressure is said to be present if it is persistently at or above 140/90 mm-Hg. This hypertension is a major risk factor for stroke, myocardial infarction (heart attacks), heart failure, aneurysms of the arteries (e.g., aortic aneurysm), peripheral arterial disease and is a cause of chronic kidney disease. According to Eq. (9.34), there are two main ways to reduce $\Delta P = P_2 - P_1$ while maintaining the volume flow rate Q. One way is to increase the inner radius R of the arteries. Other way is to decrease the viscosity η of the blood. Reducing the high blood pressure condition by using these two ways may be achieved by using proper medications.

Blood viscosity is a measure of the resistance of blood to flow, which is being deformed by either shear or extensional strain. Viscous blood can cause painful leg cramps or leg pain caused by poor circulation. This condition is called "intermittent claudication". Blood is a liquid that consists of plasma and particles, such as the red blood cells. The viscosity of blood thus depends on the viscosity of the plasma, in combination with the hematocrit. Plasma can be considered a Newtonian fluid, but blood cannot because the red blood cells add non-idealities to the fluid. The fluid must reach a shear rate of about 100 s^{-1} to be assumed Newtonian. Until this point, the term viscosity cannot be applied to blood. The viscosity of blood after reaching this shear rate is about five times as great as the viscosity of water. When the hematocrit rises to either 60 or 70, the blood viscosity can become as great as 10 times that of water. It is noted that hematocrit rises to either 60 or 70 in polycythemia. At this level of hematocrit, its flow through blood vessels is greatly retarded because of increased resistance to flow. This will lead to decreased oxygen delivery. The viscosity of blood at 37 °C is normally 3×10^{-3} to 4×10^{-3} pascal-seconds (Pa·s). The analogous unit in the centimeter gram second system of units is the poise. The viscosity of blood at 20 °C is normally 10 centipoise (cP). Plasma's viscosity is determined by water-content and macromolecular components, so these factors that affect blood viscosity are the plasma protein concentration and types of proteins in the plasma. However, these effects are not significant because they are much less than the effect of hematocrit. The elevation of plasma viscosity is correlated to the progression of coronary and peripheral vascular diseases. Also, anemia can lead to decrease blood viscosity, which may lead to heart failure. Other factors influencing blood viscosity include temperature, where an increase in temperature results in a decrease in viscosity. This is particularly important in hypothermia, where an increase in blood viscosity will cause problems with blood circulation.

SOLIDS AND FLUIDS

Exercise Problem 9.8

What diameter must a 21 m long air duct have if the pressure of a ventilation and heating system is to replenish the air in a room 9.0 m x 14 m x 4.0 m every 10 min.? Assume the pump can exert a gauge pressure of 0.71×10⁻³ atm. (**Note**: The viscosity of air is $\eta_{air} = 0.018 \times 10^{-3}$ Pa·s.)

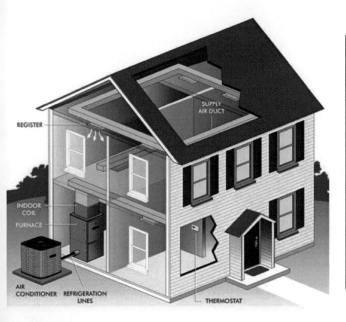

PROBLEM SOLVING STRATEGY

What do we know?

The dimensions of the room and the pressure differential are given in the problem.

What concepts are needed?

Poiseuille's equation to determine the volume flow rate is needed.

Why?

Air in an air duct is a real fluid and requires pressure differential to initiate flow.

Solution:

Since air should be considered as a real gas, we use Poiseuille's equation for the volume flow rate

$$Q = \frac{\Delta V}{\Delta t} = \frac{\pi r^4 (P_1 - P_2)}{8 \eta L}.$$

With some algebraic calculation, we can obtain the radius r as

$$r = \left(\frac{8 \eta L}{\pi (P_1 - P_2)} \frac{\Delta V}{\Delta t} \right)^{\frac{1}{4}}$$

$$= \left(\frac{8 \times 0.018 \times 10^{-3} \text{Pa} \cdot \text{s} \times 21\text{m}}{\pi \times 0.71 \times 10^{-3} \text{atm} \times (1.013 \times 10^5 \text{Pa}/1\text{atm})} \frac{9.0\text{m} \times 14\text{m} \times 4.0\text{m}}{10 \text{min} \times (60 \text{s}/1 \text{min})} \right)^{\frac{1}{4}}$$

The diameter of the air duct is

$$\cong 0.06 \text{m} \Rightarrow d = 2r = 0.12 \text{m}.$$

The biggest surprise in the application of Poiseuille's law to fluid flow is the dramatic effect of changing the radius. Suppose we have an emergency requirement for a five-fold increase in blood volume flow rate (like being chased by a big dog)? How does our body supply it? According to Poiseuille's law, a five-fold increase in blood pressure would be required if the increase were supplied by blood pressure alone! In other words, the blood pressure should increase from 120 mm-Hg to 600 mm-Hg. Of course, this is not a realizable pressure. However, our body has a much more potent method for increasing volume flow rate in the vasodilation of the small vessels called arterioles. The blood vessels may dilate and increase the radius by 50% and go from r to $1.5r$. A 50% dilation of resistance vessels means the volume flow rate will increase by a factor of 5.06. In addition since the smaller vessels provide most of the resistance to flow, the arterioles in their position just prior to the capillaries can provide a major controlling influence on the volume flow rate. This system of small vessels can constrict flow to one part of the body while enhancing the flow to another to meet changing demands for oxygen and nutrients.

SUMMARY

i. The amount of **elongation of an object depends on the applied force** exerted on the material, and on its dimension. The change in length ΔL due to the applied force F is

$$\Delta L = \frac{1}{E}\frac{F}{A}L_o$$

where L_o is the original length of the object, A is the cross-sectional area where the force is applied, and E is the elastic modulus (or Young's modulus).

ii. **Stress** is defined as force per unit area (i.e., F/A with units of N/m²). **Strain** is defined as the ratio of change in length to the original length $\Delta L/L_o$. There are following three types of stresses: (i) tensile stress, (ii) compressive stress, and (iii) shear stress. **Tensile stress** (or tension) is the stress state leading to expansion due to a force pulling down. **Compressive stress** is the stress on materials due to a force of compression leads to a smaller volume. **Shear stress** is the component of stress coplanar with a material cross section. The **shear strain** $\Delta L/L_o$ is computed from $\Delta L/L_o = (F/A)/G$ where G is the shear modulus. The maximum amount of stress that a material can absorbed without fracturing is tensile strength, compressive strength, and shear strength.

iii. If an object is subject to inward forces from all sides, its volume will decrease. The fractional amount of change in the volume may be described by

$$\frac{\Delta V}{V_o} = -\frac{1}{B}\Delta P$$

where ΔP is the change in pressure, B is the bulk modulus, ΔV is the change in volume, and V_o is the original volume.

iv. **Absolute pressure** P is zero-referenced against a perfect vacuum, so it is equal to gauge pressure P_G plus atmospheric pressure P_A,

$$P = P_A + P_G$$

Gauge pressure P_G is zero-referenced against ambient air pressure, so it is equal to absolute pressure minus atmospheric pressure. Hence, gauge pressure is the pressure over and above atmospheric pressure.

v. There are four governing principles for fluid. These governing principles are (i) Pascal's principle, (ii) Archimedes' principle, (iii) continuity principle, and (iv) Bernoulli's principle. **Pascal's law or the principle of transmission of fluid-pressure** is a principle in fluid mechanics that states that

SOLIDS AND FLUIDS

pressure exerted anywhere in a confined incompressible fluid is transmitted equally in all directions throughout the fluid such that the pressure ratio (initial difference) remains the same. **Archimedes' principle** (or **Archimedes' principle**) is a law of physics stating that the upward buoyant force F_B exerted on a body immersed in a fluid is equal to the weight of the fluid ($m_F g$) the body displaces

$$F_B = \rho_F g V = m_F g.$$

The **continuity equation** ensures that what goes into one end (region 1) of a pipe must come out at the other end (region 2). The continuity equation is written as

$$\rho_1 A_1 v_1 = \rho_2 A_2 v_2.$$

Bernoulli's principle states that for the steady laminar flow of a fluid through an enclosed tube or pipe, a place where the velocity of a fluid is high the pressure is low, and a place where the velocity is low, the pressure is high. This is the statement of energy conservation in fluids and is written as

$$P_1 + \frac{1}{2}\rho v_1^2 + \rho g y_1 = P_2 + \frac{1}{2}\rho v_2^2 + \rho g y_2.$$

vi. The real fluids have some resistance (i.e., viscosity) in between their molecules, particles or layers. **Poiseuille's equation** is applicable for a laminar flow of fluid in a round tube

$$Q = \frac{\Delta V}{\Delta t} = \frac{\pi R^4 (P_2 - P_1)}{8 \eta L}.$$

where R is the inside radius of the tube, L is the length of the tube, $P_2 - P_1$ is the pressure difference between the ends, and η is the coefficient of viscosity.

MORE WORKED PROBLEMS

Problem 9.1

A jet airplane in level flight has mass of 8.66×10^4 kg, and the two wings have an estimated total area of 90.0 m². (a) What is the pressure difference between the lower and upper surfaces of the wings? (b) If the speed of air under the wings is 225 m/s, what is the speed of the air over the wings? Assume air has a density of 1.29 kg/m³. (c) Explain why all aircraft have a "ceiling," a maximum operational altitude.

Solution:

a. Assuming the airplane is in level flight, the net lift (the difference in the upward and downward force exerted on the wings by the air flowing over them) must equal the weight of the plane

$$\left(P_{lower} - P_{upper}\right) A = mg.$$

We solve for the pressure difference and obtain

$$P_{lower} - P_{upper} = \frac{mg}{A} = \frac{8.66 \times 10^4 \text{ kg} \times 9.80 \text{ m/s}^2}{90.0 \text{ m}^2} = 9.43 \times 10^3 \text{ Pa}.$$

b. Neglecting the small difference in altitude between the upper and lower surface of the wings, and applying Bernoulli's equation, yields

$$P_{lower} + \frac{1}{2}\rho_{air}v_{lower}^2 = P_{upper} + \frac{1}{2}\rho_{air}v_{upper}^2.$$

We solve for the speed on the upper surface and obtain

$$v_{upper} = \sqrt{v_{lower}^2 + \frac{2(P_{lower} - P_{upper})}{\rho_{air}}} = \sqrt{\left(225\frac{m}{s}\right)^2 + \frac{2(9.43\times 10^3\,\text{Pa})}{1.29\,\text{kg}/\text{m}^3}} = 255\frac{m}{s}.$$

c. The density of air decreases with increasing height, resulting in a smaller pressure difference

$$\Delta P = \frac{1}{2}\rho_{air}\left(v_{upper}^2 - v_{lower}^2\right).$$

Beyond the maximum operational amplitude, the pressure difference can no longer support the aircraft.

Problem 9.2

A hospital patient receives a quick 500-cc blood transfusion through a needle with a length of 5.0 cm and an inner diameter of 1.0 mm. If the blood bag is suspended 0.85 m above the needle, how long does the transfusion take? Neglect the viscosity of the blood flowing in the plastic tube between the bag and the needle. (The viscosity and density of blood is $\eta = 1.7\times 10^{-3}$ Pa·s and 1.05×10^3 kg/m³, respectively.)

Solution:

The pressure difference between the bag and needle, obtained from $P = P_o + \rho g y$, is

$$\Delta P = \rho g h = 1.05\times 10^3\,\frac{\text{kg}}{\text{m}^3}\times 9.80\,\frac{\text{m}}{\text{s}^2}\times 0.85\,\text{m} = 8.75\times 10^3\,\text{Pa}.$$

According to Poiseuille's equation, the volume rate of flow for a real fluid is given by

$$Q = \frac{\pi r^4(\Delta P)}{8\eta L} = \frac{\pi(0.5\times 10^{-3}\,\text{m})^4 \times 8.75\times 10^3\,\text{Pa}}{8\times 1.7\times 10^{-3}\,\text{Pa·s}\times 0.05\,\text{m}} = 2.53\times 10^{-6}\,\frac{\text{m}^3}{\text{s}}.$$

We use the definition of the volume rate of fluid $Q = V/t$ (i.e., volume per unit time) and obtain

$$t = \frac{V}{Q} = \frac{500\,\text{cm}^3 \times 10^{-6}\,\text{m}^3/\text{cm}^3}{2.53\times 10^{-6}\,\text{m}^3/\text{s}} = 2.0\times 10^2\,\text{s}.$$

SOLIDS AND FLUIDS

Problem 9.3

A solid brass sphere is subjected to a pressure of 1.0×10^5 Pa due to the Earth's atmosphere. On Venus the pressure due to the atmosphere is 9.0×10^6 Pa. By what fraction $\Delta r/r_o$ (including the algebraic sign) does the radius of the sphere change when it is exposed to the Venusian atmosphere? Assume that the change in radius is very small relative to the initial radius. (The bulk modulus for brass is $B = 6.7 \times 10^{10}$ N/m².)

Solution:

The fractional change in volume is given by

$$\frac{\Delta V}{V_o} = -\frac{\Delta P}{B} = -\frac{8.9 \times 10^6 \text{ Pa}}{6.7 \times 10^{10} \text{ Pa}} = -1.33 \times 10^{-4},$$

where the B is the bulk modulus. The B for brass is 6.7×10^{10} N/m². The initial volume of the sphere is $V_o = (4/3)\pi r^3$. If we assume that the change in the radius of the sphere is very small relative to the initial radius, we can think of the sphere's change in volume as the addition or subtraction of a spherical shell of volume ΔV, whose radius is r and whose thickness is Δr. The change in volume of the sphere is equal to the volume of the shell

$$\Delta V = \left(4\pi r^2\right) \Delta r$$

From this, we may compute the fractional change in radius as

$$\frac{\Delta V}{V_o} = \frac{4\pi r^2 \Delta r}{\frac{4}{3}\pi r^3} = \frac{3\Delta r}{r} \Rightarrow \frac{\Delta r}{r} = \frac{1}{3}\frac{\Delta V}{V_o} = \frac{1}{3} \times \left(-1.33 \times 10^{-4}\right) = -4.4 \times 10^{-5}.$$

Problem 9.4

A 1.0×10^{-3} kg spider is hanging vertically by a thread that has an elastic modulus of 4.5×10^9 N/m² and radius of 13×10^{-6} m. Suppose that a 95 kg person is hanging vertically on the aluminum wire. What is the radius of the wire that would exhibit the same strain as the spider's thread, when the thread is stressed by the full weight of the spider?

Solution:

We need to use the definition of strain and write

$$\text{strain} = \frac{\Delta L}{L_o} = \frac{1}{E}\frac{F}{A},$$

where $F = mg$, $A = \pi r^2$ is the cross-sectional area where the force F is applied, and E is the elastic modulus. We now set the strain for the spider web equal to the strain for the wire as

$$\frac{1}{E}\frac{F}{A} = \frac{1}{E'}\frac{F'}{A'} \Rightarrow \frac{1}{E}\frac{F}{\pi r^2} = \frac{1}{E'}\frac{F'}{\pi r'^2}.$$

By taking the value for the elastic modulus E' for aluminum as 6.9×10^{10} N/m², we find

$$r = r\sqrt{\frac{F'}{F}\frac{E}{E'}} = 13 \times 10^{-6}\,\text{m}\sqrt{\frac{95\,\text{kg} \times 9.80\,\text{m}/\text{s}^2 \times 4.5 \times 10^9\,\text{N}/\text{m}^2}{1.0 \times 10^{-3}\,\text{kg} \times 9.80\,\text{m}/\text{s}^2 \times 6.9 \times 10^{10}\,\text{N}/\text{m}^2}} = 1.0 \times 10^{-3}\,\text{m}.$$

PROBLEMS

1. What hanging mass will stretch a 2.0-m-long, 0.50-mm-diameter steel wire by 1.0 mm?

2. A 2000 N force stretches a wire by 1.0 mm. (a) A second wire of the same material is twice as long and has twice the diameter. How much force is needed to stretch it by 1.0 mm? Explain. (b) A third wire of the same material is twice as long as the first and has the same diameter. How far is it stretched by a 4000 N force?

3. A spring which is not stretched has a length of 10 cm. It exerts a restoring force F when stretched to a length of 11 cm. (a) For what total stretched length of the spring is its restoring force $3F$? (b) At what compressed length is the restoring force $2F$?

4. A walkway suspended across a hotel lobby is supported at numerous points along its edges by a vertical cable above each point and a vertical column underneath. The steel cable is 1.27 cm in diameter and is 5.75 m long before loading. The aluminum column is a hollow cylinder with an inside diameter of 16.14 cm, an outside diameter of 16.24 cm, and unloaded length of 3.25 m. When the walkway exerts a load force of 8500 N on one of the support points, how much does the point move down?

5. The drawing shows a hydraulic chamber with a spring (spring constant = 1600 N/m) attached to the input piston and a rock of mass 40.0 kg resting on the output plunger. The piston and plunger are nearly at the same height, and each has a negligible mass. By how much is the spring compressed from its unconstrained position?

6. A copper wire has a length of 5.0 m and a diameter of 3.0 mm. Under what load will its length increase by 0.3 mm?

7. A cylinder (with circular ends) and a hemisphere are solid throughout and made from the same material. They are resting on the ground, the cylinder on one of its ends and the hemisphere on its flat side. The weight of each causes the same pressure to act on the ground. The cylinder is 0.500 m high. What is the radius of the hemisphere?

8. A 1.0×10^{-3} kg spider is hanging vertically by a thread that has a Young's modulus of 4.5×10^9 N/m² and a radius of 13×10^{-6} m. Suppose that a 95 kg person is hanging vertically on an aluminum wire. What is the radius of the wire that would exhibit the same strain as the spider's thread, when the thread is stressed by the full weight of the spider?

9. Water flows at 0.25 L/s through a 10-m-long garden hose 2.5 cm in diameter that is lying flat on the ground. The temperature of the water is 20 °C. What is the gauge pressure of the water where it enters the hose?

10. Piston 1 has a diameter of 0.25 inch; piston 2 has a diameter of 1.5 inch In the absence of friction, determine the force F necessary to support the 500-lb weight.

11. A solid brass sphere is subjected to a pressure of 1.0×10^5 Pa due to the earth's atmosphere. On Venus the pressure due to the atmosphere is 9.0×10^6 Pa. By what fraction Dr/r_o (including the algebraic

sign) does the radius of the sphere change when it is exposed to the Venusian atmosphere? Assume that the change in radius is very small relative to the initial radius.

12. A brass cube 6.0 cm on each side is placed in a pressure chamber and subjected to a pressure of 1.2×10^7 N/m² on all of its surfaces. By how much will each side be compressed under this pressure?

13. A man of mass $m = 70.0$ kg and having a density of $\varrho = 1,050$ kg/m³ (while holding his breath) is completely submerged in water. (a) Write Newton's second law for this situation in terms of the man's mass m, the density of water r_w, his volume V, and g. Neglect any viscous drag of the water. (b) Substitute $m = rV$ into Newton's second law and solve the acceleration a, canceling common factors. (c) Calculate the numeric value of the man's acceleration. (d) How long does it take the man to sink 8.00 m to the bottom of the lake?

14. As the drawing illustrates, a pond has the shape of an inverted cone with the tip sliced off and has a depth of 5.00 m. The atmospheric pressure above the pond is 1.01×10^5 Pa. The circular surface (radius = R_2) and circular bottom surface (radius = R_1) of the pond are both parallel to the ground. The magnitude of the force acting on the top surface is the same as the magnitude of the force acting on the bottom surface. Obtain (a) R_2 and (b) R_1.

15. Ethyl alcohol has been added to 200 mL of water in a container that has a mass of 150 g when empty. The resulting container and liquid mixture has a mass of 512 g. What volume of alcohol was added to the water?

16. An antifreeze solution is made by mixing ethylene glycol ($\varrho = 1116$ kg/m³) with water. Suppose that the specific gravity of such solution is 1.0730. Assuming that the total volume of the solution is the sum of its parts, determine the volume percentage of ethylene glycol in the solution.

17. A 1.0-m-diameter vat of liquid is 2.0 m deep. The pressure at the bottom of the vat is 1.3 atm. What is the mass of the liquid in the vat?

18. The pressure increases by 1.0×10^4 N/m² for every meter of depth beneath the surface of the ocean. At what depth does the volume of a Pyrex glass tube, 1.0×10^{-2} m on an edge at the ocean's surface, decreases by 1.0×10^{-10} m³?

19. The British gold sovereign coin is an alloy of gold and copper having a total mass of 7.988 g, and is 22-karat gold. (a) Find the mass of gold in the sovereign in kilograms using the fact that the number of karats = 24 × (mass of gold)/(total mass). (b) Calculate the volumes of gold and copper, respectively, used to manufacture the coin. (c) Calculate the density of the British sovereign coin.

20. In a sample of seawater taken from an oil spill, an oil layer 4.0 cm thick floats on 55 cm of water. If the density of the oil is 0.75×10^3 kg/m³, what is the absolute pressure on the bottom of the container?

21. Water is forced out of a fire extinguisher by air pressure. What gauge air pressure in the tank (above atmospheric pressure) is required or the water to have a jet speed of 30.0 m/s when the water level in the tank is 0.500 m below the nozzle?

22. A container is filled to a depth of 20.0 cm with water. One top of the water floats a 30.0-cm-thick layer of oil with specific gravity 0.700. What is the absolute pressure at the bottom of the container?

23. A hydraulic balance used to detect small change in mass. If a mass m of 0.25 g is placed on the balance platform, by how much will the height of the water in the smaller, 1.0 cm-diameter cylinder have changed when the balance comes to equilibrium?

24. A cargo barge is loaded in a saltwater harbor for a trip up a freshwater river. If the rectangular barge is 3.0 m by 20.0 m and sits 0.80 m deep in the harbor, how deep will it sit in the river?

25. A U-shaped tube, open to the air on both ends, contains mercury. Water is poured into the left arm until the water column is 10.0 cm deep. How far upward from its initial position does the mercury in the right arm rise?

26. A research submarine has a 20-cm-diameter window 8.0 cm thick. The manufacturer says the window can withstand forces up to 1.0×10^6 N. What is the submarine's maximum safe depth? The pressure inside the submarine is maintained at 1.0 atm.

27. Suppose a distant world with surface gravity of 7.44 m/s² has an atmospheric pressure of 8.04×10^4 Pa at the surface. (a) What force is exerted by the atmosphere on a disk-shaped region 2.00 m in radius at the surface of a methane ocean? (b) What is the weight of a 10.0-m deep cylindrical column of methane with radius 2.00 m? (c) Calculate the pressure at a depth of 10.0 m in the methane ocean. *Note*: The density of liquid methane is 415 kg/m³.

28. Oil having a density of 930 kg/m³ floats on water. A rectangular block of wood 4.00 cm high and with a density of 960 kg/m³ floats partly in the oil and partly in the water. The oil completely covers the block. How far below the interface between the two liquids is the bottom of the block? (F11, 9.90)

29. A spherical weather balloon is filled with hydrogen until its radius is 3.00 m. Its total mass including the instruments it carries is 15.0 kg. (a) Find the buoyant force acting on the balloon, assuming the density of air is 1.29 kg/m³. (b) What is the net force acting on the balloon and its instruments after the balloon is released from the ground? (c) Why does the radius of the balloon tend to increase as it rises to higher altitude?

30. Spherical particles of a protein density 1.8 g/cm³ are shaken up in a solution of 20 °C water. The solution is allowed to stand for 1.0 h. If the depth of water in the tube is 5.0 cm, find the radius of the largest particles that remain in solution at the end of the hour.

31. The bottom of a steel "boat" is a 5.0 m × 10 m × 2.0 cm piece of steel ($r_{steel} = 7900$ kg/m³). The sides are made of 0.50-cm-thick steel. What minimum height must the sides have for this boat to float in perfectly calm water?

32. A flat-bottomed rectangular boat is 4.0 m long and 1.5 m wide. If the load is 2000 kg (including the mass of the boat), how much of the boat will be submerged when it floats in a lake?

33. A block of wood made of oak is held under the water's surface in a swimming pool. At the instant the block is released, what is its acceleration?

34. A rectangular trough, 2.0 m long, 0.60 wide, and 0.45 m deep, is completely full of water. One end of the trough has a small drain plug right at the bottom edge. When you pull the plug, at what speed does water emerge from the hole?

35. A room measures 3.0 m by 4.5 m by 6.0 m. If the heating and air-conditioning ducts to and from the room are circular with diameter 0.30 m and all the air in the room is to be exchanged every 12 min, (a) what is the average flow rate? (b) What is the necessary flow speed in the duct? (Assume that the density of the air is constant.)

36. Water flowing through a 2.0-cm-diameter pipe can fill a 300 L bathtub in 5.0 min. What is the speed of the water in the pipe?

37. Water flows at a rate of 25.1 L/min through a horizontal 7.0-cm-diameter pipe under a pressure of 6.0 Pa. At one point, calcium deposits reduce the cross-sectional area of the pipe to 30 cm². What is the pressure at this point? (Consider the water to be an ideal fluid.)

38. The aorta carries blood away from the heart at a speed of about 40 cm/s and has a radius of approximately 1.1 cm. The aorta branches eventually into a large number of tiny capillaries that distribute the blood to the various body organs. In a capillary, the blood speed is approximately

0.07 cm/s, and the radius is about 6×10^{-4} cm. Treat blood as an incompressible fluid, and use these data to determine the approximate number of capillaries in the human body.

39. A Venturi meter is a device that is used for measuring the speed of a fluid within a pipe. The drawing shows a gas flowing a speed of u_2 through a horizontal section of pipe whose cross-sectional area is $A_2 = 0.0700$ m². The gas has a density of $\rho = 1.30$ kg/m³. The Venturi meter has a cross-sectional area of $A_1 = 0.0500$ m² and has been substituted for a section of the larger pipe. The pressure difference between the two sections is $P_2 - P_1 = 120$ Pa. Find (a) the speed u_2 of the gas in the larger, original pipe and (b) the volume flow rate Q of the gas.

THERMAL PHYSICS

CHAPTER 10

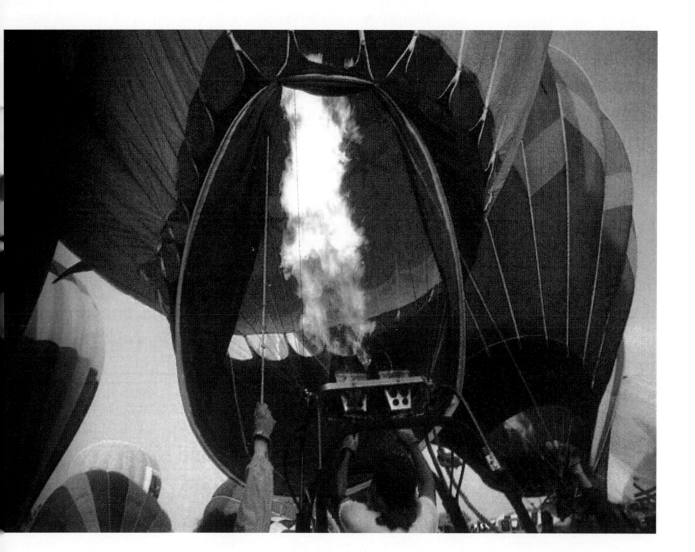

It is fun to ride on a hot air balloon on a warm breezy sunny day. The hot air balloon is the oldest successful human-carrying flight technology. It is part of a class of aircraft known as balloon aircraft. The Montgolfier brothers developed and publicly demonstrated the first flight of a hot air balloon in Annonay, Ardeche, France on June 4, 1783. This was an unmanned flight which lasted 10 minutes. Shortly later, on November 21, 1783, the first untethered manned flight of hot air balloon was performed by de Rozier and d'Arlandes in Annonay, France. Since then, many improvements had been made. Hot air balloons can be also propelled through the air rather than just being pushed along by the wind.

These are known as airships or, more specifically, thermal airships. However, the basic design has not changed much. A hot air balloon consists of a bag called the envelope which is capable of containing heated air. Suspended beneath is a gondola or wicker basket (in some long-distance or high-altitude balloons, a capsule), which carries passengers and usually a source of heat, in most cases an open flame.

How does a hot air balloon, shown above, work? A simple answer is that hot air rises when it is heated because its density decreases. The heated air inside the envelope makes it buoyant since it has a lower density than the relatively cold air outside the envelope. As with all aircraft, hot air balloons cannot fly beyond the atmosphere. This rising hot air results in lift force which can carry both the balloon and passengers upward. Making a more careful observation of the balloon, we may notice that heated air expands. This may be easily seen from the fact that when an operator fills the balloon with hot air it becomes bigger and fuller due to the expansion and can rise to high altitudes. So, hot air balloons are able to fly to extremely high altitudes. On November 26, 2005, Singhania set the world altitude record for highest hot air balloon flight, reaching 21,027 m (68,986 ft). He took off from downtown Mumbai, India, and landed 240 km (150 mi) south in Panchale.

As we will see in this chapter, there is another way to think about how hot air balloons operate. This way is based on the physical principles we discussed in previous chapters. For example, when the air inside the balloon is heated, air molecules will have greater kinetic energy indicating that they, on the average, will move faster. When these faster moving molecules are confined within a container, such as in the hot air balloon, they will collide with the walls of a container and change the direction of motion. As we learned in Chapter 6, when a moving particle changes its direction due to a collision, there is a change in momentum. The momentum change of each air molecule indicates that it exerts force on the walls of the container. When many air molecules collide within some area of the wall at the same time, these molecules will exert pressure. As we discussed qualitatively, increasing the temperature of an object can lead to many interesting phenomena. In this chapter, we will examine the effect of increasing temperature on both solid and gaseous state of matter.

10.1 ZEROTH LAW OF THERMODYNAMICS

It should be noted that whenever we deal with heat energy, as in hot air balloons, we need to consider the laws of thermodynamics. There are three laws of thermodynamics governing the heat and work energy in a given system. In addition to these three laws, there is the zeroth law which defines the equilibrium. We will postpone our discussion of these three laws to Chapter 12. However, first, we discuss the zeroth law here.

The **zeroth law of thermodynamics** states that, if two systems are in thermal equilibrium with the third system, then all three systems are in thermal equilibrium with each other. Sometimes this statement is called the **law of equilibrium**. It should be noted that two objects are in thermal equilibrium if they are in thermal contact. Whenever, two objects are in thermal contact, heat energy transfer can occur. The mechanisms for heat energy transfer are discussed in Chapter 11. However, once the two systems are thermal equilibrium, there is no net exchange of heat energy.

Due to its importance, let us reiterate the concept of thermal equilibrium according to the zeroth law of thermodynamics. It is noted that **if two objects are in thermal equilibrium with each other, then they are at the same temperature.** As shown in Fig. 10.1, when Object #1 (thermometer) is in contact with Object #2, the thermometer measures the temperature of Object #2 when they are in thermal equilibrium. Likewise, when Object #3 is in contact with Object #1 separately (i.e., Object

THERMAL PHYSICS

#2 is not in contact with Object #3), the thermometer measure the temperature of Object #3. If the temperatures of all objects are the same, then these objects are in thermal equilibrium.

In order for multiple systems to be in thermal equilibrium, heat energy must be transferred through thermal contacts. Hence, thermal contact is an important concept relating to thermal equilibrium. Multiple systems are considered to be in thermal contact if they are capable of affecting the others' temperatures. If a cup of orange juice is removed from the refrigerator and placed on a table which is at room temperature, then the table and the cup of orange juice are in thermal contact.

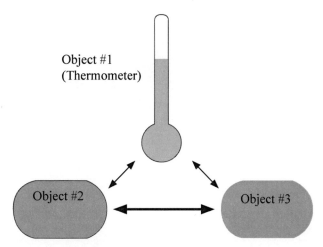

Figure 10.1: A schematic illustration of a system of three objects in thermal equilibrium. When the system is in thermal equilibrium, the temperatures of the three objects are the same.

Heat energy from the table flows to the cold cup of orange juice. Eventually, their temperatures will become equal and they will be in a state of thermal equilibrium. As we can see from this example, heat flows from the warmer object, which contains more thermal energy, to the cooler object, which contains less thermal energy, in thermal systems involving objects in thermal contact. Therefore, objects and systems can either gain or lose heat energy. In a thermal system, the warmer object loses heat energy to the cooler object until equilibrium is achieved.

10.2 TEMPERATURE AND HEAT

As we discussed above, the temperature of an object measured by using a thermometer indicates the condition (or state) of the object, namely, how hot or cold it is. This quantity is useful for knowing how much thermal energy the object has. So, we may define **temperature** as a measure of how hot or cold an object is. For example, common sense indicates that a hot oven and a frozen lake have different temperatures. The hot oven is at high temperature, and the ice in a frozen lake is at low temperature. When we know how hot or cold an object is, we may determine in which direction that thermal energy will flow. (An aspect of energy flow will be discussed in Chapter 11.) This flow of thermal energy is called **Heat**. Thermal energy usually flows naturally from a higher temperature object to a lower temperature object (i.e., energy transfer).

Thermometer

A description of how hot or cold an object is can give a qualitative description of the condition. This information is not enough to describe the state of the object accurately. To do this, we need to measure its temperature accurately by using a thermometer. Here, the **thermometer** is an instrument designed to measure temperature by using the thermal property of materials. All thermometers depend on some property of matter which changes with temperature for their operation. Some thermometers shown in Fig. 10.2 rely on thermal expansion, while others use thermal electric property for their operation.

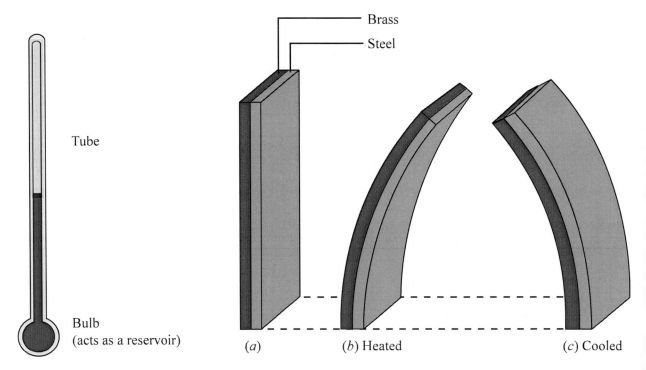

Figure 10.2: Many thermometers use the thermal expansion property of either liquids or solids to measure the temperature quantitatively. On the left panel, the thermal expansion property of mercury is used. On the right panel, however, the difference in the expansion property of both brass and steel is used.

In contrast to most other methods of temperature measurement, thermocouples are an inexpensive way to measure a wide range of temperatures by exploiting the junction thermoelectric effect. A **thermocouple** consists of two dissimilar conductors in contact, which produces a voltage when heated. The size of the voltage depends on the temperature difference across the junction to another junction in other parts of the circuit which is held at a fixed reference temperature. Thermocouples are a widely used type of temperature sensor for measurement and control. They can also be used to convert a temperature gradient into electricity.

How to measure temperatures?

Once we have a material for a thermometer, we need to determine a numerical scale for measuring the temperature. To measure temperature quantitatively, we need to define some sort of numerical scale. One can define any numerical scale for the purpose of measuring temperature, but it is convenient to use one of the most common scales in order to reduce confusion. The three most commonly used temperature scales are the following:

1. Celsius scale or centigrade scale
2. Fahrenheit scale (used widely in USA)
3. Absolute or Kelvin scale (important scale in scientific work)

These scales appear different, but they are closely related to each other. This means that a temperature measurement in one scale can be easily converted to corresponding temperature in another scale. In most cases, they differ only by a few numerical factors. This compatibility means that every temperature on the Celsius scale corresponds to a particular temperature on the Fahrenheit scale.

THERMAL PHYSICS

As we discussed above, to measure temperature quantitatively, we need to define a numerical scale. This means that we need to determine how to define a temperature scale. So, we may ask "How do we define a temperature scale?" One way to start is to assign arbitrary values to two readily reproducible temperatures. To do this, we need to choose a reference material. We choose water as the reference material since water is the most abundant material on the surface of Earth. Once we have chosen a material, we need to choose two fixed points on the temperature scale. For water, the freezing and boiling points are the two well-known points associated with the phase change. For both Celsius and Fahrenheit scale, the freezing point and the boiling point of water at atmospheric pressure are used to set the temperature scale.

	Celsius scale	Fahrenheit scale
Freezing point	0 °C	32 °F
Boiling point	100 °C	212 °F

Now, we need to divide these two marks into equal intervals. These two marks are divided into 100 equal intervals for the Celsius scale and divided into 180 equal intervals for the Fahrenheit scale.

Daniel Fahrenheit proposed his temperature scale in 1724. This scale is based on three reference points of temperature. In his initial scale (which is not the final Fahrenheit scale), the zero point is determined by placing the thermometer in brine (a solution of salt in water). He used a mixture of ice, water, and ammonium chloride (a salt) at a 1:1:1 ratio. This is a frigorific mixture which stabilizes its temperature automatically. The stable temperature was defined as 0 °F (−17.78 °C). The second point, at 32 °F, was a mixture of ice and water without the ammonium chloride at a 1:1 ratio. The third point, 96 °F, was approximately the human body temperature, which was called "blood-heat".

Celsius versus Fahrenheit

Since every temperature on the Celsius scale corresponds to a particular temperature on the Fahrenheit scale, let us see how these two scales correspond to each other. Based on dividing the two marks as described above, one Fahrenheit degree (1 F°) corresponds to 5/9 of a Celsius degree (1 C°)

$$1F° = \frac{5}{9}C° \qquad (10.1)$$

It should be noted that we use "degree Celsius" (°C) to refer to a specific temperature and use a slightly different terminology "Celsius degree" (C°) to refer to a change in temperature or temperature interval.

Conversion of temperature in Fahrenheit scale to Celsius scale may be accomplished by using

$$T(°C) = \frac{5}{9}\left[T(°F) - 32\right] \qquad (10.2)$$

while temperature in Celsius scale is converted to that in Fahrenheit scale by using

$$T(°F) = \frac{9}{5}T(°C) + 32. \qquad (10.3)$$

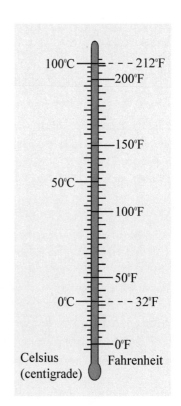

This suggests that similar equations for converting temperature either in Celsius or in Fahrenheit to other scale may also be found.

Understanding-the-Concept Question 10.1:

At what temperature are the numerical readings on the Fahrenheit and Celsius scales the same?

    a. -30°   **b.** -40°   **c.** -50°   **d.** -60°

Explanation:

Since the Fahrenheit and Celsius scale readings are the same, we may write Eq. (10.2) as

$$T = \frac{5}{9}(T - 32)$$

We solve for T and obtain

$$T - \frac{5}{9}T = -\frac{5 \times 32}{9} \Rightarrow \frac{4}{9}T = -\frac{5 \times 32}{9} \Rightarrow T = -40°$$

Answer: b

It is interesting to note that the record lowest temperature in North Dakota occurred at Parshall on February 15, 1936. The recorded temperature was -60 °F (-51 °C). This is not as bad as the climate of Antarctica which is the coldest on the whole of Earth. Antarctica has the lowest naturally occurring temperature ever recorded on the ground on Earth: −89.2 °C (−128.6 °F) at Vostok Station. This temperature value was measured on July 21, 1983.

10.3 EXPANSION OF MATERIALS

The thermal expansion of materials is exploited in thermometers. When a material is heated, it responds by changing its size. Most materials expand when temperature is increased, but some materials, such as water, contract. For example, when water is heated starting from 0 °C, it contracts as the temperature increases to 4.2 °C. From that point onwards it expands. Water has highest density at 4.2 °C. This is the reason why that fishes are perfectly fine at the bottom of a large lake during very cold winter days in North Dakota and Minnesota.

Most materials expand with temperature, but different materials do not expand in quite the same way over a wide temperature range. So, thermometers use materials with large thermal expansion coefficients. Liquids tend to expand more than solids. If a material for some standard kind of thermometer must be chosen to measure temperature accurately, then liquids may be better than solids because their thermal expansion coefficient is larger. This is the reason for using mercury in standard thermometers. The amount of expansion or contraction of a solid material is proportional to the change in temperature ΔT. The linear expansion of a solid is described by

$$\Delta L = \alpha L_0 \Delta T, \tag{10.4}$$

THERMAL PHYSICS

where α is the coefficient of linear expansion and L_o is the original length of the object. It is noted that α is a constant which accounts for the expansion property of a material.

As shown schematically in Fig. 10.3, a rod at initial temperature T_o will expand in length when the temperature is increased to $T_o + \Delta T$. The amount that the rod expands in length is ΔL. This expansion amount is expected to be different for a different material. This difference due to material is accounted for by α, the coefficient of linear expansion.

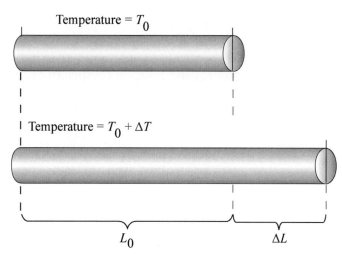

Figure 10.3: A schematic illustration of thermal expansion of a long rod. With an increase of temperature by ΔT from the initial value T_o leads to an expansion of the length L_o by ΔL.

It is interesting to note from Table 11.1 that the coefficient of volume expansion β for a solid material is roughly three times the value for the coefficient of linear expansion α. As we will see below, the factor of 3 between β and α is due to the difference in physical spatial dimensions, three dimensions for a volume and one dimension for a length. This means that, when the volume of an object expands, it increases its length in all three directions.

Table 11.1: Coefficient of expansion at 20 °C for selected materials.

Material	Coefficient of expansion	
	Linear expansion α (C°)$^{-1}$	Volume expansion β (C°)$^{-1}$
Solids		
Aluminum	25×10^{-6}	75×10^{-6}
Brass	19×10^{-6}	56×10^{-6}
Iron or steel	12×10^{-6}	35×10^{-6}
Lead	29×10^{-6}	87×10^{-6}
Glass (Pyrex)	3×10^{-6}	9×10^{-6}
Glass (ordinary)	9×10^{-6}	27×10^{-6}
Quartz	0.4×10^{-6}	1×10^{-6}
Concrete and brick	$\approx 12 \times 10^{-6}$	$\approx 36 \times 10^{-6}$
Marble	$1.4\text{-}3.5 \times 10^{-6}$	$4\text{-}10 \times 10^{-6}$
Liquids		
Gasoline		950×10^{-6}
Mercury		180×10^{-6}
Ethyl alcohol		1100×10^{-6}
Glycerin		500×10^{-6}
Water		210×10^{-6}
Gasses		
Air (& most other gasses at atmospheric pressure)		3400×10^{-6}

Understanding-the-Concept Question 10.2:

A steel bridge is 1000 m long at -20 °C in winter. What is the change in length when the temperature rises to 40 °C in summer? (The average coefficient of linear expansion of steel is 11x10⁻⁶ /C°.)

    a. 0.33 m   **b.** 0.44 m   **c.** 0.55 m   **d.** 0.66 m

Explanation:

As we discussed above, a steel bridge expands as temperature increases. The change in the length can be computed by using Eq. (10.4). We obtain

$$\Delta L = \alpha L_0 (\Delta T) = 11 \times 10^{-6} \, (\text{C}°)^{-1} \times 1000 \, \text{m} \times [40\,°\text{C} - (-20\,°\text{C})] = 0.66 \, \text{m}$$

This is the reason that all steel bridges have an expansion joint or movement joint. This is a common technique to avoid damage is to make a break in the bridge deck. The break has a small gap or flexible insert between the two parts of the bridge deck and as the deck expands and contracts, so the gap changes size. This method means that a minimal force is exerted on the bridge supports and therefore structural damage is avoided. The longer the bridge, the more expansion can take place and several expansion gaps can be used. Typically, the gaps will allow for movement of less than one inch up to as much as twelve inches.

Answer: d

Exercise Problem 10.1

An iron ring is to fit snugly on a cylindrical iron rod. At 20 °C, the diameter of the rod is 6.445 cm and the inside diameter of the ring is 6.420 cm. To slip over the rod, the ring must be slightly larger than the rod diameter by 0.008 cm. To what temperature must the ring be brought if its hole is to be large enough so it will slip over the rod?

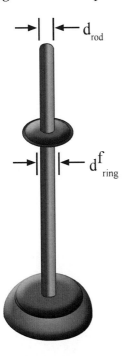

PROBLEM SOLVING STRATEGY

What do we know?

The inner diameter of the ring and the diameter of the rod at 20°C are known.

What concepts are needed?

When heated, an object expands and increases its length. The size of a hole also expands with increasing temperature

Why?

We need to compute the temperature at which the iron ring becomes large enough to slip over the rod.

THERMAL PHYSICS

Solution:

In order for an iron ring to fit, the hole in the ring must be increased to

$$d^f_{ring} = d_{rod} + \delta d$$

This increase in diameter of the hole is computed from the thermal expansion relation $\Delta L = \alpha L_o (\Delta T)$, indicating that the corresponding temperature change must be

$$\therefore \Delta T = T_f - T_i = \frac{\Delta L}{\alpha L_o} = \frac{d^f_{ring} - d^i_{ring}}{\alpha L_o}.$$

The final temperature T_f of the ring has to be

$$T_f = \frac{6.445\,\text{cm} + 0.008\,\text{cm} - 6.420\,\text{cm}}{12 \times 10^{-6}\ 1/C° \times 6.420\,\text{cm}} + 20°C = 448.3\,°C \cong 450\,°C.$$

Sometimes, thermal expansion is exploited in assembling together machine parts made from different materials. This process is called shrink-fitting. **Shrink-fitting** is a technique in which pieces of a structure are heated or cooled, employing the phenomenon of thermal expansion, to make a joint. For example, the thermal expansion of a piece of a metallic drainpipe allows a builder to fit the cooler piece to it. As the adjoined pieces reach the same temperature, the joint becomes strained and stronger. Another example is fitting the iron strip around the edge of a cartwheel. The strip will be heated and expands to the wheel's diameter, and is fitted around it. After cooling, the iron rim contracts, binding tightly in place. A common method used in industry is the use of induction shrink fitting which refers to the use of induction heating technology to pre-heat metal components between 150°C and 300°C thereby causing them to expand and allow for the insertion or removal of another component. Other methods of shrink-fitting include compression shrink fitting which uses a cryogen such as liquid nitrogen to cool the insert, while heating the housing.

Understanding-the-Concept Question 10.3:

When a material with a hole is heated, what happens to the hole?

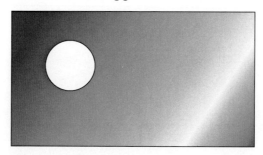

Explanation:

When an area expands, every linear dimension (i.e., length and width) increases by the same percentage with a change in temperature. This expansion also includes holes in the material. Here, we assume that the expansion of the material is uniform. We may easily extend the idea to the volume expansion. For example, if a large uniform cube of concrete with a small spherical hole in the middle is heated, then the radius of the hole becomes larger. If the side of the block increases by 4%, then the radius of the hole will change by the same percentage. So, when they heat up and expand, all the dimensions expand together by the same percentage. This uniform expansion obviously lets each part expand by the same amount in every direction. That includes the dimensions of any holes.

Volume Thermal Expansion

As we discussed above, when an object is heated, it expands. The object has volume. So, this expansion occurs in all three directions, yielding volume expansion. The change in volume of a material due to a temperature change is given by

$$\Delta V = \beta V_0 \Delta T, \qquad (10.5)$$

where β is the coefficient of volume expansion and V_o is the original volume. One example of this effect is the automobile cooling system shown in Fig. 10.4. Excess coolant fluid from the radiator transfers to the coolant reservoir when the radiator becomes hot and the fluid in the radiator expands.

For isotropic materials and for small expansions, the linear thermal expansion coefficient is one third the volumetric coefficient β

$$\beta = \alpha_V = 3\alpha. \qquad (10.6)$$

This ratio arises because the volume is composed of three mutually orthogonal directions. Thus, in an isotropic material, for small differential changes, one-third of the volumetric expansion is in a single axis. As an example, suppose we take a cube of steel that has sides of length L_o. The original volume will be $V_o = (L_o)^3$ and, after a temperature increase, the new volume V will be

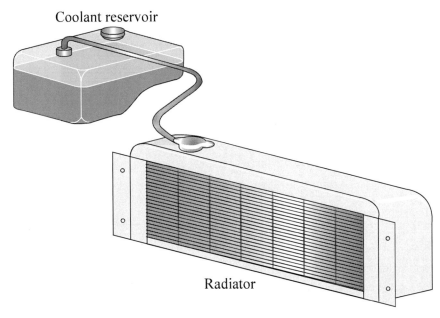

Figure 10.4: When the coolant fluid in the radiator of an automobile is heated, its volume expands. Excess fluid may be transferred to the coolant fluid reservoir.

THERMAL PHYSICS

$$V_o + \Delta V = (L_o + \Delta L)^3 = L_o^3 + 3L_o^2\Delta L + 3L_o(\Delta L)^2 + (\Delta L)^3 \approx L_o^3 + 3L_o^2\Delta L = V_o + 3V_o\frac{\Delta L}{L_o}.$$

We make a use of the relation $\Delta V = \alpha_V L_o^3(\Delta T)$ and substitute ΔV in the above equation. Also, for isotropic materials, we may use the relation $\Delta L = \alpha L_o(\Delta T)$ and write

$$L_o^3 + L_o^3\alpha_V\Delta T = L_o^3 + 3L_o^3\alpha\Delta T + 3L_o^3\alpha^2(\Delta T)^2 + L_o^3\alpha^3(\Delta T)^3 \approx L_o^3 + 3L_o^3\alpha\Delta T.$$

Since the volumetric and linear coefficients are defined only for small temperature and dimensional changes (that is, when ΔT and ΔL are small), the last two terms in the above equation can be ignored. With this simplifying approximation, we obtain Eq. (10.6), which is the relation between the coefficients α and β. If we are trying to go back and forth between volumetric and linear coefficients using larger values of ΔT then we will need to take into account the third term, and sometimes even the fourth term.

Similarly, the area thermal expansion coefficient is 2/3 of the volumetric coefficient ($\alpha_A = 2\beta/3$). This ratio can be found in the similar way to that in the linear example above, noting that the area of a face on the cube is just L^2. Also, the same considerations must be made when dealing with large values of ΔT.

Understanding-the-Concept Question 10.4: (Bursting water pipe)

Why do water pipes burst in cold weather?

Explanation:

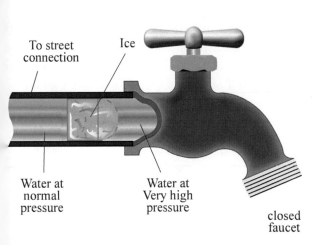
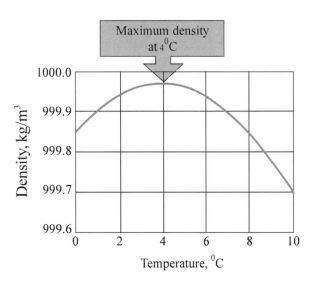

Note that the volume of water decreases with decreasing temperature until it reaches 4 °C and then it increases, yielding the maximum density at 4 °C as shown in the above graph. The reason that many water pipes burst is because water expands as it freezes. If this expansion occurs in a closed environment such as a water pipe, there could be substantial pressure placed on the inner walls of the pipe material, whether it is iron, lead or PVC plastic. Even if a pipe is strong enough to handle the occasional stress of frozen water, the cumulative effect can weaken the structural integrity of the pipe and eventually, it will burst.

Exercise Problem 10.2:

The 70 liter steel gas tank of a car is filled to the top with gasoline at 20 °C. The car is then left to sit in the sun, and the tank reaches a temperature of 40 °C (104 °F). How much gasoline do you expect to overflow from the tank?

Solution:

Since we need to estimate the volume expansion, we use the expression for the volume change

$$\Delta V = \beta V_0 \Delta T.$$

We, first, compute the expansion of the gasoline

$$\Delta V_{gas} = \beta_{gas} V_0^{gas} \left(T_f - T_i \right) = 950 \times 10^{-6} \frac{1}{C^\circ} \times 70L \times \left(40°C - 20°C \right) = 1.344L.$$

Then, we compute the expansion of the gas tank

$$V_{tank} = \beta_{tank} V_0^{tank} \left(T_f - T_i \right) = 35 \times 10^{-6} \frac{1}{C^\circ} \times 70L \times \left(40°C - 20°C \right) = 0.049L.$$

The difference in the volume change is

$$\Delta V_{tank} - \Delta V_{gas} = 0.049L - 1.344L = -1.295L.$$

So, the gasoline will spill out.

The thermal expansion of gasoline is a real effect and was a concern of the oil industry. The oil industry has known for 100 years that gasoline expands with temperature. As it warms, gasoline expands by volume but not by weight or energy content. Since the 1920's, the oil industry has taken temperature into account for wholesale transactions, and use a 60 °F standard when measuring gasoline at wholesale. However, the oil industry does not adjust for temperature in retail sales to consumers. As a result, consumers pay a hot fuel premium when gasoline temperatures exceed 60 °F, as they may do during the summer months.

Thermal Stress

Thermal stress is stress induced on a solid material due to changes in temperature. Recall that a mechanical stress F/A can produce a mechanical strain $\Delta L/L_0$ according to

$$\Delta L = \frac{1}{E} \frac{F}{A} L_0, \tag{10.7}$$

where E is Young's modulus or elastic modulus. Similarly, a temperature change ΔT can also produce a mechanical strain $\Delta L/L_0$, according to

$$\Delta L = \alpha L_0 \Delta T.$$

THERMAL PHYSICS

Equating the two expressions for ΔL, we see that a temperature change ΔT can have a similar effect to a mechanical stress F/A:

$$\alpha L_0 \Delta T = \frac{1}{E}\frac{F}{A}L_0 \Rightarrow F = \alpha E A \Delta T. \qquad (10.8)$$

Equation (10.8) indicates that the change in temperature of the object is equivalent to applying force over the cross-sectional area A (i.e., mechanical stress).

Mechanical stress is induced on a material when some or all of its parts are not free to expand or contract in response to changes in temperature. In most continuous bodies, thermal expansion or contraction cannot occur freely in all directions because of geometry, external constraints, or the existence of temperature gradients. So, stresses are produced. In the real world, problems of thermal stress arise in many practical design problems, such as those encountered in the design of steam and gas turbines, diesel engines, jet engines, rocket motors, nuclear reactors and those exposed to a wide range of temperature variations. The high aerodynamic heating rates associated with high-speed flight present even more severe thermal-stress problems for the design of spacecraft and missiles.

Exercise Problem 10.3:

A highway is to be made of blocks of concrete 10 m long placed end to end with no space in between them to allow for expansion. If the blocks were placed at a temperature of 10 °C, what force of compression would occur if the temperature reached 40 °C? The contact area between each block is 0.20 m². Will fracture occur?

Problem Solving Strategy

What do we know?

When the temperature increases, the 10 m long concrete blocks expands but there is no room for expansion.

What concepts are needed?

We need to apply the concept of thermal stress. The temperature induced expansion of the concrete blocks has the same effect as the expansion due to application of stress (F/A).

Why?

Fracturing will occur if applied stress is greater than the strength of the concrete blocks.

Solution:

We start with the expression for thermal stress

$$F = \alpha E A (\Delta T)$$

and compute the magnitude of the force exerted on the concrete blocks as

$$F = 12\times 10^{-6}\frac{1}{\text{C}°}\times 20\times 10^{9}\frac{\text{N}}{\text{m}^2}\times 0.20\text{m}^2 \times \left(40°\text{C}-10°\text{C}\right)\cong 1.4\times 10^{6}\,\text{N}.$$

According to its definition, the thermal stress on the concrete blocks is

$$\frac{F}{A}=\frac{1.4\times 10^{6}\,\text{N}}{0.20\text{m}^2}=7.0\times 10^{6}\,\frac{\text{N}}{\text{m}^2}.$$

From Chapter 9, we note that the tensile and compressive strength of concrete are given by 2×10^{6} N/m² and 20×10^{6} N/m, respectively. Also, the shear strength is 2×10^{6} N/m². This means that if the concrete is not perfectly aligned, part of the force will act in shear and fracture will occur since the shear strength is lower.

Concrete fracture is brittle. Brittle fracture means there is no deformation around the crack, but the very grains of the material are either separating, or cracking themselves. Concrete is the material of choice for many structures such as buildings, bridges, and dams because concrete offers many advantages, including low cost, good weather and fire resistance, good compressive strength, and excellent formability. However, concrete is bad in tension and shear stress. Therefore, it is prone to cracking. Also, its behavior is not so predictable. Many dams are made of concrete and have suffered from cracking, which is caused by many different forces.

It is interesting to note that concrete is not found in nature. To form concrete, a number of different elements have to be combined, and this combination is not done by nature. Unlike the way we would find aluminum, nickel, or iron in nature, concrete is formed from combining water, a special cement and rock:

Portland cement + H_2O + rock = hardened concrete + energy (heat)

Note that, in general, heat and temperature variations can cause cracking problems. Of course, lots of heat is needed if a concrete structure is big. It is important to note that cement is only a component of concrete and concrete is the structural material. So when you see a parking garage, a driveway, a sidewalk or a road remember it is made of concrete, not cement.

10.4 ATOMIC THEORY OF MATTER

We now turn our attention to thermal properties of gases. Central to this discussion is the ideal gas law. The **ideal gas law** is the equation of state of a hypothetical ideal gas. This law serves as a foundation for understanding the property of gasses and describes the relationship among three variables: pressure, volume, and temperature. The ideal gas law is considered as a good approximation to the behavior of many gasses under many conditions, although it has several limitations and does not describe real gasses accurately. The ideal gas law was first stated by Emile Clapeyron (1799-1864) in 1834 as a combination of Boyle's law, Charles's law, and Gay-Lussac's law (see below), which were obtained from phenomenological observations. The ideal gas law can also be derived from the kinetic theory. This derivation was achieved independently in 1856 by Krönig and, later, in 1857 by Rudolf Clausius (1822-1888). In the relationship among pressure, volume, and temperature of the system as described by the ideal gas law, a constant has to be introduced naturally. We call this constant the universal ga

THERMAL PHYSICS

constant because it does not depend on specific gasses. This universal gas constant R is useful because the problem of introducing a specific gas constant for each gas needing a large number of gas constants can be avoided. This universal gas constant was discovered and first introduced into the ideal gas law in 1874 by Dmitri Mendeleev (1834-1907).

A microscopic approach for understanding the property of gasses is based on the kinetic theory. The kinetic theory assumes that matter is made up of atoms and that these atoms are in continuous random motion. Note that "kinetic" is Greek for "moving". The **kinetic theory** of gasses describes a gas as a large number of small particles (atoms or molecules), and all of them are in constant, random motion. The rapidly moving particles constantly collide with each other and with the walls of the container. Kinetic theory explains macroscopic properties of gasses, such as pressure, temperature, and volume, by considering their molecular composition and motion. As discussed in Sec. 10.6, the theory posits that pressure is due to collisions between molecules moving at different velocities according to Brownian motion and **not** due to static repulsion between molecules as conjectured by Newton. While gas molecules are too small to be visible they collide with larger objects such as pollen grains or dust particle producing a jittering motion, known as Brownian motion. Brownian motion can be seen under a microscope. As pointed out by Einstein in 1905, this experimental evidence for kinetic theory is generally seen as having confirmed the existence of atoms and molecules.

Atoms are the basic building blocks of all matters. Gasses, liquids and solids are all made up of atoms, molecules, and/or ions, but the behaviors of these particles differ in the three phases. **A solid** has a definite shape and volume. A **liquid** has a definite volume but it takes the shape of a container whereas a **gas** fills the entire volume of a container. Figure 10.5 illustrates the microscopic differences. Gas, liquid, and solid are known as the **three states** of matter or material, but each of solid and liquid states may exist in one or more forms. We already know that diamond and graphite are solids made up of the element carbon. They are two **phases** of carbon, but both are solids. Thus, another term is required to describe the various forms, and the term **phase** is used. Each distinct form is called **a phase**, but the concept of **phase** defined as a homogeneous portion of a system, extends beyond a single material, because a phase may also involve several materials. For example, a homogeneous solution of any number of substances is a one-phase system. Phase is a concept used to explain many physical and chemical changes.

All elements, including naturally occuring elements as well as man-made elements, can be organized in a periodic table. Mendeleev is generally credited with the publication, in 1869, of the

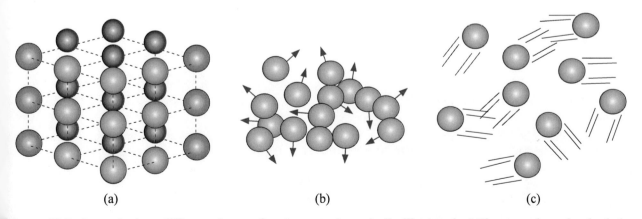

Figure 10.5: Atoms in three different phases of matter are schematically illustrated. a) The atoms have fixed relative positions in the solid phase. b) In the liquid phase, atoms are close to each other but do not have fixed positions. c) The atoms are far apart and are freely moving in the gas phase.

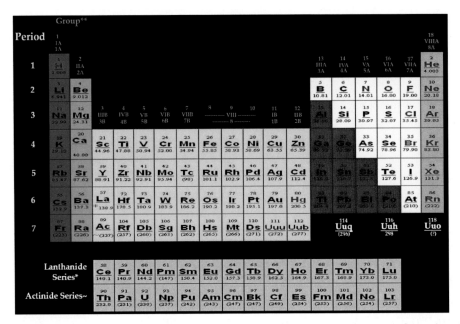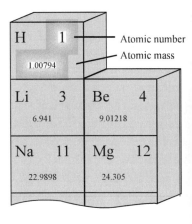

Figure 10.6: A periodic table of atomic elements. The right panel illustrates information about the atomic elements presented in a typical illustration of the periodic table.

first widely recognized periodic table. A **periodic table** as shown in Fig. 10.6 is a tabular display of the chemical elements, organized on the basis of their atomic numbers, electron configurations, and recurring chemical properties. Elements are presented in order of increasing atomic number (number of protons). The standard form of table comprises an 18 × 7 grid or main body of elements, positioned above a smaller double row of elements. The table can also be deconstructed into four rectangular blocks: the s-block to the left, the p-block to the right, the d-block in the middle, and the f-block below that. The rows of the table are called periods; the columns of the s-, d-, and p-blocks are called groups, with some of these having names such as the halogens or the noble gases. Since, by definition, a periodic table incorporates recurring trends, any such table can be used to derive relationships between the properties of the elements and predict the properties of new, yet to be discovered or synthesized, elements. As a result, a periodic table—whether in the standard form or some other variant—provides a useful framework for analyzing chemical behavior, and such tables are widely used in chemistry and other sciences.

Concept of the Atom

As we discussed above, the idea that there are lots of small moving particles is important for understanding the property of gasses. The concept of the atom, the smallest piece of matter, dates back to the ancient Greeks (i.e., atomic theory). As illustrated in Fig. 10.7, the **atom** is a basic unit of matter that consists of a dense central nucleus surrounded by a cloud of negatively charged electrons. The atomic nucleus contains a

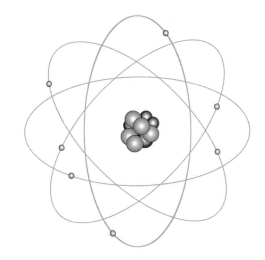

Figure 10.7: An illustration of Bohr model for an atom indicates that a nucleus composed of protons and neutrons is at the center and electrons orbit around this nucleus.

mixture of positively charged protons and electrically neutral neutrons (except in the case of hydrogen, which is the only stable nuclide with no neutrons). The electrons of an atom are bound to the nucleus by the electromagnetic force. Evidence for atomic theory is given by i) The law of definite proportions and ii) Brownian motion.

A simple model of atom, known as the **Bohr model** which was introduced by Neils Bohr (1885-1962) in 1913, depicts the atom as a small, positively charged nucleus surrounded by electrons that travel in circular orbits around the nucleus. This model is similar in structure to the solar system, but with attraction provided by electrostatic forces rather than gravity. The Bohr model is a relatively primitive model of the hydrogen atom. The model's key success lay in explaining the Rydberg formula for the spectral emission lines of atomic hydrogen.

Law of definite proportion

A crucial piece of evidence supporting the atomic theory was the "law of definite proportions", which states that when two or more elements are combined to form a compound they always do so in the same proportion by mass. This is law is a summation of experimental results collected during the half century prior to 1800. The **law of definite proportions**, sometimes called **Proust's Law**. An equivalent statement is the **law of constant composition**, which states that all samples of a given chemical compound have the same elemental composition by mass. For example, oxygen makes up about $8/9$ of the mass of any sample of pure water, while hydrogen makes up the remaining $1/9$ of the mass. Along with the law of multiple proportions, the law of definite proportions forms the basis of stoichiometry.

The law of definite proportions may appear obvious to the modern chemist because it is inherent in the very definition of a chemical compound. However, the law was novel at the end of the eighteenth century, when the concept of a chemical compound had not yet been fully developed. In fact, when first proposed, it was a controversial statement and was opposed by other chemists. The law of definite proportions contributed to and was placed on a firm theoretical basis by the atomic theory. In 1803, John Dalton (1766-1844) began to promote the law as a way to explain that matter as consisting of discrete atoms, that there was one type of atom for each element, and that the compounds were made of combinations of different types of atoms in fixed proportions.

Each atom in a periodic table has a definite mass, known as atomic mass. Atomic mass or molecular mass are relative masses of atoms and molecules, respectively. The atomic mass is defined as the total mass of the protons, neutrons, and electrons in a single atom when it is at rest. Atomic mass is measured by mass spectrometry. We can figure out the molecular mass of a compound by adding the atomic mass of its atoms. Atomic mass unit (u) is defined in terms the abundant carbon atom ^{12}C which is exactly 12.0000 u.

$$1u = 1.660538921 \times 10^{-27} \, kg \qquad (10.9)$$

Sometimes it is abbreviated as amu (i.e., atomic mass unit). Also, it is called Dalton, but this is not a commonly used teminology. Hydrogen atom's mass is 1.0078 u

Brownian movement

Another important piece of evidence for the atomic theory is the Brownian movement (e.g., motion of tiny pollen grains suspended in water). **Brownian movement** or motion is defined as zigzag, irregular motion exhibited by minute particles of matter when suspended in a fluid. The effect has been observed in all types of colloidal suspensions such as solid-in-liquid, liquid-in-liquid, gas-in-liquid, solid-in-gas, and liquid-in-gas. This movement is named for Robert Brown (1773-1853), the botanist who observed the movement of plant spores floating in water in 1827. The effect, being independent of all external factors, is ascribed to the thermal motion of the molecules of the fluid. These molecules are in constant irregular motion with a velocity proportional to the square root of the temperature. Small particles of matter suspended in the fluid are buffeted about by the molecules of the fluid. Brownian motion is observed for particles about 0.001 mm in diameter; these are small enough to share in the thermal motion, yet large enough to be seen with a microscope or ultra-microscope. The first satisfactory theoretical treatment of Brownian motion was made by Albert Einstein (1879-1955) in 1905. Jean Perrin (1870-1942) made a quantitative experimental study of Brownian motion in 1908 to understand its dependence on temperature and particle size. This study provided verification for Einstein's mathematical formulation. Perrin's work is regarded as one of the most direct verifications of the kinetic-molecular theory of gases.

The Brownian motion as shown in Fig. 10.8 is easily explained if the atoms of any substance are continually in motion. In 1905, Einstein introduced a theory to explain this phenomenon in in his doctoral dissertation to the University of Zurich. The underlying base of his theory is that matter is made up of atoms. Einstein predicted that the random motions of molecules in a liquid impacting on larger suspended particles would result in irregular, random motions of the particles, which could be directly observed under a microscope. More technically, there are two parts to Einstein's theory: the first part consists in the formulation of a diffusion equation for Brownian particles, in which the diffusion coefficient is related to the mean squared displacement of a Brownian particle, while the second part consists in relating the diffusion coefficient to measurable physical quantities. In this way Einstein was able to determine the

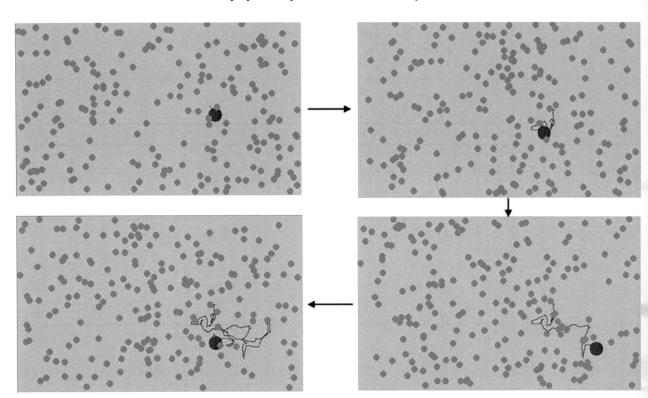

Figure 10.8: A simulation of Brownian motion for a red dot interaction with other particles (gray dots) is schematically illustrated. The random motion of the red dot with evolution of time (clockwise direction, starting from the upper left panel) may be seen from its trajectory (blue line).

THERMAL PHYSICS

size of atoms, and how many atoms there are in a mole, or the molecular weight in grams, of a gas. In accordance to Avogadro's law this volume is the same for all ideal gases, which is 22.414 liters at standard temperature and pressure. The number of atoms contained in this volume is referred to as Avogadro's number. The determination of this number is tantamount to the knowledge of the mass of an atom since the latter is obtained by dividing the mass of a mole of the gas by Avogadro's number.

Avogadro's Number

Avogadro's law (sometimes referred to as **Avogadro's hypothesis** or **Avogadro's principle**) is a gas law which states that, under the same condition of temperature and pressure, equal volumes of all gasses contain the same number of molecules. The law is named after Amedeo Avogadro (1776-1856) who hypothesized in 1811 that two given samples of an ideal gas, of the same volume and at the same temperature and pressure, contain the same number of molecules. Thus, the number of molecules or atoms in a specific volume of ideal gas is independent of their size or the molar mass of the gas.

Figure 10.9: The Hope diamond is the most famous blue diamond. It weighs 45.52 carats. The carat is a unit of mass equal to 200 mg. Assuming the crystal structure of a diamond, one may estimate the number of carbon atoms in this blue diamond.

As we can see from our earlier discussion, macroscopic amounts of materials, such as the Hope diamond shown in Fig. 10.9, contain large numbers of atoms or molecules. Since this number is so enormously large, it is convenient to express it in terms of a single unit: gram-mole or mole (mol). The number of atoms per mole is known as Avogadro's number N_A is 6.022×10^{23} per mole. The number of moles n contained in any one sample is

$$n = \frac{\text{\# of particles N in the sample}}{N_A} \quad (10.10)$$

The Avogadro constant is named after the Italian scientist Avogadro, who proposed that the volume of a gas (at a given pressure and temperature) is proportional to the number of atoms or molecules regardless of the nature of the gas. In 1909, Perrin proposed naming the constant in honor of Avogadro. Perrin won the 1926 Nobel Prize in Physics for his work in determining the Avogadro constant by several different methods.

Understanding-the-Concept Question 10.5:

A mole of diatomic oxygen molecules and a mole of diatomic nitrogen molecules at STP have

    a.   the same average molecular speeds
    b.   the same number of molecules
    c.   the same diffusion rate
    d.   all of the above

Explanation:

It should be noted that the word "mole" as a unit means that there are 6.022×10^{23} objects, regardless of whether those objects are oxygen or nitrogen molecules. So, there should be exactly the same number of molecules for both oxygen and nitrogen.

Answer: b

10.5 IDEAL GAS LAW

We learned that when temperature increases most materials expand. However, the thermal expansion property for solids and gasses is quite different. For solids, thermal expansion is usually small. The coefficient of linear and volume thermal expansion (α and β) is a material property is indicative of the extent to which a material expands upon heating. Also, different substances expand by different amounts. Over small temperature ranges, the thermal expansion of uniform linear objects is proportional to temperature change. Thermal expansion is useful for application in bimetallic strips for the construction of thermometers but can generate detrimental internal stress when a structural part is heated and kept at constant length. If the temperature of a substance is increased by ΔT, then the volume of the substance will expand by

$$\Delta V = \beta V_0 \Delta T.$$

However, this relation for the volume expansion is not valid for gasses since i) expansion of gasses can be so great and ii) gasses in a container generally expand to fill the entire space they are in. This means that a different approach is needed to describe the effects of increasing temperature on a gas. But, how was this relation obtained?

To determine the volume expansion of gasses, a number of experiments had been performed. In 1702, Guillaume Amontons (1663-1705) discovered that pressure P increased linearly with temperature T for air. He discovered that if a container of air were to be sealed at 0 °C, at ordinary atmospheric pressure of 15 pounds per square inch, and then heated to 100 °C but kept at the same volume V, the air would now exert a pressure of about 20 pounds per square inch on the sides of the container. (Of course, strictly speaking, there is a tiny correction, of about half percent for copper and even less for steel and glass, due to the container which will also have increased in size. This would lower the effect.) In other words, he found that P increased about 33% from the freezing point to the boiling point of water. Remarkably, Amontons discovered that, if the gas were initially at a pressure of 30 pounds per square inch at 0 °C, on heating to 100 °C the pressure would go to about 40 pounds per square inch, indicating that the percentage increase in pressure was the same for any initial pressure. On heating through 100 °C, the pressure would always increase by about 33%. Furthermore, the result turned out to be the same for different gasses. This observation indicates that the volume of gas depends on pressure P as well as temperature T

$$V = V(P, T). \tag{10.11}$$

This relation is called "**equation of state**". Here "state" means the physical condition of the system. An **equation of state** is a relation between state variables. More specifically, an equation of state is a thermodynamic equation describing the state of matter under a given set of physical conditions. It is this constitutive equation which provides a mathematical relationship between two or more state functions associated with the matter, such as its temperature, pressure, volume, or internal energy.

Equation of state was determined experimentally

More precisely determined experimental relationships governing the volume of gasses were obtained from combining the following three gas laws: i) Boyle's law, ii) Charles's law, and Gay-Lussac's law.

Boyle's law: Boyle's law was perhaps the first expression of an equation of state. In 1662, Robert Boyle (1627-1691) performed a series of experiments employing a J-shaped glass tube, which was sealed on one end. Mercury was added to the tube, trapping a fixed quantity of air in the short, sealed end of the tube. Then the volume of gas was carefully measured as additional mercury was added to the tube. As we discussed in Chapter 9, the pressure of the gas could be determined by the difference between the mercury level in the short end of the tube and that in the long, open end. Through these experiments, Boyle noted that the gas volume varied inversely with the pressure. In mathematical form, this finding can be stated as the volume of gas is inversely proportional to the pressure applied to it when the temperature is kept constant

$$V \propto \frac{1}{P} \quad \Rightarrow \quad PV = \text{constant at fixed T}.$$

The above relationship has also been attributed to Mariotte and is sometimes referred to as **Mariotte's law**. However, Mariotte's work was not published until 1676.

Charle's law: In 1787, Jacques Charles (1746-1823) found that oxygen, nitrogen, hydrogen, carbon dioxide, and air expand to the same extent over the same 80 kelvin interval. This finding is known as Charle's law. Hence, Charles' law describes the direct relationship of temperature and volume of a gas. Assuming that pressure does not change, a doubling in absolute temperature of a gas causes a doubling of the volume of that gas. A drop of absolute temperature sees a proportional drop in volume. The volume of a gas increases by 1/273 of its volume at 0°C for every degree Celsius rise in temperature. The volume of a given amount of gas is directly proportional to the absolute temperature when the pressure is kept constant

$$V \propto T \quad (T \text{ is in Kelvin (K)}).$$

In other words, Charles's law states that if a given quantity of gas is held at a constant pressure, its volume is directly proportional to the absolute temperature. We may think of the physics of this law in the following way. As the temperature of the gas increases, the gas molecules will begin to move around more quickly and hit the walls of their container with more force, leading to increasing volume.

Gay-Lussac's law: This gas law may be derived as a corollary to Boyle's and Charles's laws. Suppose we double the thermodynamic temperature of a sample of gas. According to Charles's law, the volume should double. Now, how much pressure would be required at the higher temperature to return the gas to its original volume? According to Boyle's law, we would have to double the pressure to halve the volume. Thus, if the volume of gas is to remain the same, doubling the temperature will require doubling the pressure. This law was first stated by Gay-Lussac (1778-1850). At constant volume, the pressure of a gas is directly proportional to the absolute temperature

$$P \propto T.$$

Gay-Lussac's law tells us that it may be dangerous to heat a gas in a closed container. The increased pressure might cause the container to explode.

Equation of state for ideal gas

Let us discuss how the equation of states for an ideal gas is obtained. The main point of three phenomenological findings that we discussed above indicates the state of an amount of gas is determined by its pressure P, volume V, and temperature T. We may now combine the three laws, Boyle's law, Charles law, and Gay-Lussac's law, to obtain

$$PV \propto T.$$

The result of careful experiments also indicates that this proportionality relation depends on the amount of gas in the system. This finding suggests that we must incorporate the effects due to the amount of gas since the experimental data show that, at constant T and P, the volume V of an enclosed gas increases in direct proportion to the mass m of gas present

$$\therefore PV \propto mT.$$

Also noting that the mass of gas present in the system may be written in terms of moles of gas by using the relation

$$n(\text{mol}) = \frac{\text{mass (grams)}}{\text{molecular mass (gm/mol)}}.$$

With the introduction of the universal constant R, we may rewrite the proportionality relation as an equation describing the relationship among P," V, and T. The most frequently used form of the ideal gas law is written as

$$PV = nRT \qquad (10.12)$$

where P is the absolute pressure, T is temperature in Kelvin scale, and R is the universal gas constant

$$R = 8.315 \frac{J}{mol \cdot K} = 0.0821 \frac{L \cdot atm}{mol \cdot K} = 1.99 \frac{cal}{mol \cdot K}. \qquad (10.13)$$

The ideal gas law equation of (10.12) is also called "**equation of state for an ideal gas**". The **ideal gas law** is use to describe the equation of state of a hypothetical ideal gas. It is a good approximation to the behavior of many gasses under many conditions, although it may not be accurate for real gasses. An ideal gas is defined as one in which all collisions between atoms or molecules are perfectly elastic and in which there are no intermolecular attractive forces. We can visualize an ideal gas as a collection of perfectly hard spheres which collide but do not otherwise interact with each other. In such a gas, all the internal energy is in the form of kinetic energy and any change in this internal energy is accompanied by a change in temperature. Note that real gases do not precisely follow Eq. (10.12).

As we will see below, the ideal gas law can be viewed as arising from the kinetic theory of gas molecules colliding with the walls of a container in accordance with Newton's laws. However, there is also a statistical element in the determination of the average kinetic energy of those molecules since there are a huge number of molecules in the container. In the context of statistical physics, the temperature is taken to be proportional to this average kinetic energy. This relation invokes the idea of kinetic temperature. Kinetic temperature is a measure of the energy of motion possessed by a substance, especially as a function of the root mean square velocity of the molecules of a gas.

THERMAL PHYSICS

Exercise Problem 10.4:

A helium party balloon, assumed to be a perfect sphere, has a radius of 18.0 cm. At room temperature (20 °C), its internal pressure is 1.05 atm. Find the number of moles of helium needed to inflate the balloon to these values.

PROBLEM SOLVING STRATEGY

What do we know?

The diameter of the spherical party balloon, temperature and internal pressure are known.

What concepts are needed?

We need to use the ideal gas law.

Why?

The ideal gas law relates the volume, temperature, and pressure to the amount of gas in the balloon.

Solution:

Assuming that the shape of the balloon as a sphere, we compute the volume as

$$V = \frac{4}{3}\pi r^3 = \frac{4}{3}\pi \times (0.18\,\text{m})^3 = 0.0244\,\text{m}^3.$$

Since the helium gas is trapped in the balloon, we use the ideal gas law $PV = nRT$ and compute the number n of mole of helium in the balloon as

$$n = \frac{PV}{RT} = \frac{1.05\,\text{atm} \times 1.013 \times 10^5 \,\text{N}/_{\text{m}^2 \times \text{atm}} \times 0.0244\,\text{m}^3}{8.315\,\text{J}/_{\text{mol} \times \text{K}} \times (273+20)\,\text{K}} = 1.065\,\text{mol}.$$

Since the molecular mass of helium is 4.00 g/mol, we compute the mass of helium gas as

$$\text{Mass} = n \times \text{molecular mass} = 1.065\,\text{mol} \times 4.00\,\frac{\text{g}}{\text{mol}} = 4.26\,\text{g}.$$

We may apply the ideal gas law to describe the property of real gas under different conditions. One commonly found condition for a gas system is that it is placed in a closed container with a fixed amount of gas. In this case, the amount of gas in the system cannot change, while the pressure, volume, and temperature may change.

For a fixed amount of gas

As we discussed above, the ideal gas of Eq. (10.12) obtained from the three gas laws describes the fundamental relationships between P and T, V and T, and V and P for a gas in a container. When there is a fixed amount of gas in the container, the cases involving changes in the pressure, temperature, and volume may be described by rewriting the ideal gas law as

$$PV = nRT \Rightarrow \frac{PV}{T} = nR. \tag{10.14}$$

Suppose the thermodynamic variables P_1, T_1, and V_1 represent the initial state of the system and the variables P_2, T_2, and V_2 represent the final state after the change, then Eq. (10.14) becomes

$$\frac{P_1 V_1}{T_1} = nR = \frac{P_2 V_2}{T_2}. \tag{10.15}$$

A few things should always be kept in mind when we work with this equation. Pressure is directly proportional to number of molecules and temperature since P is on the opposite side of the equation to n and T. However, pressure is inversely proportional to volume since P is on the same side of the equation as V.

Exercise Problem 10.5:

An automobile tire is filled to a gauge pressure of 200 kPa at 10 °C. After driving 100 km, the temperature within the tire rises to 40 °C. What is the pressure within the tire now?

PROBLEM SOLVING STRATEGY

What do we know?

The initial pressure and temperature as well as the final temperature of the tire are known.

What concepts are needed?

We need to apply the ideal gas law with the condition that the volume of the tire and the amount of gas inside it does not change.

Why?

We have a closed system in which the gas cannot leave. Also the volume of the tire will expand by a very small amount.

Solution:

When the tire temperature increases, the tire expands by only negligible amount. Since the volume will essentially remain constant, we may set $V_1 = V_2$ and simplify Eq. (10.15) to

$$\frac{P_1 V_1}{T_1} = \frac{P_2 V_2}{T_2} \Rightarrow \frac{P_1}{T_1} = \frac{P_2}{T_2}.$$

THERMAL PHYSICS

We now solve for P_2 and obtain

$$P_2 = \frac{T_2}{T_1} P_1 = (200\,\text{kPa} + 101\,\text{kPa}) \times \frac{(40+273)\,\text{K}}{(10+273)\,\text{K}} \cong 333\,\text{kPa}$$

and compute the percent increase in pressure as

$$\%\text{ increase} = \frac{(333\,\text{kPa} - 101\,\text{kPa}) - 200\,\text{kPa}}{200\,\text{kPa}} \times 100\% \cong 16\%.$$

The solution of Exercise Problem 10.5 indicates that the tire pressure can change when there are temperature fluctuations. For safety, it is important to remember that gas expands when heated and contracts when the temperature declines. In North America, the daily temperatures rise and fall between day and night, as well as seasonally. As the days get shorter and colder during fall and winter, it is especially important to check the tire pressure. Also, it is important to know that the recommended tire pressure for our vehicle (as specified in the owner's manual and the tire placard for the vehicle) is based on cold inflation pressure. This means that the tire pressure should be checked in the morning before the tire has been run, before the ambient temperature rises during the day and before the tire is exposed to direct sunlight. A good estimate to use when comparing tire pressure to air temperature is that tire pressure will adjust by 1 PSI for every 10 F°. For example, if the outside air temperature increases 10 F°, then the tire pressure will increase by 1 PSI. Conversely, if the air temperature falls 10 F°, then the tire pressure will decrease by 1 PSI.

we may write the ideal gas law differently

We may also express the ideal gas law of Eq. (10.12) in terms of the number N of molecules, rather than the number n of moles. Since the number of mole of gas molecules is $n = N/N_A$, where $N_A = 6.02 \times 10^{23}$ molecules/mol is Avogadro's number, we write the ideal gas law as

$$PV = nRT = \frac{N}{N_A} RT = N \frac{R}{N_A} T = N k_B T \tag{10.16}$$

where $k_B = R/N_A = 1.38 \times 10^{-23}$ J/K is called Boltzmann's constant. The difference between the two equations (10.12) and (10.16) is that we use the universal gas constant R when we describe the ideal gas from a macroscopic perspective, while we use Boltzmann's constant k_B when we describe it in the particle perspective. The latter perspective is useful for us in understanding how the ideal gas law is related to individual gas molecules obeying Newton's laws of motion which we discussed in Chapters 4 and 6. Of course, since there are so many gas molecules in the system, we also need to employ statistics. This indicates that we may also investigate the properties of gasses in terms of the kinetic theory.

10.6 KINETIC THEORY

As we have noted earlier, the kinetic theory is an investigation based on theory that matter is made up of atoms and that these atoms are in continuous random motion. Since a gaseous system in a container has an enormously large number of atoms or molecules which are constantly moving in random directions, we may use the kinetic theory to study the microscopic behavior of molecules. With this theory, we may derive the macroscopic relationships like the ideal gas law by accounting for the interaction between gas molecules and the walls of the container. In investigation of the property of gasses from the point of view of the kinetic theory, we make the following assumptions about the atoms or molecules in a gaseous system:

i. There are a large number of molecules N, each of mass m, moving in random directions with a variety of speeds.
ii. The molecules are, on the average, far apart from one another.
iii. The molecules are assumed to obey the laws of classical mechanics and are assumed to interact with one another only when they collide.
iv. Collisions with another molecule or the wall of the vessel are assumed to be perfectly elastic.

With these assumptions, we can see that the pressure P exerted on a wall of the container of gas is due to the constant bombardment of molecules. If the volume V is reduced by half, then the molecules are closer together and twice as many times they will be striking a given area of the wall per second. Hence, the pressure will be twice as great, indicating that

$$\therefore P \propto \frac{1}{V} \ (Boyle's\ Law).$$

Using the above four basic assumptions about the system, we now derive the equation of state for ideal gasses. This approach of study for the molecules in a gaseous system is a good example of a physical situation in which statistical methods give precise and dependable results for macroscopic manifestations of microscopic phenomena.

Derivation of ideal gas law

Suppose we calculate the pressure in a gas based on the kinetic theory, assuming that molecules are contained in a rectangular vessel.

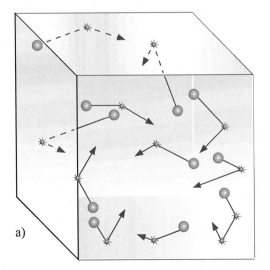
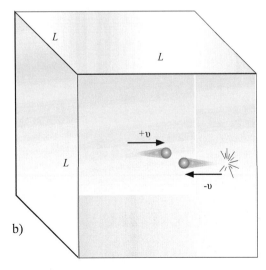

Figure 10.10: a) Gas molecules moving randomly in a cubical container is illustrated. b) A gas molecule moving in one dimension rebounds from one wall.

THERMAL PHYSICS

When one gas molecule strikes the end wall as shown in Fig. 10.10, it exerts a force on the wall, and the wall exerts an equal and opposite force back on the molecule. The magnitude of this force is equal to the rate of change of momentum. Assuming the collision is elastic, only the x-component of the molecule's momentum changes

$$\Delta(mv) = mv_x - (-mv_x) = 2mv_x.$$

Since this molecule will make many collisions with the wall, each collision at the same wall is separated by a time interval Δt

$$\Delta t = \frac{2L}{v_x}.$$

The force exerted on the wall due to this one molecule is

$$F = \frac{\Delta(mv)}{\Delta t} = \frac{2mv_x}{2L/v_x} = \frac{mv_x^2}{L}$$

Since there are many molecules in the container and these molecules also exert force on the wall, the force due to all the molecules in the box is given by

$$F = \frac{mv_{x1}^2}{L} + \frac{mv_{x2}^2}{L} + \cdots + \frac{mv_{xN}^2}{L} = \frac{m}{L}\left(v_{x1}^2 + v_{x2}^2 + \cdots + v_{xN}^2\right)$$

$$= \frac{m}{L} N \frac{v_{x1}^2 + v_{x2}^2 + \cdots + v_{xN}^2}{N} = \frac{m}{L} N \overline{v_x^2}$$

(10.17)

where

$$\overline{v_x^2} = \frac{v_{x1}^2 + v_{x2}^2 + \cdots + v_{xN}^2}{N}$$

is the average value of the square of the x-component of velocity. Since $v^2 = v_x^2 + v_y^2 + v_z^2$ and velocities of the molecules in the gaseous system are assumed to be random, there is no preference to one direction or another, we may write

$$\overline{v_x^2} = \overline{v_y^2} = \overline{v_z^2} \Rightarrow \overline{v^2} = 3\overline{v_x^2} \Rightarrow \overline{v_x^2} = \overline{v^2}/3.$$

Using this result, we may rewrite Eq. (10.17) for the force F that the gas molecules exert on the wall as

$$F = \frac{m}{L} N \frac{\overline{v^2}}{3}.$$

(10.18)

The pressure P exerted on the wall, according to the definition of pressure, may be related to the average of the velocity square as

$$P = \frac{F}{A} = \frac{(m/L)N(\overline{v^2}/3)}{L^2} = \frac{1}{3} N \frac{m\overline{v^2}}{L^3} = \frac{1}{3} \frac{N}{V} m\overline{v^2}.$$

(10.19)

We rewrite Eq. (10.19) for pressure P in a more useful form as

$$PV = \frac{2}{3}N\left(\frac{1}{2}\overline{mv^2}\right). \tag{10.20}$$

We may now write Eq. (10.20) in terms of temperature by using the **equipartition theorem**, which is a general formula relating the temperature of a system with its average energies. The equipartition theorem is also known as the law of equipartition, or equipartition of energy. The original idea of equipartition was that, in thermal equilibrium, energy is shared equally among all of its various forms. For example, the average kinetic energy per degree of freedom in the translational motion of a molecule should equal that of its rotational motions. The equipartition theorem states that the average kinetic energy may be written as

$$\frac{1}{2}\overline{mv^2} = \frac{3}{2}k_B T. \tag{10.21}$$

This suggests that we may write Eq. (10.20) as the equation of state for ideal gas

$$PV = N\frac{2}{3}\left(\frac{1}{2}\overline{mv^2}\right) = Nk_B T. \tag{10.22}$$

The average translational kinetic energy of molecules in the gas is directly proportional to the absolute temperature. According to the kinetic theory, **the higher the temperature the faster the molecules are moving, on the average.** The **root-mean-square velocity** v_{rms} is obtained by taking the square root of the mean of the square of the velocity

$$v_{rms} = \sqrt{\overline{v^2}} = \sqrt{\frac{3k_B T}{m}}. \tag{10.23}$$

Although the molecules in a gaseous system have an average kinetic energy (and therefore an average speed) the individual molecules move at various speeds and they stop and change direction according to the law of density measurements and isolation (i.e., they exhibit a distribution of speeds). Some move fast, while others relatively slowly. Collisions change individual molecular speeds, but the distribution of speeds remains the same. This equation is derived from the kinetic theory of gases using the Maxwell–Boltzmann distribution function. The higher the temperature the greater the mean velocity will be. This works well for both nearly ideal, atomic gasses like helium and for molecular gasses like diatomic oxygen. The reason for this is because despite the larger internal energy in many molecules (compared to that for an atom), $3RT/2$ is still the mean translational kinetic energy.

Understanding-the-Concept Question 10.6:

A fixed container holds oxygen and helium gases at the same temperature. Which one of the following statements is correct?

    a.   The oxygen molecules have the greater kinetic energy
    b.   The helium molecules have the greater kinetic energy
    c.   The oxygen molecules have greater speed
    d.   The helium molecules have the greater speed

THERMAL PHYSICS

Explanation:

According to Eq. (10.21), the average kinetic energy of a gas molecule is proportional to the temperature. This indicates that any two gas molecules at the same temperature will have the same kinetic energy. However, the root-mean-square velocity of Eq. (10.23) indicates that the speed is inversely proportional to the square root of the mass of a gas molecule. This suggests that a lighter molecule will have a greater speed.

Answer: d

Understanding-the-Concept Question 10.7:

According to the ideal gas law, PV = constant for a given temperature. As a result, an increase in volume corresponds to a decrease in pressure. This happens because the molecules

    a.   collide with each other more frequently.

    b.   move slower on the average.

    c.   strike the container wall less often.

    d.   transfer less energy to the walls of the container each time they strike it.

Explanation:

According to the ideal gas law at a fixed temperature, an increase in volume leads to decrease in pressure. This means the force per unit area on the walls of a container exerted by the colliding gas molecules is less. When a gas molecule strikes the container wall, it changes its direction of motion and a momentum is transferred to the wall in a very short time period. This indicates that the force exerted to the container wall is proportional to the number of collisions that the gas molecules make with the wall.

Answer: c

Example Problem 10.6:

What is the root-mean-square speed (v_{rms}) of air molecules (O_2 and N_2) at room temperature (20 °C)?

Gas	Ratio compared to Dry Air (%)		Molecular Mass - M (kg/kamol)	Chemical Symbol	Boiling Point (°C)
	By volume	By weight			
Oxygen	20.95	23.20	32.00	O_2	-196
Nitrogen	78.09	75.47	28.02	N_2	-183
Carbon Dioxide	0.03	0.046	44.01	CO_2	-78.5
Hydrogen	0.00005	~0	2.02	H_2	-252.87
Argon	0.933	1.28	39.94	Ar	-185
Neon	0.0018	0.0012	20.18	Ne	-245
Helium	0.0005	0.00007	4.00	He	-269
Krypton	0.0001	0.0003	83.8	Kr	-153.4
Xenon	9×10^{-6}	0.00004	131.29	Xe	-108.1

Solution:

To compute the root-mean-square speed of Eq. (10.23), first, we need to compute the mass of each molecule. The mass of one molecule of O_2 (32 u) and N_2 (28 u) are

$$m(O_2) = 32 \times 1.67 \times 10^{-27} \text{ kg} \cong 5.3 \times 10^{-26} \text{ kg}$$

$$m(N_2) = 28 \times 1.67 \times 10^{-27} \text{ kg} \cong 4.7 \times 10^{-26} \text{ kg}$$

Using Eq. (10.23), we compute the root-mean-square speed urms of one oxygen molecule and nitrogen molecule as

$$\therefore v_{rms}^{O_2} = \sqrt{\frac{3k_BT}{m}} = \sqrt{\frac{3 \times 1.38 \times 10^{-23} \text{ J/K} \times (273+20)\text{K}}{5.3 \times 10^{-26} \text{ kg}}} \cong 480 \frac{\text{m}}{\text{s}},$$

$$v_{rms}^{N_2} = \sqrt{\frac{3k_BT}{m}} = \sqrt{\frac{3 \times 1.38 \times 10^{-23} \text{ J/K} \times (273+20)\text{K}}{4.7 \times 10^{-26} \text{ kg}}} \cong 510 \frac{\text{m}}{\text{s}}.$$

It is noted that the speed of a nitrogen molecule is greater than that of an oxygen molecule because it has smaller mass. These high speed gas molecules are constantly colliding with our skin. This is the atmospheric pressure that we feel on the surface of Earth.

Velocity Distribution

Since molecules in a gas are assumed to be in random motion, there are many molecules with speeds less than the rms speed v_{rms} and greater than this speed. There is a wide range of speeds. The kinetic theory also predicts that the speeds of molecules in a gas are distributed according to the "**Maxwell distribution of speeds**" as shown below in Fig. 10.11.

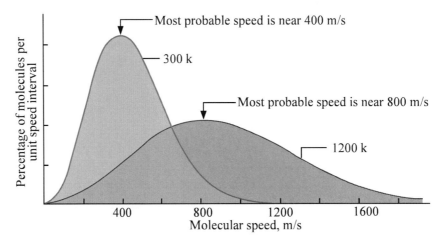

Figure 10.11: A schematic illustration of Maxwell-Boltzmann distribution of speed. Two curves indicate the distribution of speed of gas molecules in a container at 300 K (blue) and at 1200 K (red). The speed corresponding to the peak of each curve is the most probable speed at a fixed temperature.

THERMAL PHYSICS

The Maxwell-Boltzmann distribution describes the distribution of particle speeds in an ideal gas. The distribution may be characterized in a variety of ways. These ways include the average speed, most probable speed, and width of the distribution. The average speed is the sum of the speeds of all of the particles divided by the number of particles. The most probable speed is the speed associated with the highest point in the Maxwell-Boltzmann distribution. Only a small fraction of particles might have this speed, but it is more likely than any other speed. The width of the distribution characterizes the most likely range of speeds for the particles. One measure of the width is the **Full Width at Half Maximum**. To determine this value, we need to find the height of the distribution at the most probable speed (this is the maximum height of the distribution). We, then, divide the maximum height by two to obtain the half height, and locate the two speeds in the distribution that have this half-height value. One speed will be greater than the most probably speed and the other speed will be smaller. The full width is the difference between the two speeds at the half-maximum value.

In our discussion of distribution of speeds, we assumed that each gas molecule is unique. In other words, we may be able to distinguish each gas molecule by labeling it individually. This is called "classical particles". However, the microscopic world is strange in a sense that we cannot distinguish each particle separately. All gas molecules are exactly identical to each other. These indistinguishable gas molecules are called "quantum particles". Similar to classical particles, quantum particles have a speed distribution. However, these quantum particles are not distributed according to the Maxwell-Boltzmann distribution. They are distributed according to either Fermi-Dirac distribution or Bose-Einstein distribution. This is beyond the scope of this book.

10.7 INTERNAL (OR THERMAL) ENERGY

As we have seen from the kinetic theory, each gas molecule in a container has energy. The energy of each molecule is added to yield the total energy of the system. The **internal energy** is the total energy contained by a thermodynamic system. It is the energy needed to create the system, but excludes the energy to displace the system's surroundings, any energy associated with motion as a whole, or due to external force fields. Internal energy has two major components, kinetic energy and potential energy. The kinetic energy is due to the motion of the system's particles. The kinetic energy includes the energy from translational, rotational and vibrational motion. The potential energy is associated with the static rest mass energy of the constituents of matter, static electric energy of atoms within molecules or crystals, and the static energy of chemical bonds. The internal energy of a system can be changed by heating the system or by doing work on it. According to the first law of thermodynamics which we discuss in Chapter 12, the increase in internal energy is equal to the total heat added minus work done by the system on the surroundings. If the system is isolated, its internal energy cannot change. **Internal energy is the total sum of the energy of all the molecules in the system.** The internal energy U of $n = N/N_A$ moles of an ideal monatomic (one atom per molecule) gas is the sum of the translational kinetic energies of all the atoms

$$U = N\left(\frac{1}{2}m\overline{v^2}\right). \tag{10.19}$$

Since the average kinetic energy of a molecule, based on the equipartition theorem, is proportional to the temperature T and is written as

$$\overline{K.E.} = \frac{1}{2}m\overline{v^2} = \frac{3}{2}k_BT,$$

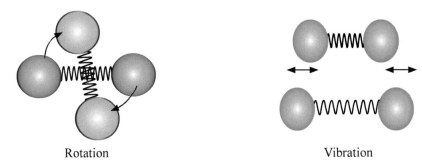

Rotation   Vibration

Figure 10.12: For a gas of diatomic molecules has larger internal energy than a monatomic gas because, in addition to the translational degrees of freedom, it has both the rotational (left panel) and vibrational (right panel) degrees of freedom which add to the kinetic energy of each molecule.

we may write the expression for the internal energy of the system as

$$U = \frac{3}{2} N k_B T \text{ or } U = \frac{3}{2} n R T. \tag{10.20}$$

Equation (10.20) may be understood from the fact that each monatomic gas molecule has three translational degrees of freedom, and the energy due to each degree of freedom yields $(1/2) k_B T$ as the corresponding kinetic energy. If the gas molecules are not monatomic, then each molecule contains more than one atom. Then, the **rotational** and **vibrational** energy of each molecule as shown in Fig. 10.12 must also be taken into account.

For a diatomic molecule, the three translation direction contribute $(3/2) k_B T$ per molecule, and for each molecule there are also two axes of rotation, contributing rotational kinetic energy in the amount of $(1/2) k_B T$ per each axis per molecule. This amounts to the total internal energy of $U = (5/2) N k_B T$ for a diatomic gas. For a polyatomic gas which has contributions from three translation direction and three axes of rotation, the internal energy is $U = (6/2) N k_B T = 3 N k_B T$.

SUMMARY

i. Three **most common scales for temperature are**
   1. Celsius scale or centigrade scale
   2. Fahrenheit scale (used widely in USA)
   3. Absolute or Kelvin scale (important scale in scientific work)

ii. The **amount of thermal expansion or contraction of a solid material** is proportional to the change in temperature ΔT. The linear expansion of a solid is described by

$$\Delta L = \alpha L_0 \Delta T,$$

where α is the coefficient of linear expansion.

iii. The **change in volume of a material** due to a temperature is given by

$$\Delta V = \beta V_0 \Delta T,$$

where β is the coefficient of volume expansion and V_0 is the original volume.

THERMAL PHYSICS

iv. **Thermal stress** is defined as the induced stress due to changes in temperature of the object which yields thermal expansion

$$\alpha L_o \Delta T = \frac{1}{E}\frac{F}{A} L_o \Rightarrow F = \alpha E A \Delta T.$$

The thermal stress equation indicates that change in temperature of an object is equivalent to applying force over a cross-sectional area A (i.e., mechanical stress).

v. Experimentally determined relations governing the pressure, temperature and volume of a gas were determined from combining the following three laws associated with gas: i) Boyle's law, ii) Charles's law, and Gay-Lussac's law. The ideal gas law equation of

$$PV = nRT = Nk_B T$$

is also called "**equation of state for an ideal gas**".

vi. The **kinetic theory of gases** is the study of the microscopic behavior of molecules and the interactions which lead to macroscopic relationships like the ideal gas law. According to kinetic theory, the higher the temperature the faster the molecules are moving, on the average. The **root-mean-square velocity** v_{rms} is obtained by taking the square root of the mean of the square of the velocity

$$v_{rms} = \sqrt{\overline{v^2}} = \sqrt{\frac{3k_B T}{m}}.$$

The kinetic theory also predicts that the speeds of molecules in a gas are distributed according to the "**Maxwell distribution of speeds**."

MORE WORKED PROBLEMS

Problem 10.1

The brass bar and aluminum bar in the drawing are attached to an immovable wall. At 28 °C the air gap between the rods is 1.3×10^{-3} m. At what temperature will the gap be closed?

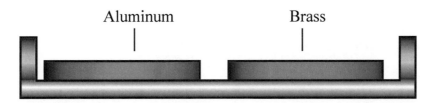

Solution:

The expression $\Delta L = \alpha L_o (\Delta T)$ may be rewritten for the expansion of the aluminum and brass as

$$\Delta L_{Al} = \alpha_{Al} L_{o,Al}(\Delta T) \text{ and } \Delta L_{Brass} = \alpha_{Brass} L_{o,Brass}(\Delta T).$$

We combine these two equations and write

$$\Delta L_{Al} + \Delta L_{Brass} = \alpha_{Al} L_{o,Al}(\Delta T) + \alpha_{Brass} L_{o,Brass}(\Delta T).$$

We now solve for ΔT and obtain

$$\Delta T = \frac{\Delta L_{Al} + \Delta L_{Brass}}{\alpha_{Al} L_{o,Al} + \alpha_{Brass} L_{o,Brass}} = \frac{1.3 \times 10^{-3}\,\text{m}}{23 \times 10^{-6}(\text{C}^\circ)^{-1} \times 1.0\,\text{m} + 19 \times 10^{-6}(\text{C}^\circ)^{-1} \times 2.0\,\text{m}} = 21\text{C}^\circ.$$

Since $\Delta T = T_f - T_i = 21\,°\text{C}$ and $T_i = 28\,°\text{C}$, we may write the final temperature as $T_f = T_i + \Delta T$. We obtain $T_f = 49\,°\text{C}$.

Problem 10.2

On a warm day (92 °F) an air-filled balloon occupies a volume of 0.20 m³ and is under a pressure of 20.0 lb/in². If the balloon is cooled to 32 °F in a refrigerator while its pressure is reduced to 14.7 lb/in², what is the volume of the air in the container? (Assume that the air behaves as an ideal gas.)

Solution:

Before starting to solve this problem, we convert temperatures in Fahrenheit to Kevin scale:

$$T_1 = 92\,°\text{F} = \frac{5}{9}(92-32)\,°\text{C} = 33.3\,°\text{C} \Rightarrow T_1 = (33.3 + 273)\,\text{K} = 306\,\text{K}$$

$$T_2 = 32\,°\text{F} = 0\,°\text{C} \Rightarrow T_2 = 273\,\text{K}.$$

We now use the ideal gas law $PV = nRT$. Since the amount of air in the balloon does not change, we may rewrite the equation of state for ideal gas as

$$\frac{PV}{T} = nR \Rightarrow \frac{P_1 V_1}{T_1} = \frac{P_2 V_2}{T_2}.$$

We solve for V_2 by doing some algebra

$$V_2 = V_1 \frac{P_1 T_2}{T_1 P_2} = 0.20\,\text{m}^3 \times \frac{20.0\,\text{PSI}}{306\,\text{K}} \times \frac{273\,\text{K}}{14.7\,\text{PSI}} = 0.24\,\text{m}^3.$$

It should be noted that we used here the units of lb/in² for pressure, rather than the units of N/m², because it is in a ratio, which results in cancellation of these units.

Problem 10.3

Two small containers, each with a volume of 100 cm³, contain helium gas at 0 °C and 1.00 atm pressure. The two containers are joined by a small open tube of negligible volume, allowing gas to flow from one container to the other. What common pressure will exist in the two containers if the temperature of one container is raised to 100 °C while the other container is kept at 0 °C?

Solution:

Suppose we let container 1 be maintained at $T_1 = T_o = 0\,°\text{C} = 273\,\text{K}$ while the temperature of container 2 is raised to $T_2 = 100\,°\text{C} = 373\,\text{K}$. Both containers have the same constant volume V, and the same initial

THERMAL PHYSICS

pressure $P_{o,1} = P_{o,2} = P_o$. As the temperature of container 2 is raised, gas flows from one container to the other until the final pressures are again equal, $P_1 = P_2 = P$. The total mass of gas is constant, so we may write

$$n_2 + n_1 = (n_o)_2 + (n_o)_1.$$

From the ideal gas law $PV = nRT$, the amount of gas in each container may be written as $n = PV/RT$. Using the expression for n, we write

$$\frac{PV}{RT_2} + \frac{PV}{RT_1} = \frac{P_o V}{RT_o} + \frac{P_o V}{RT_o} \Rightarrow P\left(\frac{1}{T_2} + \frac{1}{T_1}\right) = \frac{2P_o}{T_o}.$$

We solve for P and obtain

$$P = \frac{2P_o}{T_o} \frac{T_1 T_2}{T_1 + T_2} = \frac{2 \times 1.00 \,\text{atm}}{273 \,\text{K}} \times \frac{273 \,\text{K} \times 373 \,\text{K}}{273 \,\text{K} + 373 \,\text{K}} = 1.15 \,\text{atm}.$$

Problem 10.4

Two identical thermometers made of Pyrex glass contain, respectively, identical volumes of mercury and methyl alcohol. If the expansion of the glass is taken into account, how many times greater is the distance between the degree marks on the methyl alcohol thermometer than the distance on the mercury thermometer?

Solution:

The cavity that contains the liquid in either Pyrex thermometer expands according to

$$\Delta V_{glass} = \beta_{glass} V_o (\Delta T).$$

On the other hand, the volume of mercury expands by an amount $\Delta V_m = \beta_m V_o (\Delta T)$. The net change in volume for the mercury thermometer is

$$\Delta V_m - \Delta V_{glass} = (\beta_m - \beta_{glass}) V_o (\Delta T).$$

Similarly, the net change in volume for the alcohol thermometer is

$$\Delta V_a - \Delta V_{glass} = (\beta_a - \beta_{glass}) V_o (\Delta T).$$

In each case, this volume change is related to a movement of the liquid into a cylindrical region of the thermometer with volume $V = \pi r^2 h$, yielding height h as

$$h_m = \frac{(\beta_m - \beta_{glass}) V_o (\Delta T)}{\pi r^2} \quad \text{(for the mercury thermometer)}$$

and

$$h_a = \frac{(\beta_a - \beta_{glass}) V_o (\Delta T)}{\pi r^2} \quad \text{(for the alcohol thermometer).}$$

We now take the ratio of these two heights and obtain

$$\frac{h_m}{h_a} = \frac{\beta_m - \beta_{glass}}{\beta_a - \beta_{glass}} = \frac{1200 \times 10^{-6} (C°)^{-1} - 9.9 \times 10^{-6} (C°)^{-1}}{182 \times 10^{-6} (C°)^{-1} - 9.9 \times 10^{-6} (C°)^{-1}} = 6.9,$$

indicating that the degree marks are 6.9 times farther apart.

PROBLEMS

1. (a) The largest temperature drop recorded in the United States in one day occurred in Browning, Montana in 1916, when the temperature went from 7 °C to -49 °C. What is the corresponding change on the Fahrenheit scale? (b) On the Moon, the average surface temperature is 127 °C during the day and -183 °C during the night. What is the corresponding change on the Fahrenheit scale?

2. Temperature differences on the Rankine scale are identical to differences on the Fahrenheit scale, but absolute zero is given as 0 °R. (a) Find a relationship converting the temperatures T_F of the Fahrenheit scale to the corresponding temperatures T_R of the Rankine scale. (b) Find a second relationship converting temperatures T_R of the Rankine scale to the temperatures T_K of the Kelvin scale.

3. A wire is made by attaching two segments together, end to end. One segment is made of aluminum and the other is steel. The effective coefficient of linear expansion of the two-segment wire is 19×10^{-6} (C°)$^{-1}$. What fraction of the length is aluminum?

4. Lead has a density of 11.3×10^3 kg/m³ at 0 °C. (a) What is the density of lead at 90 °C? (b) Based on your answer to part (a), now consider a situation in which you plan to invest in gold bar. Would you be better off buying it on a warm day? Explain.

5. Concrete sidewalks are always laid in sections, with gaps between each section. For example, the drawing shows three identical 2.4-m sections, the outer two of which are against immovable walls. The two identical gaps between the sections are provided so that thermal expansion will not create the thermal stress that could lead to cracks. What is the minimum gap width necessary to account for an increase in temperature of 32 C°?

6. A steel bicycle wheel (without the rubber tire) is rotating freely with an angular speed of 18.00 rad/s. The temperature of the wheel changes from -100.0 to +300.0 °C. No net external torque acts on the wheel, and the mass of the spokes is negligible. (a) Does the angular speed increase or decrease as the wheel heats up? Why? (b) What is the angular speed at the higher temperature?

7. Show that the coefficient of volume expansion, β, is related to the coefficient of liner expansion, α, through the expression $\beta = 3\alpha$.

8. The average coefficient of volume expansion for carbon tetrachloride is 5.81×10^{-4} (°C)$^{-1}$. If a 50.0-gal steel container is filled completely with carbon tetrachloride when the temperature is 10.0 °C, how much will spill over when the temperature rises to 30.0 °C?

9. The bottom of an old mercury-in-glass thermometer has 45-mm³ reservoir filled with mercury. When the thermometer was placed under your tongue, the warm mercury would expand into a very narrow cylindrical channel, called a capillary, whose radius was 1.7×10^{-2} mm. Marks were placed along the capillary that indicated the temperature. Ignore the thermal expansion of the glass and determine how far (in mm) the mercury would expand into the capillary when the temperature changed by 1.0 C°.

THERMAL PHYSICS

10. A construction worker uses a steel tape to measure the length of an aluminum support column. If the measured length is 18.700 m when the temperature is 21.2 °C, what is the measured length when the temperature rises to 29.4 °C? (*Note*: do not neglect the expansion of the tape.)

11. A 1.5 kg steel sphere will not fit through a circular hole in a 0.85 kg aluminum plate, because the radius of the sphere is 0.10% larger than the radius of the hole. If both the sphere and the plate are always kept at the same temperature, how much heat must be put into the two so the ball just passes through the hole?

12. A brass rod has a circular cross section of radius 0.500 cm. The rod fits into a circular hole in a copper sheet with a clearance of 0.010 mm completely around it when both it and the sheet are at 20 °C. (a) At what temperature will the clearance be zero? (b) Would such a tight fit be possible if the sheet was brass and rod was copper?

13. What temperature change would cause a 0.10 % increase in the volume of a quantity of water that was initially at 20 °C?

14. If a container holds 0.865 liter of ethyl alcohol at 12.0 °C, what is the volume of the alcohol when its temperature is raised to 35.0 °C?

15. The bulk modulus of water is $B = 2.2 \times 10^9$ N/m². What change in pressure DP (in atmospheres) is required to keep water from expanding when it is heated from 15 to 25 °C?

16. On a day when the temperature is 20.0 °C, a concrete walk is poured in such a way that its ends are unable to move. (a) What is the stress in the cement when its temperature is 50.0 °C on a hot, sunny day? (b) Does the concrete fracture? Take Young's modulus to concrete to be 7.00×10^9 N/m² and the compressive strength to be 2.00×10^7 N/m².

17. A liquid with a coefficient of volume expansion of β just fills a spherical flask of volume V_o at temperature T_i. The flask is made of a material that has a coefficient of linear expansion of α. The liquid is free to expand into a capillary of cross-sectional area A at the top. (a) Show that if the temperature increases by DT, the liquid rises in the capillary by the amount $Dh = (V_o/A)(β - 3α) DT$. (b) For a typical system, such as a mercury thermometer, why is it a good approximation to neglect the expansion of the flask?

18. How many grams of water (H_2O) have the same number of oxygen atoms as 1.0 mol of oxygen gas?

19. A runner weighs 580 N (about 130 lb), and 71% of this weight is water. (a) How many moles of water are in the runner's body? (b) How many water molecules (H_2O) are there?

20. In a period of 1.0 s, 5.0×10^{23} nitrogen molecules strike a wall of area 8.0 cm². If the molecules move at 300 m/s and strike the wall head on in a perfectly elastic collision, find the pressure exerted on the wall. (The mass of one N_2 molecule is 4.68×10^{-26} kg.)

21. In 10.0 s, 200 bullets strike and embed themselves in a wall. Assume that the bullets strike the wall perpendicularly. Each bullet has a mass of 5.0×10^{-3} kg and a speed of 1200 m/s. (a) What is the average change in momentum per second for the bullets? (b) Determine the average force exerted on the wall. (c) Assuming the bullets are spread out over an area of 3.0×10^{-4} m², obtain the average pressure exerted on this region of the wall.

22. 7.5 mol of helium are in a 1.5 L cylinder. The pressure gauge on the cylinder reads 65 pounds per square inch. What are (a) the temperature of the gas in °C and (b) the average kinetic energy of a helium atom?

23. A quantity of an ideal gas is at 0 °C. An equal quantity of another ideal gas is twice as hot. What is its temperature?

24. The density of helium gas at 0 °C is $r_o = 0.179$ kg/m³. The temperature is then raised to $T = 100$ °C, but the pressure is kept constant. Assuming the helium is an ideal gas, calculate the new density r_f of the gas.

25. A 7.00-L vessel contains 3.50 mole of ideal gas at a pressure of 1.60×10^6 Pa. Find (a) the temperature of the gas and (b) the average kinetic energy of a gas molecule in the vessel. (c) What additional information would you need if you were asked to find the average speed of a gas molecule?

26. A gas fills the right portion of a horizontal cylinder whose radius is 5.00 cm. The initial pressure of the gas is 1.01×10^5 Pa. A frictionless movable piston separates the gas from the left portion of the cylinder, which is evacuated and contains an ideal spring, as the drawing shows. The piston is initially held in place by a pin. The spring is initially unstrained, and the length of the gas-filled portion is 20.0 cm. When the pin is removed and the gas is allowed to expand, the length of the gas-filled chamber doubles. The initial and final temperatures are equal. Determine the spring constant of the spring.

27. A tank having a volume of 0.100 m³ contains helium gas at 150 atm. How many balloons can the tank blow up if each filled balloon is a sphere 0.300 m in diameter at an absolute pressure of 1.20 atm?

28. A diver 50 m deep in 10 °C fresh water exhales a 1.0-cm-diameter bubble. What is the bubble's diameter just as it reaches the surface of the lake, where the water temperature is 20 °C?

29. A weather balloon is designed to expand to a maximum radius of 20 m at its working altitude, where the air pressure is 0.030 atm and the temperature is 200 K. If the balloon is filled at atmospheric pressure and 300 K, what is its radius at liftoff?

30. A driver releases an air bubble of volume 2.0 cm³ from a depth of 15 m below the surface of a lake, where the temperature is 7.0 °C. What is the volume of the bubble when it reaches just below the surface of the lake, where the temperature is 20 °C?

31. It is possible to make a thermometer by sealing gas in a rigid container and measuring the pressure. Such a constant-volume gas thermometer is placed in an ice-water bath at 0.00 °C. After reaching thermal equilibrium, the gas pressure is recorded as 55.9 kPa. The thermometer is then placed in contact with a sample of unknown temperature. After the thermometer reaches a new equilibrium, the gas pressure is 65.1 kPa. What is the temperature of this sample?

32. 10 g of liquid water is placed in a flexible bag, the air is excluded, and the bag is sealed. It is then placed in a microwave oven where the water is boiled to make steam at 100 °C. What is the volume of the bag after all the water has boiled? Assume that the pressure inside the bag is equal to atmospheric pressure.

33. The drawing shows two thermally insulated tanks. They are connected by a valve that is initially closed. Each tank contains neon gas at the pressure, temperature, and volume indicated in the drawing. When the valve is opened, the contents of the two tanks mix, and the pressure becomes constant throughout. (a) What is the final temperature? Ignore any change in temperature of the tank themselves. (*Hint*: The heat gained by the gas in one tank is equal to that lost by the other.) (b) What is the final pressure?

34. If the temperature of an ideal gas increases from 300 K to 600 K, what happens to the root-mean-square (rms) speed of the gas molecules?

35. At what temperature, in °C, is the rms speed of oxygen molecules (a) half and (b) twice its value at 0 °C?

36. For an average molecule of N_2 gas at 0 °C, what are its (a) translational kinetic energy, (b) rotational kinetic energy, and (c) total energy?

37. Helium (He), a monatomic gas, fills a 0.010-m³ container. The pressure of the gas is 6.2×10^5 Pa. How long would a 0.25-hp engine have to run (1 hp = 746 W) to produce an amount of energy equal to the internal energy of this gas? (S10, 14.42)

ENERGY TRANSFER PROCESSES

11
CHAPTER

Spending a nice fall evening with friends and family members around a warm campfire at a camp site near a lake shore, burning firewood and roasting marshmallows, is one of the pleasant moments to remember about camping. Imagine sitting downwind of a bonfire. The burning wood has lots of heat. Some of this heat is transferred to the air. The wind pushes the air toward toward where we are siting, and when the air hits our skin, it transfers some of its heat to our skin. The net result is a transfer of heat from the wood to our skin. This heat energy transfer process is called **convection**. Another example of the convection process might be the way everything in a small kitchen gets warm in a hurry when things are baking in a hot oven. Some radiative heat transfer occurs, but lots of air picks up heat from the stove and then rises to be displaced by cooler air. The hot air heats things in the upper regions of the kitchen, and then cools and sinks. It then may return to the stove to pick up more heat as hot air there continues t circulate upward. Convection currents in air transfer heat. As we discuss in this chapter, convectio is one of the heat transfer mechanisms. Convection is the process by which heat is transferred by "fluid" (which, in this case, can actually mean a liquid or a moving gas - both are considered "fluids" Heat is always transferred from an area of high heat to an area of low heat, regardless of the methoc When our hand touches a hot stove, heat moves from the stove to our hand to try to "even out" th amount of heat between the two objects.

In convection, heat is first transferred from an area of high heat to the fluid, then from the fluid t an area of (relatively) lower heat. When we stand near to a roaring bonfire, we can feel the heat fron the bonfire falling on our face. This happens because hot objects transfer heat by sending out rays This method of energy transfer is called radiation. No substance (solid, liquid, or gas) is needed so th radiation can travel through empty space. As described by these examples, heat transfer can occur i a number of different ways. In this chapter, we focus our attention to the energy transfer processes.

11.1 HEAT

As we discussed in Chapter 10, temperature is a measure of how hot or cold an object is. For example, hot oven is at high temperature and ice of a frozen lake is at low temperature. However, **heat** is define as a measure of flow of energy from a higher temperature object to a lower temperature object (i.e energy transfer). Let us focus on heat as a mechanism for a flow of energy. **Heat** is energy transferre from one body to another by thermal interactions. The transfer of energy can occur in a variet of ways, including conduction, radiation, and convection. Heat is not a property of a system o body, but instead is always associated with a process of some kind, and is synonymous with hea flow and heat transfer.

Heat flow from hotter to colder systems occurs spontaneously, and is always accompanied by an increase in entropy. The concept of entropy is discussed in Chapter 12. In a heat engine, internal energy of bodies is harnessed to provide useful work. Also, as we discuss in Chapter 12, the second law of thermodynamics prohibits heat flow directly from cold to hot systems. However, with the aid of a heat pump, external work can be used to transport internal energy indirectly from a cold to a hot body. Transfers of energy such as heat are macroscopic processes. The origin and properties of heat can be understood through the statistical mechanics of microscopic constituents such as molecules and photons. For instance, heat flow can occur when the rapidly vibrating molecules in a high temperature body transfer some of their energy (by direct contact, radiation exchange, or other mechanisms) to the more slowly vibrating molecules in a lower temperature body.

Figure 11.1: When charcoal is lighted at a grill in a par heat generated by the burning charcoals is transferred t the surroundings, since thermal energy flows from a hig temperature region to a cold temperature region.

ENERGY TRANSFER PROCESSES

There are two commonly used units for measuring heat energy. These two units are the calorie and the British thermal units. A calorie (cal) is defined as the amount of heat necessary to raise temperature of 1 g of water by 1 C°. The **calorie** is a pre-SI metric unit of energy. It was first defined by Clément in 1824 as a unit of heat. This unit for heat entered French and English dictionaries between 1841 and 1867. In most fields, its use has been replaced by the SI unit of energy, the joule. However, it remains a commonly used energy unit in the field of chemistry, and in many countries it remains in common use as a unit of food energy. On the other hand, the **British thermal unit (Btu)** is defined as the heat needed to raise the temperature of 1 lb of water by 1 F°. The **Btu** is a traditional unit of energy equal to about 1055 joules. It is approximately the amount of energy needed to heat one pound of water by one degree Fahrenheit. In scientific contexts the Btu has largely been replaced by the SI unit of energy, the joule. The unit is most often used as a measure of power (as Btu/h) in the power, steam generation, heating and air conditioning industries, and also as a measure of agricultural energy production (Btu/kg). It is still used in metric English-speaking countries (such as Canada), and remains the standard unit of classification for air conditioning units manufactured and sold in many non-English-speaking metric countries. In North America, the term "Btu" is used to describe the heat value (energy content) of fuels.

Since heat is a form of energy, it may be converted into another form, such as mechanical energy. In 1800's, a series of experiments carried out by James Joule (1818-1889) determined that a given amount of work done is always equivalent to a particular amount of heat input. This means that mechanical energy is equivalent to heat energy. It is reasonable to make this connection because both mechanical energy and heat energy are different forms of energy. Joules found out that **the mechanical equivalent of heat energy** may be expressed as

$$4.186 \text{ J} = 1 \text{ cal} \tag{11.1}$$

This equivalency has rather important implications. For example, it indicates that heat is energy which may be transferred from one body to another because of a difference in temperature. This transferred heat energy may be used to for work. As illustrated in Fig. 11.2, this equivalency relation suggests that heat energy needed to raise the temperature of water may be viewed as being equivalent to mechanical work energy needed to lift a weight vertically. This point of view allows us to consider both heat and work on an equal footing.

The concept of mechanical equivalence of heat states that motion and heat are mutually interchangeable, and that in every case, a given amount of work would generate the same amount of heat, provided the work done is totally converted to heat energy. In a classic experiment in 1843, Joule showed the energy equivalence of heating and doing work by using the change in potential energy of falling masses to stir an insulated container of water with paddles. Based on the result, Joule claimed that heat was only one of many forms of energy and only the sum of all forms was conserved. This is the basis for the theory of conservation of energy (the first law of thermodynamics) which we discuss in Chapter 12.

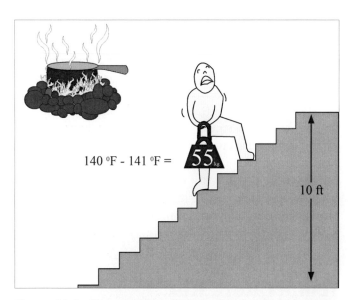

Figure 11.2: The mechanical equivalence of heat energy indicates that heat energy needed to raise the temperature of water in a pot can be measured in terms of the mechanical energy required to lift a certain massive object over some height.

Exercise Problem 11.1:

When a bullet with a mass of 3.0 g is traveling at the speed of 400 m/s passes through a rose, its speed is reduced to 200 m/s. How much heat Q is produced and shared by the bullet and the rose?

PROBLEM SOLVING STRATEGY

What do we know?
A bullet passing through a rose reduces its speed.

What concepts are needed?
We may apply conservation of energy.

Why?
Once we account for the heat energy shared by the bullet and the rose, the energy of the system is conserved.

Solution:

According to energy conservation, we may write

$$K.E._i = K.E._f + Q.$$

The heat Q produced by the bullet is $Q = K.E._i - K.E._f$. We solve for Q and obtain

$$Q = \frac{1}{2} \times 3.0 \times 10^{-3} \text{kg} \times \left[\left(400\frac{\text{m}}{\text{s}}\right)^2 - \left(200\frac{\text{m}}{\text{s}}\right)^2\right] \cong 180\text{J} \times \frac{1\text{cal}}{4.186\text{J}} \cong 43\text{cal}$$

The heat produced by the bullet goes into raising its temperature. If all of the heat goes to raising the temperature of the bullet, how can we determine its final temperature? As we will see below, how much temperature changes will depend on what material is used in the bullet as well as its mass.

As Exercise Problem 11.1 indicates, friction always creates heat. For example, brakes and tires depend on friction for their function. Heat is generated from friction between the brake pad and the rotor. This heat has a tremendous influence on the performance of the brake system. Every brake has a nice temperature range where it will work best. Understanding how the brakes react to heat is useful for determining the optimal operating conditions for the brakes to work best, to provide sufficient friction to stop a moving car. However, more friction is not always better. In the engine, friction is never good. So, engine manufacturers use everything from oil to high-tech coatings to reduce friction in order to get a little extra horsepower.

Temperature Rise

When heat is put into an object, its temperature rises. However, how much the temperature rises depends on the specific heat capacity of the object. The amount of heat Q needed to raise the temperature of a material by ΔT is given by

ENERGY TRANSFER PROCESSES

$$Q = cm\,(\Delta T), \tag{11.2}$$

where m is the mass of the material and c is the specific heat of the material. The units of specific heat may be seen easily from

$$c = \frac{Q}{m(\Delta T)}. \tag{11.3}$$

Equation (11.3) indicates the units for the specific heat c are J/kg·C° or kcal/kg·C°. It is noted that **specific heat** (or thermal capacity) is the measurable physical quantity which shows the amount of heat required to change the temperature of a substance by a given amount. Depending on situations, another quantity can be used to measure thermal capacity. Molar heat capacity and specific heat capacity are derived quantities which specify heat capacity as an intensive property (i.e., independent of the size of a sample). The **molar heat capacity** is the heat capacity per mole of a pure substance, and the **specific heat capacity** (often simply called **specific heat**) is the heat capacity per unit mass of a material. Occasionally, a volumetric heat capacity is used in engineering contexts. **Volumetric heat capacity** describes the ability of a given volume of a substance to store internal energy while undergoing a given temperature range with no phase transition. Because heat capacities of materials tend to reflect the number of particles they contain, they tend to vary within a much narrow range when intensive heat capacities of substances are expressed either directly or indirectly per particle number.

Table 11.1: Specific heat at 1 atm constant pressure and 20 °C unless stated otherwise.

Substance	Specific heat kcal/kg·C°	Specific heat J/kg·C°
Aluminum	0.22	900
Alcohol (ethyl)	0.58	2400
Copper	0.093	390
Glass	0.20	840
Iron or steel	0.11	450
Lead	0.031	130
Marble	0.21	860
Mercury	0.033	140
Silver	0.056	230
Wood	0.4	1700
Water: ice (-5 °C)	0.50	2100
Liquid (15 °C)	1.00	4186
Steam (110 °C)	0.48	2010
Human body (average)	0.83	3470
Protein	0.4	1700

Understanding-the-Concept Question 11.1:

The reason ocean temperatures do not vary drastically is that

    a.   water has a relatively high rate of heat conduction.

b. water is a good radiator.
c. water has a relatively high specific heat.
d. water is poor heat conductor.

Explanation:

To suppress a drastic temperature variation of surroundings, a medium needs to be able to absorb a large amount of heat when temperature of the surrounding increases and to release large amount of heat when the temperature decreases. Ocean temperatures do not vary drastically because water has high capacity to absorb heat energy. This means a relatively high amount of energy can be absorbed for a small increase in temperature, indicating that water has a relatively high specific heat compared to the ground.

Answer: c

Another important aspect of high heat capacity of water is that roughly 80 to 90 percent of heat from global warming may actually be going into the ocean. This means that oceans are behaving as a big heat bucket for Earth. In other words, this is where most of the heat winds up going since the oceans can absorb as much as 1000 times the amount of heat as the atmosphere without changing their temperature much. So, an increase of 1 C° in ocean temperature implies that lots of energy is trapped in the oceans. This additional energy may lead to more violent weather.

Exercise Problem 11.2:

a) How much heat is required to raise the temperature of an empty 20 kg vat made of iron from 10 °C to 90 °C? b) What if the vat is filled with 20 kg of water?

Solution:

The total amount of heat required is the sum of heat needed to raise the temperature of an empty vat and that of water inside the vat.

a. The amount of heat needed for the empty iron vat is computed from

$$c_{iron} m_{vat} \Delta T = 0.11 \frac{\text{kcal}}{\text{kg} \cdot \text{C}°} \times \frac{4.186 \times 10^3 \text{J}}{1 \text{kcal}} \times 20 \text{kg} \times (90°\text{C} - 10°\text{C}) \cong 736 \text{kJ}.$$

b. The amount of heat needed for water is

$$Q_{water} = c_{water} m_{water} \Delta T = 1.00 \frac{\text{kcal}}{\text{kg} \cdot \text{C}°} \times \frac{4.186 \times 10^3 \text{J}}{1 \text{kcal}} \times 20 \text{kg} \times (90°\text{C} - 10°\text{C}) \cong 6697 \text{kJ}.$$

The total amount of heat required is

$$Q = Q_{water} + Q_{vat} \cong 736 \text{kJ} + 6697 \text{kJ} \cong 7433 \text{kJ}.$$

Since the specific heat of water is greater than that of iron, more heat is needed to raise the temperature of the same amount of water.

Direction of heat energy flow

When different parts of an isolated system are at different temperatures, heat will flow from the part at higher temperature to the part at lower temperature. This statement is consistent with the principle of energy conservation for an isolated system which was discussed in earlier chapters. In other words, if the system is completely isolated, no energy can flow into or out of it (i.e., conservation of energy). The energy conservation principle is expressed as

$$\text{Heat lost (by one part of the system)} + \text{Heat gained (by the other part)} = 0 \qquad (11.4)$$

The conservation relation of Eq. (11.4) is a useful way to account for how much of energy is traveling in which direction within a system to achieve thermal equilibrium. Similar heat transfer happens in the human body. In order to ensure that one portion of the body is not significantly hotter than another portion, heat must be distributed evenly through bodily tissues. Blood flowing through blood vessels acts as a convective fluid and helps to prevent any buildup of excess heat inside the tissues of the body. This flow of blood through the vessels can be modeled as pipe flow in an engineering system. The heat carried by the blood is determined by the temperature of the surrounding tissue, the diameter of the blood vessel, the thickness of the fluid, velocity of the flow, and the heat transfer coefficient of the blood. We may use Eqs. (11.2) and (11.4) to compute the amount and direction of heat transfer within an isolated system.

Measurement of Heat Transfer

To check the validity of energy conservation in an isolated system, we need to be able to measure heat transfer within the system. So, we may ask the following question.

How do we measure heat transfer?

To measure heat transfer between objects, we may use calorimetry. **Calorimetry** is a technique known to measure heat quantitatively. This is a commonly used technique. The word calorimetry is derived from the Latin word *calor*, which means heat, and the Greek word *metron*, which means measure. It is noted that heat is distinct from temperature. Calorimetry is the act of measuring physical changes or the science of making such measurements. This technique is also used to measure the heat of chemical reactions. Calorimetry is performed with a calorimeter, a device used to measure heat transfer.

A **calorimeter** is an object used for calorimetry, or the process of measuring the heat of chemical reactions or physical changes as well as heat capacity. There are several types of calorimeters. The most common types include differential scanning calorimeters, isothermal microcalorimeters, titration calorimeters, and accelerated

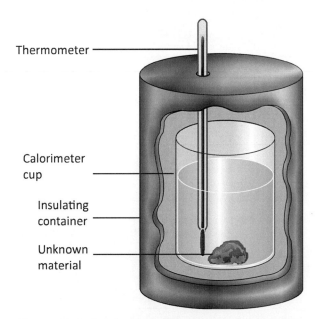

Figure 11.3: A schematic illustration of calorimeter. To measure heat transfer accurately, heat lost to the environment must be minimized. This may be achieved by having a nice thermal jacket which surrounds the calorimeter.

rate calorimeters. Although these types differ from each other by techniques, they achieve the same function as a simple calorimeter. A simple calorimeter just consists of a thermometer attached to a metal container full of water suspended in an insulating chamber as shown in Fig. 11.3. A calorimeter may be operated under constant (atmosphere) pressure, or constant volume.

Exercise Problem 11.3:

We wish to determine the specific heat of a new alloy. A 0.150 kg sample of the alloy is heated to 540 °C. It is then quickly placed in 400 g of water at 10.0 °C, which is contained in a 200 g aluminum calorimeter cup. (We do not need to know the mass of the insulating jacket since we assume the air space between it and the calorimeter cup insulates it well, so that its temperature does not change significantly.) The final temperature of the mixture is 30.5 °C. Calculate the specific heat of the alloy.

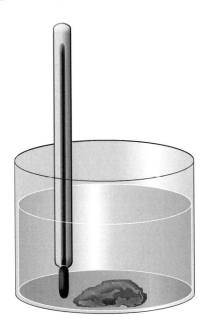

PROBLEM SOLVING STRATEGY

What do we know?

We are given the initial temperature and mass of both an unknown alloy and water. Also, the final temperature of the alloy-water mixture is known.

What concepts are needed?

We need to apply energy conservation. If the system is isolated, then heat energy lost by one part of the system equals heat energy gained by another part.

Why?

In an isolated system, heat energy flows from the high temperature region to the low temperature region until the equilibrium temperature is reached.

Solution:

Applying the law of energy conservation, we may write that the heat lost by the sample is equal to the sum of heat gained by the water and calorimeter cup as

$$c_s m_s \Delta T_s = c_w m_w \Delta T_w + c_{cal} m_{cal} \Delta T_{cal}.$$

Noting that $\Delta T_w = \Delta T_{cal} = \Delta T$, we solve for the specific heat c_s of the unknown sample as

$$c_s = \frac{(c_w m_w + c_{cal} m_{cal}) \Delta T}{m_s \Delta T_s}$$

$$= \frac{\left(4186 \, \frac{J}{kg \cdot C^\circ} \times 0.40 \, kg + 900 \, \frac{J}{kg \cdot C^\circ} \times 0.20 \, kg\right) \times 20.5 \, C^\circ}{0.15 \, kg \times (540 \, ^\circ C - 30.5 \, ^\circ C)} \cong 497 \, \frac{J}{kg \cdot C^\circ}.$$

We may determine what the unknown material is by measuring its specific heat as described in this problem and by comparing with those for known materials.

11.2 HEAT INVOLVED IN PHASE CHANGE

When a material undergoes a phase transition either from solid to liquid or from liquid to gas, a certain amount of energy is involved in this transition. When a substance changes from one state (or phase) of matter to another, we say that it has undergone a change of phase. These changes of phase always occur with a change of heat. Heat is energy which enters or leaves the material during a change of phase. However, although the energy content of the material changes, the temperature does not. For example, ice at 0 °C absorbs heat and changes its phase to water at 0 °C without changing the temperature. Similarly, water at 100 °C absorbs heat and changes its phase to steam at 100 °C without changing its temperature.

A typical example of a change of state of matter, meaning a phase transition, is the melting of ice or the boiling of water, as shown in Fig. 11.4. The **heat of fusion** L_f is the heat required to change 1.0 kg of substance from the solid to the liquid state. The heat of fusion is heat absorbed by a unit mass of a solid at its melting point in order to convert the solid into a liquid at the same temperature; "the heat of fusion is equal to the heat of solidification". On the other hand, the **heat of vaporization** L_v is the heat required to change a substance from the liquid to the vapor phase. This is the amount of heat required to convert a unit mass of a liquid at its boiling point into vapor without an increase in temperature. In genaral, latent heat is the heat released or absorbed by a body or a thermodynamic system during a process without temperature change. More specifically, the heat needed for a phase change is

$$Q = mL, \tag{11.5}$$

where L is the latent heat of that phase change (**units:** $L = J/kg$) and m is the mass of the substance. The term "latent heat" was introduced by Joseph Black (1728-1799) around 1762. It is derived from the Latin *latere* (to lie hidden). Black used the term in the context of calorimetry when referring to the heat transferred that caused a change of volume while the thermodynamic system was held at constant temperature. In contrast, when energy causes in a change of the temperature of the system, it is called "sensible heat".

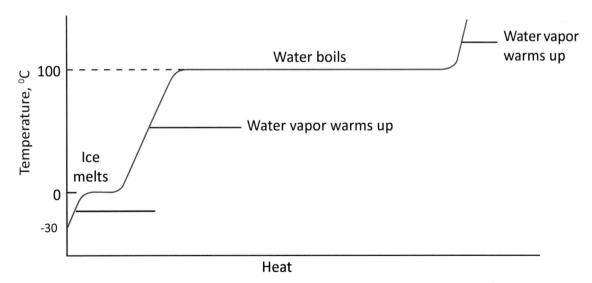

Figure 11.4: A temperature-versus-heat plot illustrates the amount of heat needed to change the phase from ice to water (at 0 °C) and from water to vapor (at 100 °C) as water warms up. For water, the latent heat of vaporization is much greater than the latent heat of fusion.

Table 11.2: Latent heat at 1 atm for selected materials.

Substance	Melting pt. (°C)	L_f kcal/kg	L_f kJ/kg	Boiling pt. (°C)	L_v kcal/kg	L_v kJ/kg
Oxygen	-218.8	3.3	14	-183	51	210
Nitrogen	-210.0	6.1	26	-195.8	48	200
Ethyl alcohol	-114	25	104	78	204	850
Ammonia	-77.8	8.0	33	-33.4	33	137
Water	0	79.7	333	100	539	2260
Lead	327	5.9	25	1750	208	870
Silver	961	21	88	2193	558	2300
Iron	1808	69.1	289	3023	1520	6340
Tungsten	3410	44	184	5900	1150	4800

Understanding-the-Concept Question 11.2:

Ice has a latent heat of fusion of 80 kcal/kg. How much work is required to melt 200 g of ice?

    a. 400 kJ   b. 160 kJ   c. 67 kJ   d. 16 kJ

Explanation:

First, we compute the heat energy needed to melt 0.200 kg of ice. The amount of energy needed may be computed as

$$Q = mL_f = 0.200\,\text{kg} \times 79.9\,\frac{\text{kcal}}{\text{kg}} = 16.0\,\text{kcal} \times 4.186\,\frac{\text{kJ}}{\text{kcal}} \approx 67\,\text{kJ}.$$

In the last step of the calculation, we used the mechanical equivalence of heat (1 cal = 4.186 J).

Answer: c

    Ice, which is considered a mineral and is frozen water at atmospheric pressure, has interesting and unusual properties. As a naturally-occurring crystalline inorganic solid with an ordered structure ice possesses a regular crystalline structure based on the structure of a water molecule, which consist of a single oxygen atom covalently bonded to from hydrogen atoms (i.e., H-O-H). Ice is solid and i approximately 8.3% less dense than liquid water. The density of ice is 0.9167 g/cm³ at 0 °C, wherea water has a density of 0.9998 g/cm³ at the same temperature. Liquid water is densest, essentially 1.0 g/cm³, at 4.2 °C and becomes less dense as the water molecules begin to form hexagonal crystals o ice as the freezing point is reached. This is due to hydrogen bonding dominating the intermolecula forces, which results in a packing of molecules less compact in the solid. The density of ice increase slightly with decreasing temperature and has a value of 0.9340 g/cm³ at -180 °C (93 K). The effect o expansion during freezing can be dramatic, and it is a basic cause of freeze-thaw weathering of roc in nature. Also, it is a common cause of the flooding of houses when water pipes burst due to th pressure of expanding water when it freezes. When ice melts, it absorbs as much energy as it woul take to heat an equivalent mass of water by 80 °C. During the melting process, the temperatur remains constant at 0 °C. While melting, any energy added breaks the hydrogen bonds between ice (water) molecules. Energy becomes available to increase the thermal energy (temperature) only after enough hydrogen bonds are broken that the ice can be considered liquid water. As with water, ice absorbs light at the red end of the spectrum preferentially as the result of an overtone of an oxygen-hydrogen (O-H) bond stretch. Compared with water, this absorption is shifted toward slightly lower energies. Thus, ice appears blue, with a slightly greener tint than for liquid water. Since absorption is cumulative, the color effect intensifies with increasing thickness or if internal reflections cause the light to take a longer path through the ice. Other colors can appear in the presence of light absorbing impurities, where the impurity is dictating the color rather than the ice itself. For instance, icebergs containing impurities (e.g., sediments, algae, air bubbles) can appear brown, grey or green. Ice was originally thought to be slippery due to the pressure. The pressure of an object coming into contact with the ice would create heat, thereby melting a thin layer of the ice and allowing the object to glide across the surface. However, now, it is commonly accepted that ice is slippery because ice molecules in contact with air cannot properly bond with the molecules of the mass of ice beneath (and thus are free to move like molecules of liquid water). These molecules remain in a semi-liquid state, providing lubrication regardless of pressure against the ice exerted by any object.

Exercise Problem 11.4:

At a party, a 0.50 kg chunk of ice at -10 °C is placed in 3.0 kg of "iced" tea at 20 °C. At what temperature and in what phase will the final mixture be? The tea can be considered as water.

PROBLEM SOLVING STRATEGY

What do we know?

A chunk of ice is placed in the iced tea at room temperature. Heat transfer will occur.

What concepts are needed?

We need to apply the principle of energy conservation. Thermal energy is involved in phase transition of matter.

Why?

When the chunk of ice is placed in the tea, it absorbs thermal energy and increases its temperature and may undergo phase transition to liquid water.

Solution:

First, we check to see if the final state will be all ice, a mixture of ice and water at 0 °C, or all water. We compute the energy release by the water at 20 °C when it reaches 0 °C, and the energy needed by the ice at -10 °C to become water at 0 °C as

$$Q_w = c_w m_w \Delta T_w = 4186 \frac{J}{kg \cdot C°} \times 3.0\,kg \times (20°C - 0°C) \cong 251\,kJ \ (for\ water)$$

$$Q_{ice} = c_{ice} m_{ice} \Delta T_{ice} = 2100 \frac{J}{kg \cdot C°} \times 0.5\,kg \times (0°C - (-10°C)) \cong 10.5\,kJ \ (for\ ice)$$

$$Q_{fusion} = m_{ice}L_f = 0.5\,\text{kg} \times 333\frac{\text{kJ}}{\text{kg}} \cong 167\,\text{kJ} \; \left(\text{from ice to water at } 0°C\right).$$

Therefore, the total energy needed to convert ice to water at 0 °C is

$$Q = Q_{fusion} + Q_{icc} \cong 167\,\text{kJ} + 10.5\,\text{kJ} \cong 177\,\text{kJ}.$$

This indicates that there is not enough energy to bring the 3.0 kg of water at 20 °C down to 0 °C. We now apply the conservation of energy and write

$$Q_{ice} + Q_{fusion} + c_w m_{ice}\left(T - 0°C\right) = c_w m_w \left(T_i^w - T\right). \tag{4a}$$

where Q_{ice} is the heat needed to raise the ice from $T_i = -10$ °C to 0 °C, Q_{fusion} is the heat needed to change the phase of the ice into water at 0 °C, $c_w m_{ice}(T - 0\,°C)$ is the heat needed to raise the mass of water from ice at 0 °C to T, and $c_w m_w (T_i^w - T)$ is the heat given off by the tea. We may rewrite the conservation equation as

$$c_w \left(m_{ice} + m_w\right)T = c_w m_w T_i^w - Q_{ice} - Q_{fusion} \tag{4b}$$

and solve for T to obtain

$$T = \frac{c_w m_w T_i^w - Q_{ice} - Q_{fusion}}{c_w \left(m_{ice} + m_w\right)} = \frac{251\,\text{kJ} - 10.5\,\text{kJ} - 167\,\text{kJ}}{4.186\,\frac{\text{kJ}}{\text{kg}\cdot\text{C}°} \times \left(0.5\,\text{kg} + 3.0\,\text{kg}\right)} = 5.0\;°\text{C}. \tag{4c}$$

Equation (4c) does not include the energy released by the container of tea. If we include this contribution, what will happen to the final temperature of the mixture and why? When the energy included in the container is accounted in the calculation, the final temperature would be higher than 5.0 °C.

Understanding-the-Concept Question 11.4:

If a glass of water is left out overnight, the water level will have fallen by morning. Some of the water will have changed to the vapor or gas phase.

- How can this happen?

Explanation:

As we know, when a glass of water is left out in the open overnight, water evaporates from the glass. Evaporation is a cooling process. Boiling and evaporation are both processes by which a liquid turns into a gas, but they are not the same. Evaporation is a slower process that occurs at the surface of a liquid below the boiling point. However, boiling occurs throughout a liquid once the temperature reaches the boiling point.

Evaporation

Realize it or not, heat transfer processes occur all around us. For example, condensation and evaporation are examples of these processes. **Evaporation** is a type of vaporization of a liquid as shown in Fig. 11.5. The other type of vaporization is boiling which occurs within the entire mass of the liquid. On average, the molecules in a glass of water do not have enough heat energy to escape from the liquid. With sufficient heat, the liquid would turn into vapor quickly, similar to at the boiling point. When the molecules collide, they transfer energy to each other in varying degrees, based on how they collide. Sometimes the transfer is so one-sided for a molecule near the surface that it ends up with enough energy to 'escape'. Evaporation is an essential part of the water cycle. The sun (solar energy) drives evaporation of water from oceans, lakes, moisture in the soil, and other sources of water. In hydrology, evaporation and transpiration (which involves evaporation within plant stomata) are collectively termed evapotranspiration. Evaporation of water occurs when the surface of the liquid is exposed, allowing molecules to escape and form water vapor; this vapor can then rise up and form clouds.

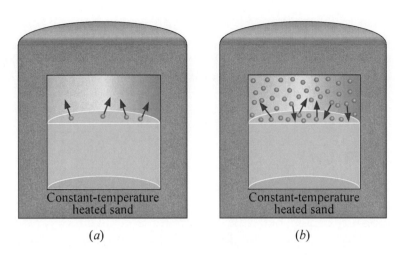

Figure 11.5: Evaporation process is schematically illustrated. (a) Molecules with high kinetic energy leave a liquid from the surface. (b) When there are large enough number of molecules in the air, the number of molecules leaving the liquid becomes the same as that returning back to the liquid via condensation.

For molecules of a liquid to evaporate, they must be located near the surface, be moving in the proper direction, and have sufficient kinetic energy to overcome liquid-phase intermolecular forces. When only a small proportion of the molecules meet these criteria, the rate of evaporation is low. Since the kinetic energy of a molecule is proportional to its temperature, evaporation proceeds more quickly at higher temperatures. As the faster-moving molecules escape, the remaining molecules have lower average kinetic energy, and the temperature of the liquid decreases. This phenomenon is also called evaporative cooling. This is why evaporating sweat cools the human body. Evaporation also tends to proceed more quickly with higher flow rates between the gaseous and liquid phase and in liquids with higher vapor pressure. For example, laundry on a clothes line will dry (by evaporation) more rapidly on a windy day than on a still day. Three key parts to evaporation are heat, atmospheric pressure (determines the percent humidity) and air movement. On a molecular level, there is no strict boundary between the liquid state and the vapor state. Instead, there is a Knudsen layer, where the phase is undermined. Because this layer is only a few molecules thick, a clear phase transition interface can be seen at a macroscopic scale.

Condensation is the change of the physical state of matter from gaseous phase into liquid phase, and is the reverse of vaporization. When the transition happens from the gaseous phase into the solid phase directly, the change is called deposition. As the molecules move, some of these molecules strike the liquid surface and again become part of the liquid phase. Condensation is initiated by the formation of atomic/molecular clusters of that species within its gaseous volume similar to rain drop or snow-flake formation within clouds. Also, condensation occurs at the contact between such gaseous phase and a (solvent) liquid or solid surface.

Saturated vapor pressure is the pressure of a vapor when it is fully in equilibrium with its liquid. The process of evaporation in a closed container will proceed until there are as many molecules returning to the liquid as there are escaping. At this point, the vapor is said to be saturated, and the pressure of that vapor (usually expressed in mm of Hg) is called the saturated vapor pressure. It is noted that if the container is large or is not closed, all liquid may evaporate before saturation is reached. The saturated vapor pressure of a liquid increases with temperature. When the temperature is raised to the point where the saturated vapor pressure at that temperature equals the external pressure, boiling occurs.

Humidity

When we refer to the weather as being dry and humid, we are referring to the water vapor content of the air. Of course, as shown in Fig. 11.6, we may clearly see and feel the difference in the amount of moisture content of the air if we are traveling in either Nevada or Louisiana during a hot summer day. The most commonly used way to indicate the water vapor content in the air is humidity.

What is humidity?

Humidity is a measure of the amount of water vapor in the air. Water vapor is the gas phase of water and is invisible. Humidity indicates the likelihood of precipitation, dew, or fog due to water vapor. We would feel uncomfortable in a hot humid environment because higher humidity reduces the effectiveness of sweating in cooling the body by reducing the rate of evaporation of moisture from the skin. This effect is calculated in a heat index table, which is used during summer months to indicate weather.

There are three main measurements of humidity: absolute, relative and specific. Absolute humidity is the water content of air. The humidity reading that is used generally by most meteorologists is relative humidity. The relative humidity of air describes the saturation of air with water vapor. It is expressed as a percent humidity (e.g., 50% relative humidity) and measures the current absolute humidity *relative* to the maximum for that air pressure and temperature. The relative humidity is the ratio of the partial pressure to the saturated vapor pressure at a given temperature

$$\text{relative humidity} = \frac{\text{partial pressure of } H_2O}{\text{saturated vapor pressure of } H_2O} \times 100\% \qquad (11.6)$$

where partial pressure refers to the pressure each gas would if it alone occupied the whole volume. This measurement allows a comparison of the amount of water vapor in the air with the maximum amount water vapor at a given temperature. **Specific humidity** is a ratio of the water vapor content of the mixture to the total air content on a mass basis.

Figure 11.6: Humidity of air for a desert climate (left panel) is less that for a swampy climate (right panel) due to a smaller amount of water vapor content.

11.3 HEAT TRANSFER MECHANISMS

Heat is defined as the transfer of thermal energy across a well-defined boundary around a thermodynamic system. It is a characteristic of a process and is never *contained* in matter. So it is an energy transfer mechanism. Heat is transferred from one place or body to another in three different ways: (i) convection, (ii) conduction, and (iii) radiation.

Convection

Convection (or convective heat transfer) is the transfer of heat from one place to another by the movement of fluids. In this process, the transfer of heat occurs essentially via mass transfer. Bulk motion of fluid enhances heat transfer in many physical situations, such as (for example) between a solid surface and the fluid. Convection is usually the dominant form of heat transfer in liquids and gases. Although convection is usually used to describe the combined effects of heat conduction within the fluid (diffusion) and heat transference by bulk fluid flow streaming, the process of transport by fluid streaming is known as *advection*. Pure advection is a term associated only with mass transport in fluids, such as advection of pebbles in a river. In the case of heat transfer in fluids, where transport is by advection, it is always accompanied by transport via heat diffusion (also known as heat conduction).

The process of heat convection is understood to refer to the sum of heat transport by advection and diffusion/conduction. As shown in Fig. 11.7, heat is transferred by the mass movement of molecules from one place to another via (i) forced convection and (ii) natural convection. **Forced convection** is a term used when the streams and currents in the fluid are induced by external means such as fans, stirrers, and pumps, creating an artificially induced convection current. An example of forced convection is a forced-air furnace in which heated air is blown into a room by a fan. **Natural convection** occurs when bulk fluid motion (steams and currents) are caused by buoyancy forces that result from density variations due to variations of temperature in the fluid. An example of this is hot air rising.

Convective heating or cooling in some circumstances may be described by Newton's law of cooling: "The rate of heat loss of a body is proportional to the difference in temperatures between the body and its surroundings." However, by definition, the validity of Newton's law of cooling requires that the rate of heat loss from convection be a linear function of (i.e., proportional to) the temperature difference

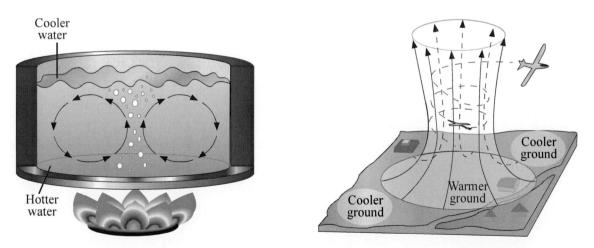

Figure 11.7: Two examples for the convection mechanism for heat transfer are schematically illustrated. Both water convection (left panel) and air convection (right panel) arise naturally because hot water and hot air rise up.

that drives heat transfer. Sometimes in convective cooling, this is not the case because convection, in general, is not linearly dependent on temperature gradients and in some cases is strongly nonlinear. In these cases, Newton's law does not apply.

Understanding-the-Concept Question 11.5:

Convection can occur
- a. only in solids
- b. only in liquids
- c. only in gases
- d. only in liquids and gases
- e. in solids, liquids, and gases

Explanation:

The convection mechanism transfers heat by moving mass (i.e., atoms or molecules). This can happen only if either atoms or molecules can move freely within the system. In the fluid phases (i.e., gases and liquids), atoms or molecules move freely, transporting heat via the convection mechanism.

Answer: d

Similar to engineering systems, the human body transfers heat. Heat is produced in the body by the continuous metabolism of nutrients which provides energy for the systems of the body. The human body must maintain a consistent internal temperature in order to maintain healthy bodily functions. Therefore, excess heat must be dissipated from the body to keep it from overheating. When a person engages in elevated levels of physical activity, the body requires additional fuel which increases the metabolic rate and the rate of heat production. The body must then use additional methods to remove the excess heat produced in order to keep the internal temperature at a healthy level. This removal of heat occurs mostly by convection. Heat transfer by convection is driven by the movement of fluids over the surface of the body. This convective fluid can be either a liquid or a gas. For heat transfer from the outer surface of the body, the convection mechanism is dependent on the surface area of the body, the velocity of the air, and the temperature gradient between the surface of the skin and the ambient air. The normal temperature of the body is approximately 37°C. Heat transfer occurs more readily when the temperature of the surroundings is significantly less than the normal body temperature. This concept explains why we feel "cold" when we do not wear enough covering and are exposed to a cold environment. Clothing can be considered an insulator which provides thermal resistance to heat flow over the covered portion of the body. This thermal resistance causes the temperature on the surface of the clothing to be less than the temperature on the surface of the skin. This smaller temperature gradient between the surface temperature and the ambient temperature will cause a lower rate of heat transfer than if the skin were not covered. Heat transfer through clothing occurs by conduction, rather than by convection.

conduction

The conduction mechanism transfers heat differently than that for the convection mechanism. In many materials, heat conduction is due to molecular collisions. When an object is heated, molecules

ENERGY TRANSFER PROCESSES

move faster and faster due to increase in their kinetic energy and collide with neighboring molecules. As they collide with their slower moving neighbors, they transfer some of their energy by collision. This collision process continues with molecules still further along the object as shown in Fig. 11.8. As a result, the energy of thermal motion is transferred by molecular collision along the object and continues as long as there is a temperature difference in the material.

On a microscopic scale, collisions occur between the vibrating molecules in materials. Hence, heat conduction occurs as rapidly moving or vibrating atoms and molecules (i.e., hot) interact with neighboring atoms and molecules, transferring some of their heat energy to these neighboring particles. In other words, heat is transferred by conduction when adjacent atoms vibrate against one another, or as electrons move from one atom to another. Conduction is the most significant means of heat transfer within a solid or between solid objects in thermal contact. Fluids, especially gases, are less conductive because the relative positions between molecules are not fixed and collisions between the vibrating molecules are much less frequent.

The transfer of energy via heat conduction occurs only if there is a difference in temperature. The rate of heat flow, ΔQ per time interval Δt, is given by

$$\frac{\Delta Q}{\Delta t} = \kappa A \frac{T_1 - T_2}{L} \qquad (11.7)$$

where A is the cross sectional area of the object, L is the distance between the two ends, and κ is the thermal conductivity [units: $J/(s \cdot m \cdot C°)$]. In general, two types of conduction are considered. **Steady state conduction** is a form of conduction that happens when the temperature difference driving the conduction is constant, so that after an equilibration time, the spatial distribution of temperatures in the conducting object does not change any further. In steady state conduction, the amount of heat entering

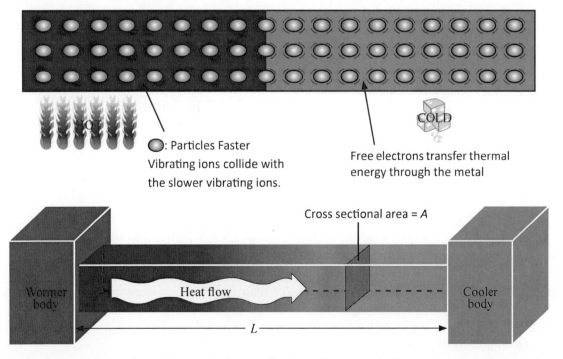

Figure 11.8: A schematic illustration of the conduction mechanism of heat transfer. Energy is transferred from the warmer region of the object to the cooler region since the atoms in the warmer region has greater vibrational (or thermal) energy than those in the cooler region.

a section is equal to amount of heat coming out. **Transient conduction** occurs when the temperature within an object changes as a function of time. Analysis of transient systems is more complex and often calls for the application of approximation theories or numerical analysis by computer.

Table 11.2: Thermal conductivity of selected materials.

Substance	Thermal conductivity κ	
	kcal/s·m·C°	J/s·m·C°
Silver	10×10^{-2}	420
Copper	9.2×10^{-2}	380
Aluminum	5.0×10^{-2}	200
Steel	1.1×10^{-2}	40
Ice	5×10^{-4}	2
Glass (typical)	2.0×10^{-4}	0.84
Brick and concrete	2.0×10^{-4}	0.84
Water	1.4×10^{-4}	0.56
Human tissues (excluding blood)	0.5×10^{-4}	0.2
Wood	$0.2\text{-}0.4\times10^{-4}$	0.08-0.16
Fiberglass insulation	0.12×10^{-4}	0.048
Cork and glass wool	0.1×10^{-4}	0.042
Wool	0.1×10^{-4}	0.040
Goose down	0.06×10^{-4}	0.025
Polyurethane foam	0.06×10^{-4}	0.024
Air	0.055×10^{-4}	0.023

The rate of heat flow in an object $\Delta Q/\Delta t$ depends on a material. This dependence in Eq. (11.7) is denoted by the thermal conductivity κ. **Thermal conductivity** is the property of a material to conduct heat. It is evaluated primarily in terms of Fourier's law for heat conduction which states that the time rate of heat transfer through a material is proportional to the negative gradient in the temperature and to the area, at right angles to that gradient, through which the heat is flowing. Heat transfer across materials of high thermal conductivity occurs at a higher rate than across materials of low thermal conductivity. Correspondingly materials of high thermal conductivity are widely used in heat sink applications and materials of low thermal conductivity are used as thermal insulation. The thermal conductivities of materials are temperature-dependent. The reciprocal of thermal conductivity is thermal resistivity.

Understanding-the-Concept Question 11.6:

Consider two neighboring rectangular house built from the same materials. One of the houses has twice the length, width, and height of the other. Under identical climatic conditions, what would be true about the rate that heat would have to be supplied to maintain the same inside temperature on a cold day? Compare to the small house, the larger house would need heat supplied at

    a.   twice the rate

    b.   4 times the rate

c.   8 times the rate
   d.   16 times the rate

Explanation:

The rate of heat flow via the conduction mechanism is described by Eq. (11.7). According to this equation, the rate of heat flow increase with the cross-sectional area. This area corresponds to the area of the wall. If the dimensions of the house increase by twice, then the area of each wall increases by four times. Also, the rate of heat flow decreases with increasing distance between the warmer inside and cooler outside. This distance corresponds to the thickness of walls. It should be noted that even when the dimensions of the house is increased by twice the thickness of the walls will be kept the same. This means the heat flow rate is affected only by the changes in the area A, indicating that the larger house will have four times the rate of heat flow than the smaller house.

Answer: b

About Thermal Conductivities

When κ is large, substances conduct heat rapidly (i.e., good thermal conductors). Examples are that most metals. They are good thermal conductors. On the other hand, when κ is small, substances conduct heat slowly (i.e., poor thermal conductors) Examples of poor thermal conductors are wool and fiberglass. These materials are also called "good insulators".

The thermal properties of building materials, particularly when considered as insulation, are usually specified by R-value (or "thermal resistance") defined for a given thickness L of material. For example, 1/8 in. of glass corresponds to $R = 1$ ft²·h·F°/Btu and 4 in. of fiberglass insulation corresponds to $R = 12$ ft²·h·F°/Btu. The **R-value** is a measure of thermal resistance used in the building and construction industry. Under uniform conditions, it is the ratio of the temperature difference across an insulator and the heat flux (heat transfer per unit area, $(1/A)\, \Delta Q/\Delta t$ through it or

$$R = (\Delta T)\frac{A\,\Delta t}{\Delta Q} = \frac{L}{\kappa}. \tag{11.8}$$

The R-value being discussed is the unit thermal resistance. This is used for a unit value of any particular material. It is expressed as the thickness of the material divided by the thermal conductivity. For the thermal resistance of an entire section of material, instead of the unit resistance, we divide the unit thermal resistance by the area of the material. For example, if we have the unit thermal resistance of a wall, then we need to divide by the cross-sectional area of the wall. The unit thermal conductance of a material is denoted by C and is the reciprocal of the unit thermal resistance (i.e., $C = A/R$). This can also be called the unit surface conductance. The higher the R, the building insulation's effectiveness is better. R-value is the reciprocal of U-value (i.e., $U = 1/R$ and sometimes incorrectly referred to as "U-value") which, is the overall heat transfer coefficient measuring how well a building element conducts heat. Note that the phrase "U-factor" is generally used to express the insulation value of windows only, while R-value is used for insulation in most other parts of the building envelope (walls, floors, roofs).

While the majority of insulation in buildings is for thermal purposes, the term also applies to acoustic insulation, fire insulation, and impact insulation (e.g. for vibrations caused by industrial

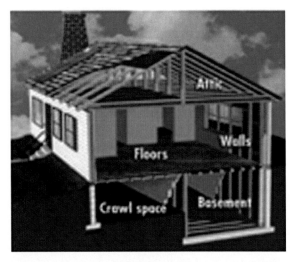

Zone	Energy Type				Ceilings		Walls/Floors				Basement	
	Gas	Heat Pmp	Fuel Oil	Elec Furn	Attic	Cath	Wall (a)	Flr	Crwl (b)	Slab Edg.	Int	Ext
1	x	x	x		R49	R38	R18	R25	R19	R8	R11	R10
1				x	R49	R60	R28	R25	R19	R8	R19	R15
2	x	x	x		R49	R38	R18	R25	R19	R8	R11	R10
2				x	R49	R38	R22	R25	R19	R8	R19	R15
3	x	x	x	x	R49	R38	R18	R25	R19	R8	R11	R10
4	x	x	x		R38	R38	R13	R13	R19	R8	R11	R4
4				x	R49	R38	R18	R25	R19	R8	R11	R10
6	x				R22	R22	R11	R11	R11	(c)	R11	R4
6		x	x		R38	R30	R13	R11	R13	R4	R11	R4
6				x	R49	R38	R18	R25	R19	R8	R11	R10

Figure 11.9: Thermal insulation regulation for houses in different regions of the United States is tabulated in terms of R-value. In general, houses in cold climate require larger R-value insulation, while those in warm climate require smaller R-value insulation. Higher R-value insulation is needed for attics since heat tends to escape from there.

applications). Often an insulation material will be chosen for its ability to perform several of these functions at once. As shown in Fig. 11.9, an initial estimate of insulation needs in the United States depends on the region but can be determined by the US Department of Energy's ZIP-Code Insulation Calculator. The **ZIP-Code Insulation Program** will tell us the most economic insulation level for our new or existing house.

Exercise Problem 11.5:

A major source of heat loss from a house is through the window. Calculate the rate of heat flow through a glass window 2.0 m × 1.5 m in area and 3.2 mm thick, if the temperatures at the inner and outer surfaces are 15 °C and 14 °C, respectively.

ENERGY TRANSFER PROCESSES

PROBLEM SOLVING STRATEGY

What do we know?

We know the temperature difference between the inside and outside of a house. Also, the dimensions of the window glass are given.

What concepts are needed?

We need to apply the heat transfer equation for the conduction mechanism.

Why?

Heat loss from the house occurs through the window.

Solution:

Since the thermal conductivity of typical glass is 0.84 J/(s·m·C°), the rate of heat loss is given by

$$\frac{\Delta Q}{\Delta t} = \kappa A \frac{T_1 - T_2}{\ell} = 0.84 \frac{J}{s \cdot m \cdot C°} \times (2.0 \text{m} \times 1.5 \text{m}) \times \frac{15°C - 14°C}{3.2 \times 10^{-3} \text{m}}$$

$$\cong 788 \frac{J}{s} \times \frac{1 \text{kcal}}{4.186 \times 10^3 \text{J}} \times \frac{3600 \text{s}}{1 \text{h}} \cong 678 \frac{\text{kcal}}{\text{h}}$$

This calculation indicates that if the temperature difference between inside and outside of the house is roughly 40 times greater (i.e., 15 °C inside and -25 °C outside) as in some winters in North Dakota, then the rate of heat loss can be as much as 2.71×10⁵ kcal/h. This is a significant amount of heat loss!

A better way to keep the heat inside is to use double-pane windows. The main advantage of the double-pane windows is better insulation. That insulation is not a result of the thickness of the glass but a result of the thickness of the resulting sandwich of materials. Double-pane windows usually use an airtight spacing between the sheets of glass of one-half inch to 1 inch. Depending on the double-pane window we choose, it will be between two and 10 times more insulated than a single-pane window. Simple double-pane windows use air between the two sheets of glass; higher-end windows use noble gas such as argon or krypton in place of air. The main disadvantage of air as a filler for that space is that when temperature differences are high, condensation can occur. Noble gases do not contain water in gaseous form, and, therefore, condensation is not a problem. Air also conducts energy better than both argon and xenon, reducing the efficiency of the window. Of the noble gas used in window making, argon is the most common and the cheapest, but is also the gas providing the least improvement over air. Krypton and xenon follow in that order for efficiency and price.

Radiative Heat Transfer

Radiation is another important heat energy transport mechanism. This process is due to either emission or absorption of electromagnetic (EM) waves. One important example of this type of heat transfer is the radiation from the Sun as shown in Fig. 11.10.

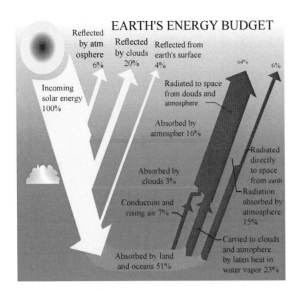
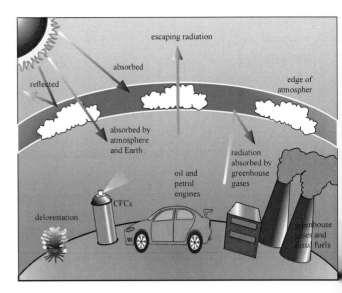

Fig. 11.10: Earth receives energy from the Sun in the form of radiation. However, as illustrated on the left panel, not all energy from the Sun is absorbed by the Earth. Also, as illustrated on the right panel, Earth emits radiation back to the space. Some of this radiation is trapped by greenhouse gases.

Thermal radiation is emitted by all matters that has a temperature above absolute zero. Thermal radiation can propagate through a vacuum: it does not need to be carried by matter. Thermal radiation is a direct result of the random movements of atoms and molecules in matter. Since these atoms and molecules are composed of charged particles (protons and electrons), their movement results in the emission of EM radiation, which carries energy away from the surface. Unlike conductive and convective forms of heat transfer, thermal radiation can be concentrated in a small spot by using reflecting mirrors, which is exploited in concentrating solar power generation. For example, the sunlight reflected from mirrors heats the PS10 solar power tower and during the day it can heat water to 285 °C (545 °F).

The flow of energy to the Earth from the Sun is constant. That constant flow is just the amount of EM radiation hitting the outside of the atmosphere. The energy from the Sun has changed by the time it hits the surface of the Earth. At the top of the atmosphere you have 29.4 MJ for every square meter for each day. We only see about 17 MJ of energy for every square meter for each day at the surface of the Earth. Something happens to the EM radiation between the top of the atmosphere and the surface of the Earth. Energy and certain types of EM radiation (**Infrared** and **Ultraviolet**) are filtered away. There are several factors which determine the amount of radiant energy hitting the Earth. Our planet is a specific distance from the Sun. Even though we have a slightly **elliptical** orbit around the Sun, we can expect a certain amount of energy to hit our planet. If we were closer there would be more energy hitting our atmosphere and if we were further away there would be less energy. We also have a Sun with special characteristics. As our Sun gets older, it will begin to expand. That increase in size is going to affect the amount of radiant energy hitting our atmosphere. The solar constant is based on the Sun's activity now. Earth maintains its temperature because it emits back into space as much energy as it absorbs from the Sun.

If Earth absorbs more energy from the Sun than it emits back to space, then gradual warming will occur. As shown in Fig. 11.10, EM radiation energy may be trapped on Earth due to an excess of greenhouse gasses in the atmosphere. The emission of excess greenhouse gases is an important contributing factor to Global warming. Roughly 72% of the totally emitted greenhouse gases are carbon dioxide (CO_2), 18% Methane and 9% Nitrous oxide (NO_x). Therefore, carbon dioxide emissions are the most important cause of global warming. Carbon dioxide is inevitably created by burning fuels

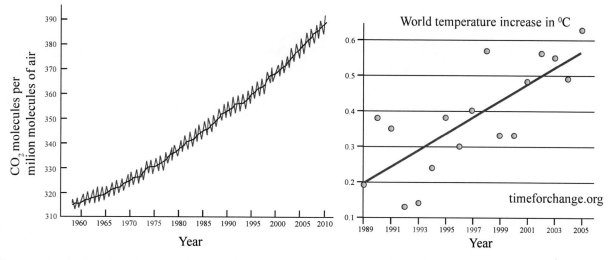

Figure 11.11: Carbon dioxide is one of the greenhouse gasses. The amount of CO_2 molecules in air has increased over several decades (left panel). As shown in the temperature versus year plot (right panel), the increase in the average temperature of the Earth may be correlated to the increase in the greenhouse gasses in air.

like such as oil, natural gas, diesel, organic-diesel, petrol, organic-petrol, and ethanol. CO_2 emissions have dramatically increased within the last 50 years and are still increasing by almost 3% each year as shown in Fig. 11.11. Some recent investigations have shown that inconceivable catastrophic changes in the environment will take place if the global temperatures increase by more than 2 °C (3.6° F). A warming of 2 °C (3.6 °F) corresponds to a carbon dioxide concentration of about 450 ppm (parts per million) in the atmosphere.

Let us see how radiative heat transfer occurs. Radiative heat transfer is concerned with the exchange of thermal radiation energy between two or more bodies. Thermal radiation is defined as EM radiation in the wavelength range of 0.1 to 100 microns (which encompasses the visible light regime), and arises as a result of a temperature difference between two bodies. No medium need exist between the two bodies for heat transfer to take place (as is needed by conduction and convection). Rather, the intermediaries are photons which travel at the speed of light. The heat transferred into or out of an object by thermal radiation is a function of several components. These include its surface reflectivity, emissivity, surface area, temperature, and geometric orientation with respect to other thermally participating objects. In turn, an object's surface reflectivity and emissivity is a function of its surface conditions (roughness, finish, etc.) and composition. The rate at which an object radiates energy is given by

$$\frac{\Delta Q}{\Delta t} = e\sigma A T^4, \tag{11.9}$$

where $\sigma = 5.67 \times 10^{-8}$ W/m²·K⁴ is the Stefan-Boltzmann constant, e is the emissivity, A is the area of the emitting object, and T is the temperature of object in Kelvin. This equation is called Stefan-Boltzmann equation. The **Stefan–Boltzmann law** is the statement that the total radiant heat energy emitted from a surface is proportional to the fourth power of its absolute temperature. In 1879, Josef Stefan (1835-1893) formulated this law as a result of his experimental studies. The same law was derived in 1884 by Ludwig Boltzmann (1844-1906) from thermodynamic considerations: if E is the radiant heat energy emitted from a unit area in one second and T is the absolute temperature (in degrees Kelvin), then $E = \sigma T^4$. The law applies only to blackbodies, which are theoretical surfaces that absorb all incident heat radiation.

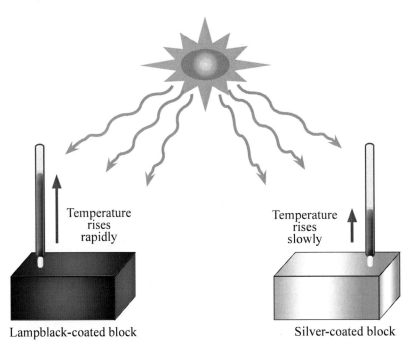

Figure 11.11: A schematic illustration of the effect of emissivity. For two identical objects, an object with higher value of emissivity (i.e., e ~ 1) absorbs more radiation than that with lower value of emissivity.

According to Eq. (11.9), the rate of energy transfer depends on emissivity. Emissivity is a dimensionless quantity. It is noted that emissivity e is close to 1 for a very black surface and is very close to 0 for a very shiny surface. The **emissivity** e of a material is the relative ability of its surface to emit energy by radiation. It is the ratio of energy radiated by a particular material to energy radiated by a black body at the same temperature. A true black body would have an $e = 1$ while any real object would have $e < 1$. In general, the duller and blacker a material is, the closer its emissivity is to 1. The more reflective a material is, the lower its emissivity. Highly polished silver has an emissivity of about 0.02. Shiny surfaces emit less radiation and absorb little of the radiation that falls upon them (most is reflected), but very dark objects absorb nearly all the radiation that falls on them.

A good absorber of EM radiation is also a good emitter. This means that any object not only emits energy by radiation, but also absorbs energy radiated by other bodies. The net rate of radiant heat flow from the object is given by

$$\frac{\Delta Q}{\Delta t} = e\sigma A \left(T_1^4 - T_2^4 \right). \tag{11.10}$$

Equation (11.10) indicates that because both the object and surroundings radiate energy, there is a net transfer of energy from one to the other unless everything is at the same temperature.

Exercise Problem 11.6:

A ceramic teapot ($e = 0.70$) and a shiny one ($e = 0.10$) each hold 0.75 liter of tea at 95 °C. (a) Estimate the rate of heat loss from each, and (b) estimate the temperature drop after 30 min for each. Consider only radiation and the surroundings are at 20 °C.

ENERGY TRANSFER PROCESSES

PROBLEM SOLVING STRATEGY

What do we know?

The emissivity and temperature of two teapots which can hold 750 cm³ of tea are known.

What concepts are needed?

We need to apply the radiation mechanism for heat transfer by using the Stefan-Boltzmann equation.

Why?

The radiant heat flow occurs from the hot teapots to the colder surrounding.

Solution:

A teapot holds 0.75L (~750 cm³ since 1L is equivalent to 100 cm³) of tea is roughly equivalent to a cube of side $(750 \text{ cm}^3)^{1/3} \sim 9.1$ cm. Since five sides are exposed, the surface area would be

$$A = 5 \times (9.1 \text{cm})^2 = 414 \text{cm}^2 = 4.14 \times 10^{-2} \text{m}^2$$

assuming that heat loss from the bottom in negligiable. The rate of heat flow is

$$\frac{\Delta Q}{\Delta t} = e\sigma A \left(T_1^4 - T_2^4\right) = e \times 5.67 \times 10^{-8} \frac{\text{W}}{\text{m}^2\text{K}^4} \times 4.14 \times 10^{-2} \text{m}^2 \left[(368\text{K})^4 - (293\text{K})^4\right]$$

We now substitute the value of emissivity and obtain

$$\frac{\Delta Q}{\Delta t} \cong 18 \text{ W} \quad (\text{ceramic teapot}; e = 0.70);$$

$$\frac{\Delta Q}{\Delta t} \cong 2.6 \text{ W} \quad (\text{shiny teapot}; e = 0.10).$$

The temperature drop can be computed using

$$\Delta Q = mc\Delta T \Rightarrow \Delta T = \frac{\Delta Q}{mc}.$$

We may divide both sides of the above equation by Δt and obtain

$$\frac{\Delta T}{\Delta t} = \frac{\frac{\Delta Q}{\Delta t}}{mc} = \frac{e \times 26 \text{W}}{0.75 \text{kg} \times 4.19 \times 10^3 \frac{\text{J}}{\text{kg} \cdot \text{C}^\circ}} = e \times 0.0083 \frac{\text{C}^\circ}{\text{s}}.$$

The result of this calculation indicates that the temperature drop after 30 minutes for the ceramic teapot is 10.45 C° and that for the shiny teapot is 1.49 C°. Based on this calculation, we may conclude that a shiny metallic teapot would be better in keeping tea longer at higher temperatures. If this is the correct conclusion, then why do we use ceramic teapots? What are we missing from this argument? Most likely, the answer is the amount of heat loss due to the conduction and convection mechanisms.

As we noted earlier, **thermal radiation** is EM radiation generated by the thermal motion of charged particles in matter. This means all matter with a temperature greater than absolute zero (i.e., 0 K) emits thermal radiation. The mechanism is that bodies with a temperature above absolute zero have atoms or molecules with kinetic energies which are changing, and these changes result in charge-acceleration and/or dipole oscillation of the charges that compose the atoms. This motion of accelerating charges produces EM radiation in the usual way. However, the spectrum of this radiation reflects the fact that there is a wide spectrum of energies and accelerations of the charges in any piece of matter even at a single temperature. A part of this radiation spectrum is used in night vision and thermography.

Night vision and thermography

Night vision is the ability to see in low light conditions. Night vision is made possible by a combination of two approaches: (i) sufficient spectral range and (ii) sufficient intensity range. Humans have poor night vision compared to many animals, in part because the human eye lacks a tapetum lucidum. The tapetum lucidum is a layer of tissue in the eye of many vertebrate animals. It lies immediately behind the retina. It reflects visible light back through the retina, increasing the light available to the photoreceptors, though blurring the initial image of the light on focus. Night vision technology enhances images in dark or low-light conditions and allows us to see in near total darkness with amazing clarity by wearing a set of night vision goggles. Night vision is commonly used by the military and police forces for nighttime missions, as well as by private investigators and security firms for surveillance. There are three types of night vision technology, each with its own principles of operation. One such technology is thermal imaging technology. Thermal imaging technology, which was developed in 1978, operates by enhancing the existing infrared light in the environment. Unlike generation zero night vision, thermal imaging is powerful enough to detect the infrared light emitted by all living things and, therefore, does not require an external source of infrared light. In addition to detecting the heat signature of living animals, it can also detect soil and plants because they emit residual heat absorbed during daylight hours. This technology was put into practical use by the US military services in the late 1980s and is still used today.

Infrared thermography (IRT), thermal imaging, and **thermal video** are examples of infrared imaging science. Thermal imaging cameras detect radiation in the infrared range of the EM spectrum (roughly from 9 to 14 μm). The produced images of that radiation are called **thermograms.** Since infrared radiation is emitted by all objects above absolute zero according to the black body radiation law, thermography makes it possible for us to see our environment with or without visible illumination. The amount of radiation emitted by an object increases with temperature; therefore, thermography allows one to see variations in temperature. When viewed through a thermal imaging camera, warm objects stand out well against cooler backgrounds; humans and other warm-blooded animals become easily visible against the environment, day or night.

As a result, thermography is particularly useful to military and other users of surveillance cameras. Thermography has a long history. Its use has increased dramatically with the commercial and industrial applications during the last fifty years. Government and airport personnel used thermography to detect suspected swine flu cases during the 2009 pandemic. Firefighters use thermography to see through smoke to find persons and to localize the base of a fire. Maintenance technicians use thermography to locate overheating joints and sections of power lines, which are a sign of impending failure. Building construction technicians can see thermal signatures that indicate

ENERGY TRANSFER PROCESSES

heat leaks in faulty thermal insulation and can use the results to improve the efficiency of heating and air-conditioning units. Some physiological changes in human beings and other warm-blooded animals can also be monitored with thermal imaging during clinical diagnostics.

As a diagnostic tool, the thermogram is used in medical imaging to detect problems in the breast by using heat radiating from person's own body. It involves no contact with the body, no compression, and is completely painless. A thermogram combines advanced digital technology with ultra-sensitive infrared camera imaging. It does not use radiation. So, it can be done as frequently as anyone thinks is necessary. Thermograms work by creating infrared images (heat pictures) that are then analyzed to find asymmetries anywhere in the chest and underarm area. Any abnormality that causes change in heat production is seen on a thermogram. Any source of inflammation such as infection, trauma to the breast, and even sun burn will cause abnormality in the thermal picture. Breast thermography detects patterns of heat generated by the increased circulation produced by abnormal metabolic activity in cancer cells. This activity occurs long before a cancer starts to invade new tissue. So, a breast thermogram has the ability to identify a breast abnormality five to ten years before the problem can be found on a mammogram.

SUMMARY

i. **The mechanical equivalent of heat energy** may be expressed as

$$4.186 \text{ J} = 1 \text{ cal}$$

The amount of heat Q needed to raise the temperature of a material via ΔT is given by

$$Q = cm(\Delta T),$$

where m is the mass of the material and c is the specific heat of the material. For water (i.e., liquid phase), the specific heat is $c = 4.186$ kJ/kg·C°.

ii. When different parts of an isolated system are at different temperatures, heat will flow from the part at higher temperature to the part at lower temperature. If the system is completely isolated, no energy can flow into or out of it (i.e., conservation of energy). So, we may write

Heat lost (by one part of the system) + Heat gained (by the other part) = 0

iii. When a material changes its phase, a certain amount of energy is involved in this phase change. The amount of heat required to convert the phase of a unit mass of a materials is called latent heat. **Latent heat** L is the value for heats of fusion (L_f) and vaporization (L_v). The heat needed to change a phase is

$$Q = mL,$$

where L denotes the latent heat of particular process and substance (units: L = J/kg), and m is the mass of the substance.

iv. Heat is transferred from one place or body to another in three different ways: i. convection, ii. conduction, and iii. radiation. These ways are called heat transfer mechanisms.

v. For the **conduction** heat transfer mechanism, then experimentally determined heat flow ΔQ per time interval Δt is given by

$$\frac{\Delta Q}{\Delta t} = \kappa A \frac{T_1 - T_2}{L},$$

where A is the cross sectional area of the object, L is the distance between the two ends, and κ is the thermal conductivity [units: J/(s·m·C°)].

vi. For the **radiation** heat transfer mechanism, the rate at which an object radiates energy is given by

$$\frac{\Delta Q}{\Delta t} = e\sigma A T^4,$$

where $\sigma = 5.67 \times 10^{-8}$ W/m²·K⁴ is Stefan-Boltzmann constant, e is the emissivity, A is the area of the emitting object, and T is the temperature of object in Kelvin. This equation is called Stefan-Boltzmann equation.

MORE WORKED PROBLEMS

Problem 11.1 :

A 750 g aluminum pan is removed from the stove and plunged into a sink filled with 10 kg of water at 20.0 °C. The water temperature quickly rises to 24.0 °C. What was the initial temperature of the pan?

Solution:

According to the conservation of energy principle, the heat lost by the aluminum pan Q_{Al} is equal to the heat gained by the water Q_{water}. The total heat change of the aluminum-water system is zero. We may write this energy conservation principle as

$$Q_{Al} + Q_{water} = 0 \Rightarrow m_{Al}c_{Al}(T_f - T_{i,Al}) + m_{water}c_{water}(T_f - T_{i,water}) = 0.$$

At the thermal equilibrium, the aluminum pan and water reach a common final temperature $T_f = 24.0$ °C. We may use this information to find the initial temperature of the pan

$$m_{Al}c_{Al}T_{i,Al} = m_{Al}c_{Al}T_f + m_{water}c_{water}(T_f - T_{i,water}).$$

We now solve for $T_{i,Al}$ and obtain

$$T_{i,Al} = T_f + \frac{m_{water}c_{water}(T_f - T_{i,water})}{m_{Al}c_{Al}}$$

$$= 24.0\,°C + \frac{10\,\text{kg} \times 4190\,\frac{J}{kg \cdot K} \times [(24.0+273)K - (20.0+273)K]}{0.750\,\text{kg} \times 900\,\frac{J}{kg \cdot K}} = 270\,°C.$$

The solution suggests that if the mass of the aluminum pan was greater than 750 g then its initial temperature would be lower than 270 °C to reach the same thermal equilibrium temperature of 24 °C.

ENERGY TRANSFER PROCESSES

Problem 11.2:

A large Styrofoam cooler has surface area of 1.0 m² and a thickness of 2.5 cm. If 5.0 kg of ice at 0 °C is stored inside and the outside temperature is a constant 35 °C, how long does it take for all the ice to melt? (Consider the conduction mechanism only.)

Solution:

First, we compute the heat required to melt the ice. This heat is

$$Q = mL_f = 5.0\,\text{kg} \times 3.3 \times 10^5 \frac{\text{J}}{\text{kg}} = 1.65 \times 10^6 \,\text{J}.$$

We now compute the rate heat energy transfer via the conduction mechanism as

$$\frac{\Delta Q}{\Delta t} = \frac{\kappa A (\Delta T)}{d} = \frac{0.042 \frac{\text{J}}{\text{m} \cdot \text{s} \cdot \text{C}^\circ} \times 1.0\,\text{m}^2 \times (35\,^\circ\text{C} - 0\,^\circ\text{C})}{0.025\,\text{m}} = 58.8 \frac{\text{J}}{\text{s}}.$$

Therefore, the time required to melt the ice is

$$\Delta t = \frac{\Delta Q}{\left(\frac{\Delta Q}{\Delta t}\right)} = \frac{1.65 \times 10^6 \,\text{J}}{58.8 \,\text{J}/\text{s}} = 2.8 \times 10^4 \,\text{s} \times \frac{1\,\text{h}}{3600\,\text{s}} = 7.8\,\text{h}.$$

It takes a long time to transfer enough heat energy through the Styrofoam cooler via the conduction mechanism because it is a good thermal insulator (i.e., κ is small).

Problem 11.3:

Liquid helium is stored at its boiling-point temperature of 4.2 K in a spherical container (r = 0.30 m). The container is a perfect blackbody radiator. The container is surrounded by a spherical shield whose temperature is 77 K. A vacuum exists in the space between the container and the shield. The latent heat of vaporization for helium is 2.1×10⁴ J/kg. What mass of liquid helium boils away through a venting valve in one hour?

Solution:

The heat needed to vaporize a mass m of any substance at its boiling point is $Q = mL_v$. This means the mass vaporized by an amount of heat Q is $m = Q/L_v$. Noting that the system continually absorbs heat through radiation, we compute the net power absorbed as

$$P_{net} = e\sigma A \left(T^4 - T_o^4\right).$$

Here, T_o is the temperature of the liquid helium, T is the temperature maintained by the shield, and the area of a sphere is $A = 4\pi r^2$. Note that we set the emissivity e = 1 since the container is considered as a perfect blackbody radiator. We compute the rate of evaporation from

$$m = \frac{Q}{L_v} \Rightarrow \frac{m}{t} = \frac{\left(Q/t\right)}{L_v} = \frac{P_{net}}{L_v} = \frac{e\sigma A\left(T^4 - T_o^4\right)}{L_v}.$$

The rate at which helium boils away is

$$\frac{m}{t} = \frac{1 \times 5.67 \times 10^{-8} \frac{J}{s \cdot m^2 \cdot K^4} \times 4\pi(0.30\,m)^2 \times \left[(77\,K)^4 - (4.2\,K)^4\right]}{2.1 \times 10^4 \; J/kg}.$$

The mass of liquid helium that boiled away in 1 hour is 1.07×10^{-4} kg/s × 3600 s = 0.39 kg.

Problem 11.4 :

The ends of a thin bar maintained at different temperatures. The temperature of the cooler end is 11 °C, while the temperature at a point 0.13 m from the cooler end is 23 °C and the temperature of the warmer end is 48 °C. Assuming that heat flows only along the length of the bar (the sides are insulated), find the length of the bar.

Solution:

Heat Q flows along the length L of the bar via conduction is given by

$$Q = \frac{\kappa A (\Delta T) t}{L} \Rightarrow L = \frac{\kappa A (\Delta T) t}{Q} = \frac{\kappa A (T_w - T_c) t}{Q}.$$

We may evaluate the heat Q transferred by recognizing that it flows through the entire length of the bar and using

$$Q = \frac{\kappa A (T - T_c) t}{D},$$

where $D = 0.13$ m and $T = 23$ °C. Substituting this result, we obtain

$$L = \frac{\kappa A (T_w - T_c) t}{Q} = \frac{\kappa A (T_w - T_c) t}{\frac{\kappa A (T - T_c) t}{D}} = \frac{(T_w - T_c) D}{T - T_c} = \frac{(48\,°C - 11\,°C) \times 0.13\,m}{23\,°C - 11\,°C} = 0.40\,m.$$

PROBLEMS

1. A fast-food hamburger (with cheese and bacon) contains 1000 Calories. (a) What is the burger's energy in joules? (b) If all this energy is used to lift a 10 kg mass, how high can it be lifted?

ENERGY TRANSFER PROCESSES

PROBLEMS

1. A fast-food hamburger (with cheese and bacon) contains 1000 Calories. (a) What is the burger's energy in joules? (b) If all this energy is used to lift a 10 kg mass, how high can it be lifted?

2. The ends of a thin bar are maintained at different temperatures. The temperature of the cooler end is 11 °C, while the temperature at a point 0.13 m from the cooler end is 23 °C and the temperature of the warmer end is 48 °C. Assuming that heat flows only along the length of the bar (the sides are insulated), find the length of the bar.

3. A 75-kg sprinter accelerates from rest to a speed of 11.0 m/s in 5.0 s. (a) Calculate the mechanical work done by the sprinter during this time. (b) Calculate the average power the sprinter must generate. (c) If the sprinter converts food energy to mechanical energy with an efficiency of 25%, at what average rate is the burning Calories? (d) What happens to the other 75% of the food energy being used?

4. A 1.5-kg copper block is given an initial speed of 3.0 m/s on a rough horizontal surface. Because of friction, the block finally comes to rest. (a) If the block absorbs 85% of its initial kinetic energy as internal energy, calculate its increase in temperature, (b) What happens to the remaining energy?

5. When a driver brakes an automobile, the friction between the brake drums and the brake shoes converts the car's kinetic energy to thermal energy. If a 1,500-kg automobile traveling at 30 m/s comes to a halt, how much does the temperature rise in each of the four 8.0-kg iron brake drums? (The specific heat of iron is 448 J/kg·°C.)

6. A locomotive wheel is 1.00 m in diameter. A 25.0-kg steel band has a temperature of 20.0 °C and a diameter that is 6.00×10^{-4} m less than that of the wheel. What is the smallest mass of water vapor at 100 °C that can be condensed on the steel band to heat it, so that it will fit onto the wheel? Do not ignore the water that results from the condensation.

7. Three liquids are at temperatures of 10 °C, 20 °C, and 30 °C, respectively. Equal masses of the first two liquids are mixed, and the equilibrium temperature is 17 °C. Equal masses of the second and third are then mixed, and the equilibrium temperature is 28 °C. Find the equilibrium temperature when equal masses of the first and third are mixed.

8. A block of material has a mass of 130 kg and a volume of 4.6×10^{-2} m³. The material has a specific heat capacity and coefficient of volume expansion, respectively, of 750 J/(kg×C°) and 6.4×10^{-5} (C°)$^{-1}$. How much heat must be added to the block in order to increase its volume by 1.2×10^{-5} m³?

9. A certain steel railroad rail is 13-yard (yd) in length and weighs 70.0 lb/yd. How much thermal energy is required to increase the length of such a rail by 3.0 mm? [*Note*: Assume the steel has the same specific heat as iron.]

10. In a passive solar house, the sun heats water stored in barrels to a temperature of 38 °C. The stored energy is then used to heat the house on cloudy days. Suppose that 2.4×10^8 J of heat is needed to maintain the inside of the house at 21 °C. How many barrels (1 barrel = 0.16 m³) of water are needed?

11. Huge icebergs are occasionally found floating on the ocean's currents. Suppose one such iceberg is 120 km long, 35 km wide and 230 m thick. (a) How much heat would be required to melt this iceberg (assumed to be at 0 °C) into liquid water at 0 °C? The density of ice is 917 kg/m³. (b) The annual energy consumption by the United States in 1994 was 9.3×10^{19} J. If this energy were delivered to the iceberg every year, how many years would it take before the ice melted?

12. It is claimed that if a lead bullet goes fast enough, it can melt completely when it comes to a halt suddenly, and all its kinetic energy is converted into heat via friction. Find the minimum speed of a lead bullet (initial temperature = 30.0 °C) for such an event to happen.

13. A 5.00-g lead bullet traveling at 300 m/s is stopped by a large tree. If half the kinetic energy of the bullet is transformed into internal energy and remains with the bullet while the other half is transmitted to the tree, what is the increase in temperature of the bullet?

14. Steam at 100 °C is bubbled into 0.250 kg of water at 20 °C in a calorimeter cup. How much steam will have been added when the water in the cup reaches 60 °C? (Ignore the effect of the cup.)

15. An aluminum spoon at 100 °C is placed in a Styrofoam cup containing 0.200 kg of water at 20 °C. If the final equilibrium temperature is 30 °C and no heat is lost to the cup itself, what is the mass of the aluminum spoon?

16. A flow calorimeter is an apparatus used to measure the specific heat of a liquid. The technique is to measure the temperature difference between the input and output points our flowing stream of the liquid while adding energy at a known rate. (a) Start with the equation $Q = mc(\Delta T)$_ and $m = rV$, and show that the rate at which energy is added to the liquid is given by expression $\Delta Q/\Delta t = rc(\Delta T)(\Delta V/\Delta t)$. (b) In a particular experiment, a liquid of density 0.72 g/cm3 flows through the calorimeter at the rate of 3.5 cm³/s. At steady state, a temperature difference of 5.8 °C is established between the input and output point when energy is supplied at the rate of 40 J/s. What is the specific heat of the liquid?

17. A 1.50-kg iron horseshoe initially at 600 °C is dropped into a bucket containing 20.0 kg of water at 25.0 °C. What is the final temperature of the water-horseshoe system? Ignore the heat capacity of the container and assume a negligible amount of water boils away.

18. Marianna really likes coffee, but on summer days she doesn't want to drink a hot beverage. If she is served 200 mL of coffee at 80 °C in a well-insulated container, show how much ice at 0 °C should she add to obtain a final temperature of 30 °C? (F08, 12.40)

19. A high-end gas stove usually has at least one burner rated at 14,000 Btu/h. (a) If you place a 0.25-kg aluminum pot containing 2.0 liters of water at 20 °C on this burner, how long will it take to bring the water to a boil, assuming all the heat from the burner goes into the pot? (b) Once boiling begins, how much time is required to boil all the water out of the pot?

20. Some ceramic materials will become superconducting if immersed in liquid nitrogen. In an experiment, a 0.150-kg piece of such material at 20 °C is placed in liquid nitrogen at its boiling point to cool in a perfectly insulated flask, which allows the gaseous N_2 immediately to escape. How many liters of liquid nitrogen will be boiled away in doing this operation? (Take the specific heat of the ceramic material to be the same as that of glass, and take the density of liquid nitrogen to be 0.80×10^3 kg/m³.)

21. The maximum amount of water an adult in temperature climates can perspire in one hour is typically 1.8 L. However, after several weeks in a tropical climate the body can adapt, increasing the maximum perspiration rate to 3.5 L/hr. At what rate, in watts, is energy being removed when perspiring that rapidly? Assume all of the perspired water evaporates. At body temperature the heat of vaporization of water is $L_v = 24 \times 10^5$ J/kg. (F08, 12.33)

22. At what average rate would heat have to be removed from 1.5 L of (a) water and (b) mercury to reduce the liquid's temperature from 20 °C to its freezing point in 3.0 min?

23. A wood stove is used to heat a single room. The stove is cylindrical in shape, with a diameter of 40.0 cm and a length of 50.0 cm, and operates at a temperature of 400 °F. (a) If the temperature of the room is 70.0 °F, determine the amount of radiant energy delivered to the room by the stove each second if the emissivity is 0.920. (b) If the room is a square with walls that are 8.00 ft high and 25.0 ft wide, determine the R-value needed in the walls and ceiling to maintain the inside temperature at 70.0 °F if the outside temperature is 32.0 °F. Note that we are ignoring any heat conveyed by the stove via convection and any energy lost through the walls (and windows!) via convection or radiation.

24. A steam pipe is covered with 1.50-cm-thick insulating material of thermal conductivity 0.200 cal/cm·°C·s. How much energy is lost every second when the steam is at 200 °C and the surrounding air is at 20.0 °C? The pipe has a circumference of 800 cm and a length of 50.0 m. Neglect losses through the ends of the pipe.

25. A Styrofoam box has a surface area of 0.80 m² and a wall thickness of 2.0 cm. The temperature of the inner surface is 5.0 °C, and the outside temperature is 25 °C. If it takes 8.0 h for 5.0 kg of ice to melt in the container, determine the thermal conductivity of the Styrofoam.

26. Three building materials, plasterboard [κ = 0.30 J/(s·m·C°)], brick [κ = 0.60 J/(s·m·C°)], and wood [κ = 0.10 J/(s·m·C°)], are sandwiched together as the drawing illustrates. The temperatures at the inside and outside surfaces are 27 °C and 0 °C, respectively. Each material has the same thickness and cross-sectional area. Find the temperature (a) at the plasterboard-brick interface and (b) at the brick-wood interface.

27. Consider two cooking pots of the same dimensions, each containing the same amount of water at the same initial temperature. The bottom of the first pot is made of copper, while the bottom of the second pot is made of aluminum. Both pots are placed on a hot surface having a temperature of 145 °C. The water in copper-bottomed pot boils away completely in 425 s. How long does it take the water in the aluminum-bottomed pot to boil away completely?

28. Seals may cool themselves by using *thermal windows*, patches on their bodies with much higher than average surface temperature. Suppose a seal has a 0.030 m² thermal window at a temperature of 30 °C. If the seal's surroundings are frosty -10 °C, what is the net rate of energy loss by radiation? Assume an emissivity equal to that of a human.

29. A solid cylinder is radiating power. It has a length that is ten times its radius. It is cut into a number of smaller cylinders, each of which has the same length. Each small cylinder has the same temperature as the original cylinder. The total radiant power emitted by the pieces is twice that emitted by the original cylinder. How many smaller cylinders are there?

30. Measurements of two stars indicate that Star X has a surface temperature of 5,727 °C and Star Y has a surface temperature of 11,727 °C. If both stars have the same radius, what is the ratio of the luminosity (total power output) of Star Y to the luminosity of Star X? Both stars can be considered to have an emissivity of 1.0.

31. A small sphere (emissivity = 0.90, radius = r_1) is located at the center of a spherical asbestos shell (thickness = 1.0 cm, outer radius = r_2). The thickness of the shell is small compared to the inner and outer radii of the shell. The temperature of the small sphere is 800.0 °C, while the temperature of the inner surface of the shell is 600.0 °C, both temperatures remaining constant. Assuming that r_2/r_1 = 10.0 and ignoring any air inside the shell, find the temperature of the outer surface of the shell.

THERMODYNAMICS

12
CHAPTER

When you go to Disneyland, you can get on a steam locomotive at the main street railroad station at the entrance and use it for transportation to other areas of the park or simply for the experience of the "Grand Circle Tour". Although, nowadays the steam locomotive is only found in museums and amusement parks, it was an important mode of transportation almost 100 years ago. Steam locomotives were first developed in Britain and dominated railway transportation until the middle of the twentieth century. These locomotives are fueled by burning some combustible material, usually coal, wood, or oil, to produce steam in a boiler, which drives steam engine. Hence, both fuel and water supplies are carried with the locomotive, either on the locomotive itself or in wagons pulled behind. From the early 1900s, steam engines were gradually superseded by electric and diesel locomotives. In this chapter, we will examine the basic principles behind steam engines and internal combustion engines, which are good example of heat engines governed by the laws of thermodynamics.

The word thermodynamics originates from the Greek words *therme* which means heat and *dynamis* which means force. Thermodynamics is a branch of physical science that describes matter and radiation in terms of macroscopic properties, and explains how these properties are related and by what laws they change with time. Within the context of this science, we will utilize many variables such as pressure, volume, density, temperature, and specific heat to facilitate the description of macroscopic systems and their relations to the surrounding environment. Thermodynamics is considered one of the most important branches of physics, because it involves fundamental laws and principles that relate to all the different fields of engineering and science. One reason thermodynamics is useful is that it applies to a wide variety of topics in science and engineering, including internal combustion engines, chemical reactions, transport phenomena, and even black holes. Results of thermodynamic calculations are essential for other fields of physics as well as for chemistry, chemical engineering, aerospace engineering, mechanical engineering, biology, and materials science. Ideas from thermodynamics are even useful in fields such as economics. The field of thermodynamics began developing in the nineteenth century when heat and its ability to produce mechanical work became of great interest.

12.1 LAWS OF THERMODYNAMICS

Thermodynamics is a branch of physical science concerned with heat and its relation to energy and work. Work is done when energy is transferred from one body to another by mechanical means. On the other hand, heat is a transfer of energy from one body to a second body at a different temperature. Thermodynamics simply describes the movement or change of a process due to heat flow. It uses macroscopic variables such as temperature, internal energy, entropy, and pressure that characterize materials and radiation. Also, thermodynamics (i) explains how these macroscopic variables are related and (ii) by what laws they change with time. As thermodynamics describes the average behavior of very large numbers of microscopic constituents, its laws can be derived from Statistical Physics, which we do not discuss here. Much of the empirical content of thermodynamics is contained in the four laws. Before we discuss these four laws, let us first define some useful terminologies.

Some useful terminology

In thermodynamics, the universe is divided into a system and its surrounding environment. The system is defined as any object or set of objects we wish to consider, and the environment is everything else other than the system under consideration. The system and environment are separated by a boundary. Both energy and matter may be exchanged between the system and the environment by transferring them through the boundary. Depending on the properties of the boundary, the system may be defined as either a closed system, an open system, or an isolated system. A closed system is defined as a system for which no mass enters or leaves (but energy may be exchanged with the environment). An open system is defined as a system for which mass as well as energy may enter

THERMODYNAMICS

or leave. Finally, an isolated system is defined as a closed system for which no energy crosses its boundaries in any form. It is important to pay particular attention to these terns, especially when solving problems.

Three Laws of Thermodynamics

There are four laws, including the zeroth law, of thermodynamics. As we discussed in Chapter 10, the zeroth law of thermodynamics states that if two systems are in thermal equilibrium with a third system, then they are in thermal equilibrium with each other. This law may be considered as the defining conditions for equilibrium, but it is as important as the defining other three laws. The **first law** of thermodynamics states that energy is conserved. Suppose we define the internal energy of a system as the sum of all the energy of the molecules of the system. The change in the internal energy of a closed system ΔU is equal to the heat Q added to the system minus the work W done by the system. This statement of internal energy change may be written as

$$\Delta U = Q - W \tag{12.1}$$

This is statement of the conservation of energy. The **second law** of thermodynamics can be stated in two equivalent ways. The first statement of this law was made by Rudolf Clausius (1822-1888) in 1850. He stated that heat flows naturally from a hot object to a cold object. This means that heat will not flow spontaneously from a cold object to a hot object. The second statement of this law focuses thermodynamic efficiency rather than the direction of heat flow. The second statement of the second law was by Lord Thomson (1824-1907) in 1851. He stated that no device can transform a given amount of heat completely into work. This statement indicates that there cannot be a perpetual motion machine, which operates continuously with no energy input. The **third law** of thermodynamics states that it is not possible to lower the temperature of any system to absolute zero in a finite number of steps. In 1918, Walter Nernst (1864-1941), a Nobel Prize winning chemist, stated that the entropy change for a chemical or physical transformation approaches zero as absolute zero temperature is approached. The statement, which is known as Nernst theorem, became the third law of thermodynamics.

Thermodynamics is also applicable in biological systems. For biological processes, thermodynamics is focused on principles of chemical dynamics. Principles of thermodynamics govern internal biochemical dynamics such as ATP hydrolysis, protein stability, DNA binding, membrane diffusion, enzyme kinetics, and other essential energy controlled pathways. Thermodynamically, the amount of energy capable of doing work during a chemical reaction is measured quantitatively by the change in a type of energy quantity known as the Gibbs free energy. Gibbs free energy is the energy associated with a chemical reaction that can be used to do work. This concept may also be applicable in studies of evolution. In 1925, Alfred Lotka (1880-1949) attempted to develop an evolutionary theory based on natural selection in his book on mathematical biology, described in terms of a thermodynamic energy called Gibbs free energy.

12.2 APPLICATION OF FIRST LAW OF THERMODYNAMICS

The changes in the state of a system are described by using thermodynamics. A system can proceed from an initial state to a final state through a thermodynamic process. A state of the system is typically described by the following three thermodynamic variables: pressure P, temperature T, and volume V. There variables are also called thermodynamic coordinates. If any one of these variables is changed, the system goes into a different state. Paths through the space of thermodynamic variables are often specified by holding certain thermodynamic variables constant. However, the state of a system is determined by the value of thermodynamic variables, and not on how those values were attained. For this reason, the state of the system is described by a state function. The state function is a function of thermodynamic variables which depend only on the current state of the system, not the path taken to reach that state.

There are two types of thermodynamic processes which are very helpful for understanding the operation of thermodynamic devices such as heat engines, refrigerators, air conditioners, and heat pumps. These processes are i) isothermal and ii) adiabatic processes. These processes are schematically shown in Fig. 12.1. An isothermal process is a change of a system in which the temperature remains constant: $\Delta T = 0$. This process typically occurs when a system is in contact with a thermal reservoir (i.e., heat bath), and the change occurs slowly enough to allow the system to continually adjust to the temperature of the reservoir through heat exchange. On the other hand, an adiabatic process is any process occurring without gain or loss of heat. This means that, during the adiabatic process, the system is thermodynamically isolated so that there is no heat transfer with the surroundings. This is the opposite of a diabatic process, where heat transfer may occur. Many rapid chemical and physical processes in thermodynamics are described or approximated as adiabatic processes.

In general, adiabatic processes can occur if the container of the system has thermally-insulated walls, or if the process happens in an extremely short time, so that there is no opportunity for significant heat exchange. Although the terms adiabatic and isocaloric can often be used interchangeably, adiabatic processes may be considered as a subset of isocaloric processes. The remaining complement subset of isocaloric processes includes processes where net heat transfer does not diverge regionally such as in an idealized case with infinite thermal conductivity or non-existent thermal capacity. For these processes, the system has an adiabatic boundary. An adiabatic boundary is a boundary that is impermeable to heat transfer. A thermally insulated wall approximates an adiabatic boundary.

As shown in Fig. 12.1, thermodynamic processes may be visualized by graphically plotting the changes to the system's state variables. Examples of these two thermodynamic processes are discussed above. Each of these processes has a well-defined start and end point in the pressure-volume (P-V) state space. In this P-V diagram, the process A-B is isothermal, whereas the process A-C is adiabatic. The P-V diagram is particularly useful for visualization of a thermodynamic process. It is also useful for extracting information about energy transfer because the area under the curve is the amount of work done by the system during that process. Thus, work is considered to be a process variable, as its exact value depends on the particular path taken between the start and end points of the process. Similarly, heat may be transferred during a process. It should be noted that it too is a process variable. In contrast, pressure and volume, as well as numerous other properties, are considered state variables because their values depend only on the position of the start and end points, not the particular path between them.

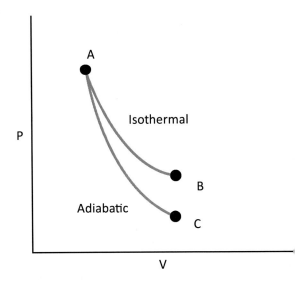

Figure 12.1: A schematic illustration of the isothermal and adiabatic process shows the difference between these two processes in the pressure P versus volume V diagram. The temperature T is kept fixed in the isothermal process, while no heat transfer is allowed in the adiabatic process.

Suppose we consider an isothermal process...

As we discussed above, an isothermal process occurs at a constant temperature. An example of this process would be a system immersed in a large constant-temperature bath. In other words, the system is thermally connected, by a thermally conductive boundary to a constant-temperature reservoir. In this case, any work energy performed by the system will be lost to the bath, but its temperature will remain constant. Now, let us consider an idealized process carried out at constant temperature by applying the ideal gas law

$$PV = \text{constant} \Rightarrow P = \frac{\text{constant}}{V}. \quad (12.2)$$

Suppose we consider an ideal gas enclosed in a container fitted with a movable piston and is in contact with a heat reservoir. Here, a heat reservoir is defined as a body whose mass is so large that its temperature does not change significantly when heat is exchanged with the system. For this system, the relation between the pressure and volume is described by Eq. (12.2). For a fixed amount of gas in the container, the value of the constant in Eq. (12.2) changes only if the temperature is changed. Hence, the curves in the P-V diagram corresponding to different temperatures do not cross, as shown in Fig. 12.2.

We assume that the process of compression or expansion is done very slowly so that all of the gas stays in equilibrium. If the gas is initially in the state represented by point A in Fig. 12.2 and an amount of heat Q is added to the system, then the system moves to another state represented by point B. Noting that this is an isothermal process, we know that heat can be added to the system. When heat is added, the gas expands even

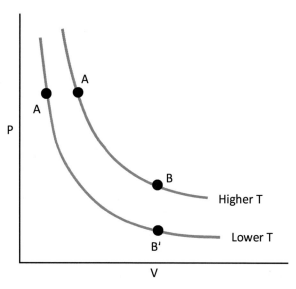

Figure 12.2: Isothermal process at two different temperatures is illustrated. These two isothermal curves do not cross at any place in the P-V diagram.

if the temperature T is fixed. This expansion will do a certain amount of work on the environment. We may determine the amount of work W by the system by using the first law of thermodynamics.

In an isothermal process, the internal energy of an ideal gas which depends only on temperature does not change,

$$\Delta U = \frac{3}{2} nR(\Delta T) = 0. \quad (12.3)$$

According to the first law, if the internal energy U of the system remains unchanged during the thermodynamic process, then we may conclude that

$$\Delta U = Q - W = 0 \Rightarrow W = Q. \quad (12.4)$$

Equation (12.4) indicates that the work W done by the gas equals the heat added to the gas. In the isothermal process, the work done by compressing the gas is considered as negative work done by the system since the work is done on the system. As a result of compression, the volume will decrease and the movable piston shown in Fig. 12.3 moves down. Also, the temperature of the gas will try to increase, but heat energy has to leave the system and enter the environment because the system is constantly in thermal contact with a bath at fixed temperature. The amount of energy entering the environment is equal to the work done while compressing the gas because the internal energy does not change. The thermodynamic sign convention is that heat entering the environment is also negative. This conclusion of $Q = W$ is consistent with that of Eq. (12.4).

The above argument shows that for an ideal gas undergoing an isothermal process (a process at constant temperature), the internal energy is also constant. However, the situation for a real gas is more complex because of the interaction between gas molecules. For a real pure substance, there is a component of internal energy corresponding to the energy used in overcoming inter-molecular forces which depend on the separation distance between molecules. This indicates that when the volume of the gas changes, the average distance between each molecule changes as well. So if the real gas undergoes an isothermal process, there is a net change in internal temperature consistent with this component of internal energy.

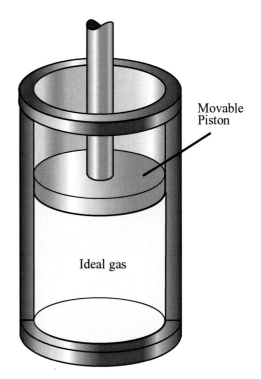

Figure 12.3: An ideal gas contained in a cylindrical container with a movable piston is shown to illustrate that the mechanical work may be either done on or by the gas as its volume shrinks or expands, respectively.

Suppose we consider an adiabatic process...

An adiabatic process is a thermodynamic process in which energy is neither added nor subtracted from the system by heating or cooling. This process occurs, in a system that is thermally insulated from its environment since its boundary is a thermal insulator. For a reversible process, this process is identical to an isentropic process. It should be noted that, an isentropic process is a thermodynamic process which does not change entropy. Entropy measures disorder in the system (see Sec. 12.4). If the entropy of the system has not yet reached its maximum equilibrium value, then the entropy will increase even though the system is thermally insulated.

Under certain conditions two states of a system may be considered adiabatically accessible. These situations can occur if the system is extremely well insulated or the process happens so quickly that heat has not time to flow in or out. For a slow adiabatic expansion (i.e., curve AC in Fig. 12.2) of an ideal gas in an inslating container, we may use the first law to conclude that

$$Q = 0 \Rightarrow \Delta U = Q - W = -W. \tag{12.5}$$

In the adiabatic process, the temperature of the system will change, indicating that the internal energy will also change. Qualitatively, equation (12.5) indicates that the internal energy decreases ($\Delta U < 0$) if the gas expands ($W > 0$). For a monatomic gas, according the decrease in the internal energy U results in the temperature T decrease for an adiabatic expansion.

$$\Delta U = \frac{3}{2} nR(\Delta T), \tag{12.6}$$

THERMODYNAMICS

However, for an adiabatic compression (i.e., curve CA in Fig. 12.1), work is done on the gas and hence the internal energy increases, which results in rising of the temperature T.

Understanding-the-Concept Question 12.1:

A monatomic ideal gas is compressed to one-half its original volume during an adiabatic process. The final pressure of the gas

    a.   increases to twice its original value.

    b.   increases to less than twice its original value

    c.   increases to more than twice its original value

    d.   does not change

Explanation:

For an ideal gas, the relation between pressure P and volume V for an isothermal process is described by $PV = nRT$, indicating that if the volume decreases by half then the pressure must increase by twice. The P-V diagram shown in Fig. 12.1 indicates that the change in the pressure P, for the same amount of change in volume, is greater for the adiabatic process than the isothermal process. This means that for an adiabatic process, the change in pressure should be more than twice when the volume decreases by half.

Answer: c

Other simple Thermodynamics Processes

There are two more useful types of thermodynamic processes. These two simple thermodynamic processes are i) isobaric and ii) isochoric processes as shown in Fig. 12.4. An isobaric process is a thermodynamic process during which the pressure stays constant: $\Delta P = 0$. The term derives from the Greek *iso-* which means "equal", and *baros* which means "weight". An isochoric process is a constant-volume process. It is also known as an isovolumetric process or an isometric process. An isochoric process is exemplified by the heating or the cooling of the contents of a sealed inelastic container. Since the container cannot deform, its volume stays constant as well as the temperature. The addition or removal of heat to the closed system change occur, in an isochoric process.

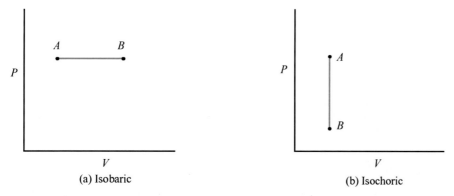

Figure 12.4: Two thermodynamic processes are illustrated. a) In an isobaric process, the pressure P is kept fixed, while b) in an isochoric process the volume V is kept fixed.

Heating a gas in a cylinder with a piston under a constant load is an example of an isobaric process. The ideal Otto cycle is an example of an isochoric process if it is assumed that the burning of the gasoline-air mixture in an internal combustion engine car is instantaneous. An Otto cycle is an idealized thermodynamic cycle which describes the functioning of a typical spark ignition reciprocating piston engine. There is an increase in the temperature and the pressure of the gas inside the cylinder while the volume remains the same.

Computing the work done

When the gas in a container as shown in Fig. 12.5 expands, it does work on its surroundings (i.e., work is done by the gas). This work, in general, is equal to the area under a curve in the P-V diagram which describes an expansion or compression process. The calculation of work is straightforward for an isobaric process. For an isobaric process, noting that the pressure $P = F/A$ (i.e., $F = PA$) is kept constant, we may compute the work done by the gas, using the definition of work, as

$$W = \vec{F} \cdot \vec{d} = PAd = P(\Delta V), \tag{12.7}$$

where the change in the volume of the system is $\Delta V = Ad$, A is the area of the movable piston, and d is the vertical displacement of the piston as shown in Fig. 12.5. It should be noted that the equation $W = P(\Delta V)$ holds also if the gas is compressed at constant pressure. In this case, ΔV is negative and W is negative since the work is done on the gas, rather than by the gas.

In an isochoric process, the volume V is kept constant. The work W done by the system is associated only with either expansion or compression (i.e., with changes of volume). So, when the volume is kept constant, the work done by the gas is zero (i.e., $W = 0$) since $\Delta V = 0$ in Eq. (12.7). This means that there is no area under the curve for this process in the P-V diagram.

The work done by the system depends on the path taken from the initial to final state. Figure 12.5 shows two different ways that the system can go from the initial state A to the final state B. The system can proceed by an isothermal process. Alternatively, the system can proceed first by an isochoric process to reach the intermediate state D, and then to the final state B via an isobaric process. The work done by the system in going from State A to State B, via these two paths as shown in Fig. 12.6 is different. The equation $W = P(\Delta V)$ cannot be used to compute the work when the pressure P changes

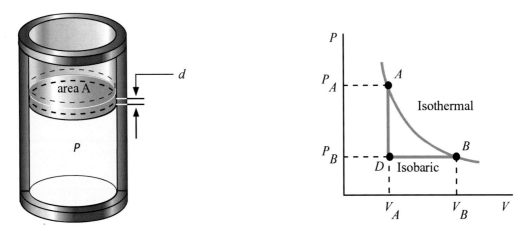

Figure 12.5: Two different ways to go from the state A to B are illustrated. One way is to proceed by an isothermal process. Other way is to go from State A to an intermediate State D by an isochoric process and then to State B by an isobaric process.

THERMODYNAMICS

during a thermodynamic process (as for the isothermal process AB in Fig. 12.7b) because the constant pressure was assumed. It should be noted that $W = P(\Delta V)$ is only applicable to an isobaric process as shown in Fig. 12.7a. One way to get around this difficulty of accounting for pressure change is to make a rough estimate of W by using an "average" value for P. This rough estimate is useful for a quick calculation, but it may not be useful if an accurate value of W is needed. For an accurate determination of W, we need to compute the area under the P-V curve as shown in Fig. 12.7. For an ideal gas, the work done by the gas during an isothermal expansion process is given by

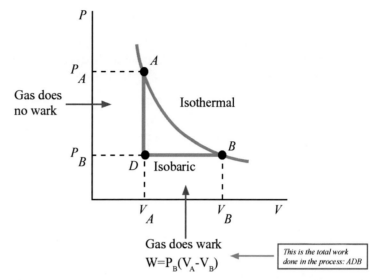

Figure 12.6: The net work done by the system to go from State A to State B thru the intermediate state D is the sum of the work done in the isochoric process and that in the isobaric process.

$$W = nRT \ln\left(\frac{V_B}{V_A}\right) \tag{12.8}$$

Equation (12.8) may be derived by using calculus since the area under a curve may be found more effective by integration.

As shown in Fig. 12.6, we can change the state of the gas from State A to State B via either path. However, the work done by the system alone the path A → D → B is less than the work done alone the isothermal path A → B. The work done by a gas depends not only on the initial and final states of the gas but also on the processes used to change the state. Different thermodynamic paths can lead to the same state, but produce different amounts of work.

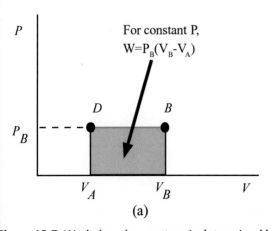
(a)

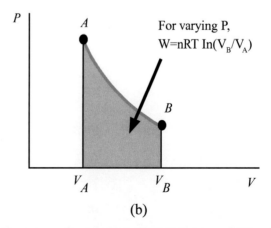
(b)

Figure 12.7: Work done by a system is determined by computing the area under a curve is the P-V diagram. a) When the pressure is kept fixed, work equals the pressure times the volume change. b) When the pressure changes with volume, the area under the curve (i.e., shaded region) needs determined to compute work.

Understanding-the-Concept Question 12.2:

Consider the P-V diagrams for a gas expanding in two ways, isothermally and adiabatically. The initial volume V_A was the same in each case, and the final volumes were the same ($V_B = V_C$). Note that $P_B \neq P_C$. In which process was more work done by the gas?

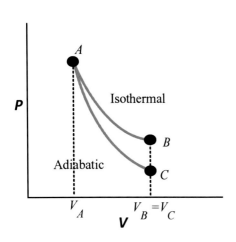

What do we know?

Both the initial and final state of the isothermal and adiabatic process in the P-V diagram are given.

What concepts are needed?

The relation between the work and curves in the P-V diagram is needed.

Why?

Pressure is related to the applied force and volume is related to the displacement. The work is the area under the curve in the P-V diagram.

Explanation:

The work done by the system for a particular thermodynamic process may be determined from the P-V diagram by computing the area under the curve. The area under the curve AB representing the isothermal process is larger than that for the curve AC representing the adiabatic process. The work done by the gas for the isothermal process is greater.

It should be noted that the work done by the gas depends not only on the process, but also depends on the heat transferred to the gas. For the adiabatic process in Understanding-the-Concept Question 12.2, the curved line from State A to State C transferred no heat to the gas. However, in the alternate isochoric and isobaric processes, the straight (vertical) line from State A to an intermediate state (isochoric) and then the straight (horizontal) line to State C (isobaric) in the P-V diagram, heat was transferred to the gas during the constant pressure process. The heat transferred to a gas depends not only on the initial and final states of the gas but also on the process used to change the state.

Understanding-the-Concept Question 12.3:

When the first law of thermodynamics, $Q = \Delta U + W$, is applied to an ideal gas that is taken through an isothermal process,

    a. $\Delta U = 0$ b. $W = 0$ c. $Q = 0$ d.   none of the above

THERMODYNAMICS

Explanation:

When an ideal gas is taken through an isothermal process, the temperature T of the system is held constant. This means there is no change in temperature. Since the internal energy of an ideal gas is $U = (3/2)\,nRT$, the change in internal energy is $\Delta U = 0$.

Answer: a

Exercise Problem 12.1:

An ideal gas is slowly compressed at constant pressure of 2.0 atm from 10.0 liters to 2.0 liters. This process is represented on a P-V diagram as the path B to D. (In this process, some heat flows out and the temperature drops.) Heat is then added to the gas, holding the volume constant, and the pressure and temperature are allowed to rise until the temperature reaches its original value. This process is D to A in the PV curve. Calculate (a) the total work done by the gas in the process BDA, and (b) the total heat flow into the gas.

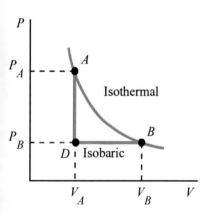

What do we know?

An ideal gas is being compressed and the system goes from state B to A either via isothermal process or via isobaric and then isochoric process.

What concepts are needed?

We need to apply the definition of work and first law of thermodynamics.

Why?

For the isothermal process, there is no change in the internal energy which is the heat added minus the work done by the system.

Solution:

a. Work is done only in the isobaric compression process BD. The work done on the system in this process is

$$W = P\Delta V = P_B(V_A - V_B)$$

$$= 2.0\,\text{atm} \times \frac{1.013 \times 10^5 \,\text{N}/\text{m}^2}{1\,\text{atm}} \times (2.0\,\text{L} - 10.0\,\text{L}) \times \frac{1.0 \times 10^{-3} \,\text{m}^3}{1\,\text{L}} \cong -1.62 \times 10^3 \,\text{J}$$

It is noted that there is no work is done on the system from State D to State A since $\Delta V = 0$. The total work done is -1.62×10^3 J is the sum of contribution from that for the isochoric and isobaric processes.

b. Since the temperature at the beginning and at the end of the process is the same, $\Delta U = 0$. Applying the first law of thermodynamics, we obtain

$$0 = \Delta U = Q - W \Rightarrow Q = W.$$

The total heat flows into the gas is $Q = -1.62 \times 10^3$ J. Since Q is negative, this means that 1620 J of heat flows out of the gas.

Exercise Problem 12.2:

In an engine, 0.25 moles of gas in the cylinder expands rapidly and adiabatically against the piston. During the process, the temperature drops from 1150 K to 400 K. How much work does the gas do? Assume the gas is ideal.

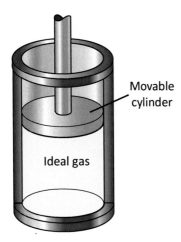

What do we know?

While a gas expands adiabatically, its temperature decreases.

What concepts are needed?

We apply the first law of thermodynamics. Also, we need to note the internal energy change is proportional to the temperature change.

Why?

For an adiabatic expansion, the heat transfer is $Q = 0$. According to the first law, this means the work done by the gas is directly related to the internal energy change.

Solution:

During an adiabatic expansion, the pressure does not remain constant, but the heat flow into the system is zero (i.e., $Q = 0$). We may apply the first law of thermodynamics and obtain

$$\Delta U = Q - W \Rightarrow W = Q - \Delta U$$

We may determine ΔU using $U = (3/2)\,nRT$ for the internal energy of an ideal monatomic gas and write

$$\Delta U = \frac{3}{2} nR\left(T_f - T_i\right) = U_f - U_i$$

$$= \frac{3}{2} \times 0.25 \,\text{mol} \times 8.315 \frac{\text{J}}{\text{mol} \cdot \text{K}} \times (400\text{K} - 1150\text{K}) = -2300\,\text{J}$$

We combine the result and compute the work done by the gas as

$$W = Q - \Delta U = 0 - (-2300\text{J}) = 2300\text{J}.$$

THERMODYNAMICS

As an example of adiabatic expansion, let us consider free expansion. Free expansion is an irreversible process in which a gas expands into an insulated evacuated chamber. For an adiabatic free expansion of an ideal gas, the gas is contained in an insulated container and then allowed to expand into a vacuum. Since there is no external pressure for the gas to expand against, the work done by or on the gas is zero. Also, the first law of thermodynamics implies that the net internal energy change of the system is zero since this process does not involve any heat transfer or work. For an ideal gas, the temperature remains constant because the internal energy only depends on temperature in that case. As we discuss below in Sec. 12.4, since the entropy is proportional to the volume at constant temperature, the entropy which measures disorder increases in this case. Therefore, this free expansion process is irreversible.

Understanding-the-Concept Question 12.4:

Work is done when a person walks or runs, or lifts a heavy object. Work requires energy. Energy is also needed for growth – to make new cells, and to replace old cells that have died. A great many energy transforming processes occur within an organism (i.e., metabolism). If we apply the first law of thermodynamics $\Delta U = Q - W$ to an organism, say the human body, work W is done by the body in various activities and heat Q flows out of the body (i.e., decreases the body's internal energy). What is the source of internal energy?

Metabolic Rates (for a 65-kg human)		
	Approximate Metabolic Rate	
Activity	kcal/h	watts
Sleeping	60	70
Sitting upright	100	115
Light activity (eating, dressing, household chores)	200	230
Moderate work (tennis, walking)	400	460
Running (15 km/h)	1000	1150
Bicycling (racing)	1100	1270

Explanation:

According to conservation of energy, heat energy that flows out of the body and work done by the body must come from somewhere. Energy that the body needs comes from food that processed by the body. When more energy is put into the body than consumed, the body stores the remaining energy as body fat.

According to Understanding-the-Concept Question 12.4, we may treat the human body as a thermodynamic system and account for its energy consumption. Although, it is not written in the table above, the human body consumes significant amount of energy intake in normal brain activities which include thinking. In 1986, researchers isolated both the "at rest" and "active" consumption of calories in the brain. Since then, we have learned quite a bit about brain activity, particularly as positron emission tomography (PET) scans have been applied to monitor glucose consumption in the brain. As a result, we know lots of things. We now know that the energy consumption in the brain is related to learning.

In other words, once we have learned something (like solving a challenging physics problem), the energy consumption goes down. Also, we know that the energy consumption in the brain is more than two times higher for children under age 4. This is no surprise because they are learning and building brain structure. The brain's energy consumption levels off at age around 10 to 12. Intelligence quotient (IQ) can affect energy consumption. After learning a task, lower IQ people have to exert more energy to complete a task than high IQ people who have learned the same task. Energy consumption by the brain is estimated as 230-247 Calories per day. This estimate is based on 17 Calories/gram and human brain sizes of 1,350-1,450 grams. During periods of peak performance, adults increase that energy consumption by up to 50%. While this may not seem an extraordinary amount of energy, the brain may use 30% of a body's total energy, while being only 2-3% of total body mass.

12.3 APPLICATION OF SECOND LAW OF THERMODYNAMICS

As we discussed in Sec. 12.1, the second law of thermodynamics can be stated in two ways. One of them is the Kelvin statement which says that it is impossible to construct an engine that will convert all the heat energy from a reservoir into work. The other one is the Clausius statement which says that it is impossible for a heat engine to transfer heat from a body of low temperature to a body of high temperature without external sources of energy. Heat is defined as the transfer of energy which arises naturally from a warmer body to a cooler body due to a temperature difference between the system and its surroundings. A heat engine is a device that converts heat into mechanical work while transferring heat from a warmer reservoir to a colder reservoir. Examples of heat engines include diesel engines and petrol engines.

Suppose we consider heat engines such as steam locomotive engines and automobile engines. In general, internal combustion engines are considered heat engines because they change thermal energy into mechanical work. The basic idea behind any heat engine is that mechanical energy can be obtained from thermal energy only when heat is allowed to flow from a high temperature to a low temperature. More spacifically, steam, gasoline, and diesel engines operate by first burning fuel(converting chemical energy to heat energy) and then using a heat engine to convert the heat energy to mechanical energy. This is different than fuel calls which go directly from chemical energy to mechanical energy, without using heat as an intermediate. We focus on the part of the engines that run in a repeating cycle (i.e., the system returns repeatedly to its starting point) and thus can run continuously.

Energy Transfers in Heat Engine

The relationship between heat and work may be seen easily in a heat engine. A heat engine is a thermodynamic device which typically uses energy provided from the high temperature reservoir in the form of heat to do work and then exhausts the heat which cannot be used to do work to another reservoir as shown in Fig. 12.8. All heat engines have two operating temperatures. Heat flows into the engine from the high temperature reservoir. The heat engine converts some of this heat into work. The unused heat is then exhausted into the lower temperature reservoir. Sometimes, the surrounding environment serves as the low temperature reservoir.

As shown in a schematic energy flow diagram of Fig. 12.8 for a heat engine, the amount of heat input to the engine is Q_H. This energy flows in from the high temperature reservoir at T_H. The amount of heat exhaust into the lower temperature reservoir at T_L is Q_L. The amount of work done by the heat engine W is related to both Q_H and Q_L by conservation of energy (i.e., $Q_H = Q_L + W$).

THERMODYNAMICS

Heat engines, including the two types of heat engines shown in Fig. 12.9, can be characterized by their specific power, which is typically given in kilowatts per liter of engine displacement (and horsepower per cubic inch in the US). The result offers an approximation of the peak-power output of an engine. This is not to be confused with fuel efficiency, since high-efficiency often requires a lean fuel-air ratio, and thus lower power density. A modern high-performance car engine makes in excess of 75 kW/L (1.65 hp/in³). Examples of everyday heat engines include the steam engine, the diesel engine, and the gasoline engine in an automobile. A common toy that is also a heat engine is a drinking bird (also known as an insatiable birdie or a dipping bird), which mimics the

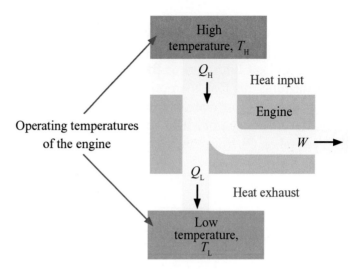

Fig. 12.8: A schematic illustration of energy transfer in operation of a heat engine. The heat engine operates between two reservoirs. The temperatures of the reservoir are T_H and T_L.

motions of a bird drinking from a water source. Also, the Stirling engine is a heat engine operating by cyclic compression and expansion of air or other gas as the working fluid at different temperature levels such that there is a net conversion of heat energy to mechanical work. This heat engine was conceived in 1816 as an industrial prime mover to rival the steam engine. The Stirling engine's practical use was largely confined to low-power domestic applications for over a century. All of these familiar heat engines are powered by the expansion of heated gases. The general surroundings are the heat sink, providing relatively cool gases which, when heated, expand rapidly to drive the mechanical motion of the engine.

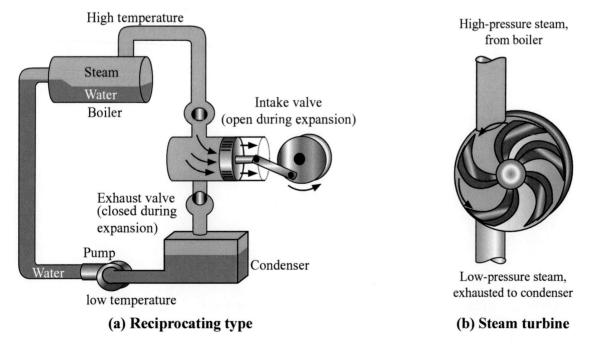

Figure 12.9: A schematic illustration of two different heat engines. a) A reciprocating engine uses one or more pistons to convert pressure into motion. b) A steam turbine extracts thermal energy stored in pressurized steam to do mechanical work.

Efficiency of a Heat Engine

Converting heat into work is the main task of a heat engine. Some heat engines are more effective in performing this task than others. In determining the effectiveness of any heat engine, we need a dimensionless performance measure for a device that uses thermal energy. This measure is the efficiency. Here, the efficiency e of a heat engine is defined as the ratio of the work W done by the heat engine to the heat taken Q_H from the hot reservoir:

$$e = \frac{W}{Q_H}. \tag{12.9}$$

Since the energy of the system is conserved (i.e., $Q_H = W + Q_L$), we may write Eq. (12.9) as

$$e = \frac{W}{Q_H} = \frac{Q_H - Q_L}{Q_H} = 1 - \frac{Q_L}{Q_H}. \tag{12.10}$$

The efficiency e of any heat engine may be described by using Eq. (12.10). One may ask the following question: Can a heat engine be 100% efficient? It turns out that the second law of thermodynamics puts a limit on the efficiency of heat engines.

A heat engine is most efficient when we have an ideal engine or Carnot engine. A Carnot heat engine is a hypothetical engine that operates on the reversible Carnot cycle. The basic model for this engine was developed by Carnot (1796-1832) in 1824. The Carnot engine model was graphically expanded upon by Clapeyron (1799-1864) in 1834 and mathematically elaborated upon by Clausius (1822-1888) in 1857 from which the concept of entropy emerged. The operation of a heat engine is described in terms of a sequence of state changes in the P-V diagram. Every single thermodynamic system exists in a particular state. A thermodynamic cycle occurs when a system is taken through a series of different states, and finally returned to its initial state. In the process of going through this cycle, the system may perform work on its surroundings, thereby acting as a heat engine. A heat engine acts by transferring energy from a warm region to a cool region of space and, in the process, converting some of that energy to mechanical work.

In a Carnot engine, every process (heat addition, exhaust, expansion, or compressional) must be done reversibly. A process is reversible if it is done slowly enough that it can be considered to pass through a series of equilibrium states, and the whole process can be done in reverse with no change in the magnitude of work done or heat exchanged. For an idealized reversible engine, the heat Q_H and Q_L are proportional to the operating temperatures T_H and T_L. Hence, we may write the efficiency e of Eq. (12.10) as

$$e_{ideal} = \frac{T_H - T_L}{T_H} = 1 - \frac{T_L}{T_H}. \tag{12.11}$$

This is called either the Carnot efficiency or the ideal efficiency. Here, e_{ideal} is the theoretical limit to the efficiency, and, according to Carnot's theorem, real engines cannot exceed this efficiency. Carnot's theorem states that no engine operating between two heat reservoirs can be more efficient than a Carnot engine operating between the same reservoirs. We may now ask the following:

THERMODYNAMICS

Why can real engines not have e_{ideal}?

In real engines, each of the processes of gas expansion or compression occur quickly. Friction and turbulence is present in the gas. A real process cannot be done precisely in reverse. In reversing the process, the turbulence would be different and the heat lost to friction would not reverse itself. So, the reversible engine is not possible in the real world. Suppose we have a Carnot engine. We ask another question: "Is 100% efficient heat engine possible?" The possibility of 100% efficient engine may be answered by using

$$e_{ideal} = 1 - \frac{T_L}{T_H}.$$

This equation indicates that 100% efficiency may be obtained if and only if $T_L = 0$ K, but it is impossible to reach absolute zero (i.e., the third law of thermodynamics). This is a practical and theoretical impossibility. The physical implication is that the only way to have all heat transfer go into doing work is to remove all thermal energy, requiring a cold reservoir at absolute zero.

Carnot Cycle

Let us examine the Carnot engine in more detail by considering the thermodynamics processes involved in this ideal heat engine. The Carnot cycle is a theoretical thermodynamics cycle proposed by Carnot in 1823 and expanded by Clapeyron in the 1830s and 40s. It can be shown that it is the most efficient

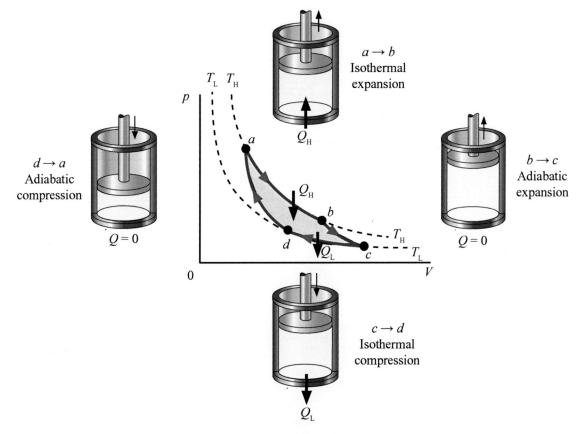

Figure 12.10: The four steps of the Carnot cycle are illustrated: reversible isothermal expansion (a→b), reversible adiabatic expansion (b→c), reversible isothermal compression (cd), and reversible adiabatic compression (d→a).

cycle for converting a given amount of thermal energy into work, or conversely, creating a temperature difference (e.g., refrigeration) by doing a given amount of work. Every thermodynamic system exists in a particular thermodynamic state. When a system is taken through a series of different states and finally returned to its initial state, a thermodynamic cycle is said to have occurred. In the process of going through this cycle, the system may perform work on its surroundings, thereby acting as a heat engine. A system undergoing a Carnot cycle is called a Carnot heat engine, although such a ‹perfect› engine is only a theoretical limit and cannot be built in practice.

The Carnot cycle when acting as a heat engine consists of the following steps as shown in Fig. 12.10: i) reversible isothermal expansion of the gas at the "hot" temperature, T_H (isothermal heat addition or absorption), ii) isentropic (reversible adiabatic) expansion of the gas (isentropic work output), iii) reversible isothermal compression of the gas at the "cold" temperature, T_L. (isothermal heat rejection), and iv) isentropic compression of the gas (isentropic work input).

The Carnot heat engine cycle described here is a totally reversible cycle. Therefore, all the processes that comprised it can be reversed, in which case it becomes the Carnot refrigeration cycle. This time, the cycle remains exactly the same, except that the directions of any heat and work interactions are reversed: Heat is absorbed from the low temperature reservoir, heat in is rejected to a high temperature reservoir, and a work input is required to accomplish all this. The P-V diagram of the reversed Carnot cycle is the same as for the Carnot cycle, except that the directions of the processes are reversed.

Understanding-the-Concept Question 12.5:

A Carnot cycle consists of

    a.   two adiabats and two isobars.

    b.   two isobars and two isotherms.

    c.   two isotherms and two isomets.

    d.   two adiabats and two isotherms.

Explanation:

As shown in Fig. 12.10, a Carnot cycle consists of i) isothermal expansion (curve ab), ii) adiabatic expansion (curve bc) iii) isothermal compression (curve cd), and iv) adiabatic compression (curve da). Hence, the Carnot cycle has two adiabats and two isotherms.

Answer: d

Although we would like to obtain the efficiency of Carnot engine, the reality is far from it. Ever since the invention of the internal combustion engine, scientists and engineers have worked to increase its efficiency. There are different ways to find the efficiency. We can rate the efficiency on different parts of an engine. Thermal efficiency is the percentage of energy taken from the combustion which is converted to mechanical work. In a typical low compression engine, the thermal efficiency is about 26%. In a highly modified engine, such as a race engine, the thermal efficiency is about 34%. Mechanical efficiency is the percentage of energy that the engine puts out after subtracting mechanical losses such as friction, compared to what the engine would put out with no power loss. Most engines are about 94% mechanically efficient. This means that for a stock engine, only 20% of the power in fuel combustion is effective. So, as it stands now, the average internal combustion engine only converts roughly 24% of its energy into power. Most of the energy is expended through heat loss in various locations. One

THERMODYNAMICS

reason for the loss of mechanical efficiency in an engine is the sheer number of moving parts at high speeds. The movement creates great amount of friction between parts, which create heat. Lubricants and cooling systems can only do so much about this problem. So, the typical internal combustion engine remains about 20-25% efficient.

Internal Combustion Engine

The internal combustion engine is an engine in which the combustion of a fuel (normally a fossil fuel) occurs with an oxidizer (usually air) in a combustion chamber that is an integral part of the fluid circuit. In an internal combustion engine the expansion of the high-temperature and high-pressure gases produced by combustion applies direct force to some component of the engine. The force is applied typically to pistons on turbine blade. This force moves the component over a distance, transforming chemical energy into useful mechanical energy. The first commercially successful internal combustion engine was created by Lenoir. The term internal combustion engine usually refers to an engine in which combustion is intermittent, such as the more familiar four-stroke and two-stroke piston engines, along with variants, such as the six-stroke piston engine and the Wankel rotary engine. A second class of internal combustion engines uses continuous combustion: gas turbines, jet engines and most rocket engines. These engines operate on the same principle as we described.

Almost all cars currently use what is called a four-stroke combustion cycle to convert gasoline into motion. The four-stroke approach is also known as the Otto cycle, in honor of Otto, who invented it in 1867. The four strokes are illustrated in Fig. 12.11. They are i) intake stroke, ii) compression stroke, iii) combustion stroke, and iv) exhaust stroke. The piston is connected to the crankshaft by a connecting rod. As the crankshaft revolves, it has the effect of "resetting the cannon." The following is what happens as the engine goes through its cycle. First, the piston starts at the top, the intake valve opens, and the piston moves down to let the engine take in a cylinderful of air and gasoline. This is the intake stroke. Only the tiniest drop of gasoline needs to be mixed into the air for this to work. Second, the piston moves back up to compress this fuel/air mixture. Compression makes the explosion more powerful. Third, when the piston reaches the top of its stroke, the spark plug emits

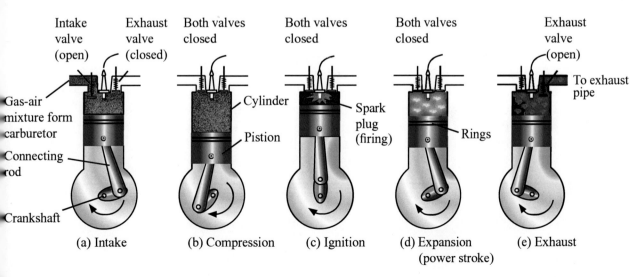

Figure 12.11: A schematic illustration of operation of a four-stroke internal combustion engine: a. intake/suction stroke, b. compression stroke, c. ignition of the fuel mixture, d. power stroke, and e. exhaust stroke.

a spark to ignite the gasoline. The gasoline charge in the cylinder explodes, driving the piston down. Finally, once the piston hits the bottom of its stroke, the exhaust valve opens and the exhaust leaves the cylinder to go out the tailpipe. Now the engine is ready for the next cycle, so it intakes another charge of air and gas.

According to Second Law of Thermodynamics

As Lord Kelvin expressed "It is impossible, by means of inanimate material agency, to derive mechanical effect from any portion of matter by cooling it below the temperature of the coldest of the surrounding objects.", no process is possible in which the sole result is the absorption of heat from a reservoir and its complete conversion into work as shown in Fig. 12.12.

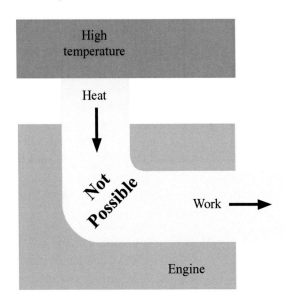

Figure 12.12: A schematic illustration of a heat transfer process which is not allowed by the second law of thermodynamics. According to Lord Kelvin, it is not possible to transform a given amount of heat completely into work.

This means that it is impossible to extract energy by heat from a high-temperature energy source and then convert all of this heat energy into work. At least some of the energy must be passed on to a low-temperature energy sink as heat. Thus, a heat engine exhibiting 100% efficiency is thermodynamically impossible. This also means that it is impossible to build solar panels that generate electricity solely from the infrared band of the electromagnetic spectrum with 100% efficiency. Although solar (photo voltaic) panels are not limited by Carnot efficiency as long as they convert light directly into electricity, the maximum theoretically possible conversion efficiency for sunlight is 85% due to the Carnot limit. This is due to the temperature of the photons emitted at the Sun's surface.

Exercise Problem 12.3:

A 100-hp car engine operates at about 15 percent efficiency. Assume the engine's water temperature of 85 °C is its cold-temperature (exhaust) reservoir and 500 °C is its thermal "intake" temperature (the temperature of the exploding gas/air mixture). (a) Calculate its efficiency relative to its maximum possible (Carnot) efficiency. (b) Estimate how much power (in Watts) goes into moving the car, and how much heat, in joules and in kcal, is exhausted to the air in 1 h.

Solution:

The efficiency of engine may be expressed as a fraction of the ideal efficiency

$$e = fe_{ideal} = f\frac{T_H - T_L}{T_H}.$$

THERMODYNAMICS

We solve for this fraction f and obtain

$$f = e\frac{T_H}{T_H - T_L} = 0.15 \times \frac{(500+273)\text{K}}{(500+273)\text{K} - (85+273)\text{K}} \cong 0.28 \Rightarrow f = 28\%.$$

The power output of the 100-hp engine is

$$P = 100\,\text{hp} \times \frac{746\,\text{W}}{1\,\text{hp}} = 7.46 \times 10^4\,\text{W}$$

Noting that power $P = W/t$, we find the intake heat flow rate from

$$e = \frac{W}{Q_H} = \frac{P}{\left(Q_H/t\right)} \Rightarrow \frac{Q_H}{t} = \frac{P}{e} = 7.46 \times 10^4\,\text{W} \times \frac{1}{0.15} \times \frac{3600\,\text{s}}{1\,\text{h}} = 1.79 \times 10^9\,\frac{\text{J}}{\text{h}}.$$

We find the discharge heat flow by using energy conservation (i.e., $Q_H = Q_L + W$)

$$\frac{Q_L}{t} = \frac{Q_H}{t} - \frac{W}{t} = 1.79 \times 10^9\,\frac{\text{J}}{\text{h}} - 7.46 \times 10^4\,\frac{\text{J}}{\text{s}} \times \frac{3600\,\text{s}}{1\,\text{h}} = 1.52 \times 10^9\,\frac{\text{J}}{\text{h}} \times \frac{1\,\text{kcal}}{4.18 \times 10^3\,\text{J}}$$

$$= 3.46 \times 10^5\,\frac{\text{kcal}}{\text{h}}.$$

Another important statement of the first and second law of thermodynamics is that it is impossible to construct a perpetual motion machine. Perpetual motion describes motion that continues indefinitely without any external source of energy. This is impossible in practice because of friction. It can also be described as the motion of a hypothetical machine which, once activated, would run forever unless subject to an external force or to wear. There is a scientific consensus that perpetual motion in an isolated system would violate the first and/or second law of thermodynamics. There are cases of apparent perpetual motion in nature, but these are either not truly perpetual or cannot be used to do work without changing the nature of the motion. For example, the motion of a planet around its star may appear «perpetual», but interplanetary space is not completely frictionless, so planets› orbital motion is very gradually slowed over time. The fly-by of a space probe past a planet can be used to speed up the probe but in doing so it alters the motion and reduces the energy of the planet in its orbit around the Sun. The flow of current in a superconducting loop can be used as an energy storage medium, but just as with a battery, using it to power a device will remove an equivalent amount of energy from the current in the loop. Machines which extract energy from seemingly perpetual sources such as ocean currents are capable of moving "perpetually" (for as long as that energy source itself endures), but they are not considered to be perpetual motion machines because they are consuming energy from an external source and are not isolated systems (in reality, no system can ever be a fully isolated system). Similarly, machines which comply with both laws of thermodynamics but access energy from obscure sources are sometimes referred to as perpetual motion machines, although they also do not meet the standard criteria for the name. Despite the fact that successful isolated system perpetual motion devices are physically impossible in terms of the current understanding of the laws of physics, the pursuit of perpetual motion is still popular.

Refrigerators, Air Conditioners, and Heat Pumps

The thermodynamic cycle of a heat engine may be reversed. The system may be worked upon by an external force, and in the process, it can transfer thermal energy from a cooler system to a warmer one, thereby acting as a refrigerator or heat pump rather than a heat engine. The Clausius statement tells us that energy (heat) will not flow from cold to hot regions without outside assistance. The devices that provide this help are called refrigeration units and heat pumps. Both types of devices satisfy the Clausius requirement of external action through the application of mechanical power. The distinction between refrigerator and heat pump is one of purpose rather than technique. The

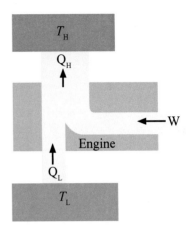

Figure 12.13: A schematic illustration of energy transfer of a thermodynamic engine which operates in the reverse direction compared to a heat engine.

refrigeration unit transfers energy (heat) from cold to hot regions for the purpose of cooling the cold region while the heat pump does the same thing with the intent of heating the hot region. We make the distinction between refrigeration and heat pumps only when it is essential to the discussion

The operating principle of refrigerators, air conditioners, and heat pumps is just the reverse of a heat engine. Now, we will focus on refrigeration. As shown in Fig. 12.13, for a refrigerator, a certain amount of energy is extracted from a colder reservoir and is exhausted into a hotter reservoir. The work W is done by a compressor motor which compresses a fluid to do this.

According to the second law of thermodynamics (i.e., Clausius statement), there can be no perfect refrigerator, similar to heat engine. We may ask the following question:

What is a perfect refrigerator?

A perfect refrigerator is a device in which no work is required to take heat from the low temperature region to the high temperature region. Of course, real refrigerators cannot do this and a certain amount of work is always needed. The effectiveness of a refrigerator is measured by using a coefficient of performance, rather than efficiency. The coefficient of performance (CoP) of a refrigerator is defined as the heat Q_L removed from the low temperature area divided by the work W done to remove the heat,

$$CoP_{refrigerator} = \frac{Q_L}{W} = \frac{Q_L}{Q_H - Q_L}. \quad (12.12)$$

Note that $Q_H = Q_L + W$, according to the energy conservation principle. For an ideal refrigerator, we may write

$$CoP_{ideal} = \frac{T_L}{T_H - T_L}. \quad (12.13)$$

Equation (12.13) is similar to the efficiency of an ideal (Carnot) engine. It can also be shown that $CoP_{cooling} = CoP_{heating} - 1$. It should be noted that these equations must use the absolute temperature (i.e. either the Kelvin or Rankine scale). Also, we note that $CoP_{heating}$ applies to heat pumps and $CoP_{cooling}$

… applies to air conditioners or refrigerators. Values for actual systems will always be less than these theoretical maxima. In Europe, ground source heat pump units are tested at $T_H = 35\,°C$ (95 °F) and $T_L = 0\,°C$ (32 °F). According to the above formula, the maximum achievable CoP would be 8.8. However, the test results of the best systems are around 4.5. When measuring installed units over a whole season, we also need to account for the energy needed to pump water through the piping systems. When this correction is included in the calculation, we find that seasonal CoP's are around 3.5 or less. This indicates that there is room for making improvement.

As Eq. (12.13) indicates, the CoP of a heat pump system can be improved by reducing the temperature gap, T_H minus T_L, at which the system works. For a heating system this would mean two things: i) reducing the output temperature to around 30 °C (86 °F) which requires piped floor, wall or ceiling heating, or oversized water to air heaters and ii) increasing the input temperature by using an oversized ground source or by access to a solar-assisted thermal bank. For an air cooler, the CoP could be improved by using ground water as an input instead of air, and by reducing the temperature drop on the output side by increasing the air flow. For both systems, increasing the size of pipes and air canals would also help to reduce noise and the energy consumption of pumps (and ventilators).

Understanding-the-Concept Question 12.5:

A refrigerator removes heat from the freezing compartment at the rate of 20 kJ and ejects 24 kJ into a room per cycle. What is the coefficient of performance?

a.   0.20   b.   0.50   c.   2.0   d.   5.0

Explanation:

Since the coefficient of performance for a refrigerator is defined as $CoP_{refrigerator} = Q_L/W$, we may rewrite it, by using conservation of energy, as $CoP = Q_L/(Q_H - Q_L) = 20\,kJ/(24\,kJ - 20\,kJ) = 5.0$

Answer: d

A household refrigerator may have a coefficient of performance of about 2.5, whereas a deep freeze unit will be closer to 1.0. When the temperature of the space is not uniform, the lower cold temperature region of space will result in lower values of CoP. Sometimes, the coefficient of performance of a refrigerator is defined as the ratio of refrigerator capacity to the work input of the refrigerator. The capacity is usually expressed in tons of refrigeration. One ton of refrigeration is 3.516 kW or 12,000 Btu/h. Here the term "ton" is derived from the fact that the heat required to melt one ton of ice is about 12,000 Btu/h.

Heat Pump

Heat pumps which are thermodynamic devices similar to refrigerators. A heat pump uses mechanical energy to transfers heat energy from a heat source to a heat sink against a temperature gradient. A heat pump may be considered similar to a refrigerator, except heat is added to a higher temperature region by doing machanical work W. It is designed to move thermal energy opposite to the direction of spontaneous heat flow. Hence, a heat pump uses some amount of external energy to accomplish the desired transfer of thermal energy from cooler outside to warmer inside of a house in winter, as schematically illustrated in Fig. 12.14. Here heat Q_L is taken from the outside at low temperature and heat Q_H is delivered to the warmer inside of the house. During the heating season, heat pumps move heat from the cool outdoors into our warm house. However, during the cooling season, heat pumps move heat from our cool house into the warm outdoors. Because they move heat rather than generate heat, heat pumps can provide up to 4 times the amount of energy they consume.

The effectiveness of a heat pump is also measured by using coefficient of performance. The coefficient of performance for a heat pump is defined as the heat Q_H delivered to inside of the house divided by the work W done to deliver the heat. We may write

$$CoP_{heat\ pump} = \frac{Q_H}{W}. \tag{12.14}$$

In general, while compressor-driven air conditioners and freezers are familiar examples of heat pumps, the term "heat pump" is used more generally and applies to HVAC devices used for space heating or space cooling. When a heat pump is used for heating, it employs the same basic refrigeration-type cycle used by an air conditioner or a refrigerator, but it operates in the opposite direction, releasing heat into the conditioned-space rather than the surrounding environment. In this use, heat pumps generally draw heat from the cooler external air or from the ground.

In heating and air conditioning (HVAC) applications, the term heat pump usually refers to easily reversible vapor-compression refrigeration devices optimized for high efficiency in both directions of thermal energy transfer. Heat spontaneously flows from warmer places to colder spaces. A heat pump can absorb heat from a cold space and release it to a warmer one, and vice-versa. "Heat" is not conserved in this process, which requires some amount of external high grade (i.e., low-entropy) energy to be expended. Heat pumps are used to provide heating because less high-grade energy is required for their operation than appears in the released heat. Most of the energy for heating comes from the external environment, and only a fraction comes from electricity (or some other high-grade energy source is required to operate a compressor). In electrically powered heat pumps, the heat transferred can be three or four times larger than the electrical power consumed, giving the system CoP of 3 or 4, as opposed to a CoP of 1 of a conventional electrical resistance heater, in which all heat is produced from input electrical energy. Heat pumps use a refrigerant as an intermediate fluid which absorbs heat and vaporizes in the evaporator. Then it releases heat in the condenser and the refrigerant condenses. The refrigerant flows through insulated pipes between the evaporator and the condenser, allowing for efficient thermal energy transfer at relatively long distances. The coefficient of performance for a heat pump itself can be improved

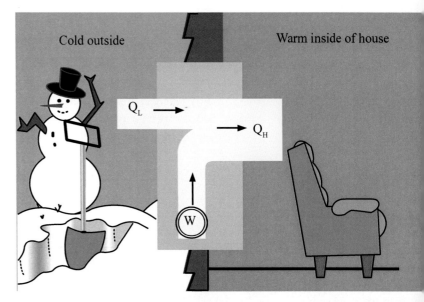

Figure 12.14: A heat pump which transfers thermal energy from the low to high temperature reservoir by using mechanical work is schematically illustrated Heat pumps use electricity to move heat from a cool space to a warm space making the cool space cooler and the warm space warmer.

THERMODYNAMICS

by increasing the size of the internal heat exchangers relative to the power of the compressor, and to reduce the system's internal temperature gap over the compressor. This latter measure, however, makes such heat pumps unsuitable to produce output above roughly 40 °C (104 °F), indicating that a separate machine is needed to produce hot tap water.

Exercise Problem 12.5:

A heat pump is used to keep a house warm at 22 °C. How much work is required of the pump to deliver 2800 J of heat into the house if the outdoor temperature is (a) 0 °C, (b) -15 °C? Assume ideal (Carnot) behavior.

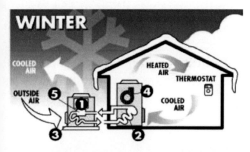

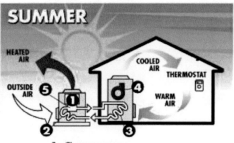

1. Compressor
2. Condenser
3. Evaporator
4. Air handler
5. Reversing valve

What do we know?

An ideal heat pump is doing work to pump heat to the house and maintains the indoor temperature at 22 °C.

What concepts are needed?

We apply the definition for the coefficient of performance for an ideal heat pump.

Why?

According to the definition, the coefficient of performance of an ideal heat pump is the heat transferred to the indoor divided by the work performed.

Solution:

a. The coefficient of performance for the ideal heat pump is

$$CoP = \frac{T_H}{T_H - T_L} = \frac{(22+273)\text{K}}{(22+273)\text{K} - (0+273)\text{K}} = 13.4 = \frac{Q_H}{W} \Rightarrow W = \frac{Q_H}{CoP}.$$

The work needed to pump the heat is

$$W = \frac{2800\text{J}}{13.4} = 2.1 \times 10^2 \text{ J}.$$

b. Similarly, the coefficient of performance for the ideal heat pump is

$$CoP = \frac{T_H}{T_H - T_L} = \frac{(22+273)\text{K}}{(22+273)\text{K} - (-15+273)\text{K}} = 7.97 = \frac{Q_H}{W} \Rightarrow W = \frac{Q_H}{CoP}.$$

and the work needed to pump the heat is

$$W = \frac{2800 \text{ J}}{7.97} = 3.5 \times 10^2 \text{ J}.$$

The result of our calculation in parts a) and b) indicates that more work energy is needed to transfer the same amount of heat energy when the temperature difference between the inside and outside is greater.

It is interesting to note that the most common type of heat pump is the air-source heat pump, which transfers heat between our house and the outside air. If electricity is used to generate heat, a heat pump can trim the amount of electricity used for heating by as much as 30% to 40%. High-efficiency heat pumps also dehumidify better than standard central air conditioners, resulting in less energy usage and more cooling comfort in summer months. However, the coefficient of performance of most air-source heat pumps drops dramatically at low temperatures, generally making them unsuitable for cold climates, although there are systems that can overcome the problem. For a heat pump operating over a moderate temperature range, the CoP can be around 4, with this value decreasing sharply as the heat pump's operating temperature range is broadened.

12.4 ENTROPY

A general statement of the second law of thermodynamics may be expressed in terms of a quantity called "entropy" (introduced by Clausius in the 1860s).

What is entropy?

Entropy is a measure of the amount of energy in a physical system not available to do work. This is related to the level of disorder in the system. As the physical system becomes more disordered, and its energy becomes more evenly distributed, that energy becomes less able to do work. For example, a car rolling along a road has kinetic energy that could do work by carrying or colliding with something. However, as friction slows it down and its energy is distributed to its surroundings as heat, the car loses its ability to do work.

The amount of entropy is often thought of as the amount of disorder in a system. We define entropy S in terms of its change, rather than the absolute amount because the absolute value of entropy is not physically meaningful. According to Clausius, the change in entropy ΔS of a system, when an amount of heat Q is added to it by a reversible process at constant temperature T (in Kelvin), is given by

$$\Delta S = \frac{Q}{T}. \tag{12.15}$$

The second law of thermodynamics states that the entropy of an isolated system never decreases. It can either stay the same or increase. Entropy can only remain the same for an idealized (reversible) process. For any real process, the change in entropy is greater than zero ($\Delta S > 0$). If the system is not isolated, then the change in entropy of the system ΔS_s plus the change in entropy of the environment ΔS_{env} must be greater than or equal to zero,

$$\Delta S = \Delta S_s + \Delta S_{env} \geq 0. \tag{12.16}$$

THERMODYNAMICS

Figure 12.15: This nursery rhyme about Humpty Dumpty is an amusing illustration of the concept of increasing entropy in a real physical process, according to the second law of thermodynamics.

The generalized statement of the second law of thermodynamics states that the total entropy of any system plus that of its environment increases as a result of any process.

Following the second law of thermodynamics, the entropy of a closed system always increases as suggested in the nursery rhyme "Humpty Dumpty", illustrated in Fig. 12.14. In heat transfer situations, heat energy is transferred from higher temperature components to lower temperature components. In thermally isolated systems, entropy runs in the direction of increasing its value for an irreversible process. Also, entropy may be used as a measure of the energy not available for work in a thermodynamic process involved in energy conversion. Thermodynamic processes and devices can only be driven by convertible energy, and have a theoretical maximum efficiency when converting energy to work. During this work, entropy accumulates in the system, which then dissipates in the form of waste heat. In classical thermodynamics, the concept of entropy is defined by the second law of thermodynamics, which states that the entropy of an isolated system always increases or remains constant. Thus, entropy is also a measure of the tendency of a process, such as a chemical reaction, to be favored to proceed in a particular direction. The direction of increasing entropy determines that thermal energy always flows spontaneously from regions of higher temperature to regions of lower temperature, in the form of heat. These processes reduce the state of order of the initial systems, and therefore entropy is an expression of disorder or randomness. This is the basis of the modern microscopic interpretation of entropy. The second law is then a consequence of this definition and the fundamental postulate of statistical mechanics.

Entropy as a Measure of Disorder

Entropy is sometimes referred to as a measure of the amount of "disorder" in a system. Lots of disorder means high entropy, while order means low entropy. For example, if water molecules are confined to a drop of water, this is orderly than if they are scattered all over the room in the form of water vapor. The water molecules in that drop are arranged in a hexagonal array (i.e., in ice) is even more orderly. Indeed, the entropy in ice is lower than that in the same amount of water. Likewise, the entropy in water is lower than that in the same amount of water vapor. Also, if all the air in our room is gathered into the same

Figure 12.16: The amount of disorder in a system is described by entropy. Lots of disorder means high entropy while order means low entropy. The level of apparent disorder in a student's room increases as time elapses.

side of the room, then this can seem more "orderly" than if they are scattered all over the place. Again, more ordered states have lower entropy. Similarly, a room with clothes and physics books strewn all over the floor has more entropy than a room in which clothes are neatly placed in drawers and physics books are neatly stacked on the bookshelves as shown in Fig. 12.16. As we may easily notice, based on our experience, it is easier to make an ordered room messy than to make a messy room ordered. It appears that a messy room is the result of a natural process while an ordered room is not. The second law of thermodynamics states that natural processes tend to move toward a state of greater of disorder.

So there is some justification to associating entropy with disorder. This is why one should never clean one's room. We could decrease the entropy in our room, but according to the second law of thermodynamics we would only be increasing it somewhere else. Of course, there is a little physics humor here! Thinking of entropy strictly as "disorder" can be misleading because while order and disorder are important to entropy, they do not take into account heat which is also very important. Occasionally we may hear that evolution violates the second law of thermodynamics. This statement is a misunderstanding because it arises from thinking of entropy as a measure of disorder. The claim is that the incredible precision with which the human body is arranged could not have evolved from less complex organisms without violating the second law of thermodynamics. However, the same argument could be used just as effectively to disprove a formation of ice! The problem with the argument is that it does not recognize the increase in entropy in the chemical processes that take place in assembling a human being because there is no visible increase in disorder. If we only think of entropy as disorder, then we may miss much of what is going on and make the wrong conclusion. In fact, living organisms may be defined as objects that tend to maintain (or decrease) their entropy, at the expense of increasing the entropy of their surroundings. Thus, animals and humans evolve into complicated forms and build complex structures while they are destroying their environment by increasing its entropy (e.g. polluting the oceans!)

THERMODYNAMICS

Understanding-the-Concept Question 12.7:

Can this happen? If not, then why not?

Explanation:

According to the second law of thermodynamics, the change in the entropy of the universe is either zero or positive, indicating that the natural processes tend to move toward a state of greater disorder. The shattered coffee mug on the floor (left panel) is in a state of greater disorder than the coffee mug resting on a table top (right panel) in one piece. Going from the state represented by the left panel to that represented by the right panel is not a natural process because the total entropy decreases. Hence, this process cannot happen in the real world.

Exercise Problem 12.6:

A sample of 50.0 kg of water at 20.0 °C is mixed with 50.0 kg of water at 24.0 °C. Estimate the change in entropy.

Solution:

Since we started with equal amounts of water, the final temperature of the mixture will be 22.0 °C.

$$Q = cm(\Delta T) = 50.0\,\text{kg} \times 1.00 \frac{\text{kcal}}{\text{kg} \cdot \text{K}} \times 2.00\,\text{C}° = 100\,\text{kcal}.$$

This heat Q flows out of the hot water and flows into the cold water. The total change in entropy may be written as

$$\Delta S = \Delta S_{hot} + \Delta S_{cold} = \frac{-Q}{T_{hot}} + \frac{Q}{T_{cold}}.$$

By using the average temperature, we obtain

$$\Delta S = \frac{-100\,\text{kcal}}{(23+273)\,\text{K}} + \frac{100\,\text{kcal}}{(21+273)\,\text{K}} = -0.338\frac{\text{kcal}}{\text{K}} + 0.340\frac{\text{kcal}}{\text{K}} = 0.002\frac{\text{kcal}}{\text{K}}.$$

This indicates that, as heat is transferred out, the hot water decreases its entropy. However, the cold water increases its entropy as heat is transferred in during the mixing process.

The second law of thermodynamics states that entropy tends to increase in an isolated system. If the universe lasts for a sufficient time, then it will asymptotically approach a state where all energy is evenly distributed. In other words, there is a tendency to dissipate mechanical energy (motion) in nature. Hence, if we extrapolate this tendency, then we may conclude that the mechanical movement of the universe will run down in time, as work is converted to heat. The idea of heat death was first proposed in loose terms in 1851 by Lord Thomson, who theorized further on the mechanical energy loss views of Nicolas Carnot in 1824, James Joule in 1843, and Rudolf Clausius in 1850. Lord Thomson's views were then elaborated on more definitively over the following decade by Herman Helmholtz (1821-1894) and William Rankine (1820-1872). The heat death of the universe was suggested to be the ultimate fate of the universe, in which the universe has diminished to a state of no thermodynamic free energy and therefore can no longer sustain processes that consume free energy, which includes supporting life. Heat death does not imply any particular absolute temperature. It only requires that there are no longer any temperature difference that may be exploited to perform work. In physics, this condition is achieved when the universe reaches thermodynamic equilibrium (i.e., maximum entropy).

SUMMARY

i. There are three laws of thermodynamics. The first law of thermodynamics states that energy is conserved, indicating that the internal energy change ΔU for a closed system is equal to the heat added to the system Q minus the work W done by the system

$$\Delta U = Q - W.$$

The second law of thermodynamics has two separate statements which are equivalent. The first statement which was made by Clausius states that heat flows naturally from a hot object to a cold object: heat will not flow spontaneously from a cold object to a hot object. The second statement which was made by Lord Thomson states that no device can transform a given amount of heat completely into work. The third law of thermodynamics states that it is not possible to lower the temperature of any system to absolute zero in a finite number of steps.

ii. Thermodynamics processes are the development of a thermodynamic system proceeding from an initial state to a final state. Useful thermodynamic processes include isothermal, adiabatic, isochoric and isobaric process. An isothermal process is a change in which the temperature remains constant: $\Delta T = 0$. An adiabatic process is any process occurring without gain or loss of heat. An isobaric process is a thermodynamic process in which the pressure stays constant: $\Delta P = 0$. An isochoric process is a thermodynamic process during which the volume of a closed system remains constant: $\Delta V = 0$.

iii. For an isobaric process, we may compute the work done by the gas as

$$W = \vec{F} \cdot \vec{d} = PAd = P(\Delta V).$$

where P is the pressure and ΔV is the change in volume of the system. This equation is useful for describing either expansion or compression of a system under a constant pressure. However, if the pressure changes during either volume expansion or compression, then we need to compute the area under the curve in the P-V diagram to determine the work done.

iv. In determining effectiveness of any heat engine, we need to examine the efficiency. The efficiency e of a heat engine is defined as the ratio of the work W done by the heat engine to the heat input Q_H from the high temperature reservoir. The efficiency e is written as

THERMODYNAMICS

$$e = \frac{W}{Q_H} = \frac{Q_H - Q_L}{Q_H} = 1 - \frac{Q_L}{Q_H}$$

noting that the energy of the system is conserved (i.e., $Q_H = W + Q_L$). Here, Q_L is the heat exhaust.

v. For an idealized reversible engine, the heat Q_H and Q_L are proportional to the operating temperatures T_H and T_L, respectively. This allows us to write the efficiency of a reversible heat engine as

$$e_{ideal} = \frac{T_H - T_L}{T_H} = 1 - \frac{T_L}{T_H}.$$

This relation is called either Carnot efficiency or ideal efficiency.

vi. The coefficient of performance (CoP) of a refrigerator is defined as the heat Q_L removed from the low temperature area divided by the work W done, and is written as

$$CoP_{refrigerator} = \frac{Q_L}{W} = \frac{Q_L}{Q_H - Q_L}.$$

For an ideal refrigerator, the coefficient of performance is

$$CoP_{ideal} = \frac{T_L}{T_H - T_L}.$$

The coefficient of performance of a heat pump is defined as the heat Q_H delivered to the hot reservoir divided by the work W done:

$$CoP_{heat\,pump} = \frac{Q_H}{W}.$$

vii According to the second law of thermodynamics, the change of entropy ΔS of a system, when an amount of heat Q is added to a reversible process at constant temperature T (in Kelvin), is given by

$$\Delta S = \frac{Q}{T}.$$

The second law of thermodynamics states that the entropy of an isolated system never decreases. Also, for any real process, the change in entropy is greater than zero ($\Delta S > 0$). If the system is not isolated, then the change in entropy of the system ΔS_s plus the change in entropy of the environment ΔS_{env} must be greater than or equal to zero, $\Delta S = \Delta S_s + S_{env} \geq 0$.

MORE WORKED PROBLEMS

Problem 12.1 :

A heat engine extracts 55 kJ from the hot reservoir and exhausts 40 kJ into the cold reservoir. What are (a) the work done and (b) the efficiency?

Solution:

a. During each cycle, the heat transferred into the engine is $Q_H = 55$ kJ and the heat exhaust is $Q_L = 40$ kJ. The thermal efficiency of the heat engine is

$$e = 1 - \frac{Q_L}{Q_H} = 1 - \frac{40\,\text{kJ}}{55\,\text{kJ}} = 0.27.$$

b. According to conservation of energy $Q_H = Q_L + W_{out}$, we compute the work done by the engine per cycle as

$$W_{out} = Q_H - Q_L = 55\,\text{kJ} - 40\,\text{kJ} = 15\,\text{kJ}.$$

Problem 12.2 :

An ideal gas expands at constant pressure. (a) Show that $P(\Delta V) = nR(\Delta T)$. (b) If the gas is monatomic, start from the definition of internal energy and show that $\Delta U = (3/2)\,W_{env}$, where W_{env} is the work done by the gas on its environment. (c) For the same monatomic ideal gas, show with the first law that $Q = (5/2)\,W_{env}$. (d) Is it possible for an ideal gas to expand at constant pressure while exhausting thermal energy? Explain.

Solution:

a. Using the ideal gas law $PV = nRT$, we may write

$$P_i V_i = nRT_i \text{ and } P_f V_f = nRT_f.$$

Thus if the pressure does not change, then $P_i = P_f = P$. We may now write

$$P_f V_f - P_i V_i = nRT_f - nRT_i \;\Rightarrow\; P(\Delta V) = nR(\Delta T).$$

b. For a monatomic ideal gas containing N gas atoms, the internal energy is

$$U = N\left(\frac{1}{2}m\overline{v^2}\right) = nN_A\left(\frac{3}{2}k_B T\right) = \frac{3}{2}nRT.$$

Thus the change in internal energy of this gas in a thermodynamic process is

$$\Delta U = \frac{3}{2}nR(\Delta T).$$

However, using the result of part a), we may write ΔU for an isobaric process involving an monatomic ideal gas as

$$\Delta U = \frac{3}{2}nR(\Delta T) = \frac{3}{2}P(\Delta V) = \frac{3}{2}W_{env}.$$

c. We recall that the work done on the gas is $W = -W_{env}$ and use the first law of thermodynamics to find that the energy transferred to the gas by heat to be

$$Q = \Delta U - W = \Delta U + W_{env} = \frac{3}{2}W_{env} + W_{env} = \frac{5}{2}W_{env}.$$

d. In an isobaric expansion (i.e., $\Delta V > 0$), the work done on the environment is

$$W_{env} = P(\Delta V) > 0.$$

Thus, from part c), the energy transfer as heat is $Q > 0$, meaning that the energy flow is into the gas. Therefore, it is impossible for the gas to exhaust thermal energy in isobaric expansion.

THERMODYNAMICS

Problem 12.3 :

When an aluminum bar is temporarily connected between a hot reservoir at 725 K and a cold reservoir at 310 K, 2.50 kJ of energy is transferred by heat from the hot reservoir to the cold reservoir. In this irreversible process, calculate the change in entropy of (a) the hot reservoir, (b) the cold reservoir, and (c) the Universe, neglecting any change in entropy of the aluminum rod. (d) Mathematically, why did the result for the Universe in part (c) have to be positive?

Solution:

The change in entropy of a reservoir is $\Delta S = Q_{res}/T$, where Q_{res} is the energy absorbed (i.e., $Q_{res} > 0$) or expelled by the reservoir, and T is the absolute temperature of the reservoir.

a. For the hot reservoir, the entropy change is

$$\Delta S_{hot} = -\frac{2.50 \times 10^3 \, J}{725 \, K} = -3.45 \frac{J}{K}.$$

b. For the cold reservoir, the entropy change is

$$\Delta S_{cold} = \frac{2.50 \times 10^3 \, J}{310 \, K} = 8.06 \frac{J}{K}.$$

c. For the universe, the entropy change is

$$\Delta S_{universe} = \Delta S_{hot} + \Delta S_{cold} = -3.45 \frac{J}{K} + 8.06 \frac{J}{K} = 4.61 \frac{J}{K}.$$

d. The magnitude of the thermal energy transfers appearing in the numerators are the same for the two reservoirs, but the cold reservoir necessarily has a small denominator. Hence its change dominates.

Problem 12.4 :

An air conditioner operating in summer extracts 100 J of heat from the interior of the house for every 40 J of electric energy required to operate it. Determine (a) the air conditioner's coefficient of performance (CoP) and (b) its CoP if it runs as a heat pump in the winter. Assume it is capable of moving the same amount of heat for the same amount of electric energy, regardless of the direction in which it runs.

Solution:

a. According to the definition, the CoP for the engine operating as an air conditioner is

$$CoP_{ref} = \frac{Q_L}{W} = \frac{100 \, J}{40 \, J} = 2.5.$$

b. When the engine operates as a heat pump, the relevant heat is the output heat, which can be calculated from the conservation of energy

$$Q_H = Q_L + W = 100\,\text{J} + 40\,\text{J} = 140\,\text{J}.$$

c. Thus, the CoP for this engine operating as a heat pump in winter, according to the definition of coefficient of performance for heat pump, is

$$CoP_{heat} = \frac{Q_H}{W} = \frac{140\,\text{J}}{40\,\text{J}} = 3.5\,\text{J}.$$

PROBLEMS

1. A piece of aluminum has a volume of 1.4×10^{-3} m³. The coefficient of volume expansion for aluminum is $\beta = 69 \times 10^{-6}$ (C°)⁻¹. The temperature of this object is raised from 20 to 320 °C. How much work is done by expanding aluminum if the air pressure is 1.01×10^5 Pa?

2. Water is heated in an open pan where the air pressure is one atmosphere. The water remains a liquid, which expands by a small amount as it is heated. Determine the ratio of the work done by the water to the heat absorbed by the water.

3. A 15 W compact fluorescent bulb and a 75 W incandescent bulb each produce 3.0 W of visible light energy. What are the efficiencies of these two types of bulbs for converting electric energy into light?

4. The label on a candy bar says 400 Calories. Assuming a typical efficiency for energy use by the body, if a 60 kg person were to use the energy in this candy bar to climb stairs, how high could she go?

5. In exercising, a weight lifter loses 0.150 kg of water through evaporation, the heat required to evaporate the water coming from the weight lifter's body. The work done in lifting weights is 1.4×10^5 J. (a) Assuming that the latent heat of vaporization of perspiration is 2.42×10^6 J/kg, find the change in the internal energy of the weight lifter. (b) Determine the minimum number of nutritional Calories of food (1 nutritional Calorie = 4186 J) that must be consumed to replace the loss of internal energy.

6. An ideal gas is compressed from a volume of $V_i = 5.00$ L to a volume of $V_f = 3.00$ L while in thermal contact with a heat reservoir at T = 295 K as in Figure P12.19. During the compression process, the piston moves down a distance of $d = 0.130$ m under the action of an average external force of $F = 25.0$ kN. Find (a) the work done on the gas, (b) the change in internal energy of the gas, and (c) the thermal energy exchanged between the gas and the reservoir. (d) If the gas is thermally insulated so no thermal energy could be exchanged, what would happen to the temperature of the gas during the compression?

7. An ideal monatomic gas expands isothermally from 0.500 m³ to 1.25 m³ at a constant temperature of 675 K. If the initial pressure is 1.00×10^5 Pa, find (a) the work done on the gas, (b) the thermal energy transfer Q, and (c) the change in the internal energy.

8. One gram of water changes to ice at a constant pressure of 1.00 atm and a constant temperature of 0 °C. In the process, the volume changes from 1.00 cm³ to 1.09 cm³. (a) Find the work done on the water and (b) the change in the internal energy of the water.

THERMODYNAMICS

9. A 100-kg steel support rod in a building has a length of 2.0 m at a temperature of 20 °C. The rod supports a hanging load of 6,000 kg. Find (a) the work done *on* the rod as the temperature increases to 40 °C, (b) the energy Q added to the rod (assuming the specific heat of steel is the same as that for iron), and (c) the change in internal energy of the rod.

10. A monatomic ideal gas is heated while at a constant volume of 1.00×10^{-3} m^3, using a ten-watt heater. The pressure of the gas increases by 5.0×10^4 Pa. How long was the heater on?

11. A sample of helium behaves as an ideal gas as it is heated at constant pressure from 273 K to 373 K. If 20.0 J of work is done by the gas during this process, what is the mass of helium present?

12. One mole of an ideal gas initially at a temperature of $T_i = 0$ °C undergoes and expansion at a constant pressure of 1.00 atm to four times its original volume. (a) Calculate the new temperature T_f of the gas. (b) Calculate the work done *on* the gas during the expansion.

13. 10 J of heat are removed from a gas sample while it is being compressed by a piston that does 20 J of work. What is the change in the thermal energy of the gas? Does the temperature of the gas increase of decrease?

14. In one cycle a heat engine absorbs 500 J from a high-temperature reservoir and expels 300 J to a low temperature reservoir. If the efficiency of this engine is 60% of the efficiency of a Carnot engine, what is the ratio of the low temperature to the high temperature in the Carnot engine?

15. A nuclear-fueled electric power plant utilizes a so-called "boiling water reactor." In this type of reactor, nuclear energy causes water under pressure to boil at 285 °C (the temperature of the hot reservoir). After the steam does the work of turning the turbine of an electric generator, the steam is converted back into water in a condenser at 40 °C (the temperature of the colder reservoir). To keep the condenser at 40 °C, the rejected heat must be carried away by some means – for example, by water from a river. The plant operates at three-fourths of its Carnot efficiency, and the electrical output power of the plant is 1.2×10^9 watts. A river with a water flow rate of 1.0×10^5 kg/s is available to remove the rejected heat from the plant. Find the number of Celsius degrees by which the temperature of the river rises.

16. A Carnot engine uses hot and cold reservoirs that have temperatures of 1684 and 842 K, respectively. The input heat for this engine is $|Q_H|$. The Work delivered by the engine is used to operate a Carnot heat pump. The pump removes heat from the 842-K reservoir and puts it into a hot reservoir at a temperature T'. The amount of heat removed from the 842-K reservoir is also $|Q_H|$. Find the temperature T'.

17. A heat engine operates in a Carnot cycle between 80.0 °C and 350 °C. It absorbs 21000 J of energy per cycle from the hot reservoir. The duration of each cycle is 1.00 s. (a) What is the mechanical power output of this engine? (b) How much energy does it expel in each cycle by heat?

18. An electrical power plant has an overall efficiency of 15%. The plant is to deliver 150 MW of electrical power to a city, and its turbines use coal as fuel. The burning coal produces steam at 190 °C, which drives the turbines. The steam is condensed into water at 25 °C by passing through coils that are in contact with river water. (a) How many metric tons of coal does the plant consume each day (1 metric ton = 1×10^3 kg)? (b) What is the total cost of the fuel per year if the delivery price is $8 per metric ton? (c) If the river water is delivered at 20 °C, at what minimum rate must it flow over the cooling coils so that its temperature doesn't exceed 25 °C? (Note: The heat of combustion of coal is 7.8×10^3 cal/g.)

19. Find the maximum possible coefficient of performance for a heat pump used to heat a house in a northerly climate in winter. The inside is kept at 20 °C while outside is -20 °C.

20. A freezer has a coefficient of performance of 6.30. The freezer is advertised as using 457 kW-h/y. (a) On average how much energy does the freezer use in a single day? (b) On average, how much thermal energy is removed from the freezer each day? (c) What maximum amount of water at 20.0 o C could the freezer freeze in a single day? (One kilowatt-hour is an amount of energy equal to running a 1-kW appliance for one hour.)

21. The surface of the Sun is approximately at 5700 K, and the temperature of the Earth's surface is approximately 290 K. What entropy change occurs when 1000 J of energy is transferred by heat from the Sun to the Earth? (S11, 12.51)

22. Every second at Niagara Falls, approximately 5000 m^3 of water falls a distance of 50.0 m. What is the increase in entropy per second due to the falling water? Assume the mass of the surroundings is so greater that its temperature and that of the water stay nearly constant at 20.0 °C. Also assume a negligible amount of water evaporates.

Appendix

EXAMINATIONS

EXAMINATION 1

Instructor: Dr. Ju H. Kim

Name (please print): _____ UID number: _____

Please do all **12** problems of this examination. The first **10** problems are multiple choices and are worth **5** points each. Indicate your answer by circling the best answer to each question. The last 2 problems are worth **25** points each and require fully explain solutions similar to the worked examples in the text book. **You may not receive full credit on each problem if you do not provide a full explanation on the problem solving part.** *Draw free-body diagram where appropriate.*

1. An average human has a heart rate of 70 beats per minute. If someone's heart beats at that average rate over a 70-yr lifetime, how many times would it beat?
   a. 7.4×10^5
   b. 2.2×10^6
   c. 1.8×10^7
   d. 2.6×10^9

2. A thick-walled metal pipe of length 20.0 cm has an inside diameter of 2.00 cm and an outside diameter of 2.40 cm. What is the total surface area of the pipe, counting the ends, in cm²?
   a. 276
   b. 277
   c. 278
   d. 279

3. A can, after an initial kick, moves up along a smooth hill of ice. It will
   a. travel at constant velocity.
   b. have a constant acceleration up the hill, but a different constant acceleration when it comes back down the hill.
   c. have the same acceleration, both up the hill and down the hill.
   d. have a varying acceleration along the hill.

4. An object is moving in a straight line with constant acceleration. Initially it is traveling at 16 m/s. Three seconds later it is traveling at 10 m/s. How far does it move during this time?
   a. 26 m
   b. 30 m
   c. 39 m
   d. 48 m

5. A ball is thrown downward from the top of a building with an initial speed of 25 m/s. It strikes the ground after 2.0 s. How high is the building?
   a. 20 m
   b. 30 m
   c. 50 m
   d. 70 m

6. A ball is thrown straight up with an initial speed of 30 m/s. What is its speed after 4.2 s?
   a. 11 m/s
   b. 30 m/s
   c. 42 m/s
   d. 72 m/s

7. A 400-m tall tower casts a 600-m long shadow over a level ground. At what angle is the sun elevated above the horizon?
   a. 34°
   b. 42°
   c. 48°
   d. Cannot be found; not enough information.

8. If vector **A** = (-3.0, -4.0), and vector **B** = (+3.0, -8.0), what is the magnitude of vector **C** = **A** − **B**?
   a. 13
   b. 16
   c. 144
   d. 7.2

9. A projectile is launched with an initial velocity of 60.0 m/s at an angle of 30.0° above the horizontal. What is the maximum height reached by the projectile?
   a. 23 m
   b. 46 m
   c. 69 m
   d. 92 m

10. A ball thrown horizontally from a point 24 m above the ground, strikes the ground after traveling horizontally a distance of 18 m. With what speed was it thrown?
    a. 6.1 m/s
    b. 7.4 m/s
    c. 8.1 m/s
    d. 8.9 m/s

11. **(25 points)** A ball is thrown upward from the top of a building at an angle of 30.0° to the horizontal and with an initial speed of 20.0 m/s, as in the figure below. The point of release is 45.0 m above the ground. (a) How long does it take for the ball to hit the ground? (b) Find the balls' speed at impact. (c) Find the horizontal range of the stone. Neglect air resistance.

12. **(25 points)** A model rocket is launched straight upward with an initial speed of 60.0 m/s. It accelerates with a constant upward acceleration of 2.50 m/s² until its engines stop at an altitude of 200 m. (a) What can you say about the motion of the rocket after its engines stop? (b) What is the maximum height reached by the rocket? (c) How long after liftoff does the rocket reach its maximum height? (d) How long is the rocket in the air?

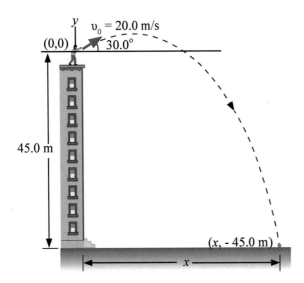

EXAMINATION 2

Instructor: Dr. Ju H. Kim

Name (please print): _____ UID number: _____

Please do all **12** problems of this examination. The first **10** problems are multiple choices and are worth **5** points each. Indicate your answer by circling the best answer to each question. The last 2 problems are worth **25** points each and require fully explain solutions similar to the worked examples in the text book. **You may not receive full credit on each problem if you do not provide a full explanation on the problem solving part.** *Draw free-body diagram where appropriate.*

1. The force that keeps you from sliding on an icy sidewalk is
   a. weight.
   b. kinetic friction.
   c. static friction.
   d. Normal force.

2. A person of weight 480 N stands on a scale in an elevator. What will the scale be reading when the elevator is accelerating upward at 4.00 m/s²?
   a. 480 N
   b. 676 N
   c. 284 N
   d. 196 N

3. An object of mass 6000 kg rests on the flatbed of a truck. It is held in place by metal brackets that can exert a maximum horizontal force of 9000 N. When the truck is traveling 15 m/s, what is the minimum stopping time if the load is not to slide forward into the cab?
   a. 5.0 s
   b. 10 s
   c. 13 s
   d. 23 s

4. An object with a mass m slides down a rough 37° inclined plane where the coefficient of kinetic friction is 0.20. If the plane is 10 m long and the mass starts from rest, what will be its speed at the bottom of the plane?
   a. 12 m/s
   b. 11.0 m/s
   c. 9.7 m/s
   d. 9.3 m/s

5. Matthew pulls his little sister Sarah in a sled on an icy surface (assume no friction), with force of 60.0 N at an angle of 37.0° from the horizontal. If he pulls her distance of 12.0 m, what is the work done by Matthew?
   a. 185 J
   b. 433 J
   c. 575 J
   d. 720 J

6. A container of water is lifted vertically 3.0 m, then returned to its original position. If the total weight is 30 N, how much work was done?
   a. 45 J
   b. 90 J
   c. 180 J
   d. No work was done.

7. A 5.00-kg object is moved from a height of 3.00 m above a floor to a height of 7.00 m above the floor. What is the change in gravitational potential energy?
   a. zero
   b. 147 J
   c. 196 J
   d. 343 J

8. A 30.0-N stone is dropped from a height of 10.0 m, and strikes the ground with a velocity of 7.00 m/s. What average force of air friction acts on it as it falls?
   a. 22.5 N
   b. 75.0 N
   c. 225 N
   d. 293 N

9. A ball of mass 0.10 kg is dropped from a height of 12 m. Its momentum when it strikes the ground is
   a. 1.5 kg·m/s
   b. 1.8 kg·m/s
   c. 2.4 kg·m/s
   d. 4.8 kg·m/s

10. A small object with momentum of magnitude 5.0 kg m/s approaches head-on a large object at rest. The small object bounces straight back with a momentum of magnitude 4.0 kg m/s. What is the magnitude of the small object's momentum change?

a. 9.0 kg·m/s
b. 5.0 kg·m/s
c. 4.0 kg·m/s
d. 1.0 kg·m/s

11. **(25 points)** A block is given an initial speed of 6.0 m/s up the 22.0° plane shown in the figure below. Assuming the coefficient of kinetic friction is 0.10, (a) draw the free-body diagram for the mass going up and going down the incline. (b) Calculate how far up the plane will the block go? (c) How much time elapses before it returns to its starting point?

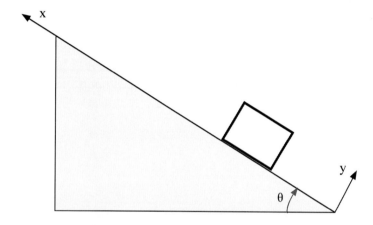

12. **(25 points)** A physics student wishes to bungee-jump from a hot-air balloon 75.0 m above a carnival midway. He will use a piece of uniform elastic cord tied to a harness around his body to stop his fall at a point 15.0 m above the ground. Model this body as a particle and the cord as having negligible mass and a tension force described by Hooke's force law. In a preliminary test, hanging at rest from a 6.00-m length of the cord, the jumper finds that his body weight stretches it by 1.50 m. He will drop from rest at the point where the top end of a longer section of the cord is attached to the stationary balloon. (a) What length of cord should he use? (b) What maximum acceleration will he experience?

EXAMINATION 3

Instructor: Dr. Ju H. Kim

Name (please print): _____ UID number: _____

Please do all **12** problems of this examination. The first **10** problems are multiple choices and are worth **5** points each. Indicate your answer by circling the best answer to each question. The last 2 problems are worth **25** points each and require fully explain solutions similar to the worked examples in the text book. **You may not receive full credit on each problem if you do not provide a full explanation on the problem solving part.** *Draw free-body diagram where appropriate.*

1. A 0.50-kg mass is attached to the end of a 1.0-m string. The system is whirled in a horizontal circular path. If the maximum tension that the string can withstand is 350 N. What is the maximum speed of the mass if the string is not to break?
    a. 700 m/s
    b. 26 m/s
    c. 19 m/s
    d. 13 m/s

2. What minimum banking angle is required for an Olympic bobsled to negotiate a 100-m radius turn at 35 m/s without skidding? (Ignore friction.)
    a. 31°
    b. 41°
    c. 51°
    d. 61°

3. Is it possible for an object moving around a circular path to have both centripetal and tangential acceleration?
    a. No, because then the path would not be a circle.
    b. No, an object can only have one or the other at any given time.
    c. Yes, this is possible if the speed is constant.
    d. Yes, this is possible if the speed is changing.

4. How many newtons does the weight of a 100-kg person change when he goes from sea level to an altitude of 5000 m? (The mean radius of the earth is 6.38×10^6 m.)
    a. 0.6 N
    b. 1.6 N
    c. 2.6 N
    d. 3.6 N

5. Let the average orbital radius of a planet be R. Let the orbital period be T. What quantity is constant for all planets orbiting the sun?
   a. T/R
   b. T/R^2
   c. T^2/R^3
   d. T^3/R^2

6. Two objects move on a level frictionless surface. Object A moves east with a momentum of 24 kg·m/s. Object B moves north with momentum 10 kg·m/s. They make a perfectly inelastic collision. What is the magnitude of their combined momentum after the collision?
   a. 14 kg·m/s
   b. 26 kg·m/s
   c. 34 kg·m/s
   d. Cannot be determined without knowing masses and velocities.

7. A car of mass m, traveling with a velocity v, strikes a parked station wagon, whose mass is $2m$. The bumpers lock together in this head-on inelastic collision. What fraction of the initial kinetic energy is lost in this collision?
   a. 1/2
   b. 1/3
   c. 1/4
   d. 2/3

8. A cable car at a ski resort carries skiers a distance of 6.8 km. The cable which moves the car is driven by a pulley with diameter 3.0 m. Assuming no slippage, how fast must the pulley rotate for the cable car to make the trip in 12 minutes?
   a. 9.4 rpm
   b. 30 rpm
   c. 60 rpm
   d. 720 rpm

9. A small bomb, of mass 10 kg, is moving toward the north with a velocity of 4.0 m/s. It explodes into three fragments: a 5.0-kg fragment moving west with a speed of 8.0 m/s; a 4.0-kg fragment moving east with a speed of 10 m/s; and a third fragment with a mass of 1.0 kg. What is the velocity of the third fragment? (Neglect air friction.)
   a. zero
   b. 40 m/s north
   c. 40 m/s south
   d. None of the above.

10. A car is negotiating a flat circular curve of radius 50 m with a speed of 20 m/s. The maximum centripetal force (provided by static friction) is 1.2 x 10⁴ N. What is the mass of the car?

   a. 0.50×10^3 kg
   b. 1.0×10^3 kg
   c. 1.5×10^3 kg
   d. 2.0×10^3 kg

11. **(25 points)** Consider a frictionless track as shown in the figure below. A block of mass $m_1 = 5.00$ kg is released from A. It makes a head-on completely inelastic collision at B with a block of mass $m_2 = 10.0$ kg that is initially at rest. Calculate the maximum height to which m_1 and m_2 rises after the collision.

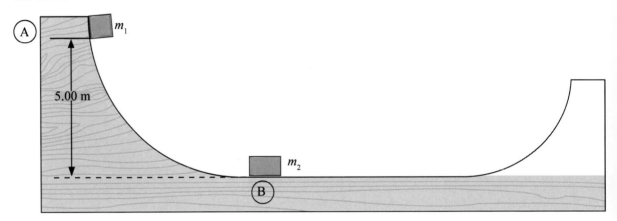

12. **(25 points)** The figure below shows a roller-coaster car moving around a circular loop of radius R. (a) What speed must the car have so that it will just make it over the top without any assistance from the track? (b) What speed will the car subsequently have at the bottom of the loop? (c) What will be the normal force on a passenger at the bottom of the loop if the loop has a radius of 10.0 m?

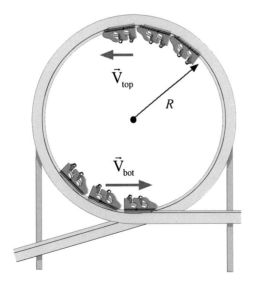

EXAMINATION 4

Instructor: Dr. Ju H. Kim

Name (please print): _____ UID number: _____

Please do all **12** problems of this examination. The first **10** problems are multiple choices and are worth **5** points each. Indicate your answer by circling the best answer to each question. The last 2 problems are worth **25** points each and require fully explain solutions similar to the worked examples in the text book. **You may not receive full credit on each problem if you do not provide a full explanation on the problem solving part.** *Draw free-body diagram where appropriate.*

1. Consider a bicycle wheel to be a ring of radius 30 cm and mass 1.5 kg. Neglect the mass of the axle and sprocket. If a force of 20 N is applied tangentially to a sprocket of radius 4.0 cm for 4.0 s, what linear speed does the wheel achieve, assuming it rolls without slipping?
   a. 3.0 m/s
   b. 5.9 m/s
   c. 7.1 m/s
   d. 24 m/s

2. A hoop of radius 0.50 m and a mass of 0.20 kg is released from rest and allowed to roll down an inclined plane. How fast is it moving after dropping a vertical distance of 3.0 m?
   a. 2.2 m/s
   b. 3.8 m/s
   c. 5.4 m/s
   d. 7.7 m/s

3. A 1.53-kg mass hangs on a rope wrapped around a frictionless disk pulley of mass 7.07 kg and radius 66.0 cm. What is the acceleration of the mass?
   a. 9.26 m/s²
   b. 2.96 m/s²
   c. 6.29 m/s²
   d. zero

4. A person weighing 800 N stands with one foot on each of two bathroom scale. Which statement is definitely true?
   a. Each scale will read 800 N.
   b. Each scale will read 400 N.
   c. If one scale reads 500 N, the other will read 300 N.
   d. None of the above is definitely true.

5. A triatomic molecule is modeled as follows: mass m is at the origin, mass 2m is at $x = 2a$, and mass 3m is at $x = 2a$. What is the moment of inertia about the origin?
   a. $2\,ma^2$
   b. $3\,ma^2$
   c. $2\,ma^2$
   d. $14\,ma^2$

6. A figure skater rotating at 5.00 rad/s with arms extended has a moment of inertia of 2.25 kg·m². If the arms are pulled in so the moment of inertia decreases to 1.80 kg·m², what is the final angular speed?
   a. 2.25 rad/s
   b. 4.60 rad/s
   c. 6.25 rad/s
   d. 0.81 rad/s

7. In a hydraulic garage lift, the small piston has a radius of 5.0 cm and the large piston has a radius of 15 cm. What force must be applied on the small piston in order to lift a car weighing 20,000 N on the large piston?
   a. 6.7×10^3 N
   b. 5.0×10^3 N
   c. 2.9×10^3 N
   d. 2.2×10^3 N

8. Water flows through a horizontal pipe of cross-sectional area 10.0 cm² at a pressure of 0.20 atm. The flow rate is 1.00×10^{-3} m³/s. At a valve, the effective cross-sectional area of the pipe is reduced to 5.00 cm². What is the pressure at the valve?
   a. 0.112 atm
   b. 0.157 atm
   c. 0.200 atm
   d. 0.235 atm

9. Which has the greatest effect on the flow of the fluid through a narrow pipe? That is, if you made a 10% change in each of the quantities below, which would cause the greatest change in the flow rate?
   a. The fluid viscosity
   b. The pressure difference
   c. The length of the pipe
   d. The radius of the pipe

EXAMINATIONS

10. Liquid flows through a 4.0 cm diameter pipe at 1.0 m/s. There is a 2.0 cm diameter restriction in the line. What is the velocity in this restriction?
    a. 0.25 m/s
    b. 0.50 m/s
    c. 2.0 m/s
    d. 4.0 m/s

11. **(25 points)** Calculate the tension F_T in the cable that supports the 30 kg beam, and the force F_H exerted by the wall on the beam (give magnitude and direction) if a person weighing 6.0×10^2 N stand 1.50 m from the wall, as shown in the figure.

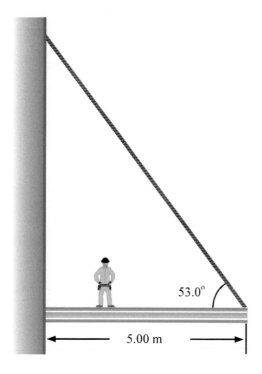

12. **(25 points)** A 1.00×10^3-kg cube of aluminum is placed in a tank. Water is then added to the tank until half the cube is immersed. (a) What is the normal force on the cube? (See Fig. a.) b) Mercury is now slowly poured into the tank until the normal force on the cube goes to zero. (See Fig. b.) How deep is the layer of mercury? (The specific gravity of Al and Mg are 2.70 and 13.6, respectively.)

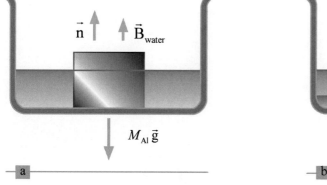
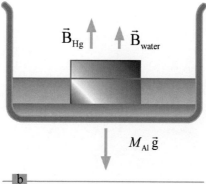

APPENDIX

Index

A

Absolute pressure   285, 289, 308, 313, 338, 354
Accelerating   28, 32, 40, 43, 59, 107, 110, 120, 131, 155, 206-7, 210
Accelerating frames   94
Accelerating objects   30, 207
Acceleration   28-32, 34, 37, 40-2, 45-8, 56-8, 83-4, 95-7, 113-17, 120-3, 162-3, 196, 198-9, 206-8, 226-7, 267-8
  center seeking   207
  centrifugal   210-11
  free-fall   217, 227, 229
  instantaneous   28, 30, 40
  linear   198, 200-1, 211, 223, 237, 251
  net   199
  total   199, 227-8
  uniform   32, 45, 47
Acceleration vectors   56-7, 64-5, 207
Action force   85, 99
Addition and Subtraction of Vectors   51
Adiabatic boundary   392
Adiabatic expansion   394, 400-1, 405-6
Adiabatic process   392, 394-5, 398, 418
Advection   369
Air   34-6, 62-4, 66, 86, 284-7, 295-7, 299-300, 304-5, 307, 309-10, 314, 336-7, 354-5, 367-8, 375-7, 407-8
  heated   318, 369
  hot   318, 356, 369
Air conditioners   392, 410-12, 414, 421
Air density   34, 39-40, 310, 314
Air flow   299-300, 304
Air molecules   35-6, 318, 345
Air pressure   40, 287, 292, 305, 313, 354, 368, 422
Air resistance   35-6, 58, 61-4, 66, 68, 73, 93, 122-3, 140, 145, 155, 157, 159
Air temperature   34, 341
Analysis of transient systems   372
Analysis, dimensional   10
Angstrom   7
Angular acceleration   196, 198-203, 223, 228, 232-3, 237-8, 247, 250-2, 257

Angular displacement   196-8, 200, 202, 207-8, 223, 226, 255
Angular kinetic energy   252, 262
Angular momentum   178, 247-8, 255-63, 268-9
  change of   256, 262
  law of   258, 260
  law of conservation of   256, 261-2
  total   257-8, 261
Angular velocity   196, 198-203, 205, 223, 226, 238, 247, 250-1, 255, 257, 261, 268-9
  final   228, 268
  rate of change of   198, 223
  uniform   199, 201
Angular velocity changes   199
Angular velocity times   252
Application of First Law of Thermodynamics   391
Applied force   91, 95-6, 112-13, 118, 127, 130, 132, 135, 137, 148, 232-3, 236, 239, 273-4, 276-7, 308
  direction of   276
Applied torque   239, 247
Approximate Metabolic Rate   401
Artificial gravity   210-11, 229
Asteroid   122, 220-1, 260
Asteroid belt   221
Atmosphere   35, 282, 284-6, 292, 311-12, 314, 318, 353, 360, 362, 376-7, 422
Atmospheric pressure   185, 285, 287, 289, 292, 308, 313-14, 321, 323, 346, 354, 364, 367
Atomic Theory of Matter   330
Atoms   2, 18, 91-2, 117, 128-9, 258, 274, 283-4, 287, 331-5, 338, 341-2, 347-8, 370-1, 376, 380
Average acceleration   28, 30, 40-1, 47
Average angular velocity   198, 223
Average force   121, 139, 157, 189-90, 353
Average frictional force   156
Average kinetic energy   338, 344-5, 347, 353-4
Average net force   121-2
Average power   148, 155, 158-9, 385
Average surface temperature   352, 387
Average temperature   377, 417
Average velocity   25-8, 31, 40, 44, 77, 155, 298
  definition of   26, 30, 32, 41
Avogadro's Number   335

B

Balance point 180
Ballistics 175-6
Basic units for Physical Quantities 6
Bernoulli's equation 304
Bernoulli's principle 290, 300-2, 304-5, 308-9
Bohr model 258, 332-3
Boiling point 321, 345, 363, 366-7, 383, 386
Bonds 128-9, 274, 365
  molecular 91-2
Bone loss 211
Bose-Einstein 347
Boyle's law 330, 337-8, 349
Brake pads 147, 268, 358
Brakes 43-4, 46-7, 147-8, 171, 210, 291, 358
Brass 277, 281, 311, 320, 323, 349-50, 353
Bridge 48, 242, 271-2, 324, 330
Bridle point 243
British thermal units 357
Brownian motion 331, 333-4
Brownian particles 334
Buoyant force 293-6, 314

C

Cables 47, 91, 108-10, 123, 159, 242, 267, 272
Calories 7, 357, 384, 401-2, 422
Calorimeter 361-2, 386
Calorimetry 361, 363
Carnot cycle 405-6, 423
Carnot engine 404-6, 423
Celsius scale 320-1, 348
Center of gravity 178-80, 182, 186, 243-6, 262, 267
Center of mass 101, 135, 178-84, 186, 188, 192, 204-6, 224, 227, 238, 243-4, 246, 252, 262, 264-5, 268
Center of mass trajectory 184-5
Center of motion 183-4, 237
Centrifugal force 94, 194, 199, 209-11
Centripetal acceleration 199, 206-9, 220, 223, 227-8
Centripetal force 132, 144-5, 194, 199, 206-7, 209-10, 216, 219, 228, 263
Charles's law 330, 337, 349
Chemical energy 127, 129, 134, 402, 407
Chemical reactions 129, 149, 361, 390-1, 415
Circular motion 193-6, 198-200, 202, 208-9, 211-13, 216, 219-21, 223, 258
  non-uniform 198-9
Circular orbit 77, 87, 132, 208, 219-22, 229, 333
Clausius statement 402, 410
Coefficient of expansion 323
Coefficient of friction 113

Coefficient of restitution see CoR
Coefficient, volumetric 326-7
Collision 35, 46, 73, 86, 93, 161-2, 165-78, 185-6, 188-91, 209, 221, 256-7, 259-60, 269, 342-3, 371
  energetic 173
  molecular 370-1
  two-dimensional 176-7
Combustion 165, 300, 406-7, 423
Compressed spring 91, 139
Compressibility 283, 288
Compression 108, 171, 271-2, 276-80, 381, 393, 396, 405, 407, 418, 422
  force of 272, 276, 308, 329
Compressive strength Shear 277-8
Compressor 412-13
Condensation 285, 366-7, 375, 385
Condenser 403, 412, 423
Conduction 130, 356, 369-71, 376-7, 379, 381, 384
Conduction mechanism 371, 373, 375, 383
Conservation 127, 141-2, 147-8, 152, 154, 167, 172, 258, 260-1, 298, 411, 420
Conservation laws 127, 162, 166
Conservation of angular momentum 256-8, 261-2
Conservation of energy 127, 133-4, 141, 143-5, 152, 172-3, 175, 177, 298, 300, 302, 309, 357-8, 361-2, 365-6, 401-2
  law of 128, 134, 162, 253, 302, 362
Conservation of momentum 98, 127, 162, 165-6, 168, 170-7, 187, 259
  law of 165-8, 185
Conservation of momentum principle 168, 172, 188
Conservative forces 140-1, 152
Constant acceleration 28, 32-4, 41, 46, 58, 63, 95-6, 156, 159
Constant angular acceleration 226, 228, 237
Constant angular velocity 237, 247, 256
Constant force 95, 130, 152, 156, 268
Constant pressure 337, 396-7, 399, 418, 420, 422-3
Constant pressure process 398
Constant temperature 363, 393-4, 401, 414, 419, 422
Constant velocity 45, 47, 59, 73, 78-9, 86, 107, 123, 131, 136, 159, 187, 227, 237
Contact force 83, 88-92, 104, 106
Convection 130, 355-6, 369-70, 377, 381, 386
  forced 369
Convection mechanism 369-70, 379
Conversion factors 7-8, 10-11, 16, 285
Conversion of temperature in Fahrenheit 321
CoP (coefficient of performance) 410-14, 419, 421-4
COP (coefficient of performance) 410-14, 419, 421-4
CoR (coefficient of restitution) 170, 173, 178
Crumple zones 87

INDEX

D

Decomposition 56-7
Definite proportions 333
   law of 333
Deformation 90, 139, 146, 200, 237, 274-8, 330
   relative 280
Defueling 245
Density 19, 38, 40, 63-4, 133, 283-4, 287-8, 294-6, 298, 309-10, 313-15, 318, 327, 352-3, 364, 386
Device 27, 128, 140, 150-1, 172, 174-5, 291-2, 315, 361, 391, 402, 404, 409-10, 415, 418
   isolated system perpetual motion 409
Dimensional Kinematics 23, 49
Dimensionless parameter 297
Dimensions 4, 10-11, 16, 24, 50, 55-7, 60, 69-70, 176-7, 234-5, 274-5, 281, 307-8, 323, 326, 373
Direction of heat energy flow 361
Direction of motion 25, 27, 30, 40, 71, 86, 93, 132-3, 162, 194, 199, 206-7, 318, 345
Displacement vectors 51-2, 77, 130, 136, 233
Dissipation, frictional 147-8
Dissipative forces 144, 146-7, 252-3
Dissipative systems 146
Distribution of mass 68, 178, 247, 249
Drag forces 93, 159, 300

E

Earth 38, 78, 85, 89, 158, 192, 230, 260, 268, 272, 376
   gravitational acceleration of 38, 244
   planet 19, 180
   surface of 34-5, 38-9, 58, 90, 102, 104, 137-8, 212, 217, 260, 321, 346
Earth orbits 218-19
Earth's gravitational field 40, 96
Earth's gravity 34, 38-41, 50, 58, 106, 212, 217
Earth's rotation 39, 50, 133, 219, 260-1
Earth's surface 34, 38-40, 58, 73, 106, 128, 221-2, 229-30, 376, 424
Effect of Dissipative Forces 146
Effects of Air Resistance 35
Effects of Work on Kinetic Energy 136
Elastic collision 170, 173, 176, 185, 189-90, 353
Elastic energy 127-9, 139, 158-9
Elastic modulus 275-6, 308, 311-12, 328
Elastic modulus Shear 281
Elastic potential energy 137, 139, 145, 152-3
Electric Charge 6, 94
Electric Current 6
Electrical Conductance 6
Electrical energy 126, 128, 148-9, 412
Electricity 2, 4, 126, 138, 151, 320, 408, 412, 414

Electromagnetism 82, 94, 234
Electrons 94, 97, 174, 194, 283, 333, 376
Elevation, angle of 20, 62
Elevator 55, 105, 107, 210
EM radiation 376-8, 380
Emission 375-6
Emissivity 377-9, 382-3, 386-7
Energy changes 146
Energy conservation principle 145, 177, 193, 361, 365, 382, 410
Energy consumption 401-2, 411
Energy content 328, 357, 363
Energy conversion 147, 415
Energy losses 128, 142-3, 148, 387
Energy of motion 87, 93, 134-5, 338
Energy principle 143, 153, 253, 382
Energy release 177, 365
Energy sources 151, 409
   high-temperature 408
Energy transfer 130, 148, 319, 356, 371, 378, 390, 392, 402-3, 410, 420
   rate heat 383
Energy transfer mechanisms 129-30, 369
Energy Transfers in Heat Engine 402
Energy transformation 143, 147
Energy transforming processes 401
Entropy 356, 390, 394, 401, 404, 414-19, 421, 424
Entropy change 391, 419, 421, 424
Equation of state 337-8
Equilibrium 91, 108-9, 180, 231, 237-8, 240-1, 246, 261, 267, 293, 313, 318-19, 354, 367, 391, 393
   mechanical 139, 237-8, 261
   static 237, 240-1
Equilibrium temperature 362, 385
Equipartition theorem 344, 347
Equivalence, mass-energy 19
Evaporation 366-7, 422
Expansion of Materials 322
External forces 97, 129, 137-8, 166, 168, 179, 183, 190, 237-8, 258, 409-10, 422

F

Fahrenheit scale 320-1, 348, 352
First Law of Thermodynamics 391
Flow 71, 79, 99, 138, 278, 283, 297-9, 301-2, 304-8, 314, 319, 350, 356, 361, 381, 384
   laminar 64, 297-8, 300, 305-6, 309
   smooth 305
   turbulent 64, 297
   volume rate of 299, 306, 310
Flow direction 300, 305
Flow rate 152, 297, 300, 307, 367
   volumetric 299-300
Flow velocity 297, 302

Fluid   40, 68, 93, 105, 151, 271-3, 280-1, 283-4, 287-8, 290-4, 297-302, 304-6, 308-10, 334, 356, 369
  compressible   291
  confined   290-1
  enclosed   290, 292
  movement of   369-70
  volume of   288, 299
Fluid density   298
Fluid dynamics   2, 4, 64, 93, 297, 304
Fluid flow   93, 104, 297-300, 307
Fluid friction   93, 111
Fluid pressure   287-8, 293
  static   290
Fluid velocity   93, 283
Fluid-pressure   308
Force constant   146, 157
Force vectors   98, 130, 135-6, 233, 276
Force-sensitivity   82
Formula, quadratic   12, 15, 76
Frequency   201, 208, 223
Friction   85-6, 92-3, 100, 106, 111-13, 122-3, 140-4, 146-7, 157-9, 188-91, 214-15, 228-9, 254, 358, 385, 405-7
  coefficient of   111, 113, 200, 206
  controlled   148
  dry   111
  law of   112
Friction force   99-100, 111-13, 116-18, 123, 136, 158, 200, 214-15, 240
  static   112, 118
Frictionless   90-1, 107, 115, 120, 141, 191, 240, 267-9, 354, 409
  stationary   258-9
Fundamental laws   2-3, 134, 298, 390
Fusion   149, 363-4, 381

G

Gas constants   331
  universal   331, 338, 341
Gas laws   335, 337, 340
Gas molecules   35, 331, 337-8, 341-3, 345-8, 354, 394
Gas pressure   354
Gauge pressure   285, 289-90, 292, 302-3, 307-8, 312, 340
Gay-Lussac's law   330, 337-8, 349
Geostationary orbit   222
Geosynchronous orbit   219, 222
Governing principles   84, 290, 308
Gravitation   39, 82-3, 102-4, 132, 170, 217
  universal law of   85, 102-3, 132, 216-21
Gravitational acceleration   37-8, 59, 66, 90, 137, 209, 212, 217, 244, 290
  local   90
  sea-level   39
Gravitational attraction   38, 217

Gravitational field   39, 137-8, 244
  uniform   66, 106, 179
Gravitational force   83-4, 87, 96, 102-4, 106-7, 122, 132-3, 140, 179, 216, 219-21, 225, 261
  non-uniform   66
  universal   222
Gravitational force definition   105
Gravitational mass   96
  active   96
Gravitational pull   90, 133, 181, 217, 246
Gravity   34-5, 37-41, 58-9, 81-2, 89-90, 94, 102, 104, 131-3, 137-8, 178-82, 216-18, 243-6, 262, 284, 294-6
  acceleration due to   216
  acceleration of   59, 94, 122
  action of   58-9, 61, 73
  apparent   39, 211
  effective   39
  force of   59, 81-2, 94, 104, 106, 110, 113, 137, 179, 186, 212, 216-18, 230, 246, 303
  moon's   217
  pull of   36, 133
  strength of   102
  zero   106, 211
Greenhouse gases   376
Ground reaction force   92, 106

H

Head-on collision   87, 173-4
  elastic   189-90
Heat   93, 128-30, 146-8, 153, 171, 173, 318-19, 330, 353-67, 369-72, 374-5, 381-6, 390-5, 398-402, 404-8, 410-24
  excess   361, 370
  frictional   130
  given amount of   391, 408, 418
  high   356
  large amount of   360
  latent   363-4, 381
  mechanical equivalence of   357, 364
  move   412
  rejected   423
  sum of   360, 362
  total   347, 399-400
  total amount of   360
Heat capacity   358-9, 361, 385-6
  molar   359
  volumetric   359
Heat conduction   129, 359, 369-72
Heat death   418
Heat energy   130, 147, 318-19, 357-8, 360, 362, 364, 367, 371, 381, 383, 393, 401-3, 408, 414-15
  lose   319
  measuring   357
  radiant   377

INDEX

    total radiant 377
    transferred 357
    transfers 411
Heat energy flows 361-2
Heat energy transfer 318
Heat energy transfer process 355
Heat engine 356, 390, 392, 402-4, 406, 408, 410, 418-19, 423
Heat exchange 392
Heat exhaust 402-3, 419
Heat flow 130, 319, 356, 370-1, 384-5
    radiant 378-9
    rate of 371-4, 379
Heat input 357, 402-3
Heat loss 374-5, 379, 392, 406, 418
    rate of 369, 375, 378
Heat of chemical reactions 361
Heat of fusion 363
Heat production 370, 381
Heat pump system 411
Heat pumps 356, 392, 410-14, 419, 421-3
    air-source 414
    ideal 413
Heat reservoir 393, 404, 422
Heat sink 403, 411
Heat transfer 130, 355-6, 361, 369-73, 375-7, 379, 392, 400-1, 405
    radiative 355, 375, 377
Heat transfer coefficient 361, 373
Heat transfer equation 375
Heat transfer mechanisms 356, 369, 381
Heat transfer processes 366, 408
High temperature reservoir 402, 406, 412, 418
Hooke's law 91, 273-5
Horsepower 148, 151, 159, 403
Hubble Space Telescope 218
Humidity 367-8
    relative 368
Humidity of air 368
Hydroelectricity 138
Hypotenuse 12, 14-15, 18, 20

I

Ice fusion 366
ICF (inertial confinement fusion) 149
Ideal gas 335, 338, 340-1, 344, 347, 349-50, 353-4, 393-5, 397-401, 420, 422-3
Ideal gas law 330-1, 336, 338-42, 345, 349, 393
Impact 78, 87, 153, 162, 166, 170-1, 189-91, 260
Impact force 87, 92, 162
Impulse 162, 167-9, 189
Inelastic collision 170, 173-5, 177, 186, 191, 256
Inertia 39, 59, 132-3, 162-3, 169, 179, 194, 209-10, 237-8, 247-50, 254-5, 257, 259-61, 263-4, 268-9
    moment of 247, 262

Inertia force 179, 247, 262
Inertial confinement fusion (ICF) 149
Intermolecular forces 94, 364
    liquid-phase 367
Internal energy 170, 336, 338, 344, 347-8, 354, 356, 359, 385, 390-1, 393-5, 399-401, 420, 422-3
    component of 394
Internal energy change 391, 400-1, 418
Internal heat exchangers 413
Internal temperature 370, 394
Isobaric 395-9
Isobaric process 395-9, 418, 420
Isochoric 395-6, 398-9, 418
Isolated system 127, 134, 164-6, 168, 185, 260, 361-2, 381, 390-1, 409, 414-15, 418-19
Isothermal 392, 396-9, 418
Isothermal process 392-6, 398-9, 418
Isotherms 406

J

Joules 7, 130, 148, 152-3, 357, 384, 408

K

Kepler's laws 219
Kinematic equations 50, 57-8, 60, 65, 67, 75, 115-16, 143, 201-3
    circular 202
    position-versus-time 65, 74
    velocity-versus-position 135
Kinematic Equations for Projectile Motion 60
Kinematics 24, 28-9, 33, 55-7, 65, 69, 83, 96, 117, 126-7
    circular 202
    two-dimensional 56, 58
Kinematics equations 31-2, 37-8, 41-3, 96, 114, 120, 202, 226
Kinetic 112, 126, 134, 141, 146, 171, 175, 193, 331
Kinetic energy 126, 128-9, 134-7, 139-40, 142-3, 147, 157-8, 163, 170-7, 190-1, 252-3, 262, 301-2, 344-5, 347-8, 367
    average translational 344
Kinetic energy contributions 252
Kinetic friction 92, 111-14, 118, 123
    coefficient of 112-13, 118-19, 121, 123, 157, 188, 268
Kinetic temperature 338
Kinetic theory 330-1, 338, 341-2, 344, 346-7, 349

L

Latent 364, 381
Latent heat of fusion 363-4

Latent heat of vaporization   363, 383, 422
Launch point   184-5
Law of definite proportion   333
Laws of motion   81, 84
Laws of Thermodynamics   391
LEO (low-Earth orbit)   219
Light   5, 19, 43, 117, 126, 129, 146, 148, 150, 153, 174, 237, 272-3, 294, 365, 380
  infrared   380
Linear expansion, coefficient of   323, 348, 353
Linear momentum   162, 178, 185, 199, 255-6, 262
  conservation of   162, 165

M

Magnitude   25-6, 39-40, 50-5, 69-73, 77-9, 94-7, 102-4, 107-9, 121-2, 156, 162-4, 189, 206-9, 227-8, 232-5, 266-8
Mass flow rate   298, 300
Mass trajectory   184-5
Mass units   7
Massless   90-1, 110, 245
Materials of low thermal conductivity   372
Matter   5, 83-4, 96, 99, 102, 126, 128, 166, 173-4, 185, 217-18, 283, 288, 330-4, 376, 380
  state of   283-4, 336, 363
Maxwell-Boltzmann distribution   346-7
Measure heat transfer   361
Measure of Disorder   415
Measurement of Heat Transfer   361
Mechanical advantage   109-10, 129, 239, 291-2
Mechanical energy   127, 129-30, 132, 134, 137, 144, 146, 253, 357, 385, 402, 407, 411, 418
  conservation of   141, 144, 175
  potential   139, 153
  principle of conservation of   141, 152
  total   134, 141, 146, 152-3
Mechanical work   139, 151, 236, 385, 390, 394, 402-4, 406, 412
Mechanisms, heat energy transport   375
Metabolic Rates   370, 401
Minimum centripetal acceleration   225-6
MKS system   5, 8, 16
Molar   6, 359
Momentum   83, 94, 96, 98, 127, 161-77, 183, 185-7, 189-90, 260-1, 318, 345, 353
  bat's   162
  changing   167
  definition of   163-4
  object›s   169
  rate of change of   163, 185, 343
Momentum change   163-4, 318
Momentum principle   168, 188
Monatomic   348, 395, 420, 423
motion   24-5, 27-30, 34-5, 55-8, 69-73, 83-8, 96-8, 111-13, 117-18, 131-5, 144-7, 178-9, 182-6, 194-6, 199-200, 409

Motionless   85, 236
Moving fluid   300-1, 305

N

Net force   83-4, 89-90, 94, 96-101, 108, 110, 113, 117, 121-2, 131-2, 134-6, 163, 207, 214, 236-7, 257-8
Net gravitational force   229
Net torque   236-7, 240-1, 244, 246-7, 251, 256-60, 262, 266
Newtonian fluids   298, 306
Newton's cradle 5-ball system   171
Newton's laws   20, 84, 89-90, 103-4, 110, 114, 116-17, 185, 214, 216-17, 220, 223, 251, 338, 369-70
Newton's Principal   103
Newton's Second Law   96
Newton's theory   117
Newton's Third Law   98
Non-conservative forces   140-2, 146, 152-3
Normal force   90-2, 98, 104-8, 112-13, 118-19, 123, 228, 240, 268
Nova laser   149

O

Ocean temperatures   360
Orbit   39, 85, 103, 106, 126, 133, 165, 179, 183-4, 211-12, 216, 218-22, 229-30, 258, 261, 268
Orbital motion   219
Orientation   51-2, 82, 278
Outward force   209-10

P

Parabola   32, 59, 62-3, 66, 73
Parallel Force   112
Particles   103, 108-9, 122, 135-7, 165-7, 177-9, 182-3, 185, 196, 199-200, 208, 216, 255-6, 262, 334-5, 347
Pascal's principle   290-3, 308
Pendulum   174-6
  ballistic   174-6
Perpetual motion machines   113, 391, 409
PET (positron emission tomography)   401
Phase   283, 331, 363, 365-7, 381
  gaseous   367
Phase change   321, 363, 381
Phase transition   359, 363, 365
Physical laws   2-4, 8, 15, 84, 103, 117, 293
Physical quantities   4, 6-7, 9-11, 24, 32, 56, 126-7, 198, 233, 275, 293

INDEX

Physical world 2-4, 8, 15, 232
Physics 2-4, 6, 10-11, 50, 82-4, 89, 94, 107, 117, 126-7, 134, 143, 162-3, 173-4, 257, 390
  laws of 2-4, 73, 193, 220, 409
Piston 291-2, 312, 354, 396, 400, 403, 407-8, 422-3
  movable 354, 393-4, 396
Plane Geometry 11
Planetary motion 85, 90, 216, 261
  laws of 85, 219-20
Planets 79, 85, 90, 96, 103-4, 106, 123, 128, 133, 150, 170, 216-21, 229-30, 261, 376, 409
  mass of 218
  motion of 219-20
Plastic Region 273
Point masses 103-4, 117, 178, 181, 216, 223, 248
Point particles 178, 196, 256, 268
Poiseuille's equation 305-7, 309-10
Polyurethane 372
Potential energy 126, 128, 134, 137-43, 145-7, 152, 154, 170-1, 193, 252-4, 347,
  large quantity of 143, 147
Pound-force 93, 96, 120, 165
Power 6, 9, 11-12, 82, 148-53, 159, 264, 306, 357, 377, 403, 406, 408-9, 413
  definition of 148-9, 151, 155, 264
Power stroke 407
Precision 8-9, 15, 82, 96
Precision grip 82
Pressure 165, 272-3, 280-93, 300, 302, 304-5, 307-9, 311-14, 330-1, 335-45, 349-50, 353-4, 364-5, 367-8, 390-3, 395-400
Pressure difference 287, 289-90, 305-6, 309-10, 315
Pressure gauges 272, 285, 353
Pressure Pascal 7
Principles of thermodynamics 391
Projectile 50, 58-64, 66-8, 73, 75, 85-6, 174-6, 229
Projectile motion 50, 58-60, 63, 66, 73, 90, 156-7, 217
  ideal 66
Projectile weapons 50
Pulley system 109-10
Pump Condenser 403
Pyrex 313, 323, 351
Pythagorean equation 14-15, 18, 52, 54, 70-2, 176, 240

R

Radians 196-8, 201, 203, 226
Radiation 5, 93, 130, 173, 356, 369, 375-6, 378, 380-1, 383, 386-7, 390
Radiator 326, 360
  perfect blackbody 383
Random motions 331, 334, 346
  continuous 331, 341

Rankine scale 352, 410
Reaction 85, 99, 167
Reaction forces 85-6, 93, 99, 101, 106, 209
Real fluids 283, 305-7, 309-10
Refrigeration 406, 410-11
Regular Process Pressure 286
Relative motion 50, 64, 68-73, 92-3, 111
Resistance 35-6, 41, 64, 82, 112, 146, 179, 212, 247, 305-6, 308-9, 330
Restitution, coefficient of 170, 178
Resultant vector 51-5, 70
Revolutions 19, 196-7, 201-2, 204, 208, 221, 223, 226, 228, 266, 269
  complete 201, 204, 208, 223, 227
Reynolds number 64, 297
Rocket engine 47, 93, 106, 168, 407
Roller coaster 128, 143, 147, 229, 269
Rolling friction 119, 200
Rolling motion 200, 203, 223, 231, 253
Rolling objects 200, 204, 252, 254
Rotation Vibration 348
Rotational analog 233, 247, 255
Rotational dynamics 194, 231-2, 247, 255
Rotational inertia, body's 255, 262
Rotational kinematics 24, 194-6, 198
Rotational motion 24, 40, 83, 101, 135-6, 178, 193-6, 198, 200, 203-4, 223, 231-3, 236, 238, 240, 252
R-value 373-4, 386

S

Satellite 87, 133, 170, 195, 218-23, 227, 229-30, 232, 258, 261
Scalars 25-6, 40, 50, 53, 73, 140, 247
Scattering, inelastic 173-4
Scientific International Units 5
Second law of motion 96, 163, 185
Second law of thermodynamics 113, 356, 391, 402, 404, 408-10, 414-19
Sensation of Weight 104
Shear 276-9, 306, 330
Shear modulus 278, 308
Shear strength 277-8, 308, 330
Shear stress 276-80, 297, 308, 330
Solar system 133, 218, 221, 261, 284, 333
South Pole 229
Speedometer 26-8, 70
Speed-versus-time 32, 41
Sphere 19-20, 39, 103, 116, 229-30, 253-4, 268, 311, 339, 353-4, 383, 387
Spring constant 16, 138-9, 145-6, 158-9, 267, 274, 312, 354
State conduction 371
State functions 336, 392
Static friction 92, 111-13, 118, 200, 206, 224, 228
  coefficient of 112, 116, 118, 123, 227, 229, 263, 266-7

Stefan-Boltzmann equation   377, 379, 382
Stethoscope Open-tube   291
Strain   139, 275, 280-1, 308, 311-12
Stress forces   104
Stretch   110, 139, 158-9, 238, 250, 273-4, 276, 312
Styrofoam   383, 386-7
Substances conduct heat   373
Suction cups   287
Sun   40, 73, 77, 85, 90, 103, 150, 218-20, 222, 261, 269, 284, 375-6, 409, 424
Superposition principle   89, 94
Surface force parallel   300
Surface gravitational acceleration   218
Surface gravity   218, 314
Surface of displacement   273
Surface reflectivity, object›s   377
Surface temperature   370, 387
Surfaces Coefficient of static friction   113
Suspension bridges   242, 271-2

T

Tail Heaviness   243
Tangent, definition of   13, 17
Tangential   199, 223, 226, 252
Tangential acceleration   207-8, 227-8
Temperature gradients   320, 329, 370, 411
Tension   90-1, 94, 98, 106, 108-11, 120-3, 156, 159, 210-13, 224, 228, 241-2, 251, 265-9, 271-2, 276-9
   single   109
Tension force   91, 109-10, 156, 158, 272
Thermal conductivity   371-3, 375, 382, 386-7
Thermal conductors   373
Thermal energy   127-8, 154, 319, 365, 369, 385, 402, 404-6, 411, 415, 420, 422-4
Thermal energy transfers   369, 411-12, 421-2
Thermal equilibrium   318-19, 344, 354, 361, 382, 391
Thermal expansion   319, 322-3, 325, 328-9, 336, 348-9, 352
Thermal imaging   380-1
Thermal radiation   376-7, 380
Thermal resistance   370, 373
Thermocouples   320
Thermodynamic processes   391-5, 397-8, 415, 418, 420
Thermodynamic system   347, 363, 369, 401, 406, 418
Thermodynamics   2, 4, 113, 130, 318, 347, 356-7, 389-92, 399, 402, 404, 408-1
Thermography   380
Thermometer   318-22, 336, 351-2, 354, 361-2
Tire pressure   285, 292, 341
   recommended   285, 341
Torricelli's law   304
Transient   372

Translational kinematics   24, 196, 198, 202
Translational kinetic energy   135, 152, 252-4, 262, 344, 347, 354
Translational motion   24, 40, 83, 135, 178-9, 186, 194, 196, 198, 200-2, 204, 233, 236, 240, 243, 250
Trigonometric   12
Trigonometric functions   12-15, 70-1
   inverse   14-15
Trigonometry   11-12
Trochanteric Bursitis   278

U

Uniform circular motion   199, 206-8, 223
Unit thermal resistance   373
units   4-8, 10, 18, 25-6, 40, 96, 148, 151-4, 197-8, 281, 285, 306, 350, 357, 359, 381-2
Units English Units   7
Units of meter   7, 11
Universal gravitation, law of   4, 20, 104, 117, 216-17, 223
Universal gravitational constant   103, 118, 216, 220, 223
Universal law of gravitation   217
Universe   3, 72, 82-4, 103-4, 117, 127-8, 151, 216, 218, 223, 283, 390, 417-18, 421

V

Valves   354, 407
Vapor   363, 366-8
Vapor pressure, saturated   367-8
Vaporization   363, 366-7, 381, 383, 386, 422
Variables, thermodynamic   391-2
Vector algebra   53, 55
Vector decomposition   52, 70
Vector quantities   20, 50-1, 55-6, 84, 95, 198, 201
Vector thruster   57
Vectors   15, 25-6, 28, 40, 46, 50-7, 69-70, 73, 77, 83-4, 88, 107, 130, 206-7, 233-5, 261-2
Velocity   25-6, 28-31, 40, 45-8, 56-7, 63-5, 68-72, 77, 82-4, 86-8, 93-4, 161-3, 166-9, 175-7, 187-91, 204-7
Velocity distribution   297, 346
Velocity equation   183
Velocity graph   45
Velocity vector   64, 73, 78, 184, 206-7
Velocity-versus-position   58, 60, 73
Velocity-versus-time   58, 60, 73
Vertical velocity magnitude   78
Viscosity   64, 146, 290, 302, 305-7, 309-10
Volume Deformation   280
Volume flow rate   298, 300, 305-8, 315
Volumetric   327

INDEX

W

Water gauge pressure   289, 302
Water molecules   272, 353, 364, 415
Water pressure   273, 288, 290, 302
Water slides   141-2
Water temperature   354, 357, 382
   engine's   408
Water vapor   165, 283, 368, 376, 385, 415
Water vapor content   368
Weight density   16, 19
Weightless   39, 90, 104, 106, 179, 230
Work energy   318, 393, 414
   mechanical   357
Work-energy principle   135-6, 152, 188, 264, 301

X

x-rays   174

Y

Young's modulus   274-5, 279, 308, 312, 328

Z

Zero potential energy position   134
Zeroth Law of Thermodynamics   318

Chapter 1: Units and Physical Quantities

TOPIC OBJECTIVE:	NAME:
	DATE:

ESSENTIAL QUESTION:

QUESTIONS:	NOTES:

SUMMARY:

Chapter 2: One Dimensional Kinematics

TOPIC OBJECTIVE:

NAME:

DATE:

ESSENTIAL QUESTION:

QUESTIONS:	NOTES:

SUMMARY:

Chapter 9: Solids and Fluids

TOPIC OBJECTIVE:

NAME:

DATE:

ESSENTIAL QUESTION:

QUESTIONS:	NOTES:

SUMMARY:

Chapter 10: Thermal Physics

TOPIC OBJECTIVE:

NAME:

DATE:

ESSENTIAL QUESTION:

QUESTIONS: **NOTES:**

SUMMARY:

Chapter 11: Energy Transfer Processes

	TOPIC OBJECTIVE:	NAME:
		DATE:

ESSENTIAL QUESTION:

QUESTIONS:	NOTES:

SUMMARY:

Chapter 12: Thermodynamics

TOPIC OBJECTIVE:

NAME:

DATE:

ESSENTIAL QUESTION:

QUESTIONS: **NOTES:**

SUMMARY: